T0178491

Examination Questions and Answers in Basic Anatomy and Physiology

Martin Caon

Examination Questions and Answers in Basic Anatomy and Physiology

2900 Multiple Choice Questions and 64 Essay Topics

Third Edition

Martin Caon
College of Nursing and Health Sciences
Flinders University
Clarence Park, SA, Australia

ISBN 978-3-030-47316-7 ISBN 978-3-030-47314-3 (eBook)
https://doi.org/10.1007/978-3-030-47314-3

This Springer imprint is published by the registered company Springer Nature Switzerland AG
The registered company address is: Gewerbestrasse 11, 6330 Cham, Switzerland

Preface

Two thousand nine hundred multiple-choice questions (MCQs) that could be asked of a student of introductory human anatomy and physiology are presented in 20 chapters and to more descriptively describe the contents and to facilitate their access are subdivided into 68 categories. In addition, there are 64 topics for a written assignment (essay topics) that may be used in such a course and as an assessment task for such students.

It is assumed that users of these questions are teachers or students who have completed at least part of an anatomy and physiology course that might be offered in the first year of a university degree program. It is also assumed that they would have access to one of the Anatomy and Physiology textbooks (or similar) listed in the bibliography below. Each question category has an introduction containing a summary of useful knowledge pertinent to that category of question. However, not all possible information is provided within these introductions, so a textbook is indispensable. The summary introductions are composed of vocabulary that may be unfamiliar to the beginning student but which should be known in order to understand the questions. You will need to look up the meaning of many unfamiliar words as your studies progress.

All questions have been used at least once, during the author's teaching career, by the end of semester examinations of a university first-year undergraduate introductory anatomy and physiology course or a physical science course for health sciences students to support their anatomy and physiology study. Consequently they reflect the author's choice of content. Students enrolled in the courses for which these questions were written include nursing, midwifery, paramedic, physiotherapy, occupational therapy, nutrition and dietetics, health science students, exercise science students and students taking the course as an elective. Many of the students do not have an extensive background in science from their secondary schooling. Some knowledge of physical science is required to understand physiology; hence, physical science questions are included. Students without some background knowledge in chemistry and physics will find such questions challenging and will need to work a little harder to develop their background knowledge. The boundary between chemistry and biochemistry is not distinct; nevertheless, chemistry is implicit in physiology. Furthermore, the physics of the body becomes physiology, so gradually that sometimes the boundary between the two is only noticed after it has been crossed.

Some questions were difficult to categorise and may span two (or more) categories. Furthermore, in order to answer some questions, you may need knowledge drawn from other "sections" of anatomy different from the name of the section in which the question appears. This is not a bad thing as it emphasises the connected nature of human anatomy and physiology. Each question is unique (there are no duplicates). However, many questions will be examining the same (or similar) material albeit with a differently worded question or a different selection of answers. If the questions are to be used to compile an examination, then care should be taken to exclude questions that are too similar to already selected ones. On the contrary, if the questions are to be used for instruction or study purposes, I would suggest including several similar questions in consecutive order to emphasise the point and to give the students practice.

Advice to the Exam Candidate

The correct choice of answer for each question is provided. Accompanying the correct choice is a justification for the choice or an explanation of the correct answer and sometimes of why the other choices are incorrect. The degree of difficulty varies, but not by intentional design. The perception of difficulty depends on that part of science that the question examines, the level of scientific background brought to the course by the student and their level of studious preparation for the examination.

There is only one best correct answer for each of the multiple-choice questions among the four choices presented. However, there may be more than one correct answer. You must choose the **best** one. In marking multiple-choice questions, I suggest that 1 mark be allocated for a correct answer and that a quarter of a mark be deducted for a wrong answer or an unanswered question. Deducting a quarter mark will reduce the score that would be gained by selecting an answer from the four choices purely at random (i.e. guessing), from about 25% to about 6%. Not to deduct a quarter mark is, in my opinion, unsound. Hence, in an examination, the students should never leave a question unanswered. If you cannot decide on an answer, guess at it (after eliminating any choices that you deem to be incorrect). That is, you will be rewarded for the ability to decrease the number of choices from which you are guessing, from 4 to 3 or 2.

Be aware of questions that are asked in the negative, that is, have NOT true or FALSE or INCORRECT or EXCEPT one in the stem. In this case, you are seeking a statement that is wrong in order to answer the question. Do not be intimidated by arithmetical calculations. The calculation itself will be simple. Deciding what to add, multiply or divide with what is the tricky part.

Some questions have been published before in the book: Caon, M. & Hickman, R. (2003) *Human Science: Matter and Energy in the Human Body* 3rd edn.,

Crawford House Australia Publishing, Belair South Australia, and are used with the authors' permission.

Some Thoughts on Writing Good MCQs and on Answering Poorly Prepared MCQ Quizzes

Ten Pieces of Advice for Writing Good Multiple-Choice Questions

1. Make all the choices of answer about the same length.
2. Do not write choices that use "all of the above", "none of the above", "both A and B", "never", "all", etc. (If you cannot think of sufficient choices for distractors, then discard that question.)
3. Use plausible distractors (do not use funny, absurd or cute choices).
4. If the choices are all numbers, list them in order of increasing magnitude.
5. Avoid choices where two are the opposite of each other (one might be guessed to be true).
6. Make the stem ask a question. Do not include irrelevant material in the stem. Do not unintentionally provide a clue in the question.
7. Spread the correct answer evenly (and randomly) among the choices. In questions with four choices of answer, about 25% of the correct choices should be "A", about 25% "B", etc. Do not avoid having two or three consecutive answers that are the same letter choice.
8. Limit the number of questions "asked in the negative", that is, where a false statement is the correct choice.
9. Be grammatically correct when writing the question and the choices. Do not be ambiguous.
10. If only one choice is meant to be the best correct answer, make sure that it is so.

Five Ways to Score More Highly on a Poorly Prepared Multiple-Choice Question Test

Knowing the subject matter is the best way to score well in a multiple-choice test, but if you do not know the answer, always guess at it after crossing out the obvious wrong answers first. Your guess will then be an educated guess.

1. Eliminate the obvious wrong answers first!!!
   (a) If marks are deducted for incorrect answers but NOT deducted for unanswered questions, do not answer the questions you are sure that you do not know the answer to.
   (b) If one of the choices is "none of the above" or "all of the above", choose that answer.

(c) Look at the answers to the preceding and following questions. If you are guessing, do not select a choice that is the same as the previous or the next choice. (This only works if you have chosen those answers correctly!)
(d) Choose the longest answer.
(e) Eliminate the choices with absolute statements such as never, always and all.

Some Thoughts on the Marking of MCQ Tests (Where There Are Four Choices of Answer, One of Which Is the Best Correct)

Testing for knowledge is an imprecise science. Using multiple-choice questions (MCQs) for the testing simplifies the marking but also introduces additional uncertainties and some unfairness.

I award 1 mark for each correct answer. This would mean that someone may score 25% without any study simply by guessing (assuming that correct choices are spread evenly among the four choices). Hence, I also deduct ¼ of a mark for each incorrect answer or unanswered question. With this deduction, it follows that in a 100 question quiz, the mark that a total guesser will score is approximately: 25 correct – (75 incorrect) × ¼ = 25 – 18¾ = 6¼% rather than about 25% if marks were not deducted for incorrect answers.

My reasoning is as follows. If you randomly choose the answers for four questions that each have a choice of four answers, the probability of guessing one correct answer from the four questions is: ¼ + ¼ + ¼ + ¼ = 1 and you would be awarded 1 mark out of 4. This would be undeserved as you did not know any answers. By deducting ¼ for each wrong answer, your score for guessing the answers of these four questions would become 1 – ¾ = ¼ mark. The score is still undeserved but more reasonable. Nevertheless, I advise my students to guess at the answer if they do not know it, after eliminating the obviously erroneous choices. If the student can reduce the potentially correct answers from 4 to 3 or 2 before guessing, the probability of guessing correctly from the remaining choices is higher, and they will score more marks. For example the probability of guessing four answers correctly after eliminating one or two obviously incorrect choices may be: $\frac{1}{3} + \frac{1}{2} + \frac{1}{3} + \frac{1}{2} = 1.67$. Hence, on average, you would be awarded 1.67 of the 4 marks (minus the deduction for wrong answers). This is reasonable as the student deserves some credit for knowing that some of the choices were wrong.

Should a ¼ mark be deducted for each *unanswered* question? Before I answer this, let us consider four possible strategies for awarding marks to a multiple-choice question quiz with 100 questions.

Strategy 1: award 1 mark for a correct answer.
Strategy 2: award 1 mark for a correct answer and deduct a $\frac{1}{3}$ mark for wrong answers.
Strategy 3: award 1 mark for a correct answer and deduct a ¼ mark for wrong answers.

Strategy 4: award 1 mark for a correct answer and deduct a ¼ mark for wrong answers AND for unanswered questions.

Given that there are four choices to each question and only one is correct and that the correct choice is evenly allocated between choices A, B, C and D, which strategy is fairer? Clearly, the more you get correct, the higher the score. It is also clear that students who bring their knowledge to bear on answering the quiz, that is, are not merely selecting choices at random, will choose far more than 25% of the answers correctly.

Consider strategy 1. If a student attempts all questions, the lowest probable score (by random guessing) is 25%, not zero. Hence, 25% is equivalent to zero (no knowledge), and the range of possible scores in a four-choice MCQ quiz is from 25 to 100, rather than from 0 to 100. This strategy suffers from rewarding lack of knowledge with 25% of the marks and also constricts the range of marks to about three quarters of the total range. To account for marks obtained by guessing, the examiner may choose to set as a pass mark, a number greater than 50/100 as the passing score for the quiz, for example 60 or 70 or 75/100. If another student leaves some questions unanswered, perhaps because this student does not know the answers, then his or her maximum possible score is reduced by the number of unanswered questions. The scenario for such students remains largely as described above. However, it is possible for both the students to answer the same number of questions correctly and so attain the same score despite the second student leaving some questions unanswered (Table 1, column 4). The examiner may consider that this outcome is fair.

It seems reasonable to me to deduct marks for an incorrect answer when the answer is chosen from four possibilities, as is the case for the type of multiple-choice questions being considered here. It also seems too great a penalty to deduct a mark (or half a mark) for an incorrect choice as the result would be a negative score when less than 50% (or 33%) of questions are answered correctly. Would deducting a ⅓ mark or a ¼ mark produce a fairer result?

Consider strategy 2. In a 100-question quiz, when a ⅓ mark is deducted for incorrect answers only, Student 1 who answers 50 questions correctly and 50 incorrectly is awarded 33.3 (see Table 2, column 6). Furthermore, Student 3 who chooses not to answer ten questions but still answers 50 questions correctly (and 40 incorrectly) is awarded a higher score (36.7) than Student 1. Is this an intended consequence?

Compare this with strategy 3 where a ¼ mark (rather than ⅓ mark) is deducted. The same scenarios above result in Students 1 and 3 being awarded 37.5 and 40, respectively, for their 50 correct answers (see Table 1, column 6), instead of 33.5 and 36.7 (if ⅓ of a mark were deducted). Hence there is more reward for effort when only a ¼ mark is deducted. However, both the strategies will result in students scoring more highly if they are able to strategically omit answering questions that they are sure they do not know the answer to. Thus students are rewarded for knowing what they do not know—or for omitting to study a section of the course and avoiding the questions on that part of the course. This is the same as inviting students to choose which questions they wish to answer and rewarding them for answering fewer questions. It is for this reason that I deduct a ¼ mark for unanswered questions. When marks are deducted for wrong answers (but not for unanswered questions), even for the same number of correct answers (50 in Tables 1 and 2), the more MCQs you leave unanswered (between 0 and 50), the higher will be the score. Hence, students would be encouraged to leave answers to questions that they are unsure about (or have not studied) blank.

Consider strategy 4. When a ¼ mark is deducted for wrong answers and also for unanswered questions, students are compelled to answer all the questions. In a 100-question quiz, Student 1, who answers 100 questions—50 correctly and 50 incorrectly—is awarded 37.5 (see Table 1, column 7). Student 3 who chooses not to answer ten questions but still answers 50 questions correctly (and 40 incorrectly) is also awarded the score of 37.5 (rather than the higher score of 40 if strategy 3 was used to encourage the student to guess at the answers to the ten unanswered questions). If the second student had, instead of leaving ten MCQs unanswered, simply guessed at the ten answers, they would probably have scored another 2 or 3 marks (Table 1, column 8). Indeed, if they had guessed the answers after first eliminating any choices they knew to be incorrect, they may have scored more than 2.5 extra marks (on average).

This marking strategy rewards students for correctly guessing at answers instead of leaving some questions unanswered. This is compensated for by the ¼ mark deduction for incorrect answers. However, students are penalised if they do not answer (or do not guess at) questions on some parts of the course. Furthermore, students who guess from fewer choices are rewarded for having the knowledge to eliminate some choices prior to guessing from the remaining choices. Such students will probably guess correctly more than 25% of the time. This is a more searching test of their knowledge of the course and is why I deduct ¼ for each unanswered question.

Deducting ¼ mark for incorrect and blank answers also advantages the better students—those who answer more questions correctly—by increasing their score. Table 3 displays the result of four students who all answer 90 questions (and leave ten unanswered) and score different numbers of correct answers. If strategy 1 is used, the students' scores would range from 90 to 50 (Table 3, column 4). Strategy 4 would result in a spread of scores between 87.5 and 37.5 (column 7) when ten MCQs are left unanswered. The score would likely increase 2.5 or more if the students had guessed at these ten answers, rather than leaving them blank, and the highest scoring student has his or her mark "restored" to 90. Hence the student marks would be spread out over a larger range of scores (90–40) than for strategy 1.

When ¼ marks are deducted for wrong answers and also for blank answers, the lowest possible score (by random guessing) is close to 6%, not zero. Hence, 6% is equivalent to zero, so the range of possible scores is from 6 to 100 (see Table 4). The examiner may wish to neglect this discrepancy from zero and use a score of 50% as the passing score for the quiz. Note also from Table 4 that the student who gets 80/100 answers correct has his or her score adjusted down to 75 due to the guessing deduction, while the student who gets only 40/100 answers correct has his or her score adjusted more severely to 25 due to the guessing deduction.

Table 1 Four students who all answer 50 questions correctly but choose to leave different numbers of questions unanswered

	No. of MCQs answered	Unanswered MCQs	Correctly answered MCQs	Incorrectly answered MCQs	Score when ¼ subtracted for incorrectly answered MCQs	Score when ¼ subtracted for incorrect and for unanswered MCQs	Extra score if the unanswered MCQs were guessed at
Student 1	100	0	50/100	50	37.5	37.5	Na
Student 2	95	5	50/95	45	38.75	37.5	+1.25
Student 3	90	10	50/90	40	40	37.5	+2.5
Student 4	50	50	50/50	0	50	37.5	+12.5

Two scenarios are considered where a ¼ mark is deducted for wrong answers (column 6) and for wrong answers and also for unanswered questions (column 7)

Table 2 Four students who all answer 50 questions correctly but choose to leave different numbers of questions unanswered

	No. of MCQs answered	Unanswered MCQs	Correctly answered MCQs	Incorrectly answered MCQs	Score when ⅓ deducted for incorrect answers
Student 1	100	0	50/100	50	33.3
Student 2	95	5	50/95	45	35
Student 3	90	10	50/90	40	36.7
Student 4	50	50	50/50	0	50

A ⅓ mark is deducted only for wrong answers

Table 3 Four students who all answer the same number of questions (and choose to leave ten questions unanswered), but who answer different numbers of questions correctly

	No. of MCQs answered	Unanswered MCQs	Correctly answered MCQs	Incorrectly answered MCQs	Score when ¼ deducted for incorrectly answered MCQs	Score when −¼ also for unanswered MCQs	Extra score if unanswered MCQs were guessed
Student 1	90	1c0	90/90	0	90	87.5	+2.5
Student 2	90	10	80/90	10	77.5	75	+2.5
Student 3	90	10	70/90	20	65	62.5	+2.5
Student 4	90	10	50/90	40	40	37.5	+2.5

Table 4 When 1 mark is awarded for a correct answer and ¼ marks are deducted for wrong answers AND also for questions that are not answered, column 2 of this table displays the score that would be awarded by answering correctly the number of questions in column 1

Correct answers (out of 100)	Awarded score (%)	Correct answers (out of 100)	Awarded score (%)	Correct answers (out of 100)	Awarded score (%)
100	100	73	66	46	33
99	99	72	65	45	31
98	98	71	64	44	30
97	96	70	63	43	29
96	95	69	61	42	28
95	94	68	60	41	26
94	93	67	59	40	25
93	91	66	58	39	24
92	90	65	56	38	23
91	89	64	55	37	21
90	88	63	54	36	20
89	86	62	53	35	19
88	85	61	51	34	18
87	84	60	50%	33	16
86	83	59	49	32	15
85	81	58	48	31	14
84	80	57	46	30	13
83	79	56	45	29	11
82	78	55	44	28	10
81	76	54	43	27	9
80	75%	53	41	26	8
79	74	52	40	25	6%
78	73	51	39	24	5
77	71	50	38	23	4
76	70	49	36	22	3
75	69	48	35	21	1
74	68	47	34	20	0

Clarence Park, SA, Australia Martin Caon

Bibliography[1]

Caon, Martin & Hickman, Ray (2003) Human Science: Matter and Energy in the Human Body 3rd ed, Crawford House Australia Publishing, Belair South Australia. ISBN 0 8633 3255 3

Drake, R.L., Vogl, A.W. and Mitchell, A.W.M. (2019) GRAYS Anatomy for Students 4th ed. Churchill Livingstone, Philadelphia.

Hall, J.E. (2015) Guyton and Hall Textbook of Medical Physiology 13th ed. W.B.Saunders, Philadelphia

Herman Irving P. (2016) Physics of the Human Body, 2nd ed, Springer

Hewitt Paul G. (2015) Conceptual Physics, 12th ed. Pearson Education Ltd.

Hobbie, Russell K., & Roth, Bradley (2015) Intermediate Physics for Medicine and Biology, 5th ed, Springer

Marieb, E.N & Hoehn, K.N. (2015) Human Anatomy & Physiology 10th ed, Pearson

Martini, F.H., Nath, J.L. & Bartholomew, E. F. (2018) Fundamentals of Anatomy & Physiology 11th ed, Pearson

Matta, Michael S.; Wilbraham, Antony C.; Staley, Dennis D (1996) Introduction to General, Organic and Biological Chemistry, D C Heath & Co.

McKinley, M.P., Oloughlin, V.D. & Bidle, T.S. (2013) Anatomy & Physiology An Integrative Approach, McGraw Hill

Nave Carl R. &. Nave Brenda C. (1985) Physics for the Health Sciences, 3rd ed W. B. Saunders

Netter, F.H. (2014) Atlas of Human Anatomy 6th ed. Saunders, Philadelphia

Patton, K.T. & Thibodeau, G.A. (2016) Anatomy & Physiology 9th ed, Elsevier

Saladin, K.S. (2014) Anatomy & Physiology: The unity of form and function 7th ed, McGraw Hill

Sherwood L (2015) Human physiology from cells to systems 9th ed., Thomson, Belmont

Spitzer, Victor M. and Whitlock David G. (1998) Atlas of the Visible Human Male. Reverse Engineering of the Human Body. Jones and Bartlett ISBN 0-7637-0273-0

Timberlake Karen C. (2015) Chemistry: An Introduction to General, Organic, and Biological Chemistry, 12th Ed. Pearson

Tortora, G.J. & Derrickson, B. (2014) Principles of Anatomy & Physiology 14th ed, Wiley

Van De Graff, K.M. & Fox, S.I. (1999) Concepts of Human Anatomy & Physiology 5th ed, WCB

VanPutte, C. Regan, A. Russo, A. & Seeley, R. (2016) Seeley's Anatomy & Physiology 11th ed, McGraw Hill

Widmaier E.P, Raff H. and Strang K.T. (2013) Vander Sherman & Luciano's Human Physiology: the mechanism of body function 14th ed. McGraw Hill, New York

[1] Textbooks suitable for use in an introductory anatomy and physiology course. Later editions may exist, and earlier editions will suffice.

Contents

Chapter 1
Organisation of the Body

A large part of beginning the study of anatomy and physiology is learning the specialised words that are used. This new terminology may seem daunting, but the challenge lies in its unfamiliarity rather than its difficulty of comprehension. You must expect to encounter a lot of new words and be prepared to learn them over the course of your study. Most of the words contain information as the words are constructed with a prefix and a suffix or a stem that identifies the word as referring to a specific part of anatomy or physiology. Many anatomical and physiological terms are in fact descriptions. For example extensor carpi radialis longus refers to a muscle that extends (extensor) the hand at the wrist (the carpals) that lies over the radius bone (radialis) and is the longer (longus) of the two muscles. The name also implies that there is a similar muscle that is not "longus"—it is the extensor carpi radialis brevis. Deoxyribonucleic acid (DNA) refers to a molecule that contains units of a ribose sugar with an oxygen atom removed, attached to a base to form a nucleoside and also attached to a phosphoric acid. This sometimes makes the words rather long or unusual.

You should know what the anatomical position of the body is and in what direction the transverse, sagittal and coronal planes of the body lie. Directional terms such as proximal/distal; deep/superficial; superior/inferior; lateral/medial; anterior/posterior and caudal/cephalic allow the location of one anatomical feature to be placed relative to another. The dorsal and ventral body cavities are located on different sides of the body and contain different organs. For ease of communication, the abdomen is divided into nine regions: right hypochondriac, epigastric, left hypochondriac, right lumbar, umbilical, left lumbar, right inguinal, hypogastric (or pubic) and left inguinal regions. Furthermore, you should be aware that superficial anatomical landmarks are referred to by regional names such as popliteal, calcaneal, cephalic, axillary and acromial. You should know the difference between physiology and anatomy and the definitions of metabolism, anabolism and catabolism.

M. Caon, *Examination Questions and Answers in Basic Anatomy and Physiology*, https://doi.org/10.1007/978-3-030-47314-3_1

1. Which of the listed term is described as "All the chemical processes that take place in the organelles and cytoplasm cells of the body"?
   A. Metabolism
   B. Cellular respiration
   C. Homeostasis
   D. Physiology

Answer is A: The quoted statement is the definition of metabolism.

2. Which of the following is the best definition of physiology?
   A. The microscopic study of tissues and cells
   B. The study of how the body works
   C. All the chemical processes that take place in the organelles of the body cells
   D. The body's automatic tendency to maintain a relatively constant internal environment

Answer is B: Physiology is indeed the study of how the (healthy) body functions.

3. What is the study of how body parts function called?
   A. Histology
   B. Physiology
   C. Homeostasis
   D. Metabolism

Answer is B: Physiology refers to function.

4. What does the process known as anabolism refer to?
   A. The use of energy for producing chemical substances
   B. The breaking down phase of metabolism
   C. All the chemical processes that take place in the organelles of the cells
   D. The supply of nutrients to the body cells

Answer is A: Anabolism refers to the process of constructing/building molecules (think anabolic steroids). B refers to catabolism. C is metabolism.

5. Which major organ lies deep to the right hypochondriac region?
   A. Stomach
   B. Spleen
   C. Liver
   D. Duodenum

Answer is C: Hypochondriac = below the rib cartilage (chondra = cartilage); liver is located mostly on the right side.

6. Which plane divides the body into dorsal and ventral regions?
   A. Transverse
   B. Axial
   C. Coronal
   D. Sagittal

Answer is C: Dorsal and ventral = front and back—a coronal section so divides the body into these sections.

7. The "anatomical position" could be described as which of the following?
   A. Lying down prone
   B. Lying down supine
   C. Standing displaying the ventral surface of the body
   D. Standing with arms and legs abducted

Answer is C: This is the best answer. Standing is required, as is having the arms hanging parallel to the sides, with palms facing forward

8. To what does the term "hypochondriac" refer?
   A. A condition of having too few chondria
   B. The region of abdomen inferior to the ribs
   C. A person who often complains of an ailment
   D. Having insufficient cartilage in the knees

Answer is B: In this case, "hypo-" means below, while "-chondr" refers to the cartilage joining the ribs to the sternum (the costal cartilages). The regions of the abdomen immediately inferior to these rib cartilages (on the left and right sides of the body) is what is being referred to.

9. If a medical image displays internal anatomy in mid-sagittal section, which of the following describes the section?
   A. A vertical section through the nose and umbilicus that divides the body into right and left halves
   B. A cross-section through the midriff at about the level of the liver
   C. A cross-section through the upper chest at about the level of the shoulders
   D. A vertical section through the midpoint of the clavicle and through either the right or left thigh

Answer is A: A sagittal section divides the body into left and right portions. A mid-sagittal section means that the dividing line is in the vertical midline of the body, so that the halves are equal.

10. Which of the following best describes the "anatomical position"?
    A. Standing vertically, arms held horizontally, legs apart so that the tips of the head, hands and feet lie on an imaginary circle, drawn around the body
    B. Standing "to attention", with hands held so that the thumbs are ventral while the fifth digit is dorsal
    C. Standing "at ease" with hands clasped behind your back while adjacent and dorsal to the sacrum
    D. Standing vertically, arms parallel and lateral to the ribs with hands inferior to the elbows and supinated

Answer is D: The anatomical position is achieved when standing with feet comfortably apart while displaying the ventral surface of the head, body, and forearms to the same direction (forwards).

11. Which of the following terms is NOT used to identify a region of the abdomen?

    A. Left hypochondriac
    B. Hypogastric
    C. Epigastric
    D. Right sacral

Answer is D: Right sacral is not a region on the anterior surface of the abdomen.

12. When the body is standing in the "anatomical position", which of the following is true?

    A. The radius is lateral to the ulna.
    B. The radius is medial to the ulna.
    C. The radius is proximal to the ulna.
    D. The radius is distal to the ulna.

Answer is A: In the anatomical position, the palms are displayed ventrally. The radius is further from the body's midline than is the ulna; hence, it is lateral to the ulna.

13. How does a coronal section divide the body?

    A. Into many transverse slices
    B. Into a ventral part and a dorsal part
    C. Into a left and right section
    D. Into superior and inferior portions

Answer is B: An imaginary cut that divides the body into a front half (or section) and back half is termed coronal. Choice C is sagittal, while choice D is a transverse section.

14. By what anatomical term is the head region known?

    A. Plantar
    B. Cephalic
    C. Hypochondriac
    D. Axillary

Answer is B: The cephalus is the head; the plantar region is the base of the foot; the hypochondriac region is inferior and deep to the rib cartilages of ribs 7–10; the axillary region is the "armpit".

15. Which organ would be found in the left hypochondriac region?

    A. Appendix
    B. Urinary bladder
    C. Liver
    D. Stomach

Answer is D: Left hypochondriac region is deep to the cartilages of the lower ribs on our left hand side. The stomach is closest to this region.

16. Which body region does "popliteal" refer to?
    A. The region around each eye
    B. The region anterior to the elbow, between arm and forearm
    C. The region dorsal to the knee
    D. The region of the anterior crease between thigh and abdomen

Answer is C: Here the popliteal artery and popliteus tendon may be located. Choice A refers to orbital; B refers to antecubital; D refers to inguinal.

17. Which region of the body is known as the acromial region?
    A. The elbow region
    B. The heel region
    C. The medial ankle region
    D. The shoulder region

Answer is D: The superior part of the shoulder, at the distal end of the clavicle is known as acromial. Here the acromion of the scapula articulates with the clavicle at the "ac" or acromio-clavicular joint. Choice A is the olecranal region; B the calcaneal; C (the medial malleolus) is not usually ascribed a region name.

18. What part of the body is known as the popliteal region?
    A. The fold of the knee
    B. The fold of the elbow
    C. The area around the ears
    D. The medial sides of the ankles

Answer is A: Behind the knee, opposite to the patella is the popliteal region. Here is found the popliteal pulse and popliteus tendon.

19. Which organs are likely to be found in the epigastric region of the abdomen?
    A. Aorta, vena cava and trachea
    B. Oesophagus and stomach
    C. Urinary bladder and some of the large intestine
    D. The spleen and left kidney

Answer is B: The epigastric region is above/upon "epi-" the stomach. The organs of choice A are in the chest; C are in the hypogastric region; D in the left hypochondriac or left lumbar regions.

20. Which region of the body is the "sural" region?
    A. The dorsal surface of the leg
    B. The dorsal surface of the thigh
    C. The regions left and right of the lumbar vertebrae
    D. The medial region of the arm

Answer is A: Anatomically the "leg" is between the knee and ankle. Its dorsal region, also known as the calf, is the sural region.

21. Where is the inguinal region of the body?
    A. On the ventral surface of each knee
    B. On the abdomen, immediately superior to each thigh
    C. It is the region between the two lungs
    D. It is superior to the heart and inferior to the larynx

Answer is B: Inguinal refers to two of the nine regions that the abdomen surface is usually divided into. In this case, to the two most inferior, on either side of the pubic (or hypogastric) region.

22. What part of the body is referred to as the plantar region?
    A. The crease (anterior surface) of the elbows
    B. The backs of the hands
    C. The palms of the hands
    D. The soles of the feet

Answer is D: The inferior surface of the feet that make contact with the ground are the plantar region. Plantar warts occur on the soles. Plantar flexion of the foot is the act of "pointing the toes".

23. The directional term "superior" in anatomy means which of the following?
    A. Cephalic
    B. Ventral
    C. Closer to the top of the head
    D. Closer to the skin surface

Answer is C: Cephalic refers to the head region. While superior refers to being closer to the head than is the other anatomical structure in question.

24. Which of the stated relationships is correct?
    A. The heart is inferior to the clavicle.
    B. The shoulder is distal to the carpals.
    C. The phalanges are proximal to the metacarpals.
    D. The eye is medial to the eyebrows.

Answer is A: The heart is indeed below (inferior) the clavicle. All other choices are wrong.

25. Which of the stated relationships is correct?
    A. The heart is superior to the large intestine.
    B. The shoulder is distal to the metacarpals.
    C. The phalanges are proximal to the carpals.
    D. The eye is medial to the nose.

Answer is A: The heart is indeed above (superior) the intestine. All other answers are wrong.

26. Which of the following correctly describes the two named body parts?
    A. The elbow is proximal to the shoulder.
    B. The phalanges are distal to the carpals.
    C. The ribs are proximal to the sternum.
    D. The elbow is distal to the knee.

Answer is B: Phalanges (finger bones) are indeed further from the trunk along the arm, than are the carpals (wrist bones).

27. Complete the sentence correctly: "Cervical vertebrae are …".
    A. Superior to the rib cage.
    B. Inferior to the thoracic vertebrae.
    C. Located between the thoracic and sacral vertebrae.
    D. Fused into a single bone called the sacrum.

Answer is A: Cervix refers to "neck". The cervical vertebrae are in the neck hence are above (superior) the rib cage.

28. Which term describes the location of the adrenal glands with reference to the kidneys?
    A. Proximal
    B. Distal
    C. Superior
    D. Inferior

Answer is C: The adrenal glands are on the cephalic side of the kidneys. Being closer to the head, they are termed "superior to the kidneys".

29. Which bones are located distal to the elbow and proximal to the wrist?
    A. Carpals
    B. Radius and ulna
    C. Tarsals
    D. Humerus

Answer is B: Distal to the elbow means further along the arm towards the hand—this eliminates the humerus. Proximal to the wrist means closer to the body than the wrist—this eliminates the carpals. The tarsals are in the ankle.

30. Imagine an image of a transverse section of the upper arm. What tissues may be identified there located from the most superficial to the deepest?
    A. Skin, subcutaneous fat, muscle, hypodermis, bone
    B. Epidermis, dermis, hypodermis, muscle, bone
    C. Integument, muscle, superficial fascia, bone, marrow
    D. Hypodermis, subcutaneous fat, muscle, marrow, bone

Answer is B: Choice A is incorrect as hypodermis is more superficial than muscle. Choice C is wrong again because superficial fascia (which is a synonym for hypodermis) is more superficial than muscle. Choice D is wrong as marrow lies within bone, and also hypodermis and subcutaneous fat are almost synonyms.

31. Which of the following describes the position of the pinna of the ear with respect to the nose and chin?
    A. The ear is superior to the nose and distal to the chin.
    B. The ear is proximal to the nose and distal to the chin.
    C. The ear is anterior to the nose and superior to the chin.
    D. The ear is lateral to the nose and superior to the chin.

Answer is D: The ear is lateral (further from the body's midline) to the nose and superior (closer to the top of the head) to the chin.

32. Which of the following describes the position of the elbow with respect to the wrist and shoulder?
    A. The elbow is lateral to the shoulder but medial to the wrist.
    B. The elbow is dorsal to the shoulder but anterior to the wrist.
    C. The elbow is distal to the shoulder but proximal to the wrist.
    D. The elbow is inferior to the shoulder but superficial to the wrist.

Answer is C: Distal and proximal are terms used to describe position on a limb. The elbow is distal (further away from the torso) to the shoulder but proximal (closer to the torso) to the wrist.

33. Which of the following describes the position of the thoracic vertebrae with respect to the sternum and the kidneys?
    A. Vertebrae are posterior (or dorsal) to the sternum and medial to the kidneys.
    B. Vertebrae are superficial to the sternum and deep to the kidneys.
    C. Vertebrae are superior to the sternum and inferior to the kidneys.
    D. Vertebrae are lateral to the sternum and medial to kidneys.

Answer is A: Vertebrae are posterior (or dorsal) to sternum, which means they are closer to the back surface and medial to the kidneys, which means that the vertebrae are closer to the body's midline.

34. Which choice best describes the location of the majority of the musculoskeletal system?
    A. It is in the dorsal cavity.
    B. It is in the ventral cavity.
    C. It is in the abdominopelvic cavity.
    D. It is not located in a body cavity.

Answer is D: The musculoskeletal system is located in the arms and legs, and surrounding, but outside of the abdominopelvic, thoracic and the dorsal cavities.

35. Which of the following is/are the contents of the dorsal body cavity?
    A. Heart and lungs
    B. Brain and spinal cord
    C. Viscera
    D. Gut, kidneys, liver, pancreas, spleen, bladder, internal reproductive organs

Answer is B: Dorsal refers to the back, the cavity enclosed by the skull and vertebrae.

36. Which of the following is/are the contents of the ventral cavity?
    A. Heart and lungs
    B. Brain and spinal cord
    C. Viscera
    D. Gut, kidneys, liver, pancreas, spleen, bladder, internal reproductive organs

Answer is C: This is the best answer. It is a collective term for all organs in the thoracic and abdominopelvic cavities.

37. Which one of the following statements is correct?
    A. The diaphragm separates the brain and the spinal cord.
    B. The ventral cavity contains the male and female reproductive system.
    C. The abdominopelvic cavity contains the spinal cord.
    D. The dorsal cavity contains the brain and spinal cord.

Answer is D: Dorsal means back and that is the cavity with spinal cord and brain. B is incorrect as the genitalia are outside the ventral cavity.

38. The dorsal body cavity contains which of the following organs?
    A. Brain
    B. Brain and spinal cord
    C. Brain, spinal cord and heart
    D. Brain, spinal cord, heart and kidneys

Answer is B: Dorsal refers to the back and is opposite to ventral. Only the brain and spinal cord occupy the dorsal cavity. All other answers are incorrect.

39. What structure separates the thoracic cavity from the abdominal cavity?
    A. Mediastinum
    B. Diaphragm
    C. Peritoneum
    D. Pylorus

Answer is B: The muscular diaphragm physically separates these two ventral cavities.

40. What structure separates the abdominal and pelvic cavities?
    A. There is no separating structure.
    B. Diaphragm
    C. Peritoneum
    D. Dura mater

Answer is A: The pelvic cavity is not physically separated from the abdominal cavity. For example parts of the small intestine are located in both "cavities".

41. In which cavity/ies does the digestive system lie?
    - A. Abdominal cavity
    - B. Abdominal and pelvic cavities
    - C. Thoracic, abdominal and pelvic cavities
    - D. Dorsal, thoracic, abdominal and pelvic cavities

Answer is C: The oesophagus is within the thoracic cavity, while the remainder is in the abdominopelvic cavity.

42. What structure separates the thoracic and abdominal cavities and what is it made of?
    - A. The mediastinum, made of serous membranes
    - B. The diaphragm, made of skeletal muscle
    - C. The diaphragm, made of smooth muscle
    - D. The pleural cavity, made of visceral and parietal layers

Answer is B: The diaphragm physically separates these two ventral cavities. It is made of skeletal (not smooth) muscle as it is under our conscious control. However, most of the time, we delegate control to the autonomic nervous system.

43. Which organs do NOT lie within a body cavity?
    - A. Heart
    - B. Brain and spinal cord
    - C. The muscles and bones of the legs
    - D. The prostate and urinary bladder

Answer is C: Each muscle and bone is an organ and the musculoskeletal system is within neither the dorsal nor ventral body cavities.

44. What is the movement called when the arms are moved from the anatomical position by sweeping them through 90° in the coronal plane, so that they are held horizontally (parallel to the ground)?
    - A. Pronation
    - B. Circumduction
    - C. Abduction
    - D. Rotation

Answer is C: Moving the straight arms away from the body in this fashion is called abduction.

45. What is meant by the term flexion (or to flex)?
    - A. Flexion is where the angle between two long bones is decreased by muscle action.
    - B. Flexion is an action performed to stretch (extend) a muscle.
    - C. Flexion is where the angle between two long bones is increased by muscle action.
    - D. Flexion is caused by the action of contracting a muscle.

Answer is A: To flex an arm is to decrease the angle between the humerus and radius (by contracting the biceps brachii). Choice D is wrong as contracting the triceps brachii causes extension of the forearm.

46. To what movement is the term "extension" applied?
    A. Extension is where the angle between two long bones is decreased by muscle action.
    B. Extension is an action performed to stretch (extend) a muscle.
    C. Extension is where the angle between two long bones is increased by muscle action.
    D. Extension occurs when an antagonistic muscle is allowed to contract.

Answer is C: To extend a body part is to increase the angle between the moving bone and the stationary bone. For example when the fingers of a clenched fist are allowed to straighten, the angle between the proximal phalanges and the metacarpals increases.

47. What is the collective term used for the contents of the body's ventral cavity?
    A. The omentum
    B. The peritoneum
    C. The internal organs
    D. The viscera

Answer is D: The peritoneum is the membrane that surrounds the abdominal cavity, and the omentum is a portion of that. The internal organs are close but also include the brain which is not in the ventral cavity.

48. What is meant by the term "retroperitoneal"?
    A. On the dorsal side of the lungs.
    B. In the space between the spinal cord and the bodies of the vertebrae.
    C. Within the body wall but not enclosed by the peritoneum.
    D. It is a small bone of the facial skeleton.

Answer is D: Retroperitoneal refers to organs inferior to the diaphragm but not enclosed by the peritoneum. For example the kidneys, pancreas, rectum and part of the duodenum.

49. What are the terms cortex and medulla used to describe?
    A. The cortex is the outer part of an organ or bone while the medulla is the inner part.
    B. The cortex is the inner part of an organ or bone while the medulla is the outer part.
    C. The cortex is the deeper part of an organ or bone while the medulla is the more superficial part.
    D. The medulla refers to the fibrous capsule around an organ, while the cortex is the tissue of an organ.

Answer is A: The cortex of the kidney, for example, is the deep (inner part) while the cortex is the more superficial, outer part.

50. What exists in the "potential space" between the visceral and parietal layers of a membrane?
    A. Serous fluid
    B. Nothing
    C. Air
    D. Synovial fluid

Answer is A: Serous membranes have a deeper visceral layer and a more superficial parietal layer. Between them is a small amount of serous fluid to lubricate their movement past each other.

51. One of the images taken for mammography of the compressed breast is known as "craniocaudal". What direction is this?
    A. Compression from the medial and lateral sides.
    B. A left to right (sideways) view.
    C. The breast is flattened against the rib cage for imaging.
    D. When standing, the breast is compressed from above and below.

Answer is D: From above (the cranial direction) and below (the caudal or tail direction).

52. What is the collective name for the contents of our ventral cavity?
    A. The giblets
    B. The abdominopelvic organs
    C. The internal organs
    D. The viscera

Answer is D: The giblets is a term used for the internals of edible fowl such as chicken. The internal organs is close but also includes the brain which is not in the ventral cavity. Abdominopelvic organs exclude the thoracic organs, which are in the ventral cavity.

53. What is/are the main function(s) of the serous membranes of the body's ventral cavities and the potential space that is between their two layers?
    A. To contain the enclosed organs and restrict them to their cavity.
    B. To secrete fluid which allows the organs and two layers to slide over each other without friction.
    C. To secure the organs in position by attaching them to the skeleton.
    D. To house the nerve supply and blood supply to the organs.

Answer is B: The serous fluid secreted allows the parietal membrane to slide over the visceral membrane without friction. This sliding occurs as we breathe and as we walk and run and as food moves through the digestive tract. It also allows the organs to shift against their neighbour without friction. The membranes surround the organs and are attached to the body wall, so another function is to secure the organs to the body wall (so choice C is not wrong).

54. What is the clinical condition that develops when air is able to enter the "potential space" between the parietal and visceral pleura?
    A. Pneumothorax
    B. Pneumonia
    C. Collapsed lung
    D. Dyspnoea

Answer is A: Pneumo refers to air and thorax to the chest. Sometimes it is called collapsed lung, but atelectasis more correctly is applied to that. Dyspnoea refers to difficult or laboured breathing.

55. What is the clinical condition called when the membrane of the abdominal cavity is inflamed?
    A. Meningitis
    B. Pleurisy
    C. Gastroenteritis
    D. Peritonitis

Answer is D: The peritoneum is the membrane around the abdominal cavity. The suffix "-itis" is used to describe inflammation. Gastroenteritis describes the illness triggered by the infection and inflammation of the digestive system. Pleurisy refers to inflammation of the pleural membranes that surround the lungs. Meningitis refers to inflammation of the meninges that surround the brain and spinal cord.

56. Which organs do NOT have regions that are referred to as the cortex and the medulla?
    A. Kidneys
    B. Adrenals
    C. Bones
    D. Arteries

Answer is D: Medulla is the inner (deep) part of an organ while the cortex is the more superficial. Arteries walls are divided into "tunics" but do not have sections that are referred to as cortex or medulla.

57. Which of the following does the "tissue level" of structural organisation refer to?
    A. Atoms, ions, molecules and electrolytes
    B. Mitochondria, ribosomes, nucleus, endoplasmic reticulum
    C. Nephron, alveolus, villus, lobule
    D. Muscle, nervous, connective, epithelial

Answer is D: The listed structures are the four major tissue types.

58. How does an organ differ from a tissue?
    A. An organ contains more than one type of tissue and performs a particular function.
    B. An organ contains more than two types of tissues and performs a specialised function.
    C. An organ contains more than three types of tissues and performs a relatively limited number of functions.
    D. An organ contains more than four types of tissues and performs or one or more specific functions.

Answer is A: "More than one type" implies at least two; and organs perform a function. The term specialised function is usually used for the work that a tissue does. Choice D is wrong as there are only four main types of tissues. Choice C is wrong as more than three implies at least four, but the brain is an organ which does not contain any muscle tissue.

59. When a pianist is in a seated position with forearms making an angle of 90° with the arms, ready to play the piano, what would describe the position of her forearms?
    A. Pronated
    B. Supinated
    C. Lateral to the abdomen
    D. Medial to the thorax

Answer is A: To play the piano, the palms of the hands are facing the piano keys and the dorsal side of the hands are facing the ceiling. In this position, the forearms are in the pronated position.

60. The following is a list of several levels of organisation that make up the human body: (1) tissue; (2) cell; (3) organ; (4) molecule; (5) organism; (6) organ system. When listed in order from the smallest to the largest level, what sequence would the numbers be?
    A. 2, 1, 4, 3, 6, 5
    B. 4, 2, 1, 3, 6, 5
    C. 4, 2, 1, 6, 3, 5
    D. 4, 2, 3, 1, 6, 5

Answer is B: Molecule (4) must come first and organ system second to last. That eliminates A and C. Cell (2) and tissue (1) are next in increasing order and this coincides with choice B.

Chapter 2
Cells and Tissues

2.1 Cell and Membrane Structure

Cells are composed of their cytoplasm, which includes the cytosol and organelles, the nucleus and the surrounding plasma membrane. You should know that the plasma membrane is a double layer of phospholipid molecules and that these molecules have a hydrophilic end and a hydrophobic end. The plasma membrane contains proteins including the ATPase (the sodium–potassium pump) which moves sodium ions out of the cell while moving potassium ions into the cell. Other proteins act as receptors and as pores or gateways into the cell. You should know the names and function of some of the organelles within the cells. For example you should know that mitochondria produce ATP and that ribosomes synthesise proteins from amino acids.

1. Which structure within the cell produces ATP (adenosine triphosphate)?
   A. Mitochondria
   B. Nucleus
   C. Peripheral proteins
   D. Endoplasmic reticulum

Answer is A: This is a basic function of mitochondria. All other answers are wrong.

2. Which of the following is **NOT** a component of the cell plasma membrane?
   A. Cholesterol
   B. Proteins
   C. Microfilaments
   D. Phospholipids

Answer is C: Microfilaments occur inside the cell.

M. Caon, *Examination Questions and Answers in Basic Anatomy and Physiology*, https://doi.org/10.1007/978-3-030-47314-3_2

3. Except for one, the following are types of cells. Which one is **NOT** a type of cell?
   A. Platelets
   B. Leucocytes
   C. Macrophages
   D. Osteoblasts

Answer is A: Platelets are fragments of a cell (a megakaryocyte) bound by a membrane.

4. In which part of a cell does the process of making ATP from oxygen and glucose take place?
   A. Lysosomes
   B. Ribosomes
   C. Mitochondria
   D. Golgi apparatus

Answer is C: ATP production is the function of mitochondria.

5. Which of the following is a function of membrane proteins?
   A. To process lipids and proteins for secretion through the plasma membrane
   B. To act as receptors for hormones
   C. To synthesise proteins from amino acids
   D. To act as a cytoskeleton to support and shape the cell

Answer is B: One function of membrane proteins is to receive (amino acid based) hormones that cannot pass through the plasma membrane.

6. What is the difference between simple squamous cells and simple columnar cells?
   A. Squamous cells are flattened while columnar cells are taller than they are wide.
   B. Simple squamous cells are one layer thick while simple columnar cells are several layers thick.
   C. Simple squamous cells are epithelial tissue while simple columnar cells are connective tissue.
   D. Squamous cells are flattened while columnar cells are cuboidal.

Answer is A: The name of the cells contains a description of their shape: either flat or like columns. Simple refers to a single layer of cells

7. Which of the following is **NOT** an example of a cell?
   A. Macrophages
   B. Lysosomes
   C. Plasmocytes
   D. Chondroblasts

Answer is B: The suffix "-some" refers to an organelle within a cell. The other suffixes all indicate a type of cell.

8. Which cell organelles contain an acidic environment capable of digesting a wide variety of molecules?
   A. Lysosomes
   B. Ribosomes
   C. Centrosomes
   D. Golgi complex

Answer is A: The prefix “lyso-” refers to the ability to dissolve or destroy molecules or cells.

9. Which statement about the plasma membrane is **INCORRECT**?
   A. It is selectively permeable.
   B. It is composed of two layers of glycoprotein molecules.
   C. It contains receptors for specific signalling molecules.
   D. The plasma membranes of adjacent cells are held together by desmosomes.

Answer is B: The PM is indeed made of two layers, but they are phospholipid (not glycoprotein) molecules.

10. What is the name of the mechanism that ensures that there is a higher concentration of sodium ions in the extracellular fluid than in the intracellular fluid?
    A. Facilitated diffusion
    B. The sodium–potassium pump
    C. Secondary active transport
    D. Osmosis

Answer is B: The “pump” (or ATPase) transports Na^+ out and K^+ into the cell.

11. What are lysosomes, centrosomes, and ribosomes example of?
    A. Stem cells
    B. Organelles within a cell
    C. Sensory receptors in the dermis
    D. Exocrine glands

Answer is B: The suffix “-some” refers to small body or organelle within a cell.

12. What is the function of phospholipids in the plasma membrane?
    A. To maintain the intracellular fluid at a similar composition to that of the interstitial fluid.
    B. To form channels to selectively allow passage of small molecules.
    C. To act as receptors for signalling chemicals.
    D. To present a barrier to the passage of water-soluble molecules.

Answer is D: Molecules that are soluble in water cannot pass through lipid (fat). So the phospholipids are a barrier. The functions described by B and C are performed by other molecules in the plasma membrane.

13. Which of the following is **NOT** a part of the plasma membrane of a cell?
    A. Integral proteins
    B. Glycoproteins
    C. Plasma proteins
    D. Peripheral proteins

Answer is C: As the name implies, plasma proteins are found in the blood plasma. Not to be confused with the plasma membrane.

14. A major role for mitochondria is to:
    A. Transcribe the information in DNA (deoxyribonucleic acid)
    B. Produce ATP (adenosine triphosphate)
    C. Synthesise proteins from amino acids
    D. Use enzymes to lyse molecules

Answer is B: ATP is only produced within the mitochondria.

15. What is a role performed by mitochondria?
    A. Contain enzymes capable of digesting molecules
    B. Produce ATP
    C. Synthesise proteins
    D. Synthesise fatty acids, phospholipids and cholesterol

Answer is B: Mitochondria produce ATP. The other tasks are performed by lysosomes, ribosomes and endoplasmic reticulum, respectively.

16. Which of the following is **NOT** found in the plasma membrane?
    A. Proteins
    B. Cholesterol
    C. Endoplasmic reticulum
    D. Phospholipids

Answer is C: Endoplasmic reticulum is an organelle and found within the cell.

17. Which of the following is **NOT** a part of the plasma membrane of a cell?
    A. Phospholipid
    B. Glycoprotein
    C. Chromatin
    D. Cholesterol

Answer is C: Chromatin makes up chromosomes.

18. A major role for mitochondria is to:
    A. Synthesise fatty acids, phospholipids and steroids
    B. Deliver lipids and proteins to plasma membrane for secretion
    C. Synthesise proteins from amino acids
    D. Produce ATP (adenosine triphosphate)

Answer is D: Mitochondria produce ATP from glucose

19. What is the purpose of mitochondria?
    A. To store the nucleolus and chromatin
    B. To produce adenosine triphosphate
    C. To support and shape the cell
    D. They produce enzymes to break down molecules

Answer is B: Mitochondria are the site of ATP production

20. The plasma membrane of a cell contains molecules that have a hydrophobic end and a hydrophilic end. What are they called?
    A. Phospholipids
    B. Cholesterol
    C. Integral proteins
    D. Glycoproteins

Answer is A: The phosphate end is hydrophilic (water soluble) while the lipid end is hydrophobic (insoluble in water).

21. What is the role of mitochondria?
    A. Function in cell division
    B. Synthesise proteins
    C. Form part of the plasma membrane
    D. Synthesise fatty acids, phospholipids and steroids

Answer is C: Mitochondria produce ATP.

22. What do fibroblasts, chondroblasts, osteoblasts and haemocytoblasts have in common?
    A. They are all types of white blood cell.
    B. They are all macrophages.
    C. They are all immature cells.
    D. They are all types of epithelial cell.

Answer is C: The suffix "-blast" implies that these cells have not yet finished their differentiation, that is, are immature.

23. Active transport across the plasma membrane may be described by which statement?
    A. Active transport requires energy from ATP.
    B. Active transport is also known as endocytosis.
    C. Active transport moves molecules along their concentration gradient.
    D. Active transport is the movement of lipid-soluble molecules through the plasma membrane.

Answer is A: This is the only correct answer. The others are not true.

24. Which of the following cell types denotes an immature cell?
    A. Macrophages
    B. Monocytes
    C. Osteoblasts
    D. Ribosomes

Answer is C: The suffix "-blast" indicates that the cell is immature.

25. Which organelle is the site of ATP production?
    A. Nucleus
    B. Endoplasmic reticulum
    C. Mitochondria
    D. Golgi apparatus

Answer is C: The mitochondria is where ATP is produced.

26. Which of the following is **NOT** one of the organelles within a cell?
    A. Desmosome
    B. Endoplasmic reticulum
    C. Mitochondrion
    D. Golgi apparatus

Answer is A: Desmosome (despite having the suffix "-some") are not within the cell. They are structures that join adjacent plasma membranes to each other.

27. The process of "diffusion" through a membrane may be described by which of the following?
    A. The movement of ions and molecules away from regions where they are in high concentration towards regions where they are in lower concentration.
    B. The use of energy from ATP to move ions and small molecules into regions where they are in lower concentration.
    C. The plasma membrane engulfs the substance and moves it through the membrane.
    D. The use of energy from ATP to move water molecules against their concentration gradient.

Answer is A: The choices with ATP are nonsense. While choice C refers to endocytosis.

28. The process of "active transport" through a membrane may be described by which of the following?
    A. The movement of ions and small molecules away from regions where they are in high concentration.
    B. The use of energy from ATP to move ions and small molecules into regions where they are in lower concentration.

C. The plasma membrane engulfs the substance and moves it through the membrane.
D. The use of energy from ATP to move ions and small molecules against their concentration gradient.

Answer is D: Energy (ATP) is required to force molecules against their concentration gradient.

29. Which of the following is the smallest living structural unit of the body?

A. Atom
B. Molecule
C. Organelle
D. Cell

Answer is D: Cell is the smallest structural unit that is deemed to be alive.

30. Which of the following enables ions such as sodium to cross a plasma membrane?

A. Phospholipid bilayer
B. Peripheral proteins
C. Integral proteins
D. Desmosomes

Answer is C: One function of integral protein in the PM is to form channels to allow for the passage of ions.

31. Cell membranes can maintain a difference in electrical charge between the interior of the cell and the extracellular fluid. What is this charge difference called?

A. Excitability
B. Membrane potential
C. Action potential
D. Sodium–potassium pump

Answer is B: The inside of a cell is negative while the exterior side of the membrane is positive. This difference in charge constitutes a difference in electrical potential (or voltage), known as the resting membrane potential. An action potential is generated when the membrane is stimulated and the potential reversed.

32. The resting membrane potential of a cell is the consequence of which of the following concentrations of ions?

A. High K^+ and Cl^- outside the cell and high Na^+ and large anions inside the cell.
B. High K^+ and Na^+ outside the cell and high Cl^- and large anions inside the cell.
C. High Cl^- and Na^+ outside the cell and high K^+ and large cations inside the cell.

D. High Ca^{2+} and Na^+ outside the cell and high K^+ and large cations inside the cell.

Answer is C: These ionic species are largely responsible for the membrane potential (cations are negative ions). While there is a higher concentration of Ca outside the cell than inside, there are fewer Ca than Cl ions.

33. What is one function of mitochondria?
    A. Produce enzymes to break down molecules
    B. Produce molecules of ATP
    C. Hold adjacent cells together
    D. Allow passage of molecules through the plasma membrane

Answer is B: Mitochondria are organelles within which ATP is made.

34. Membrane proteins perform the following functions **EXCEPT** one. Which one?
    A. Form glycocalyx
    B. Act as receptor proteins
    C. Form pores to allow the passage of small solutes
    D. Behave as enzymes

Answer is A: Glycocalyx refers to molecules in the plasma membrane that have a carbohydrate chain attached (prefix "glyco-").

35. Facilitated diffusion differs from active transport because facilitated diffusion:
    A. Requires energy from ATP
    B. Moves molecules from where they are in lower concentration to higher concentration
    C. Moves molecules from where they are in higher concentration to lower concentration
    D. Involves ions and molecules that pass through membrane channels

Answer is C: Diffusion always refers to movement from high to low concentration (without energy expenditure). Facilitated refers to the assistance provided by a transport molecule that is designed for the purpose.

36. Which one of the following terms best describes the structure of the cell membrane?
    A. Fluid mosaic model
    B. Static mosaic model
    C. Quaternary structure
    D. Multilayered structure

Answer is A: "Fluid" implies the structure can move and change (not like a brick wall); mosaic refers to the presence of proteins scattered among the glycolipids.

37. Which one of the following terms best describes a phospholipid? It consists of a:
    A. Polar head and polar tail
    B. Non-polar head and a polar tail
    C. Polar head and non-polar tail
    D. Non-polar head and a non-polar tail

Answer is C: Polar = hydrophilic head of phosphate (which can dissolve in the aqueous extracellular solution because water molecules are polar); non-polar = hydrophobic tails of lipid, which being non-polar cannot dissolve in aqueous solutions.

38. One of the functions of integral proteins in cell membranes is to:
    A. Maintain the rigid structure of the cell
    B. Support mechanically the phospholipids
    C. Interact with the cytoplasm
    D. Form channels for transport functions

Answer is D: Some proteins form channels which allow molecules and ions to enter the cell.

39. Which one of the following best describes what a cell membrane consists of?
    A. Lipids, proteins, ribosomes
    B. Lipids, cholesterol, proteins
    C. Cholesterol, proteins, cytoplasm
    D. Lipids, proteins, cytoplasm

Answer is B: These are the three major constituents. Ribosomes and cytoplasm are found inside the cell.

40. Which one of the following organelles is considered as the "energy producing" centre of the cell?
    A. Rough endoplasmic reticulum
    B. Golgi apparatus
    C. Mitochondria
    D. Ribosomes

Answer is C: Mitochondria are where ATP molecules are produced from glucose.

41. What is the major function of lysosomes?
    A. Package proteins
    B. Detoxify toxic substances
    C. Catalyse lipid metabolism
    D. Digest unwanted particles within the cell

Answer is D: The prefix "lys-" refers to the ability to alter molecules by dividing them into smaller pieces.

42. What is the purpose of the "sodium/potassium pump"?
    A. To perform endocytosis
    B. To move sodium and potassium by facilitated diffusion
    C. To perform bulk transport through the plasma membrane
    D. To produce a concentration gradient for sodium ions

Answer is D: A concentration gradient is set up by the use of energy to move Na ions to where they are in greater concentration. This requirement for energy means choice B is wrong.

43. Which of the following is **NOT** a type of cell?
    A. Ribosome
    B. Haemocytoblast
    C. Neutrophil
    D. Phagocyte

Answer is A: A ribosome is a cell organelle, not a cell type.

44. What is the name of the mechanism that ensures that there is a higher concentration of sodium ions in the extracellular fluid than in the intracellular fluid?
    A. Facilitated diffusion
    B. The sodium–potassium pump
    C. Secondary active transport
    D. Osmosis

Answer is B: The "pump" exchanges Na for K and uses energy from ATP to function.

45. What is the name given to the type of transport where glucose or an amino acid binds to a receptor protein on the plasma membrane, which then moves the molecule into the cell without the expenditure of energy?
    A. Facilitated diffusion
    B. Bulk transport
    C. Secondary active transport
    D. Active transport

Answer is A: The membrane protein facilitates the entry into the cell. No energy is expended so it is not active transport.

46. What is the name given to the movement of glucose or amino acids from the gut into the cells lining the gut, when they bind to a transport protein that has also bound a sodium ion. The sodium ion is entering the cell along its concentration gradient.
    A. Facilitated diffusion
    B. Sodium–potassium pump
    C. Active transport
    D. Secondary active transport

Answer is D: The sodium ion was transported out of the cell with the use of energy in order to set up the sodium concentration gradient. This gradient then allows other molecules to enter the cell along with the re-entry of sodium. This is active (because energy used), but secondary as it occurs as a result of the previous active transport event.

47. Mitochondria produce which of the following?
    A. ATP
    B. DNA
    C. RNA
    D. Proteins

Answer is A: Adenosine triphosphate (ATP)

48. Why does the plasma membrane of a cell present a barrier to the movement of electrolytes through it?
    A. There are no channels in the membrane for the passage of electrolytes.
    B. Electrolytes are not soluble in the lipid of the membrane.
    C. Electrolytes are too large to pass through membrane channels.
    D. Membrane proteins electrically repel charged particles.

Answer is B: Electrolytes, being charged particles, are not able to dissolve their way through the lipid plasma membrane (which is non-polar). Hence it is a barrier to them.

49. Which of the following statements about "leak channels" in the plasma membrane is correct?
    A. Proteins that form these channels bind to solutes to allow them to pass into the cell.
    B. They are passageways formed by proteins to allow water and ions to move passively through the membrane.
    C. They allow small ions and molecules to move between adjacent cells.
    D. They are formed by glycoprotein and proteoglycans to allow hormones to enter cells.

Answer is B: This is the definition of leak channels. They may be "gated" which means shut until stimulated to open. Choice A refers to facilitated diffusion.

50. Which of the following is a component of the plasma membrane of a cell?
    A. Plasma
    B. Glycolipid
    C. Plasma proteins
    D. Cholesterol

Answer is D: Despite the term "plasma", A and C are wrong. And it is phospholipids, not glycolipids that occur in the membrane.

51. What term is used to describe the movement of dissolved particles along (or down) their concentration gradient?
    A. Endocytosis
    B. Active transport
    C. Osmosis
    D. Diffusion

Answer is D: Following the concentration gradient is a passive process. Choice C applies only to water molecules.

52. Which of the following molecules cannot pass through the plasma membrane?
    A. Water molecules
    B. Non-polar molecules
    C. Amino acid-based hormones
    D. Fat-soluble molecules

Answer is C: These hormones are not lipid soluble and too large to pass through channels.

53. What is the major component of the plasma membrane of a cell?
    A. Phospholipid
    B. Glycolipid
    C. Integral protein
    D. Cholesterol

Answer is A: Cholesterol and proteins are also present in the plasma membrane but as more minor components.

54. Which one of the following is **NOT** a function of membrane proteins?
    A. They form a structure called a glycocalyx.
    B. They attach cells to each other.
    C. They form passageways to allow solutes to pass through the membrane.
    D. They from receptors which can bind messenger molecules.

Answer is A: The glycocalyx is thought of as being membrane carbohydrates.

55. Facilitated diffusion through a membrane involves which of the following scenarios?
    A. The diffusion of water through a selectively permeable membrane along its concentration gradient.
    B. The movement of a molecule against its concentration gradient with the expenditure of energy
    C. The plasma membrane surrounding (engulfing) the molecule and the molecule moving into the cell
    D. A molecule binding to a receptor which moves the molecule through the membrane without the expenditure of energy

Answer is D: Facilitation is by binding to a membrane protein.

56. The diffusion of water through a membrane is referred to as:
    A. Secondary active transport
    B. Bulk transport
    C. Osmosis
    D. Endocytosis

Answer is C: Osmosis is a word that is reserved for the movement of water through a membrane.

57. The cell membrane's resting potential (about −70 mV inside with respect to the outside) is mainly due to which of the following mechanisms?
    A. The sodium–potassium pump
    B. The diffusion of cations and anions through the membrane along their concentration gradients
    C. The diffusion of sodium and potassium across the cell membrane
    D. The presence inside the cell of anions too large to passively cross the cell membrane

Answer is A: The ATPase pump shifts three Na^+ out of the cell and two K^+ into the cell. This disparity in positive charge is the major influence on the resting potential.

58. Which of the following cells are uni-nucleate (have a single nucleus)?
    A. Red blood cells
    B. Epithelial cells
    C. Skeletal muscle cells
    D. Osteoclasts

Answer is B: Red blood cells have no nucleus, while osteoclasts and skeletal muscle cells are multinucleate.

59. What is the function of the cell plasma membrane?
    A. It maintains an intracellular environment that is hypotonic compared to the extracellular fluid.
    B. It protects the cell from dehydration by limiting water flow from the cell.
    C. It regulates the passage of molecules and ions into and out of the cell.
    D. It provides the supportive medium for membrane proteins.

Answer is C: The plasma membrane is made of lipid and separates the aqueous intracellular fluid from the aqueous extracellular fluid. Substances cannot enter or leave the cell unless they are lipid soluble or by endo/exo-cytosis or have a membrane protein pore or transport mechanism that can move them through the membrane.

60. Which of the following cells are found in the lungs?
    A. Pneumocytes
    B. Leucocytes

C. Keratinocytes
D. Sertoli cells

Answer is A: "Pneumo" comes from the Greek language and refers to lung or air. The other cells are found in the blood, epidermis and testicles, respectively.

61. Which of the following cells are found in the liver?
    A. Myocytes
    B. Pericytes
    C. Podocytes
    D. Hepatocytes

Answer is D: "Hepato" comes from the Greek language and refers to liver.

62. What are generally referred to as organelles?
    A. Small exocrine organs such as salivary glands and sudoriferous glands
    B. Small endocrine glands such as the pituitary and adrenal glands
    C. Structures of the general senses such as lamellated corpuscles and muscle spindles
    D. Structures within a cell such as ribosomes and endoplasmic reticulum

Answer is D: Organelles are not glands. They are structures within a cell.

63. What does the term "integral protein" refer to?
    A. Proteins that lie within the plasma membrane of a cell
    B. Proteins that must be included in the diet as they cannot be manufactured by the body
    C. Proteins found within the central nervous system
    D. Plasma proteins that exist in the blood but not in the interstitial fluid

Answer is A: "Integral" refers to being within the cell membrane. The other choices are not even close to being correct.

64. What is the term used to describe the bulk movement of a large number of molecules out of a cell?
    A. Lymphocytosis
    B. Exocytosis
    C. Thrombocytosis
    D. Endocytosis

Answer is B: "Exo-" refers to movement from inside to the outside (-cytosis is a suffix that means in reference to cells).

65. In which organelle of the cell do most aerobic respiration reactions happen?
    A. In the nucleus
    B. In the mitochondria
    C. In the ribosomes
    D. In the lysosomes

Answer is B: Mitochondria are the sites of aerobic respiration. That is, the production of ATP from pyruvic acid (via glucose) and oxygen.

66. Which of the following is a list of four of the organelles found within a cell?
    A. Lysosomes, ribosomes, centrosomes, Golgi complex
    B. Ribosomes, centrosomes, Golgi complex, desmosomes
    C. Endoplasmic reticulum, mitochondria, nucleus
    D. Centrosomes, Golgi complex, peroxisomes, chromosomes

Answer is A: Desmosomes are not within a cell, they are structures that hold adjacent cells together. The nucleus is not regarded as an organelle. Chromosomes while within a cell are DNA and are not regarded as an organelle.

67. Which of the following describes the structure of the plasma membrane (cell membrane)?
    A. A bilayer of lipoproteins with cholesterol molecules and plasma proteins forming pores
    B. A double layer of glycolipid molecules, with cholesterol molecules and plasma proteins
    C. A double layer of phospholipid molecules, with cholesterol molecules and membrane proteins
    D. A bilayer of cholesterol molecules and plasma proteins, with pores of phospholipid

Answer is C: Phospholipid molecules each with a hydrophilic end and a hydrophobic end arrange themselves as a double wall (a bilayer). Cholesterol molecules are interspersed among the phospholipids and give the membrane additional strength. Protein molecules form the pores or channels through which water-soluble materials may enter and leave the cell. "Plasma proteins" exist in the blood, not in the cell membrane.

68. What roles do proteins play in a cell's plasma membrane?
    A. They catalyse reactions and perform endocytosis.
    B. Receptors for hormones and selectively allow entry to some solutes via the channels they form.
    C. They allow cells to adhere to each other and allow lipid-soluble molecules to pass through the lipid bilayer.
    D. Glycoproteins act as recognisers or identification tags and are responsible for movement of the cell.

Answer is B: Some proteins are receptor sites for hormones, others form pores and selectively allow some solutes to pass through the channels they form. They do not perform endocytosis or move the cell around. Lipid-soluble molecules can pass through the lipid bilayer on their own.

69. What is active transport when applied to a cell?
    A. It is the movement of ions or molecules through a membrane against their concentration gradient using energy from ATP.
    B. It is the rapid movement of water molecules through channels called aquaporins.
    C. It is the process of bringing large substances (e.g. bacteria, proteins, polysaccharides) into cells by engulfing them in a vesicle enclosed by a membrane.
    D. It is the passage of blood cells through the intact walls of the capillaries.

Answer is A: "Active" means that energy is expended to accomplish the movement. This energy is required to shift particles from a solution where they are in a lower concentration to a solution where they are at a greater concentration. Choice B is osmosis; Choice C is endocytosis; Choice D is diapedesis.

70. Which of the following is correct?
    A. Lipids form channels in the cell membrane.
    B. Sodium and potassium are found at equal concentration in the cytoplasm.
    C. Hydrophobic tails of phospholipids in the cell membrane contact the cytoplasm.
    D. Hydrophilic heads of phospholipids in the cell membrane contact the cytoplasm.

Answer is D: Hydrophilic means water loving, so the phospholipid head dissolves in the aqueous solution of the cytoplasm. It is proteins that form the channels. And K is in much greater concentration in the cytoplasm than is Na.

71. Which is NOT a feature of a mitochondrion?
    A. It contains DNA and RNA.
    B. Its outer membrane contains pores.
    C. It produces ATP.
    D. It packages and stores protein.

Answer is D: Protein is produced and stored by the endoplasmic reticulum and Golgi apparatus. Mitochondria do have DNA (derived from the mother) and RNA.

72. Which statement about endoplasmic reticulum (ER) and Golgi apparatus (GA) is incorrect?
    A. Rough ER contains structures called ribosomes.
    B. ER packages newly synthesised proteins.
    C. GA packages enzymes for use in the cytoplasm.
    D. GA is a site for ATP synthesis.

Answer is D: ATP is produced in the mitochondria, not the Golgi apparatus.

73. What type of transport utilises a carrier protein to transport molecules across the cell membrane?
    A. Facilitated diffusion
    B. Diffusion
    C. Active diffusion
    D. Active transport

Answer is A: Facilitated diffusion is spontaneous passive transport of large molecules across a biological membrane via specific transmembrane integral proteins, from where they are in high concentration to where they are in lower concentration.

2.2 Types of Tissues

You will become familiar with the names of many cells. Often a word can be recognised as the name of a cell because it ends with "-cyte", or if it is an immature cell, it ends with "-blast". Four major types of tissues are identified in the body: epithelial, connective, muscle and neural tissues. Of course there are many subtypes within these categories. For example epithelial tissue may be squamous, cuboidal, columnar or glandular. Muscle may be skeletal, smooth or cardiac. Connective tissue is quite varied and you should be aware of the many different examples of tissues that are categorised as "connective". For example blood, bone, dermis, cartilage and tendon are all connective tissues.

1. Which list below contains the four types of tissues?
    A. Extracellular fluid, skeletal tissue, glandular tissue, connective tissue
    B. Extracellular fluid, muscle tissue, glandular tissue, cartilaginous tissue
    C. Neural tissue, skeletal tissue, epithelial tissue, cartilaginous tissue
    D. Neural tissue, muscle tissue, epithelial tissue, connective tissue

Answer is D: These are the four types. Extracellular fluid is not a tissue. Cartilage is a type of connective tissue.

2. Which of the tissue type below consists of a single layer of cells?
    A. Stratified squamous epithelial tissue
    B. Glandular epithelium
    C. Areolar connective tissue
    D. Simple columnar epithelial tissue

Answer is D: The word "simple" indicates a single layer of cells. Stratified means several layers (or strata) of cells.

3. Which one of the following is **NOT** a serous membrane?
   A. Pleura
   B. Peritoneum
   C. Mucosa
   D. Pericardium

Answer is C: Mucosa is a mucus membrane (and secretes mucus)

4. Which of the following is **NOT** made predominantly from epithelial tissue?
   A. Dermis
   B. Exocrine glands
   C. Endocrine glands
   D. Endothelium of blood vessels

Answer is A: The dermis contains connective tissue, nervous tissue and muscle as well as epithelial tissue.

5. What are tendons and ligaments composed of?
   A. Dense connective tissue
   B. Liquid connective tissue
   C. Muscular tissue
   D. Epithelial tissue

Answer is A: Tendons and ligaments are dense CT. This is strong as there is a high proportion of fibres.

6. What is the composition of the intercellular matrix in connective tissue?
   A. Cells and fibres
   B. Serous and mucus membranes and lamina propria
   C. Protein fibres and ground substance
   D. Interstitial fluid

Answer is C: "Intercellular" means between cells. So matrix is fibres and ground substance (but no cells).

7. Which of the following is **NOT** epithelial tissue?
   A. Epidermis
   B. Glandular tissue
   C. Internal lining of blood vessels
   D. Dermis

Answer is D: The dermis contains some of all four types of tissues.

8. Which of the following is **NOT** a cell found in connective tissue?
   A. Adipocytes
   B. Chondroblasts
   C. Keratinocytes
   D. Osteoblasts

Answer is C: Keratinocytes are in the epidermis which is an epithelial tissue. The other cell types occur in fat, cartilage and bone.

9. What tissue has cells that are closely packed and that have one surface attached to a basement membrane and the other free to a space?
   A. Epithelial tissue
   B. Muscle tissue
   C. Connective tissue
   D. Nervous tissue

Answer is A: This is a definition of epithelial tissue.

10. What does simple columnar epithelial tissue refer to? Tissue with:
    A. A single layer of cells longer than they are wide
    B. A single layer of cells whose length, breadth and depth are about the same size
    C. Several layers of cells, all of the same type
    D. Several layers of cells but without a basement membrane

Answer is A: Simple = one layer. Columnar means oblong or shaped like a column.

11. Which of the following is **NOT** an example of connective tissue?
    A. Blood
    B. Bone
    C. Tendon
    D. Epidermis

Answer is D: The epidermis (on top of the dermis) is an epithelial tissue.

12. Which one of the following cell types is found in epithelial tissue?
    A. Plasma cells
    B. Leucocytes
    C. Keratinocytes
    D. Chondroblasts

Answer is C: Keratinocytes produce keratin, the protein of the epidermis, which is an epithelial tissue.

13. Choose the tissue below that is one of the four primary types of body tissue.
    A. Epidermal tissue
    B. Epithelial tissue
    C. Interstitial tissue
    D. Osseous tissue

Answer is B: Epithelial is a major tissue type (as is muscle, nervous and connective)

14. What are the primary types of tissue in the body?
    A. Muscle, nervous, connective and epithelial
    B. Muscle, nervous, connective, osseous and epithelial
    C. Muscle, nervous, connective, osseous, blood and epithelial
    D. Muscle, nervous, connective, glandular and epithelial

Answer is A: There are four major types (not five or six). Osseous and blood are also connective, while glandular tissue is also epithelial.

15. What is the name of the membrane that surrounds the lungs?
    A. Visceral peritoneum
    B. Parietal peritoneum
    C. Visceral pleura
    D. Dura mater

Answer is C: Pleura is around the lung, while visceral refers to the layer of the pleura that is attached to the lung surface.

16. Which one of the following cell types is found in epithelial tissue?
    A. Mast cells
    B. Adipocytes
    C. Chondroblasts
    D. Keratinocytes

Answer is D: These cells produce keratin, the protein of the stratum corneum.

17. Choose the tissue below that is **NOT** one of the four primary types of body tissue.
    A. Connective tissue
    B. Muscular tissue
    C. Nervous tissue
    D. Osseous tissue

Answer is D: Osseous tissue (or bone) is a connective tissue.

18. Adipocytes are found in which type of tissue?
    A. Muscle tissue
    B. Epithelial tissue
    C. Nervous tissue
    D. Connective tissue

Answer is D: Adipocytes are found in fat (adipose tissue) which is a type of connective tissue.

19. Which one of the following cell types is found in epithelial tissue?
    A. Mast cells
    B. Adipocytes

C. Chondroblasts
D. Melanocytes

Answer is D: Melanocytes produce melanin to protect the skin from ultraviolet radiation, and it results in tanning of the skin.

20. What is the difference between "loose" connective tissue (CT) and "dense" connective tissue?

A. Fibres occupy most of the volume in dense CT.
B. Dense CT includes cartilage, loose CT does not.
C. Loose CT has a good blood supply while dense CT does not.
D. Loose CT has no fibres (and dense CT does).

Answer is A: The preponderance of fibres is what makes the CT "dense". Cartilage is classified as supportive CT.

21. Which is **NOT** true of connective tissue (CT)?

A. The cells are closely packed.
B. The tissue contains protein fibres and ground substance.
C. Types include loose CT, dense CT and liquid CT.
D. CT contains white blood cells.

Answer is A: Being closely packed is a property of epithelial tissue. In CT the cells are widely spaced, being separated by the ground substance.

22. Choose the membrane that is **NOT** a serous membrane.

A. Pleura
B. Peritoneum
C. Pericardium
D. Lamina propria

Answer is D: The lamina propria is a "basement membrane" attached to epithelial tissue. The others are serous membranes.

23. Which of the following is **ONE** major function of epithelial cells?

A. Movement
B. Secretion
C. Support of other cell types
D. Transmit electrical signals

Answer is B: Glandular tissues are one type of epithelial tissue, and their function is to produce material to secrete.

24. What are the major types of tissue in the body?

A. Nervous, muscle, epithelial, connective
B. Squamous, cuboidal, columnar, transitional
C. Osteocytes, chondrocytes, leucocytes, adipocytes
D. Protein, adipose, cartilage, osseous

Answer is A: Choice C refers to cell types; B is a list of epithelial tissue. Protein is applied to molecules.

25. Which list contains the main body tissue types?
    A. Glandular, connective, osseous, nervous
    B. Epithelial, nervous, connective, muscle
    C. Endothelial, connective, muscle, cartilaginous
    D. Epithelial, cartilaginous, muscle, glandular

Answer is B: The terms osseous, glandular and cartilaginous disqualify the other choices.

26. Which of the following is **NOT** a connective tissue?
    A. Blood
    B. Mesothelium
    C. Fat
    D. Tendon

Answer is B: Mesothelium is simple squamous epithelium that is found in serous membranes.

27. What are the cells that are found in tendons called?
    A. Osteocytes
    B. Adipocytes
    C. Haemocytoblasts
    D. Fibroblasts

Answer is D: Tendons are connective tissues and contain fibroblasts—the most common type of cell in connective tissue. Osteocytes occur in bone; adipocytes in adipose tissue, and haemocytoblasts in bone marrow.

28. What are the primary types of body tissue?
    A. Connective tissue, blood, muscle tissue, nervous tissue, epithelial tissue
    B. Muscle tissue, osseous tissue, epithelial tissue, nervous tissue, blood, connective tissue
    C. Nervous tissue, epithelial tissue, muscle tissue, connective tissue
    D. Epithelial tissue, connective tissue, adipose tissue, muscle tissue, nervous tissue

Answer is C: These are the four primary types. Blood is not a "type" of tissue.

29. Epithelial and connective tissues differ from each other in which of the following characteristics?
    A. Epithelial tissue contains fibres, but connective tissue does not.
    B. Connective tissue is avascular, but epithelial tissue is well-vascularised.
    C. Cells in epithelial tissue are closely packed, whereas in connective tissue they are not.

D. Connective tissue includes tissue that makes up glands, but epithelial tissue does not occur in glands.

Answer is C: The other choices are not correct.

30. Which of the following is a connective tissue?

  A. Pancreas
  B. Spinal cord
  C. Muscle
  D. Blood

Answer is D: Blood contains cells separated by a liquid matrix. Choices A and B are epithelial and nervous tissues.

31. Which of the following is an epithelial tissue?

  A. Adipose tissue
  B. Adrenal gland
  C. Heart
  D. Blood

Answer is B: The adrenal gland is a glandular epithelial tissue.

32. What is the tissue that covers the body surface and lines the internal tubes called?

  A. Epithelial tissue
  B. Connective tissue
  C. Glandular epithelium
  D. Muscle tissue

Answer is A: Epithelial tissue has one surface "open" to the exterior or to the contents of the tube.

33. Which of the following is true for connective tissue?

  A. It consists of cells, a basement membrane and intercellular matrix.
  B. Its cells are closely packed and held together by protein fibres.
  C. It has a high rate of cell division and no blood supply.
  D. It is made of cells, protein fibres and ground substance.

Answer is D: Connective tissue includes fibres and cells which are not closely packed.

34. Which of the following cells would be found in connective tissue?

  A. Mucous cells
  B. Goblet cells
  C. Chondrocytes
  D. Neurons

Answer is C: Chondros refers to cartilage, so chondrocytes are cells found in cartilage. Cartilage is a connective tissue. The other choices are found in epithelial tissue or nervous tissue.

35. Which of the following cells would be found in nervous tissue?
    A. Dendrocytes
    B. Microglia
    C. Microphages
    D. Erythrocytes

Answer is B: Microglia is a term that refers to the support cells of the nervous system. Dendrocytes are found in the epidermis, and there are no erythrocytes in the CNS.

36. Which of the four types of tissue do all organs contain?
    A. Connective, epithelial and nervous tissues
    B. Muscle, epithelial and nervous tissues
    C. Organs contain all four types of tissues
    D. Muscle, connective and epithelial tissues

Answer is C: All organs have capillaries and a blood supply. Blood is a connective tissue, and capillaries are made of epithelial tissue. The arterial end of capillaries has a pre-capillary sphincter which is made of smooth muscle cells. All organs are innervated so contain nervous tissue.

37. What is "deep fascia"?
    A. Connective tissue that surrounds muscle
    B. Epithelial tissue that lies under the skin
    C. Connective tissue of the hypodermis
    D. Epithelial tissue of the abdominal serous membranes

Answer is A: Deep fascia is a fibrous connective tissue that surrounds and penetrates into the muscles, as well as surrounding bones, nerves and blood vessels. Muscle epimysium is an example of deep fascia. Choice C refers to superficial fascia.

38. Which of the following cells are found in cartilage?
    A. Osteocytes
    B. Chondrocytes
    C. Lymphocytes
    D. Monocytes

Answer is B: "Chondros" means cartilage in Greek.

39. Which of the following cells are found in bone?
    A. Dendrocytes
    B. Erythrocytes
    C. Osteocytes
    D. Pneumocytes

Answer is C: "Osteo" is ancient Greek for bone.

40. Which of the following is NOT an example of connective tissue?
    A. Blood
    B. Bone
    C. Smooth muscle
    D. Superficial fascia

Answer is C: There are four basic types of tissue. Connective tissue and muscle are two of them, so C is not a connective tissue; the other three choices are.

41. A particular tissue contains cells that have a process that extends for 20 or 30 cm from the cell body. What type of tissue is it likely to be?
    A. Neural tissue
    B. Muscle tissue
    C. Epithelial tissue
    D. Connective tissue

Answer is D: The long process is called an axon and belongs to a neurone. So the tissue is neural tissue.

42. The cells of a tissue are long and cylindrical and contain a great number of nuclei. In what type of tissue will such a cell be found?
    A. Neural tissue
    B. Muscle tissue
    C. Epithelial tissue
    D. Connective tissue

Answer is B: The cell is a skeletal muscle fibre, so the tissue is the muscle tissue.

43. What type of cells are found in a single layer lining the inner surface of the tubes in the body and are surrounded by a non-cellular membrane?
    A. Neural tissue
    B. Muscle tissue
    C. Epithelial tissue
    D. Connective tissue

Answer is C: The description indicates epithelial cells and their basement membrane.

44. A particular tissue contains cells that are widely separated from each other and surrounded by a tissue component that is not cellular. What type of tissue conforms with this description?
    A. Neural tissue
    B. Muscle tissue
    C. Epithelial tissue
    D. Connective tissue

Answer is D: Connective tissue consists of a non-cellular matrix in which cells are distributed so that they are not adjacent to each other.

45. Which of the following is a description of nervous tissue?
    A. Covers body surfaces, lines hollow organs, body cavities and ducts and forms glands
    B. Binds organs together, stores energy as fat and provides immunity
    C. Contains cells that are specialised for contraction
    D. Is specialised for the propagation of electrical impulses around the body

Answer is D: Nervous tissue allows the sending of signals as electrical impulses around the body. Choices A, B and C describe epithelial, connective and muscle tissues, respectively.

46. Which of the following is a description of muscle tissue?
    A. Almost all of it occurs in the brain and spinal cord.
    B. Binds organs together, stores energy as fat and provides immunity.
    C. Contains cells that are specialised for contraction.
    D. The cells are closely packed and tightly bound together.

Answer is C: The cells of skeletal muscle, cardiac muscle and smooth muscle are all able to shorten their length (to contract). Choices A, B and D describe nervous, connective and epithelial tissues, respectively.

47. Which of the following is a description of epithelial tissue?
    A. Covers body surfaces, lines hollow organs and body cavities and forms glands.
    B. Forms basement membranes, bone, fat, blood, tendons and cartilage.
    C. Contains cells that are specialised for contraction.
    D. Some of their cells are the longest in the body and are referred to as fibres.

Answer is A: "Epi-" means on top of epithelial tissue that includes the epidermis and luminal surface of hollow organs. Choices B, C and D describe connective, muscle and nervous tissues, respectively.

48. Which of the following is a description of connective tissue?
    A. Their cells have one surface not attached to other tissue, that is exposed to "space".
    B. Their cells possess an axon and dendrites.
    C. Their cells contain actin and myosin.
    D. It contains a matrix of fibres and "ground substance".

Answer is D: Connective tissue consists of cells widely separated by the non-cellular matrix. Choices A, B and C describe epithelial, nervous and muscle tissues, respectively.

49. What are some of the functions of epithelial tissue?
    A. They facilitate communication within the body; enable awareness of our environment; and allow us to learn.
    B. Physically protect underlying cells; they secrete substances; they provide sensation when stimulated.

C. They store energy; they provide a structural framework for the body; they transport dissolved materials.
D. They contribute energy to maintain body temperature; they move the contents of the gut; they dilate and constrict blood vessels.

Answer is B: Epithelial cells physically protect the underlying cells from abrasion and chemicals. They secrete onto the respiratory tract and into the digestive system. Neuroepithelium (e.g. on tongue, on retina) contains sensory nerve endings. Choices A, C and D are functions of nervous, connective tissue and muscle tissue, respectively.

50. What is the difference between "loose" connective tissue (CT) and "dense" connective tissue?
    A. Dense CT forms the superficial fascia, while loose CT forms the deep fascia.
    B. Dense CT is mostly fibres, while loose CT surrounds blood vessels and nerves.
    C. Dense CT forms tendons and ligaments, while loose CT forms aponeuroses and deep fascia.
    D. Dense CT fills the spaces between organs, while loose CT forms tendons and ligaments.

Answer is B: Dense CT is mostly fibres and forms tendons and ligaments, aponeuroses and deep fascia. Loose CT surrounds vessels, nerves and muscles, forms the superficial fascia, fills the spaces between organs and surrounds blood vessels and nerves.

51. Which list has cells that would all be found in connective tissues?
    A. Leucocytes, astrocytes, ependymal cells and Schwann cells
    B. Keratinocytes, melanocytes, dendrocytes and Merkel cells
    C. Leucocytes, chondrocytes, osteoblasts and adipocytes
    D. Astrocytes, Schwann cells, ependymal cells and microglia

Answer is C: Leucocytes occur in blood, chondrocytes in cartilage; osteoblasts in bone; and adipocytes in fat. All of which are types of connective tissue. The cells in choices B and D occur in skin and nervous tissue, respectively.

52. Which of the following is NOT a function of at least some epithelial tissue?
    A. Providing physical protection
    B. Providing cushioning
    C. Providing sensations
    D. Producing specialised secretions

Answer is B: Cushioning is not a function provided by epithelial tissue. All the rest are.

53. What type of epithelium is found lining blood vessels?
    A. Simple squamous epithelium
    B. Stratified squamous epithelium
    C. Simple cuboidal epithelium
    D. Stratified cuboidal epithelium

Answer is A: Simple squamous, that is, flattened and one-cell thick, cells line blood vessels.

2.3 Cell Cycle (Mitosis and Protein Synthesis)

The cell nucleus contains chromosomes which are composed of molecules of DNA. DNA is composed of units called nucleotides which consist of a sugar (deoxyribose) attached to a phosphoric acid group $PO(OH)_3$ and one of four bases (adenine, guanine, cytosine and thymine). Chromosomes contain the code for the sequence of amino acids used to construct different proteins. Each amino acid is coded for by a particular sequence of three of the four bases. This sequence is called a "codon". mRNA "transcribes" this code then moves from the nucleus to a ribosome in the cytoplasm where it is "translated", and the protein is assembled by joining the required amino acids in the appropriate sequence.

Mitosis is the process by which a somatic cell divides to produce two cells with identical DNA. In this way an organism can grow. Before mitosis, the DNA must be duplicated. Hence, the chromosomes (consisting of one strand or "chromatid") synthesise a second strand to become a chromosome of two chromatids. Then during mitosis the two chromatids separate and move into the two daughter cells.

Meiosis occurs only in the gonads. This process results in four daughter cells. Human cells have two copies of each of 23 chromosomes, one copy being inherited from the father and the other copy from the mother. The gametes need to have only one copy of each of the 23 chromosomes, so that when the sperm fuses with the ovum, the "diploid" number of 46 (two copies of each chromosome) is restored. Meiosis is the process by which cells reduce their number of chromosomes from 46 to 23 different chromosomes. Of the 23 chromosomes in a sperm (or ovum), some (between 0 and 23) will have come from the sperm owner's mother and the rest from the sperm owner's father. The same can be said of the 23 chromosomes in the ovum. In this way the resulting children will be genetically different from each of their parents (and siblings) as each sperm/ovum will have a different assortment of the 23 available chromosomes.

1. The term "chromatin" would be used in reference to which of the following?
    A. Genetic substance
    B. Cellular energy
    C. Membrane support
    D. Nuclear membrane

Answer is A: Chromatin is DNA and the associated proteins so pertains to genetic material.

2. In protein synthesis, where dose translation occur?
   A. Cytoplasm between ribosomes, tRNA and mRNA
   B. Nucleus between ribosomes, tRNA and mRNA
   C. Nucleus between DNA and mRNA
   D. Cytoplasm between DNA and mRNA

Answer is A: Translation occurs in the cytoplasm (transcription occurs in the nucleus). DNA does not exist in the cytoplasm.

3. If the DNA strand sequence of bases is CTT AGA CTA ATA, what would the tRNA read?
   A. GAA TCT GAT TAT
   B. CUU AGA CUA AUA
   C. GAA UCU GAU UAU
   D. GUU ACA GUA AUA

Answer is C: Guanine (G) must be matched to cytosine (C) and vice versa. Adenine (A) must match with thymine (T). In RNA, uracil (U) replaces thymine; hence, U must be matched to A.

4. Which one of the following statements best describes DNA?
   A. Single-stranded, deoxyribonucleic acid
   B. Single-stranded, ribonucleic acid
   C. Double-stranded, deoxyribonucleic acid
   D. Double-stranded, ribonucleic acid

Answer is C: DNA is double stranded, while the "D" refers to "deoxy-".

5. In which phase of mitosis would chromosomes line up at the centre of the spindle?
   A. Anaphase
   B. Interphase
   C. Prophase
   D. Metaphase

Answer is D: Remember the metaphase plate occupies the middle of the cell.

6. In a cell cycle, which phase takes the longest time to complete?
   A. Anaphase
   B. Interphase
   C. Prophase
   D. Telophase

Answer is B: Interphase is the time when the cell is performing its normal function and not dividing.

7. What is the purpose of meiosis? To produce:
   A. DNA
   B. Somatic cells
   C. Diploid cells
   D. Haploid cells

Answer is D: Meiosis produces sperm or egg, so these must have half the complement of chromosomes (be haploid) to allow for the full complement to be present (and not more!) when sperm combines with egg.

8. What results from the events that occur during metaphase of mitosis?
   A. The nuclear membranes form around two nuclei.
   B. The chromosomes are aligned on a plane in the centre of the cell.
   C. The chromosomes become visible and attach to the spindle fibres.
   D. The chromatids from each chromosome separate and move to opposite sides of the cell.

Answer is B: During metaphase, chromosomes are arranged on a plane (the metaphase plate) in the middle of the cell, attached to microtubules of the spindle.

9. What is the name of the process of division of a somatic cell's nucleus into two daughter nuclei?
   A. Prophase
   B. Cytokinesis
   C. Mitosis
   D. Meiosis

Answer is C: Mitosis involves somatic cells. Meiosis refers to the production of the sex cells.

10. In a strand of DNA, what is the combination of deoxyribose and phosphate and base known as?
    A. A ribosome
    B. A chromatid
    C. A codon
    D. A nucleotide

Answer is D: Three nucleotides form a codon, and many codons form a chromatid.

11. What happens during anaphase of mitosis?
    A. Spindle fibres pull each chromatid to opposite sides of the cell.
    B. The sense and non-sense strands "unzip" along their hydrogen bonds.
    C. RNA polymerase forms a complementary strand by reading the sense strand.
    D. The cell cytoplasm divides into two cells.

Answer is A: Separation of the two chromatids of a chromosome occurs at anaphase. Choice D is cytokinesis and begins in late anaphase and continues into telophase.

12. The process by which information is read from DNA, encoded and transported outside the nucleus is known as:
    A. Translation
    B. Transcription
    C. Encoding
    D. Catalysis

Answer is B: To "transcribe" is to record the information from a source and to record it at another place (onto mRNA). Then messenger RNA moves out of the nucleus.

13. How many nucleotides are required to code for a single amino acid?
    A. Twenty
    B. Five
    C. Three
    D. One

Answer is C: A sequence of three nucleotides constitute a codon. Each codon is specific for one of the 20 amino acids.

14. The combination of a sugar, a base and at least one phosphate group is given the general term of:
    A. Nucleoside
    B. Amino acid
    C. Polypeptide
    D. Nucleotide

Answer is D: A nucleoside is a nucleotide without a phosphate group.

15. The nucleus of the cell contains the master nucleic acid:
    A. DNA
    B. RNA
    C. mRNA
    D. tRNA

Answer is A: DNA exists in the nucleus. The other three are ribonucleic acids.

16. Which of the following is the correct combination of the components for the nucleic acid DNA?
    A. Phosphate, ribose, uracil
    B. Phosphate, deoxyribose, proline
    C. Phosphate, ribose, thymine
    D. Phosphate, deoxyribose, adenine

Answer is D: DNA has the sugar deoxyribose, proline is an amino acid that does not occur in DNA.

17. In the ribosome of a cell, the mRNA is read to produce the particular amino acid sequence for the formation of a protein. What is this process called?

    A. Translation
    B. Transcription
    C. Transportation
    D. Transmutation

Answer is A: Translation occurs in the cytoplasm of a cell with a ribosome. It is when the information in mRNA is read to produce the sequence of amino acids needed to form a protein.

18. Which of the base pairings in DNA would be correct?

    A. A–T pair
    B. A–G pair
    C. C–T pair
    D. C–A pair

Answer is A: A pairs with T, while C pairs with G.

19. The combination of a sugar and a base is given the general term of:

    A. Nucleoside
    B. Amino acid
    C. Polypeptide
    D. Nucleotide

Answer is A: A nucleotide is formed from a nucleoside and a phosphate group.

20. The nucleic acid which carries the information for protein synthesis from the cell nucleus to the ribosomes is:

    A. DNA
    B. RNA
    C. mRNA
    D. tRNA

Answer is C: "Messenger" RNA carries the data (the message) from the chromosomes in the nucleus to the ribosomes in the cytoplasm.

21. Which of the following is the correct combination of the components for the nucleic acid RNA?

    A. Phosphate, ribose, uracil.
    B. Phosphate, deoxyribose, proline
    C. Phosphate, ribose, thymine
    D. Phosphate, deoxyribose, adenine

Answer is A: RNA must have the sugar ribose. Thymine exists in DNA but not RNA, where it is replaced with uracil.

22. In the nucleus of the cell, DNA is used as a template to form mRNA. What is the process called?
    A. Translation
    B. Transcription
    C. Transportation
    D. Transmutation

Answer is B: Transcription refers to the conversion of information on DNA into the form of mRNA.

23. Which of the following describes the translation step of protein synthesis?
    A. DNA unwinds to expose the "sense" strand.
    B. mRNA reads the sense strand.
    C. Amino acids are split from tRNA and join to make the protein.
    D. Amino acids are split form mRNA and join to form the protein.

Answer is C: The tRNA molecule carries an amino acid to the mRNA molecule. As each amino acid splits from its transfer RNA molecule, the amino acids assemble to form the protein. Choice B describes transcription.

24. Which statement about the triplets of bases on a nucleic acid is correct?
    A. A gene is a triplet found on DNA.
    B. A codon is a triplet found of DNA.
    C. An anticodon is a triplet found on mRNA.
    D. A codon is a triplet found on tRNA.

Answer is B: A triplet of three consecutive bases along a strand of DNA is called a codon. An anticodon is a triplet found on the anti-sense strand of DNA (and on tRNA). The three bases on mRNA are also called a codon but with the U in DNA replaced with T in RNA.

25. If the following code "GTA CGT GAG AAG CAG" is found on the DNA sense strand, what is found on the DNA anti-sense strand?
    A. GTA CGT GAG AAG CAG
    B. CAT GCA CTC TTC GTC
    C. CAU GCA CUC UUC GUC
    D. GUA CGU GAG AAG CAG

Answer is B: Looking at the first codon, G binds to C so either choice B or C is correct. T binds to A and A binds to T, so the anticodon we are looking for is CAT, hence choice B. (The amino acid sequence is: valine, arginine, glutamic acid, lysine, glutamine.)

26. From the DNA codon sequence of question 25 above, what would be the sequence of the tRNA?
    A. GTA CGT GAG AAG CAG
    B. CAT GCA CTC TTC GTC

C. CAU GCA CUC UUC GUC
D. GUA CGU GAG AAG CAG

Answer is D: As we are now focussing on RNA, the T is replaced with U, which eliminates choices A and B. mRNA TRANSCRIBES the DNA so where G and T appears on DNA, C and A (respectively) will appear on RNA and where C and A appear on the DNA, G and U will be on the RNA. Hence, GTA will be CAU on RNA so choice C is how the mRNA will appear. The tRNA TRANSLATES so is complementary to the mRNA and will be GUA.

27. What is the sequence of amino acids corresponding to the following codon sequence "GTA CGT GAG AAG CAG"?
    A. his, ala, phe, val
    B. thr, arg, phe, cys
    C. val, arg, glu, lys, gln
    D. ser, thr, arg, lys, thr

Answer is C: By consulting a DNA codon chart: GTA = valine; CGT = arginine; GAG = glutamic acid; AAG = lysine; GAG = glutamine.

28. During which phase of the cell cycle does DNA replication occur?
    A. G1 stage
    B. G2 stage
    C. S phase
    D. The resting phase

Answer is C: The cell cycle passes through phases G0, G1, S, G2, mitosis. S for synthesis phase is when DNA replicates. At this stage, chromosomes that consist of one chromatid become two chromatids.

29. Which of the following events occur in the stated phase of mitosis?
    A. Chromosomes shorten and thicken in interphase.
    B. Chromosomes line up at the cell equator at prophase.
    C. Chromosomes split into single chromatids at anaphase.
    D. Chromatin material duplicates a second chromatid at telophase.

Answer is C: This ensures that each daughter cell has one chromatid of each chromosome. Shortening and thickening happens in early prophase; assembling at the equator (metaphase plate) happens in metaphase; the cell begins to cleave at interphase which is the end of mitosis; chromosome duplication occurs if the S phase of interphase, not during mitosis.

30. What is the end result of meiosis?
    A. Two haploid cells
    B. Two diploid cells
    C. Four diploid cells
    D. Four haploid cells

Answer is D: Meiosis prepares the sex cells which must have half the number of chromosomes of somatic cells so that when they sperm fuses with the egg, the correct (diploid) number are present. In humans, meiosis results in four haploid sperm cells and one oocyte, with three polar bodies.

31. Which of the following is NOT found in DNA?
    A. Thymine
    B. A double helix
    C. Phosphate
    D. Ribose

Answer is D: Deoxyribose (not ribose) is found in DNA. The ribose sugar is found in RNA (ribonucleic acid).

Chapter 3
Measurement, Errors, Data and Unit Conversion

All measurements involve a number, a unit and a level of uncertainty. The number is usually expressed in scientific notation (with a power of 10) while the units should be metric units and be part of the standard international system (SI) of units. Examples are metre, second, kilogram, ampere as well as derived units such as litre, newton and pascal. However, sometimes the units that are commonly used are not SI and you should be aware of these. The metric units will have standard prefixes (kilo, milli, etc.) to denote known multiples of the standard unit. You should be able to convert between different prefixes of the same unit and to convert between one unit and another. No measurement of a continuous variable (like height, weight, temperature, blood pressure) is known with absolute accuracy so its level of uncertainty is usually stated. Uncertainty is sometimes stated as the "error", but this does not imply that a mistake has been made, just that the true value lies within a known upper and lower boundary.

Large amounts of data are handled by using a statistic (a number) to summarise the data. The mean and the standard deviation are two such statistics. Many biological data are "normally distributed", that is, are symmetrically distributed with most data clustered about a central value with progressively fewer data points the further you move away from the centre. For such data, the "mean" (or average) is an indication of where the middle value of the group of data lies, while the "standard deviation" describes how closely around the mean value the data are clustered.

3.1 Measurement and Errors

1. Say that someone's body temperature is measured by four different devices and the resulting four measurements are given below. Which reading has an absolute error of ±0.05 °C?

M. Caon, *Examination Questions and Answers in Basic Anatomy and Physiology*, https://doi.org/10.1007/978-3-030-47314-3_3

A. 38 °C
B. 37.8 °C
C. 37.85 °C
D. 37.855 °C

Answer is B: Absolute error is plus or minus half the smallest scale interval. Two times 0.05 = 0.1, so the smallest scale interval is 0.1 of a degree Celsius which applies to the 37.8 °C value.

2. If someone's height is measured while the person is wearing shoes, the height will be overestimated. This type of error is known as which of the following?
   A. Absolute error
   B. Parallax error
   C. Calibration error
   D. Zeroing error

Answer is D: Zeroing error because in this example, the object being measured is not aligned with the start of the measuring scale.

3. A bathroom scales displays a mass reading of 68.4 kg. Which one of the following could **NOT** be the true mass of the person standing on the scales?
   A. 68.40 kg
   B. 68.44 kg
   C. 68.47 kg
   D. 68.37 kg

Answer is C: A reading of 68.4 (that is, stated to the nearest 0.1 kg) means that the actual value is between 68.35 and 68.44, so only 68.47 (which is closer to 68.5) is outside this range.

4. Look at the figure of a thermometer. What is the temperature reading?

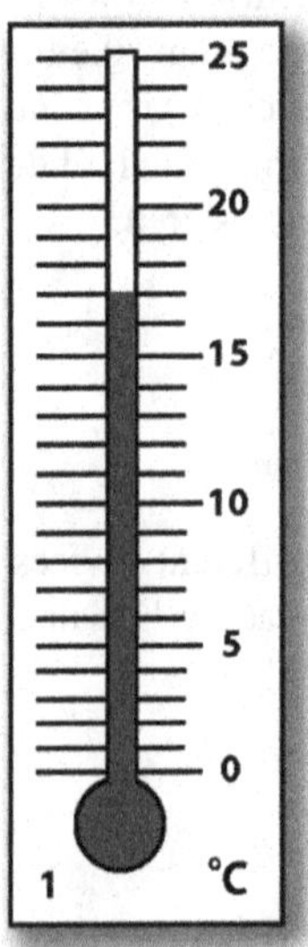

A. 15 °C
B. 15.4 °C
C. 17 °C
D. 20 °C

Answer is C: Every 5 scale intervals is labelled with a number, and each interval corresponds to 1 °C. As the reading is two intervals above 15: 15 + 2 = 17.

5. Look again at the figure of a thermometer. What is the absolute error of the temperature reading?

   A. ±0.05 °C
   B. ±0.5 °C
   C. ±1.0 °C
   D. ±5.0 °C

Answer is B: Absolute error is plus or minus half of the smallest scale interval (which is 1 °C), half of one is 0.5.

6. On a clinical thermometer where the smallest scale interval is 0.1 °C, a person's temperature is measured to be 37.7 °C. Which of the listed temperatures could **NOT** be the person's true temperature?

   A. 37.72 °C
   B. 37.76 °C
   C. 37.67 °C
   D. 37.685 °C

Answer is B: An actual value of 37.76 °C would be seen on a scale that has 0.1 as its smallest interval, as 37.8. All other values are closer to 37.7 than they are to 37.8 or to 37.6.

7. What is meant when a person's mass is stated as 73.6 kg?

   A. The mass is closer to 73.6 than it is to 73.7 or 73.5.
   B. The mass is closer to 73.6 than to any other value.
   C. The mass is between 73.5 and 73.7.
   D. The mass is 73.6 ± 0.1 kg.

Answer is A: Because the mass is stated to one decimal place, the absolute error is ±0.05.

8. Which of the following measurements is a semi-quantitative one?

   A. A blood pressure of 120/80 mmHg
   B. A blood glucose level of +++
   C. A state of anxiety measure of "calm"
   D. The patient's name is Tim Cruise

Answer is B: The number of "+" signs indicates a level of magnitude that is semi-quantitative, but there is no unit of magnitude. C is "qualitative", while D is "nominal".

9. A baby's mass measurement is 3.8 ± 0.05 kg. What is the absolute error in the measurement?

   A. ±0.05/3.8
   B. ±(0.05/3.8) × 100%
   C. ±0.05 kg
   D. 0.05 kg

Answer is C: By definition, absolute error is half the smallest scale interval (0.1 in this case) above and below the measured value.

10. 1 mL of water has a mass of 1.00 g. Which of the following sets of three measurements of the mass of 1 mL of water is the most precise set?

    A. 0.98 g, 1.00 g, 1.02 g
    B. 0.99 g, 0.99 g, 0.99 g
    C. 1.00 g, 1.01 g, 1.02 g
    D. 0.99 g, 0.99 g, 1.00 g

Answer is B: Precision refers to the repeatability of the measurement. In choice B, all measurements are the same so are precise.

11. Systematic errors arise from some inadequacy of equipment or technique. Which of the following is **NOT** an example of systematic error?

    A. Parallax error
    B. Calibration error
    C. Random error
    D. Zeroing error

Answer is C: As the words suggest, random error is not systematic as it is unpredictable.

12. What is the number 0.028 when correctly expressed in scientific notation?

    A. 28×10^{2}
    B. 2.8×10^{2}
    C. 2.8×10^{-2}
    D. 28×10^{-2}

Answer is C: Scientific notation requires one number to the left of the decimal point (choices B and C). The decimal point must be shifted to two places to achieve this. As 0.028 is less than one, the power of 10 is negative.

13. Which of the following numbers has four significant figures?

    A. 3300.0
    B. 37.60
    C. 0.008
    D. 0.0540

Answer is B: Any zero to the left of the first non-zero digit—when approached from the left—are not significant (zeros on the right are).

14. Which of the following statements involves a nominal measurement?
    A. James has a height of 170 cm.
    B. Barry's blood pressure is elevated.
    C. Gino was born in Italy.
    D. More than 5% of Australians receive a pension.

Answer is C: Gino's birthplace is "named" so the information is nominal, but no other information is available.

15. Millimetre of mercury is a unit commonly used for the measurement of blood pressure. Which of the following statements about this unit is true?
    A. It is part of the Australian Metric System but not part of the SI system.
    B. It is part of the SI system but not part of the Australian Metric System.
    C. It belongs to both the SI system and the Australian Metric System.
    D. It does not belong to either the SI system or the Australian Metric System.

Answer is D: Both the SI system and the Australian Metric System are "metric", and millimetre of mercury is not (despite having millimetre in its name).

16. A volume of urine is measured to be 325 mL using a measuring cylinder with a scale of smallest interval 5 mL. What is the absolute error in the measurement?
    A. ±0.5 mL
    B. ±1 mL
    C. ±2.5 mL
    D. ±5 mL

Answer is C: The absolute error is plus or minus half of the smallest scale interval (this means that the maximum error is this, but the actual error may be less). The interval is 5 mL, so 2.5 mL is half of this and equal to the absolute error.

17. A breath analyser unit to measure % blood alcohol content is accurate to ±0.005%. Which of the following displayed values will ensure that the reading is indeed greater than the legal driving limit of 0.05%?
    A. 0.06%
    B. 0.055%
    C. 0.05%
    D. 0.045%

Answer is A: An accuracy of ±0.005% means that for a displayed reading of 0.055%, the actual value may be as low as $0.055 - 0.005 = 0.05\%$ or as high as $0.055 + 0.005 = 0.06\%$. Hence will be between 0.05% and 0.06%. To ensure that the actual % blood alcohol concentration is in fact above 0.05%, the reading should be more than 0.005 above the limit of 0.05%.

18. What distinguishes between absolute error and relative error?
    A. Absolute error is relative error divided by the magnitude of the measurement.
    B. Relative error of a measurement is half of the smallest scale division on the measuring instrument.
    C. Absolute error of a measurement is equal to the smallest scale division on the measuring instrument.
    D. Absolute error is a feature of the scale on the measuring instrument while relative error depends on the magnitude of the measurement.

Answer is D: The absolute error of a measurement is half of the smallest scale division on the measuring instrument. Hence, it is a feature of the scale on the measuring instrument. Relative error is absolute error divided by the magnitude of the measurement. Hence, the larger is the magnitude of the measurement, the smaller will be the relative error.

19. Four anatomists have their mass measured on a scale with a smallest interval of 100 g (0.1 kg). Their masses are displayed below. Which one will have the smallest relative error for their mass?
    A. Dr. Corti, mass = 68.2 kg
    B. Dr. Schlemm, mass = 73.8 kg
    C. Dr. Eustachio, mass = 81.3 kg
    D. Dr. Fallopio, mass = 87.2 kg

Answer is D: Relative error is determined by dividing the smallest scale interval by the magnitude of the measurement. The smallest error will result from whoever has the largest mass. For Dr. Fallopio's mass: RE = (0.1/87.2) × 100 = 0.11%.

20. What is the number 37,000 when expressed in scientific notation?
    A. 37×10^3
    B. 3.7×10^4
    C. 37.0×10^3
    D. 3.7×10^{-4}

Answer is B: Scientific notation requires one number to the left of the decimal point (choices B and D). The decimal point must be shifted four places to achieve this. As 37,000 is greater than 1, the shift is to the right and the power of 10 is positive.

21. How many significant figures do the numbers (a) 3300, (b) 37.60, (c) 0.008, (d) 0.0540 have?
    A. (a) 2 (b) 4 (c) 3 (d) 3
    B. (a) 2 (b) 3 (c) 4 (d) 5
    C. (a) 4 (b) 4 (c) 1 (d) 3
    D. (a) 4 (b) 3 (c) 1 (d) 2

Answer is C: Any zero that appears before a non-zero digit is not significant (the three zeros in 0.008 are not). Zeros that appear after the first non-zero digit are significant (the third zero in 0.0540 is).

22. What is (a) the absolute error and (b) the relative error of the measurement: 49.7 cm when made with a scale whose smallest scale interval is 0.1 cm?

    A. (a) ±0.05 cm (b) (0.05/49.7) × 100
    B. (a) ±0.025 cm (b) (0.05/49.7) × 100
    C. (a) ±0.05 cm (b) (0.025/49.7) × 100
    D. (a) ±0.025 cm (b) (49.7/0.05) × 100

Answer is A: Absolute error is "plus or minus" the smallest scale interval. So half of 0.1 is 0.05. Relative error (expressed as a %) is the absolute error divided by the magnitude of the measurement, then multiplied by 100.

23. Two length measurements are: L1 = 39.25 ± 0.005 km; and L2 = 2745.2 ± 0.05 m. Which measurement is the most accurate and why?

    A. L1 has a smaller absolute error and smaller relative error.
    B. L1 has smaller absolute error, but larger relative error.
    C. L2 has larger absolute error, but smaller relative error.
    D. L2 has a smaller absolute error and smaller relative error.

Answer is D: The smaller the relative error, the more accurate is the measurement. RE of L1 = 0.005/39.25 = 0.000127 (0.0127%), while RE of L2 = 0.05/2745.2 = 0.0000182 (0.00182%). In addition, the absolute error of L1 is 5 m while that of L2 is 5 cm.

24. The acceleration due to the earth's gravity in Adelaide may be badly written as: 0979.711 × 10^{-2} m/s^2. How many significant figures are there in this number?

    A. 3
    B. 4
    C. 6
    D. 7

Answer is C: All digits except for the first zero are significant.

25. Which of the following is 0979.711 × 10^{-2} m/s^2 when written in standard (scientific) form?

    A. 0.979711 × 10 m/s^2
    B. 9.79711 m/s^2
    C. 9.8 m/s^2
    D. 97,971.1 m/s^2

Answer is B: A number written in scientific form has one digit before the decimal place followed by the remaining significant figures and multiplied a power of 10 (if required). ×10^{-2} requires a decimal point shift to the left of two places. The leading zero is not significant and should be left off. Choice C has unnecessarily eliminated some significant figures.

26. A snooker ball has a mass of 250.2 g as measured on a balance that can read to the nearest tenth of a gram. (a) What is the absolute error (AE) of the measurement? AND (b) what is the relative error (RE) of the measurement?

    A. AE = ±0.05, RE = 0.00002
    B. AE = ±0.01, RE = 0.0002
    C. AE = ±0.025, RE = 0.002
    D. AE = ±0.1, RE = 0.02

Answer is A: Absolute error is half of the smallest scale interval of 0.1 g so is 0.05. Relative error is AE/measurement = 0.05/250.2 = 1.998×10^{-4} = 0.00002 = 0.002%

27. Body mass index (BMI) is a number obtained by dividing a person's mass (in kg) by the height (in meter) squared. If mass = 71.4 ± 0.05 kg and height is 179 ± 0.5 cm, calculate the BMI and its relative error.

    A. BMI = 0.00223 ± 0.0063 kg/m^2
    B. BMI = 22.3 ± 0.0063 kg/m^2
    C. BMI = 39.9 ± 0.0035 kg/m^2
    D. BMI = 0.399 ± 0.0035 kg/m^2

Answer is B: BMI = 71.4/(1.79)^2 = 71.4/3.2 = 22.3 kg/m^2. RE of mass = 0.05/71.4 = 0.0007; RE of height = 0.5/179 = 0.0028. To determine the RE of a result obtained by multiplying or dividing measurements, the relative errors of each measurement are added together. Hence RE of BMI = 0.0007 + 0.0028 + 0.00 28 = 0.0063, or 0.63%

28. A syringe containing 50 mL of drug solution is fitted to an infusion pump that has a scale marked with 1 mL intervals. State the absolute error (AR) AND calculate the percentage errors (PE) associated with volumes for this syringe.

    A. AR = ±0.05 mL, PE = 0.01%
    B. AR = ±0.5 mL, PE = 0.01%
    C. AR = ±0.05 mL, PE = 1%
    D. AR = ±0.5 mL, PE = 1%

Answer is D: Absolute error is plus or minus half of the smallest scale interval. So half of 1.0 is 0.5 mL. Percentage error is relative error × 100: PE = (0.5/50) × 100 = 1%.

29. Six babies have their mass measured at: 4.4, 5.1, 5.2, 5.2, 5.8 and 6.1 kg. They have all been weighed on scales marked in 10 g divisions. What is the mean mass of the group? AND, which baby's mass would have the largest relative error?

    A. 5.2 kg largest relative error for 4.4 kg baby
    B. 5.3 kg largest relative error for 4.4 kg baby
    C. 5.2 kg largest relative error for 6.1 kg baby
    D. 5.3 kg largest relative error for 6.1 kg baby

Answer is B: The sum of the six masses is 31.8 kg and when divided by 6, gives the mean = 5.3 kg. Relative error is absolute error divided by the measurement. As the absolute error for all six measurements is the same (=5 g), the highest relative error will be for the smallest baby. RE = 5/4400 = 0.0011 (or 0.11%).

30. What is the number 472.93 when written in scientific notation?

A. 472.93×10^2
B. 4.7293×10^{-2}
C. 4.7293×10^2
D. 4.7293×10^4

Answer is C: Scientific notation requires on digit ahead of eth decimal point, followed by the other significant digits then a power of 10. As the decimal point was shifted two places to the left to achieve this, the power of 10 is 2.

3.2 Units and Conversion

1. Which of the following metric prefixes is used to denote one thousandth of a gram?

A. Micro
B. Milli
C. Centi
D. Kilo

Answer is B: "Milli" refers to thousandth or 10^{-3} (nothing to do with million!).

2. Which of the following does **NOT** describe a milligram?

A. 1×10^3 g
B. 1×10^{-3} g
C. One thousandth of a gram
D. 0.001 g

Answer is A: This (1×10^3 g) is 1000 g = 1 kg.

3. How many micrograms are there in 5 mg?

A. 0.005
B. 0.5
C. 500
D. 5000

Answer is D: 1 mg is 1000 μg, so 5 mg = 5000 μg.

4. Given that a milligram 1×10^{-3} g, what is a microgram?

A. 1×10^3 g
B. 1000 mg

C. 1×10^{-6} g
D. 0.001 g

Answer is C: A microgram is one millionth (0.000001 or 10^{-6}) of a gram. A = 1 kg; B = 1 g; D = 1 mg.

5. How many micrograms are there in 1 mg?
   A. 0.001
   B. 0.1
   C. 100
   D. 1000

Answer is D: 1 μg = $10^{-3} \times 1$ mg, so 1000 μg is the same as 1 mg.

6. How many milligrams are there in 1 μg?
   A. 0.001
   B. 1000
   C. 0.1
   D. 1,000,000

Answer is A. 1 mg = $10^3 \times 1$ μg, so one thousandth of a milligram is the same as 1 μg.

7. Which of the following units is **NOT** part of the Australian Metric System of units?
   A. Millimetre of mercury for measuring blood pressure
   B. Degree Celsius for measuring temperature
   C. Pascal for measuring pressure
   D. Second for measuring time

Answer is A: Pascal (Pa) is the SI unit for pressure.

8. In the Australian Metric System of units, what does the prefix micro stand for?
   A. 1000
   B. One thousandth
   C. 1,000,000
   D. One millionth

Answer is D: One millionth = 10^{-6}

9. What is 3400 cm^2 converted to square metres?
   A. 0.0034 m^2
   B. 0.34 m^2
   C. 3.4 m^2
   D. 34 m^2

Answer is B: A square metre has sides that are 100 cm long, so $100 \times 100 = 10{,}000$ cm^2 in a square metre. So 3400/10,000 = 0.34.

10. In the Australian Metric System of measurement, what does the prefix "milli" stand for?
    A. One thousandth
    B. 1000
    C. One millionth
    D. 1,000,000

Answer is A: Milli = one thousandth = 10^{-3}

11. What is 120 mg expressed as grams?
    A. 0.12 g
    B. 1.2 g
    C. 12 g
    D. 12,000 g

Answer is A: 1 mg is a thousandth of a gram (0.001 g), so 120 mg is 120 thousandths = 0.120. (Milligram to gram: shift the decimal point three places to the left.)

12. How many milligrams are in 0.75 g?
    A. 0.00075 mg
    B. 7.5 mg
    C. 75 mg
    D. 750 mg

Answer is D: 1 g is 1000 mg. So three quarters of a gram is three quarters of a thousand milligrams = 750 mg. (Gram to milligram: shift the decimal point three places to the right.)

13. What is 1.25 g converted to micrograms?
    A. 125
    B. 1250
    C. 12,500
    D. 1,250,000

Answer is D: Micro refers to "a millionth of". There are 1.25×10^6 millionths of a gram in 1.25 g. So 1,250,000 millionths of a gram = 1,250,000 µg. (Gram to microgram: shift the decimal point six places to the right.)

14. How many micrograms are there in 0.25 g?
    A. 250
    B. 25,000
    C. 250,000
    D. 25,000,000

Answer is C: Micrograms are smaller than grams, so there will be more of them. A million times more, so 0.25 × 106 = 250,000 µg. (Gram to microgram: shift the decimal point six places to the right.)

15. When 2.25 mg is converted to micrograms, how many micrograms are there?
    A. 22.5
    B. 2250
    C. 225,000
    D. 22,500,000

Answer is B: There are 1000 μg in each milligram. Hence 2.25 thousands is 2250. (Milligram to microgram: shift the decimal point three places to the right.)

16. What is the result of converting 650 μg to milligrams?
    A. 0.650
    B. 6.50
    C. 65
    D. 65,000

Answer is A: 1000 μg are needed to make a milligram, so we have less than one. (Microgram to milligram: shift the decimal point three places to the left.)

17. Given that 1 mmHg = 0.133 kPa, how can a systolic BP measurement of 120 mmHg be converted to a measurement in kilopascal?
    A. Divide 120 by 0.133
    B. Multiply 120 by 0.133
    C. Divide 0.133 by 120
    D. Add 0.133–120

Answer is B: We know that 1 mmHg is the same as 0.133 kPa, so two of them would be $2 \times 0.133 = 0.266$ kPa. Hence 120 of them would be $120 \times 0.133 = 16$ kPa

18. A newborn baby weighs 7 lb 8 oz (seven pounds and eight ounces). If 1 lb = 0.454 kg, what is done to convert the baby's weight to kilograms?
    A. Multiply 7.8 by 0.454
    B. Divide 7.8 by 0.454
    C. Divide 7.5 by 0.454
    D. Multiply 7.5 by 0.454

Answer is D: You need to know that 8 oz is half a pound and that "a half" is the same as 0.5! Hence, 7 lb 8 oz = 7.5 pounds = $7.5 \times 0.454 = 3.4$ kg

19. By Internet search, you discover that 1 in. (1″) is the same as 2.54 cm. How would you convert a man's waist girth (circumference) from 40″ to centimetres?
    A. Multiply 40 by 2.54
    B. Divide 40 by 2.54
    C. Divide 2.54 by 40
    D. Multiply 40.1 by 2.54

Answer is A: You know that 1″ is the same as 2.54 cm, so ten of them would be $10 \times 2.54 = 25.4$ cm. Hence, 40 of them would be $40 \times 2.54 = 110.6$ cm.

20. A clinical thermometer is used to measure someone's body temperature to be 101 °F (Fahrenheit). When this temperature is converted to degrees Celsius using the conversion formula: T (°C) = (T (°F) – 32) × 5/9, what is the correct value?

    A. −6.5
    B. 38.3
    C. 82
    D. 124

Answer is B: Centigrade temperature values above zero are always smaller than Fahrenheit values. T (°C) = (T (°F) – 32) × 5/9 = (101 – 32) × 5/9 = 69 × 5/9 = 34 5/9 = 38.3 °C

21. Match the following "power of 10" notations: 10^{-2}, 10^{-3}, 10^{3}, 10^{-6}, 10^{6} with the appropriate SI unit prefixes given below:

    A. Milli, centi, mega, kilo, micro
    B. Centi, milli, kilo, micro, mega
    C. Centi, kilo, milli, micro, mega
    D. Milli, centi, kilo, mega, micro

Answer is B: One hundredth is "centi" and one thousandth is "milli".

22. What is meant by the "Systeme Internationale" (SI) units that are used to measure quantities?

    A. They are the British Imperial units of: foot, pound, gallon, degree Fahrenheit, mile, etc.
    B. They are the USA customary units of: nautical mile, hundredweight, furlong, bushel, etc.
    C. They are the technical units such as: atmospheres, calorie, teaspoon, thou of an inch, etc.
    D. They are the standard international system based on the metre, kilogram, ampere, kelvin, etc.

Answer is D: Systeme Internationale is a unit system based on metre, kilogram and second. There are seven base units and many units are derived from the base. It employs a universal system of prefixes (e.g. milli, mega, pico, etc.) to denote multiples of 10 of the SI units.

23. Which of the following is a base unit within the SI unit system?

    A. Second
    B. Pascal
    C. Joule
    D. Kilometre per hour

Answer is A: The second is one of the seven base units in the SI system, the rest are all derived SI units.

24. Which of the following is a derived unit within the SI unit system?
    A. Hertz
    B. Candela
    C. Kelvin
    D. Mole

Answer is A: The unit for frequency hertz (Hz) is derived. The others (B, C, D) are SI base units for luminous intensity, temperature and amount of substance, respectively.

25. What is meant by a "scalar" quantity?
    A. One that can be measured using a scale.
    B. One that is defined by a magnitude.
    C. One that needs both a magnitude and a direction to be defined.
    D. One that is known exactly without error.

Answer is B: A scalar quantity needs only a magnitude to be defined. Examples are time, mass, speed, volume, temperature.

26. What is meant by a "vector" quantity?
    A. One that can be measured using a scale.
    B. One that is defined by a magnitude.
    C. One that needs both a magnitude and a direction to be defined.
    D. One that is known exactly without error.

Answer is C: A vector quantity requires both a magnitude and a direction to be defined completely. Examples are velocity, displacement, force, acceleration, momentum, electric field strength.

27. The "Systeme Internationale" (SI) units have a set of standard prefixes. Which of the following list four of them correctly?
    A. Micro = 10^{-6}; hecto = 10^{2}; kilo = 10^{3}; mega = 10^{6}
    B. Milli = 10^{-3}; centi = 10^{-1}; deka = 10^{1}; giga = 10^{9}
    C. Nano = 10^{-8}; micro = 10^{-5}; deci = 10^{-1}; tera = 10^{9}
    D. Deci = 10^{-1}; deka = 10^{1}; hecto = 10^{2}; kilo = 10^{4}

Answer is A: Centi = 10^{-2} (not 10^{-1}); micro = 10^{-6} (not 10^{-5}); tera = 10^{12} (not 10^{9}); kilo = 10^{3} (not 10^{4}).

28. Which of the following is the unit of a base unit in the SI system of units?
    A. Kilometre (km)
    B. Kilogram (kg)
    C. Kilopascal (kPa)
    D. Degrees Celsius (°C)

Answer is B: Kilogram (not gram) is the base unit. Metre (not km) is a base unit. Pascal is a derived unit, while kelvin (not degree Celsius), is the base unit of temperature—even though 1 K has the same magnitude as 1 °C.

29. Which of the following is the unit of a derived unit in the SI system of units?
    A. Millisecond
    B. Gram
    C. Metres/second
    D. Kelvin

Answer is C: This unit of speed is derived from distance and time. Millisecond and gram are just multiples of the base units second and kilogram.

30. Poiseuille's law for flow rate may be written in symbols as: $V = \frac{r^4 \Delta P}{2.55 \eta l}$ where r = radius of pipe, P is pressure, l is length of pipe and η is viscosity in units of Pa s. What is the unit of flow rate, V?
    A. m^4 N/m^2
    B. N/m^2 s m
    C. m^3/s
    D. Pa s m

Answer is C: Replacing the symbols in Poiseuille's law with their units gives: (m^4 Pa)/(Pa s m). Now rewrite Pa as N/m^2 to give: (m^4 N/m^2)/(N/m^2 s m); now cancel out units that appear in the numerator and the denominator within a bracket to give (m^2 N)/(N/m s); now cancel terms that appear in both brackets to give m^3/s. This is just a volume per second which is just what flow rate (V) is.

31. A healthy human body temperature is 37 °C, what is this temperature converted to kelvin?
    A. 37 K
    B. −236 K
    C. 236 K
    D. 300 K

Answer is D: 0 K = −273 °C. To convert Celsius to kelvin, add 273, hence 37 °C + 273 = 300 K.

32. What is the (gauge) pressure in a car tyre at 28 psi (pounds per square inch) when converted to kilopascals?
    A. 193 kPa
    B. 193,000 kPa
    C. 34.8 kPa
    D. 4.1 kPa

Answer is A: 1 psi = 6.89 kPa, hence 28 psi in kilopascal is 28 × 6.89 = 193 kPa.

33. How many micrograms are there in 150 mg?
    A. 150,000
    B. 0.15
    C. 0.015
    D. 0.00015

Answer is A: There are 1000 mg in a microgram. So 150 mg = 150,000 μg.

34. Which of the following is equivalent to 100 μm?
    A. 0.1 m
    B. 0.01 m
    C. 0.01 mm
    D. 0.1 mm

Answer is D: There are 1000 μm in a millimetre. So 100 μm = 0.1 mm.

35. Which of the following is equivalent to 0.01 L?
    A. 100.0 mL
    B. 0.1 mL
    C. 1.0 mL
    D. 10.0 mL

Answer is D: There are 1000 mL in 1 L. So 100 mL in 0.1 L and 10 mL in 0.01 L.

3.3 Data and Statistics

1. What can be correctly said of data that are "normally distributed"?
    A. The upper and lower values of the distribution describe the healthy range of physiological values.
    B. The standard deviation characterises the dispersion of data, and the variance characterises the central tendency of the data.
    C. The mean and range are statistics that are strictly only applicable to normally distributed data.
    D. Sixty eight percent of all data values will be within 1 SD from the mean.

Answer is D: Normally distributed data have this predicable relationship between their mean and the spread of values around the mean.

2. Median is a measure of central tendency. It may be defined as:
    A. The value that has half the values greater than it and half less than it
    B. The value that occurs most often
    C. The distribution of values that has the mode, mean and average equal to each other
    D. The sum of all values divided by the number of values.

Answer is A: Median is the mid-point of the number of measured values. The value that appears most often in a set of data is called the mode.

3. What is the standard deviation used for?
    A. As a measure of central tendency
    B. As a measure of dispersion
    C. As a measure of spread of data that are normally distributed
    D. As a measure of the error of the mean value

Answer is C: B is also correct but is not as good an answer as choice C.

4. What information does the "standard deviation" of a mean value tell us?
   A. It gives us the healthy range of values for the measured physiological quantity.
   B. It is the range within which 68% of measured values are found.
   C. It tells us that the measured values are normally distributed.
   D. It tells us the number of values that were used to calculate the mean.

Answer is B: C is also correct but choice B is the better answer.

5. What does the standard deviation of the mean represent? For values that are normally distributed, it represents:
   A. The value above and below the mean that includes 68% of all data values
   B. The difference between the highest data value and the lowest data value
   C. The average of the difference between each data value and the mean value
   D. The spread of the normal distribution

Answer is A: The term "standard" in standard deviation of a distribution of measured values, means that it may be relied upon to encompass 68% of all measured values.

6. Which of the following statements applies to the statistic known as the "standard deviation"?
   A. It is a measure of central tendency.
   B. It is only applicable to qualitative measurements.
   C. Standard deviation is also known as the "variance".
   D. 95% of all data lie within 2 SD of the mean.

Answer is D: This is the only true statement for normally distributed data. Standard deviation is the square root of the variance.

Note: The following three questions about the "central tendency" of data, rely on the following information. Consider the weekly earnings in dollars for ten workers to be: 400, 475, 475, 475, 500, 500, 525, 620, 630 and 660. These ten wages add to the total: $5260.

7. What is the "average" wage—technically referred to as the arithmetic mean?
   A. $475
   B. $500
   C. $526
   D. $5260

Answer is C: The arithmetic mean is the sum of the ten wages, divided by the number of wages (10). So $5260/10 = $526.

8. What is the "median" wage of these ten?
   A. $475
   B. $500
   C. $526
   D. $600

Answer is B: The median is the middle number. Half of the ten numbers are larger than (or equal to) the median and half are smaller. The wages have been arranged in ascending order: the fifth is $500 and the sixth is also $500. So half of the numbers are $500 or less and half are $500 or more. The median is midway between these two numbers and is $500.

9. What is the "mode" value of these ten wages?
   A. $475
   B. $500
   C. $526
   D. $620

Answer is A: Mode is the most commonly occurring value. $475 appears three times so is the mode.

10. The mean June midday temperature in Desertville is 36 °C and the standard deviation (SD) is 3 °C. Assuming the temperature data for June to be normally distributed, how many days (to the nearest whole day) in June would you expect the midday temperature to be between 39 and 42 °C? (Hint: how many SDs are these two temperatures from the mean?)
    A. 3
    B. 14
    C. 7
    D. 4

Answer is D: 39 °C is 1 SD above the mean and 42 °C is 2 SD above the mean. Sixty-eight percent of values lie within 1 SD of the mean and 95% lie within 2 SD of the mean. Therefore, 27% of days (95 − 68 = 27%) would have a temperature between 30 and 33 °C or 39 and 42 °C. Hence we would expect the temperature to be between 39 and 42 °C on half of 27% (i.e. 13.5%) of the days in June. There are 30 days in June. So 13.5% of 30 = 0.135 × 30 = 4.05 = 4 to the nearest day.

11. For a distribution of data that is "normally distributed" what can be said of the mean, median and the mode?
    A. They have the same value.
    B. The mean is lower than the median and mode.
    C. The mean is higher than the median and the mode.
    D. They cannot take on the same value.

Answer is A: In a perfectly normal distribution, the mean, median and mode have the same value.

12. The healthy male range for the physiological variable blood haemoglobin concentration is about 130–160 g/L. How is this range determined?
    A. It is ±1 SD from the mean value.
    B. It is ±2 SD from the mean value.

C. It is ±3 SD from the mean value.
D. It is the minimum and maximum value found for healthy males.

Answer is B: The reference interval comprises a range of ±2 SD from the mean. It indicates the limits that should cover 95% of normal subjects. That is 5% of healthy individuals will fall outside this range.

13. Many measured human physiological variables are found to have a normal distribution. For example mean fasting blood glucose is 4.7 mmol/L with range: 3.8–5.5 mmol/L. What does the healthy range signify?
    A. Most healthy individuals will have a blood glucose within this range.
    B. All healthy individuals will have a blood glucose within this range.
    C. Individuals whose blood glucose is outside this range are unhealthy.
    D. Five percent of healthy individuals will have a blood glucose outside this range.

Answer is D: With "normally distributed" data there is a quantitative relationship between the mean value and the spread of values about the mean. The "healthy range" is taken to be ±2 SD from the mean value. It is well known that 95% of values will fall into this range. And also that 5% of healthy folk will have a value outside of this range.

Chapter 4
Chemistry for Physiology

4.1 Atoms and Molecules

There are 90 naturally occurring simplest substances called chemical "elements". An additional 28 have been produced artificially. Thirty-four of the 118 elements are radioactive, while there also exist some radioactive forms of the stable elements, such as ^{14}C, ^{40}K and ^{3}H. This trio of naturally occurring radioactive elements are incorporated into our bodies. The smallest particle of an element is called an atom of that element. The names of these elements are shortened to one- or two-letter symbols that are displayed on the "periodic table" of elements. Metal elements (e.g. Li, Na, Ca, K) appear on the left while non-metal elements (e.g. Cl, O, N) appear on the right hand side of this table. A metal element may react with a non-metal element to form a new substance which will be a type of "ionic" compound. A non-metal element may react with another non-metal element to form a new substance which will be a type of "covalent" compound. Ionic compounds in solid form are continuous lattice structures, which, when they dissolve, allow the particles to move about separately as positive ions if they have lost an electron(s) or negative ions if they gained electron(s). Covalent compounds exist as groups of atoms (known as molecules), with a fixed ratio of different atoms. The atoms in these molecules stay together. Examples are H_2O (water), $C_6H_2O_6$ (glucose), CO_2 (carbon dioxide), CH_3COOH (acetic acid). Ions and small molecules such as these and amino acids and lipid molecules are able to move into and out of cells through pores in the plasma membrane during normal cell functioning.

On a macroscopic scale, humans are a skinful of solids and liquids with no spaces (molecules and ions are "shoulder to shoulder"). However on the sub-atomic scale, atoms are mostly empty space! If the central nucleus (which consists of two types of particle called protons and neutrons) of an atom was enlarged to the size of a tennis ball, the whole atom, enlarged by the same factor would have a diameter of about 600 m with just a few electrons occupying the mostly empty space which surrounds the "tennis-ball-sized" nucleus. Therefore, humans are mostly empty

M. Caon, *Examination Questions and Answers in Basic Anatomy and Physiology*, https://doi.org/10.1007/978-3-030-47314-3_4

space surrounding the nuclei of our atoms. This is the reason that some X-ray photons will pass through the body without being deflected by any interaction with the electrons or nuclei.

1. The chemical formula $C_6H_{12}O_6$ contains much information. However, what information is **NOT** provided by the formula?
   A. The number of atoms in a molecule
   B. The name of the substance
   C. The elements that make up the substance
   D. Whether the substance is covalent or ionic

Answer is B: The name may be guessed at, but it is possible for two substances to have the same formula but a different structure. Glucose and fructose both have this formula.

2. Which of the particles listed below is the smallest?
   A. An atom
   B. A proton
   C. An ion
   D. A molecule

Answer is B: A proton is a sub-atomic particle so is smaller than all of the others

3. The chemical for sodium is which of the following?
   A. Na
   B. So
   C. K
   D. Si

Answer is A: Na is short for natrium which is the Latin name for sodium.

4. A molecular compound may be defined by which of the following?
   A. Atoms from non-metal elements covalently bonded
   B. Atoms from metal elements covalently bonded
   C. Atoms from metal elements and non-metal elements covalently bonded
   D. Atoms from non-metal elements ironically bonded

Answer is A: Only non-metal elements are involved in covalent (molecular) bonding.

5. Many drugs are neutralised to form salts and administered in this form. What is the main advantage of administering the salt form of the drug? It is usually:
   A. Less toxic
   B. More soluble in water
   C. More pleasant to taste
   D. More soluble in lipid

Answer is B: Salts, having particles that are electrically charged, are more likely to be soluble in water.

6. Which one of the following is **INCORRECT**?
   A. Metal atoms will form compounds with non-metal atoms.
   B. Metal atoms will form compounds with metal atoms.
   C. Non-metal atoms will form compounds with non-metal atoms.
   D. Metal atoms will not form compounds with metal atoms.

Answer is B: Metal atoms can only form (ionic non-molecular) compounds with non-metal atoms.

7. Choose the correct statement about hydrogen bonds.
   A. They are stronger than covalent bonds.
   B. They act between the H in one –OH or –NH group, and the O or N in another.
   C. They operate within molecules.
   D. They act between the H in one –OH or –NH group, and the H in another.

Answer is B: The slightly positive H atom in one molecule is attracted to the adjacent slightly negative O (or N).

8. Given that the atomic mass of nitrogen is 14 and of hydrogen is 1, what is the mass in grams of 1 mol of ammonia (NH_3)?
   A. 15
   B. 16
   C. 17
   D. 18

Answer is C: From the formula, there is one N and three H atoms. So $(1 \times 14) + (3 \times 1) = 17$.

9. Choose the ending that will correctly complete the sentence: When atoms of a metal element and atoms of a non-metal element react, the result is:
   A. A covalent compound consisting of molecules
   B. A covalent compound consisting of ions
   C. An ionic compound consisting of ions in a lattice
   D. An ionic compound consisting of molecules in a lattice

Answer is C: Atoms of a metal element react with atoms of a non-metal element to form an ionic compound (ions held within a lattice when in solid form).

10. In a water molecule, the bond between oxygen (O) and hydrogen (H) is:
    A. A covalent bond and a polar bond
    B. An ionic bond
    C. A covalent bond
    D. An ionic bond and forms an electrolyte

Answer is A: As both atoms are non-metals, the bonding is covalent. As the atoms differ in their attraction for the bonding electrons, the bond is polar (the electrons are more closely attracted to O).

11. Some atoms of potassium, K, contain 19 protons and 20 neutrons in their nuclei. What is the correct symbol for these atoms?
    A. $^{39}_{19}K$
    B. $^{19}_{20}K$
    C. $^{39}_{19}K$
    D. $^{20}_{39}K$

Answer is C: The number of protons (the atomic number) is written as a subscript. The sum of the number of protons and neutrons (the mass number) is written as a superscript.

12. The chemical elements can be divided into metal elements and non-metal elements. Which of the statements about metals and non-metals is correct?
    A. Metals lose electrons to become charged particles called cations.
    B. Most of the elements are non-metals.
    C. Non-metals are located at the left hand side of the periodic table.
    D. Metals have low melting points and are good conductors of heat.

Answer is A: Metal ions are positively charged (having lost an electron/s), consequently they are attracted to a cathode (a negatively charged electrode).

13. Twenty-four of the chemical elements are essential to the human body. Four bulk elements, seven are macrominerals and 13 are trace elements. Which are the four bulk elements?
    A. Calcium, carbon, hydrogen and oxygen
    B. Nitrogen, carbon, hydrogen and oxygen
    C. Calcium, nitrogen, carbon and oxygen
    D. Carbon, oxygen, phosphorus and iron

Answer is B: These elements make up the bulk of proteins, carbohydrates and lipids.

14. Most non-molecular compounds are formed by the chemical combination of:
    A. Molecules with molecules
    B. Non-metals with non-metals
    C. Metals with metals
    D. Metals with non-metals

Answer is D: Ionic compounds are formed when metal atoms react with non-metal atoms. The structures the form are not molecular.

15. Molecules are relatively easy to separate from one another. This means that the bonds between them are:
    A. Ionic
    B. Covalent

C. Relatively weak
D. Relatively strong

Answer is C: Ease of separation implies relatively weak bonds. Both ionic and covalent bonds are strong.

16. What happens when a sodium atom reacts to form a compound? The atom will:
   A. Gain one electron
   B. Lose one electron
   C. Gain two electrons
   D. Lose two electrons

Answer is B: Sodium is a metal so will lose an electron. As it occurs in period IA of the periodic table, we know it loses a single electron.

17. When nitrogen combines with hydrogen to form a compound, its formula will be:
   A. NH_3
   B. N_3H
   C. NH_4
   D. NH

Answer is A: N requires three electrons to complete its outer shell (it occurs in column VA of the periodic table), while H requires one. Hence three H must attach to a single N.

18. Which of the following type of bond between atoms is the weakest?
   A. Ionic bond
   B. Polar bond
   C. Covalent bond
   D. Hydrogen bond

Answer is D: The hydrogen bond is the weakest. Almost all covalent bonds are polar bonds.

19. Ionic, non-molecular compounds are likely to have which of the following sets of properties?
   A. High melting point, often soluble in organic liquids, in pure form do not conduct electricity.
   B. Low melting point, no strong odour, soluble in water, electrical conductivity in solution.
   C. High melting point, electrical conductivity in solution, no strong odour, often soluble in water.
   D. Low melting point, strong odour, soluble in organic liquids, in pure form do not conduct electricity.

Answer is C: High melting points and electrical conductivity are indicative of ionic compounds.

20. Which of the following best describes a molecule?
    A. The simplest structure in an ionic compound
    B. One thousandth of a mole
    C. The particles of which covalent compounds are composed
    D. The smallest particle of an element

Answer is C: Molecules are the particles of covalent compounds.

21. Which of the following is true of metal elements?
    A. The four most common metal elements in the body have the symbols: C, H, O and N.
    B. They form compounds with non-metals.
    C. When they form compounds, they gain electrons.
    D. In the body, ions of metal atoms have a negative charge.

Answer is B: All other choices are wrong.

22. Which of the following symbols represents a chemical element?
    A. O
    B. CO
    C. CO_2
    D. H_3O^+

Answer is A: O symbolises oxygen, element number 16.

23. What is the name given to the particles that make up a covalent compound?
    A. Ions
    B. Molecules
    C. Anions and cations
    D. Metal atoms

Answer is B: This is a definition of molecules.

24. Compounds may be described by which of the following sentences?
    A. Pure substances that contain two or more elements.
    B. Materials that are composed of particles called molecules.
    C. Materials composed of more than one part, and the parts may be present in any proportion.
    D. Substances that appear on the right hand side of the periodic table.

Answer is A: Choice B refers only to covalent compounds.

25. The symbol for potassium is which of the following?
    A. PO_4
    B. Po
    C. P
    D. K

Answer is B: K is from kalium which is the Latin word for potassium. The letter P is used of phosphorus and Po for polonium.

26. "Covalent" is the term applied to which of the following bonds?
    A. Those between an ion and all the surrounding oppositely charged ions
    B. The bond between an electrolyte and the surrounding water molecules in a solution
    C. Those between non-metal atoms
    D. Those between atoms on the left hand side of the periodic table

Answer is C: Non-metals share electrons when they react with each other (rather than losing or gaining), hence the prefix "co-".

27. Given that 1 mol is 6×10^{23} particles, how much is a millimole?
    A. 6×10^{20} particles
    B. 10^{6} mol
    C. 10^{-6} mol
    D. 6×10^{-3} particles

Answer is A: A millimole is one thousandth of a mole, so 1 mmol = $10^{-3} \times 6 \times 10^{23}$ = 6×10^{20} particles.

28. Which of the following is a definition of a molecule? The particle that composes:
    A. Covalent compounds
    B. Non-metal elements
    C. Electrolytes
    D. Ionic compounds

Answer is A: This is one definition of a molecule.

29. What is the smallest particle of a non-metal element known as?
    A. A molecule
    B. An atom
    C. An ion
    D. A neutron

Answer is B: An atom is the smallest particle of any element—not just non-metals!

30. What is the difference between ions and molecules?
    A. Ions have an electrical charge, whereas molecules do not.
    B. Ions are from metal elements only, whereas molecules contain only non-metal elements.
    C. Ions arise from compounds between non-metal elements, whereas molecules arise from metal and non-metal elements.

D. An ion may be formed from a single atom, but molecules always involve more than one atom.

Answer is D: This is the only correct choice. However, molecules may also form ions—polyatomic ions.

31. Of the four different types of matter listed below, which is not an example of an element?

A. Hydrogen
B. Oxygen
C. Water
D. Gold

Answer is C: Water is a compound of H and O.

32. A certain pure substance, A, when heated is changed into two quite different pure substances, C and D. Which of the following statements must be true?

A. A is a compound.
B. C and D are not elements.
C. A, C and D are all compounds.
D. C and D are elements.

Answer is A: Only this choice must be true. The others are just possible.

33. Which of the following statements about atoms is **FALSE**?

A. They are mostly empty space.
B. Nearly all their mass is concentrated in the nucleus.
C. In a neutral atom protons and electrons are equal in number.
D. The nucleus contains equal numbers of protons and neutrons.

Answer is D: For the elements with smaller atoms, usually this is true. For heavier elements it is not.

34. In which of the following sequences are particles listed in order of increasing size from left to right?

A. Electron, atom, proton, molecule
B. Molecule, atom, proton, electron
C. Atom, proton, electron, molecule
D. Electron, proton, atom, molecule

Answer is D: An electron so far is immeasurably small, while a molecule must have at least two atoms and the proton is a sub-atomic particle.

35. Which of the following is the name of a subatomic particle?

A. Anion
B. Cation
C. Molecule
D. Neutron

Answer is D: A neutron is one of the constituents of the atomic nucleus.

36. Two atoms have the same mass number but different atomic numbers. Which of the following statements concerning these atoms is **TRUE**?

    A. Each has the same number of neutrons in its nucleus.
    B. They are isotopes.
    C. They are atoms of different elements.
    D. Each has the same number of protons in its nucleus.

Answer is C: Different atomic numbers means different numbers of protons and hence different elements.

37. What is the atomic number of the element occupying Group VA and Period 4 of the periodic table?

    A. 33
    B. 34
    C. 51
    D. 52

Answer is A: This is arsenic (As)—you need to consult a periodic table to answer this.

38. Some atoms of iodine, I, contain 53 protons and 78 neutrons in their nuclei. A correct symbol for these atoms would be:

    A. $^{131}_{53}I$
    B. $^{78}_{53}I$
    C. $^{131}_{78}I$
    D. $^{53}_{78}I$

Answer is A: The convention is: the number of protons is the subscript; the sum of the number of protons and neutrons (131) is the superscript.

39. Which of the following atoms normally forms ions having a single, positive charge?

    A. Mg
    B. S
    C. Cl
    D. K

Answer is D: This is because potassium is a metal and from group IA of the periodic table.

40. Which of the following statements concerning isotopes is **FALSE**?

    A. They contain the same number of protons in their atoms.
    B. They contain the same number of electrons in their atoms.
    C. They contain the same number of neutrons in their atoms.
    D. They have very similar chemical properties.

Answer is C: Having a **different** number of neutrons (while having the same number of protons) is what defines them to be isotopes.

41. The element nitrogen exists as molecules, N_2. Which of the following representations of the bonding in a molecule of nitrogen is correct?
    A. $N^+ N^-$
    B. N–N
    C. N=N
    D. N≡N

Answer is D: Nitrogen is placed in period VA of the periodic table and so needs to share three electrons. That is, needs to form three covalent bonds.

42. Which of the following properties is least likely to be possessed by a covalent, molecular substance?
    A. Strong odour
    B. High solubility in water
    C. Melting point above 400 °C
    D. Low electrical conductivity

Answer is C: A high melting point is characteristic of ionic substances. Some covalent compounds are soluble in water.

43. The diagrams below display the covalent bonds present in a series of simple molecules. In which case is the <u>number</u> of bonds surrounding an atom in the diagrams below **INCORRECT**?
    A. Carbon dioxide, **O–C–O**
    B. Carbon tetrachloride, 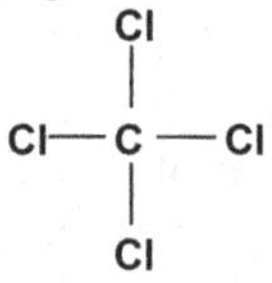 Cl–C(–Cl)(–Cl)–Cl (Cl above, below, left and right of C)
    C. Ammonia, 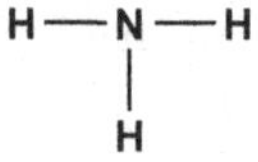 H–N(–H)–H (H left, right and below N)
    D. Water, **H–O–H**

Answer is A: Carbon must form four bonds (not two). It is from group IV A in the periodic table.

44. The formula for glucose is $C_6H_{12}O_6$. How many atoms of each type of element are there in a molecule of glucose?
    A. 6 atoms of calcium; 12 atoms of helium, 6 atoms of osmium
    B. 1 atom of carbon; 6 atoms of hydrogen, 12 atoms of oxygen
    C. 1 atom of carbon; 18 atoms of hydrogen, 6 atoms of oxygen
    D. 6 atoms of carbon; 12 atoms of hydrogen, 6 atoms of oxygen

Answer is D: The number of atoms, written as a subscript, immediately follows the symbol for the element and C, H, O are the symbols for carbon, hydrogen and oxygen, respectively.

45. The formula for oleic acid (a fatty acid) may be written as: $CH_3(CH_2)_7CH{=}CH(CH_2)_7CO_2H$. How many atoms of each type of element are there in a molecule of oleic acid?
    A. 18 atoms of carbon; 11 atoms of hydrogen, 1 atom of oxygen
    B. 12 atoms of carbon; 10 atoms of hydrogen, 2 atoms of oxygen
    C. 18 atoms of carbon; 34 atoms of hydrogen, 2 atoms of oxygen
    D. 20 atoms of carbon; 23 atoms of hydrogen, 1 atom of oxygen

Answer is C: The number of atoms written as a subscript, immediately follows the symbol for the element, hence there are 2 atoms of oxygen (O). The number 7 outside the parentheses means that there are 7 lots of the atoms that are inside the parentheses. Hence, 14 + 14 H inside the parentheses, with a further 6 H outside them, which totals 34 H.

46. The formula for leucine (an amino acid) may be written as: $(CH_3)_2C_3H_4NH_2COOH$. How many atoms of each type of element are there in a molecule of leucine?
    A. 6 atoms of carbon; 13 atoms of hydrogen, 2 atoms of oxygen, 1 atom of nitrogen
    B. 5 atoms of carbon; 8 atoms of hydrogen, 2 atoms of oxygen, 1 atom of sodium
    C. 6 atoms of carbon; 13 atoms of hydrogen, 2 atoms of oxygen, 1 atom of natrium
    D. 8 atoms of carbon; 10 atoms of hydrogen, 2 atoms of oxygen, 1 atom of nitrogen

Answer is A: The number of atoms immediately follows the symbol for the element as a subscript. A number following a parenthesis multiplies the atoms inside the parenthesis. Hence there are 6 atoms of carbon (including 2 inside the parenthesis). By the same reasoning, there are 13 atoms of hydrogen (including 6 inside the parenthesis). An amino acid always has an atom of nitrogen.

47. What may be stated about a chemical bond between atoms that is polar? They occur:
    A. Between a metal and a non-metal atom
    B. Between two non-metal atoms
    C. Between two non-metal atoms that are different to each other
    D. When the electrons in the bond are shared equally

Answer is C: Choice B is wrong if the two atoms are the same e.g. if both are oxygen. Choice D is wrong as it implies a non-polar bond.

48. What may be stated about a molecule that is polar?
    A. It is probably soluble in polar liquids such as lipids.
    B. The molecule has a highly symmetrical shape.
    C. It is likely to be soluble in non-polar liquids such as lipids.
    D. One end of the molecule will contain different atoms from the opposite end.

Answer is D: A polar molecule has opposite ends (poles) that are different from each other. That is, the molecule is not symmetrical. Choice A is wrong as lipids are not polar liquids.

49. Which of the following best describes the electrolytes that are dissolved in blood?
    A. Positive ions
    B. Any charged particle
    C. Molecules with an electric charge
    D. Negative ions

Answer is B: The term electrolyte is applied to any species of dissolved particle that has an electric charge.

50. The formula for the bone mineral calcium hydroxyapatite is $Ca_{10}(PO_4)_6(OH)_2$. How many atoms of each element are there in this molecule?
    A. 10 atoms of calcium, 6 atoms of phosphorus, 26 atoms of oxygen and 2 atoms of hydrogen
    B. 10 atoms of calcium, 6 atoms of phosphorus, 5 atoms of oxygen and 2 atoms of hydrogen
    C. 10 atoms of cadmium, 4 atoms of phosphorus, 8 atoms of oxygen and 2 atoms of hydrogen
    D. 10 atoms of calcium, 6 atoms of polonium, 1 atom of oxygen and 2 atoms of hydrogen

Answer is A: The number of atoms is written as a subscript and immediately follows the symbol for the element. Hence there are 10 atoms of calcium. When the subscript is outside the parenthesis, it refers to all of the atoms enclosed by the parentheses. Hence there are 6 atoms of phosphorus, 2 of hydrogen and $(4 \times 6) + (1 \times 2) = 26$ of oxygen. Ca is the symbol for calcium. Cadmium has the symbol Cd. Polonium has the symbol Po (not PO).

51. Urea has the formula CH_4N_2O. Identify the elements, and how many atoms of each element are present in a molecule of urea.
    A. 1 molecule of methane and 1 of di-nitrogen oxide
    B. 1 of carbon, 4 of hydrogen, 2 of nitrogen, 1 of oxygen
    C. 1 of calcium, 4 of hydrogen, 2 of ammonia, 1 of oxygen
    D. 1 of carbon, 1 of hydrogen, 4 of nitrogen, 2 of oxygen

Answer is B: The elements are: carbon, hydrogen, nitrogen and oxygen. The subscript number immediately after the symbol states how many of each atom there are. Hence 4 atoms of hydrogen.

52. Which of the following is the ion of a metal element?

    A. Na^+
    B. NH_4^+
    C. Cl^-
    D. HCO_3^-

Answer is A: A metal element loses an electron in chemical reactions so that its number of protons outnumber the electrons which leaves it with a positive charge. Choice B, while being a positive ion, is a compound, not an element.

53. Which of the following is the ion of a non-metal element?

    A. H_3O^+
    B. NH_4^+
    C. Cl^-
    D. HCO_3^-

Answer is C: While all of the elements represented by symbols in the four choices are non-metals, only choice C represents an element. All others are compounds.

54. Which of the following is the ion of a non-metal element?

    A. K^{2+}
    B. Mg^{2+}
    C. Na^+
    D. Cl^-

Answer is D: A non-metal element gains one (or more) electrons in a chemical reaction so that its number of electrons outnumber the protons which gives it with a negative charge. The other three choices are metal elements, which result in ions with a positive charge.

55. What is the definition of a chemical element?

    A. One of earth, air, water and fire.
    B. A simple substance that cannot be made into simpler substances by physical means.
    C. A pure substance that cannot be made into simpler substances by chemical means.
    D. A pure substance that cannot be made into simpler substances by ordinary physical or chemical means.

Answer is D: An element is "pure", so it contains only one type of substance. It is not a combination of simpler substances, so that it cannot be made into simpler substances. Nuclear reactors and high energy particle accelerators can convert elements into other things, but they are not considered to be "ordinary physical or chemical means". Choice A does not refer to science.

56. What is the definition of an atom?
    A. They are the particles of which an element is composed.
    B. They are the particles of which a compound is composed.
    C. They are the particles of which an ionic compound is composed.
    D. They are the particles of which a covalent compound is composed.

Answer is A: An atom is the smallest particle of an element. They are the particles of which an element is composed. The atoms of an element are different from the atoms of all other elements.

57. What is the definition of a proton?
    A. They are the subatomic particles within the nucleus of an atom.
    B. They are subatomic particles with a positive electrical charge within the nucleus of an atom.
    C. They are subatomic particles with a negative electrical charge within the nucleus of an atom
    D. They are subatomic particles within the nucleus of an atom but without an electrical charge.

Answer is B: Protons are subatomic particles (smaller than an atom). They have a positive electrical charge. They are, along with neutrons, one of the types of particle that is within the nucleus (the tiny but dense central structure) of an atom.

58. What is the definition of a neutron?
    A. They are the subatomic particles within the nucleus of an atom.
    B. They are subatomic particles with a positive electrical charge within the nucleus of an atom.
    C. They are subatomic particles within the nucleus of an atom but without an electrical charge.
    D. They are subatomic particles with a negative electrical charge within the nucleus of an atom

Answer is C: Neutrons are subatomic particles (smaller than an atom). They do not have an electrical charge (they are "neutral"). They are, along with protons, one of the type of particle that is within the nucleus (the tiny but dense central structure) of an atom.

59. What is the definition of an electron?
    A. They are the subatomic particles within the atom.
    B. They are the subatomic particles within the atom but not in the nucleus.
    C. They are the subatomic particles within the atom but outside the nucleus and possess a negative electrical charge.
    D. They are the subatomic particles within the atom but outside the nucleus and possess a positive electrical charge.

Answer is D: Electrons are subatomic particles (smaller than an atom). They have a negative electrical charge (i.e. they are attracted to positively charged objects). They exist within the atom but outside of the nucleus and so occupy most of the volume but only a tiny amount of the mass of an atom.

60. What is the definition of a metal element?
    A. The atoms of a metal element lose electrons when they participate in chemical reactions.
    B. The atoms of a metal element gain electrons when they participate in chemical reactions
    C. The atoms of a metal element share electrons when they participate in chemical reactions with atoms of other metal elements.
    D. The atoms of a metal element share electrons when they participate in chemical reactions

Answer is A: The atoms of a metal element lose (one or more) electrons when they participate in chemical reactions with non-metal atoms. This transforms them into "ions" with a positive charge. People who are not chemists use the term "electrolyte" as a synonym for "ion". Metal elements are listed on the "left hand side" of the periodic table. (The atoms of a non-metal element gain electron(s) when they participate in chemical reactions with metal atoms.)

61. What is the definition of a non-metal element?
    A. The atoms of a non-metal element lose electrons when they participate in chemical reactions.
    B. The atoms of a non-metal element gain electrons when they participate in chemical reactions.
    C. The atoms of a non-metal element share electrons when they participate in chemical reactions.
    D. The atoms of a non-metal element gain electrons when they participate in chemical reactions with atoms of metal elements.

Answer is D: The atoms of a non-metal element gain (one or more) electrons when they participate in chemical reactions with metal atoms. This transforms them into "ions" with a negative charge. People who are not chemists use the term "electrolyte" as a synonym for "ion". Non-metal elements are listed on the "right hand side" of the periodic table. (In addition, the atoms of a non-metal element share electrons with atoms of other non-metal atoms when they participate in chemical reactions.)

62. What is the definition of an ion?
    A. An ion is an atom that has lost an electron so has a positive electrical charge.
    B. An ion is an atom or molecule that has lost or gained and electron so has either a positive or negative electrical charge (respectively).
    C. An ion is an atom that has lost or gained and electron so has either a positive or negative electrical charge (respectively).
    D. An ion is an atom that has gained an electron so has a negative electrical charge.

Answer is B: Both atoms or molecules can become ions. That is, can possess an electrical charge, and that charge may be either positive (if an electron has been lost) or negative (if an electron has been gained).

63. What is the definition of a molecule?

    A. A molecule is a small independent group of two or more ions joined together.
    B. A molecule is a grouping of ions such that an ion of one charge, is completely surrounded and joined to, a group of ions of the opposite charge.
    C. A molecule is a small independent group of two or more atoms joined together.
    D. A molecule is a small independent group of two or more atoms of different elements joined together.

Answer is C: A molecule is a group of two or more atoms joined together. The atoms may be from the same element, e.g. O_2, or from different elements, e.g. $C_6H_{12}O_6$. This group of (2 or 24 in the examples given) atoms remain joined together as they move about.

64. What is the definition of a chemical compound?

    A. A pure substance formed when atoms from two or more elements join together.
    B. A pure substance formed when atoms from one or more elements join together.
    C. A pure substance formed when two or more molecules join together.
    D. A mixture that contains atoms of different elements not chemically combined.

Answer is A: A compound is a pure substance (contains only one type of substance). They form when atoms from two or more elements join together. Hence O_2 and H_2 while being molecules, are not compounds as they contain only atoms of the same element.

65. What is the definition of a covalent bond? It is the link that joins two:

    A. Non-metal atoms together in a molecule
    B. Metal atoms together in a molecule
    C. Non-metal atoms together in a lattice structure
    D. Metal atoms together in a lattice structure

Answer is A: It is the link (in fact a shared pair of electrons) that joins two non-metal atoms together in a molecule. The "co-" part refers to sharing of electrons, and it is between non-metal atoms. The joining of metal atoms to non-metal atoms is something else.

66. How may an ionic bond be described? It is the attraction between:
   A. Two or more adjacent molecules.
   B. An ion and the surrounding ions of opposite charge.
   C. Atoms within a molecule.
   D. Two ions of opposite charge.

Answer is B: When a metal atom reacts with a non-metal atom, it transfers an electron (or more) to the non-metal atom. In this way the metal atom becomes positively charged while the non-metal atom acquires an equal but opposite charge. It is the attraction between an ion and all of the immediately adjacent surrounding ions of opposite charge (which may number 6) that causes a "continuous lattice" structure to form.

67. What is the definition of an ionic (non-molecular) compound?
   A. The compound formed when atoms of a metal element chemically react with atoms of a different metal element.
   B. The compound formed when atoms of a non-metal element react with atoms of a different non-metal element.
   C. The compound formed when atoms of a metal react with atoms of a non-metal substance.
   D. The structure that results when ions form a lattice rather than molecules.

Answer is C: The compound formed when atoms of a metal element react with atoms of a non-metal element. In such compounds, the metal atoms lose electrons to become cations while the non-metal atoms gain electrons to become anions. These oppositely charged ions then attract each other to form a lattice structure rather than molecules. Choice B refers to a covalent (molecular) compound.

68. What is the definition of a molecular compound?
   A. The compound formed when atoms of a metal element chemically react with atoms of a different metal element.
   B. The compound formed when atoms of a non-metal element react with atoms of a different non-metal element.
   C. The compound formed when atoms of a metal react with atoms of a non-metal substance.
   D. The structure that results when ions form a lattice rather than molecules.

Answer is B: When atoms of a non-metal element chemically react with atoms of a different non-metal element, they share their electrons to form a covalent bond. The resulting group of atoms (which may contain two or three atoms or 20,000–30,000 atoms) is called a molecule and is the smallest particle of the new compound substance. Choice C refers to an ionic (non-molecular) compound.

69. Which of the following may be used to distinguish a metal element from a non-metal element?
    A. Metal ions have a negative electrical charge, while non-metal ions have a positive charge.
    B. Metals are listed on the "left hand side" of the periodic table, while non-metals are on the "right hand side".
    C. Metals gain an electron(s) when they react with non-metals, while non-metals lose an electron(s).
    D. Metals ions are called anions, while non-metal ions are called cations.

Answer is B: Metals are listed on the "left hand side" of the periodic table, while non-metals (except for hydrogen) are on the "right hand side". All other choices are wrong, but if stated "in reverse" would be correct.

70. Which of the following may be used to distinguish an electron from a proton?
    A. An electron resides in the nucleus of the atom, a proton is in the surrounding atomic volume.
    B. An electron has a mass of 1860, while a proton has a mass of one.
    C. An electron has a negative electrical charge, a proton has a positive charge.
    D. An electron has a negative electrical charge, a proton has no electrical charge.

Answer is C: An electron has a negative charge (and no linear dimensions!) and exists in the space within an atom but outside of the nucleus, a proton has a positive charge, exists in the nucleus of an atom, and is 1860× the mass of an electron.

71. Which of the following may be used to distinguish an atom from an ion?
    A. An atom has equal numbers of protons and electrons so is neutral: an ion has a +ve or −ve electrical charge.
    B. An atom is the smallest part of a molecule; an ion is the smallest part of an element.
    C. An atom may the combination of two or more elements; an ion is the smallest part of an element.
    D. Atoms may be found on the left hand side of the periodic table; ions are located on the right hand side.

Answer is A: An atom is the smallest piece of an element and has equal numbers of protons and electrons so is neutral: an ion has a +ve or −ve electrical charge. Ions may consist of one atom (e.g. Na^+ or Cl^-) two or more atoms of different elements (e.g. HCO_3^- or H_3O^+), in which case they are also molecules.

72. Which of the following may be used to distinguish a cation from an anion?
    A. A cation is repelled from a cathode, while an anion is attracted to the cathode.
    B. A cation is repelled from a cathode, while an anion is repelled from the anode.

C. A cation is an ion with a −ve charge, anion has a positive charge.
D. A cation is an ion with a +ve charge, anion has a negative charge.

Answer is D: A cation is an ion with a +ve charge (and so is attracted to the negatively charged electrode—called the cathode), while an anion has a negative charge (and is attracted to an anode).

73. Which of the following may be used to distinguish an atom from a molecule?
    A. An atom is the smallest part of a metal element, a molecule is the smallest part of a non-metal element.
    B. An atom is the smallest part of a covalent compound, a molecule is the smallest part of an element.
    C. An atom is the smallest part of an element, a molecule is the smallest part of a covalent compound.
    D. An atom is the smallest part of a non-metal element, while a molecule is composed of two or more metal elements.

Answer is C: An atom is the smallest part of an element, while a molecule is the smallest part of a covalent compound (it is also a group of two or more atoms that may be from the same element (e.g. a molecule of oxygen gas O_2) or from different elements (e.g. H_2O)).

74. Which of the following may be used to distinguish an element from a compound?
    A. A compound is a simple substance that cannot be broken down into a simpler substance.
    B. A compound may only be formed from non-metal elements.
    C. An element may be found on the left hand side of the periodic table; while compounds are located on the right hand side.
    D. An element is a simple substance that cannot be broken down into a simpler substance.

Answer is D: An element is a simple substance that cannot be broken down into a simpler substance. A compound is the result of a chemical reaction between two (or more) elements, so that it can be broken down into simpler substances (either its component elements or other compounds).

75. Which of the following may be used to distinguish an ionic compound (IC) from a molecular compound (MC)?
    A. An IC involves a metal element and a non-metal element. An MC involves only non-metal elements.
    B. An IC involves only non-metal elements. An MC involves a metal element and a non-metal element.
    C. An IC involves only metal elements. An MC involves only non-metal elements.
    D. An IC involves only non-metal elements. An MC involves only metal elements.

Answer is A: An ionic compound involves a metal element (which loses an electron/s to become a +ve ion) and a non-metal element (which gains an electron/s to become a −ve ion). A molecular compound involves only non-metal elements which share their electrons (shared electrons form a covalent bond) and results in a cluster of atoms called a molecule.

76. Which of the following may be used to distinguish an ionic bond from a covalent bond?
    A. A covalent bond is the attraction between two metal atoms. An ionic bond is the attraction between an ion and the surrounding adjacent ions of the same charge.
    B. A covalent bond is the attraction between two non-metal atoms. An ionic bond is the attraction between an ion and the surrounding adjacent ions of the opposite charge.
    C. A covalent bond is the attraction between two non-metal atoms. An ionic bond is the attraction between an ion and the surrounding adjacent ions of the same charge.
    D. A covalent bond is the attraction between two metal atoms. An ionic bond is the attraction between an ion and the surrounding adjacent ions of the opposite charge.

Answer is B: A covalent bond is the attraction between two **non-metal** atoms. A non-metal atom may form one, two, three or four such bonds (depending on which element we are considering), and the bond may be a single one (involving a pair of electrons) or be double bonds or even triple ones. An ionic bond is the attractions between an ion and the 6–8 surrounding adjacent ions of the **opposite** charge. Such attractions result in the formation of a crystal with a continuous lattice structure (rather than molecules).

77. Which list contains metallic elements only?
    A. Sodium, potassium, calcium, hydrogen
    B. Oxygen, nitrogen, calcium, phosphorus
    C. Iron, sodium, potassium, calcium
    D. Magnesium, zinc, cobalt, argon

Answer is C: These four elements are located on the left hand side of the periodic table (and do not include hydrogen).

78. Which elements AND how many atoms of each element are in CH_3COOH?
    A. One methyl and one carboxylic acid
    B. Three hydrogen, one oxygen, two carbon, one helium
    C. Two oxygen, two carbon and two hydrogen
    D. Two carbon, two oxygen, four hydrogen

Answer is D: In this case, the symbol for each element appears twice in the formula. The given chemical formula for acetic acid indicates its structure. The equivalent molecular formula is $C_2H_4O_2$ from which it is easier to see how many atoms of each element there are, but it does not indicate how the atoms are joined together.

79. Which statement about $Ba^{2+}SO_4^{2-}$ is correct?
    A. It is an ionic substance with one atom of barium, one of sulphur and four of oxygen.
    B. It is an ionic substance with one atom of barium, four atoms of sulphur and four of oxygen.
    C. It is a covalent substance with one atom of barium, one of sulphur and four of oxygen.
    D. It is a covalent substance with one atom of boran, one of sulphur and four of oxygen.

Answer is A: It is ionic as it is written with the symbols for elements superscripted with +ve and −ve signs to indicate ions. Also because Ba is a metal and S and O are non-metals. The sulphate ion has 1 atom of S and 4 of O. "Boran" is not a real element.

80. What features are required for a molecule to be "polar", that is to have small but opposite electrical charges at different ends of the molecule?
    A. The molecule must have polar bonds and be asymmetrical in shape.
    B. The molecule must have non-polar bonds and be asymmetrical in shape.
    C. The molecule must have polar bonds and be symmetrical in shape.
    D. The molecule must have non-polar bonds and be symmetrical in shape.

Answer is A: The molecule must have polar bonds and be asymmetrical in shape. Only covalent bonds between two atoms of the same element are not polar.

81. Which of the following statements about the atom and its nucleus is correct?
    A. The nucleus is 10^{-4} times the size of the atom.
    B. The nucleus contains neutrons with a positive charge and protons with no charge.
    C. Most of the volume of the atom is occupied by the nucleus, but most of the mass is due to the electrons.
    D. The majority of the atom's mass is due to the electrons.

Answer is A: The nucleus is one ten thousandth the size of the atom, so is mostly empty space. Protons have a +ve charge, electrons have a mass that is insignificant when compared to the atom mass.

82. What charge will an ion of magnesium have?
    A. Minus 1
    B. Minus 2
    C. Plus 1
    D. Plus 2

Answer is D: Magnesium is a metal so will have a positive ion. It is also in column two of the periodic table so will have a double positive charge. Alternatively, magnesium has 12 electrons and 12 protons (atomic number 12) so needs to lose two electrons in a chemical reaction to achieve the noble gas configuration of ten electrons like neon. When two electrons are lost, the resulting charge is plus 2.

83. What is the reason that a polar bond between two atoms is so named?
    A. The atoms share the electrons equally.
    B. The atoms share electrons unequally.
    C. The atoms share a pair of electrons.
    D. The bond is ionic and occurs between an anion and a cation.

Answer is B: When a pair of electrons are unequally shared, they are located closer to one of the bonding atoms. This makes that atom slightly negative compared to the other atom, hence the molecule has two poles.

4.2 Solutions

Water, because its molecules are electrically polar, is a very good solvent, and the substances that are dissolved in a sample of water are known as the solutes. The combination of solvent with the dissolved solutes is known as the solution. The term "% concentration" is a statement about how much solute is dissolved in a known volume of solvent. Hence a solution of concentration x% has x g of solute dissolved per 100 mL of solution. Solution concentration may also be expressed as: specific gravity; molarity in millimole per litre; osmolarity in milliosmole per litre and osmotic pressure in millimetre of mercury.

Solids that dissolve in water to produce ions are known as "electrolytes". However, many books refer to the dissolved ions themselves as electrolytes. If the concentration of a particular ion in the body's plasma is too low for healthy function, the prefix "-hypo" is used. For example: hyponatremia. If the concentration of a particular ion in the plasma is too high the prefix "-hyper" is used. For example: hyperkalemia.

1. Which item from the following list is an electrolyte?
    A. Ca^{2+}
    B. Oxygen gas (O_2) dissolved in water
    C. Table salt (solid form of Na^+Cl^-)
    D. Glucose (solid form of the sugar)

Answer is C: "An electrolyte is a substance that when dissolved in water, will produce ions in solution". Glucose is a molecular compound so does not ionise.

2. Given that 1 mol of Na^+Cl^- has a mass of 58.5 g, how many grams of Na^+Cl^- are dissolved in a solution of 0.1 mol/L?
    A. 0.585 g
    B. 5.85 g
    C. 58.5 g
    D. 0.1 g

Answer is B: 0.1 mol = 1/10 of 58.5 g = 5.85 g

3. How many grams of sodium chloride are there in (0.9%) normal saline?
   A. 0.09 g/100 mL
   B. 0.09 g/L
   C. 9 g/100 mL
   D. 9 g/L

Answer is D: 0.9% means 0.9 g/100 mL of solution = 9 g/1000 mL (i.e. per litre).

4. How many grams of sodium chloride are there in a 1 L bag of 0.9% saline?
   A. 0.09 g
   B. 0.9 g
   C. 9 g
   D. 90 g

Answer is C: 0.9% means 0.9 g/100 mL of solution = 9 g/1000 mL (i.e. per litre).

5. What is the percentage concentration of glucose if 80 g of glucose is dissolved in 1 L of solution?
   A. 0.8%
   B. 5%
   C. 8%
   D. 80%

Answer is C: 80 g/1000 mL = 8 g/100 mL = 8%.

6. Given that the concentration of hydronium ions in a solution with pH of 2 is 0.19%, what would be the concentration of hydronium ions in a solution with pH of 3?
   A. 0.13%
   B. 0.29%
   C. 0.019%
   D. 1.9%

Answer is C: A pH change of one corresponds to a change in hydronium ion concentration by a factor of 10. As the pH has risen, the acidity has decreased, so there are fewer hydronium ions. That is 1/10 of 0.19% = 0.019%.

7. What does the term "electrolyte" refer to?
   A. An uncharged dissolved particle
   B. The smallest particle of an element
   C. A substance that will conduct electricity when dissolved in water
   D. Negatively charged sub-atomic particles

Answer is C: This is the definition of an electrolyte.

8. What is meant by referring to a solution concentration of 0.18%?
   A. 0.18 g of solute in 100 mL of solution
   B. 1.8 g of solute in 100 mL of solution
   C. 0.18 g of solute in 1 L of solution
   D. 0.18 mol of solute in 100 mL of solution

Answer is A: 0.18% means 0.18 g/100 mL of solution.

9. Which of the following is **NOT** an electrolyte (or does not contain electrolytes)?
   A. Cl^-
   B. Acetic acid
   C. Glucose
   D. A 0.9% solution of sodium chloride

Answer is C: Glucose will dissolve in water, but being a molecular compound, does not produce ions.

10. If the concentration of a solution is 5%, which of the following is true?
    A. There is 0.5 g of solute per 100 mL of solution.
    B. There is 5 g of solute per 100 mL of solution.
    C. There is 5 g of solute per 1000 mL of solution.
    D. There is 50 g of solute per 100 mL of solution.

Answer is B: 5% = 5 per cent = 5 per hundred = 5 g/100 mL of solution.

11. What does the term "electrolyte" refer to?
    A. The minor component of a solution
    B. A substance that will conduct electricity when dissolved in water
    C. The smallest particle of an element
    D. Negatively charged sub-atomic particles

Answer is B: This is a definition of electrolyte.

12. The solution concentration 0.9% means that there are:
    A. 0.9 g of solute in 100 mL of solution
    B. 9.0 g of solute in 100 mL of solution
    C. 0.9 g of solute in 1 L of solution
    D. 0.9 mol of solute in 100 mL of solution

Answer is A: 0.9% = 0.9 per cent = 0.9 per hundred = 0.9 g/100 mL of solution.

13. Which of the following is **NOT** an electrolyte (or does not contain electrolytes)?
    A. K^+
    B. Nitrate ions
    C. Haemoglobin

D. A 1% solution of sodium chloride

Answer is C: Haemoglobin is an uncharged protein found within red blood cells.

14. Which is the best definition of an electrolyte?
    A. An atom that dissociates into ions
    B. A substance that will conduct electricity when it is dissolved in water
    C. Molecules of solid, liquid or gas that will conduct electricity in solution
    D. A substance that will conduct electricity

Answer is B: This is a definition of electrolyte. Choice C is restricted to molecules; and there must be a solution involved, so D is wrong.

15. What is the condition known as hyperkalemia characterised by? A:
    A. Higher than normal concentration of potassium in the blood
    B. Lower than normal concentration of potassium in the blood
    C. Serum sodium concentration greater than 150 mmol/L
    D. Serum sodium concentration greater than 5 mmol/L

Answer is A: Kalaemia refers to potassium. "Hyper-" refers to more or an excess.

16. A solution of 5% glucose is used for an infusion. Over a 2 h period, 300 mL of the solution were used. How much glucose in grams was infused?
    A. 5 g
    B. 12.5 g
    C. 15.0 g
    D. 50 g

Answer is C: 5% means 5 g/100 mL. So in 300 mL there would be 3×5 g = 15 g infused.

17. A solution of glucose is used for an infusion. Over a 3 h period 250 mL of solution is used containing a total of 5 g of glucose. What is the concentration of the glucose solution used?
    A. 2%
    B. 20%
    C. 0.2%
    D. 5%

Answer is A: 5 g/250 mL = 20 g/1000 mL (multiply by 4) = 2 g/100 mL = 2%.

18. A sudden and severe loss of potassium due to diuretic abuse is likely to result in:
    A. Hypothermia
    B. Hyponatremia
    C. Hypokalemia
    D. Hypoventilation

Answer is C: Kalium means potassium, the prefix "hypo-" means a lack or decrease in.

19. Which of the following statements relating to a patient with severe loss of potassium due to diuretic abuse is **TRUE**?
    A. The serum levels of potassium are >3 mmol/L.
    B. An ECG is probably not warranted.
    C. The condition may be treated by administering oral glucose and potassium.
    D. One course of action is to decrease the intake of potassium and to undergo ion-exchange resin treatment.

Answer is C: A loss of potassium may be treated by administering potassium. Hypokalemia refers to a blood concentration of <3 mmol/L, and such a level could affect the heart so an ECG **IS** warranted.

20. Which strategy would be most effective in dealing with a severe case of dehydration?
    A. Oral administration of a hypertonic solution
    B. Intravenous administration of distilled water
    C. Intravenous administration of isotonic glucose
    D. Intravenous administration of hypotonic sodium chloride

Answer is C: IV solutions should be isotonic (oral solutions should be hypotonic). Glucose would be absorbed by cells leaving the water behind in the blood which would reduce blood osmolarity. Hence water would redistribute itself by osmosis through the body.

21. A common IV solution is the combination of 0.18% sodium chloride and 4% glucose (also called "4% and a fifth"). How many grams of each solute will be in a 1 L bag?
    A. 0.18 g of sodium chloride and 4 g of glucose
    B. 0.9 g of sodium chloride and 5 g of glucose
    C. 1.8 g of sodium chloride and 40 g of glucose
    D. 18 g of sodium chloride and 40 g of glucose

Answer is C: 0.18% sodium chloride and 4% glucose means 0.18 g of sodium chloride per 100 mL and 4 g of glucose per 100 mL. So in 1 L there would be 1.8 g and 40 g, respectively.

22. The "extracellular" fluid compartment of the body includes which of the following?
    A. Interstitial, transcellular and connective tissue fluids
    B. Vascular, connective tissue and interstitial fluids
    C. Intracellular and transcellular fluid
    D. Vascular and connective tissue fluid

Answer is B: Extracellular must include the blood (vascular) and the fluid between cells (interstitial).

23. Fluid and electrolyte balance in the body is maintained by which of the following?
    A. The hormone aldosterone
    B. Keeping accurate account of the patient's fluid balance chart
    C. The nephron of the kidney
    D. The hypothalamus of the brain

Answer is C: The nephron is the functional unit of the kidney that does this job.

24. Which general description of the components of a solution is correct?
    A. Solvent and solute
    B. Solvent and liquid
    C. Solute and solder
    D. Liquid and solid

Answer is A: The solvent dissolves (e.g. water), and the solute (e.g. salt) is dissolved.

25. Which term below is **NOT** suitable to describe the concentration of a solution?
    A. 5 g/L
    B. 5%
    C. 2 Molar
    D. 0.5 mol

Answer is D: This states an amount of substance only, without reference to the volume of solvent involved.

26. Given that the healthy range of sodium ion concentration in the blood is 137–145 mmol/L, if the measured concentration of a blood sample was 130 mmol/L, what would the condition be called?
    A. Hyperkalemia
    B. Hypokalemia
    C. Hypernatremia
    D. Hyponatremia

Answer is D: Natrium is sodium. As 130 is less than 137, the prefix "hypo-" is the correct one.

27. Extracellular fluid includes which of the following liquids?
    A. Blood plasma
    B. Blood plasma and interstitial fluid
    C. Blood plasma and interstitial fluid and connective tissue fluid
    D. Blood plasma and interstitial fluid and connective tissue fluid and liquid inside cells

Answer is C: This choice includes more than choices A and B. D is wrong as liquid inside cells is not extracellular.

28. What can be said about a solution that conducts electricity?
    A. The solute is a polar molecule.
    B. The solvent is a non-polar liquid.
    C. The solution contains dissolved ions.
    D. The solution contains dissolved molecules.

Answer is C: Dissolved ions must be present for electrical conduction to occur.

29. Which of the body's fluid compartments does the vascular compartment form part of?
    A. Interstitial fluid
    B. Extracellular fluid
    C. Intracellular fluid
    D. Transcellular fluid

Answer is B: Vascular = blood which is almost synonymous with extracellular.

30. To what condition does the term "hypokalemia" refer?
    A. Too little phosphorus in the blood
    B. Too much sodium in the blood
    C. Too little potassium in the blood
    D. Too little sodium in the blood

Answer is C: Hypo- refers to too little and kalium is the Latin word for potassium.

31. Which of the following is correct for intracellular fluid (ICF) and extracellular fluid (ECF)?
    A. The ECF is part of the ICF.
    B. The majority of the body's water is in the ECF.
    C. The ICF contains more sodium than the ECF.
    D. The ICF contains more potassium than the ECF.

Answer is D: There is more K within cells than outside cells (the reverse is true for sodium). Most body water is inside the cells.

32. A 1 L IV bag contains 0.18% sodium chloride and 4% glucose. What mass of solutes would be dissolved in 100 mL of the solution?
    A. 0.18 g of sodium chloride and 4 g of glucose
    B. 1.8 g of sodium chloride and 4 g of glucose
    C. 1.8 g of sodium chloride and 40 g of glucose
    D. 18 g of sodium chloride and 40 g of glucose

Answer is A: 0.18% sodium chloride and 4% glucose mean 0.18 g of sodium chloride per 100 mL and 4 g of glucose per 100 mL.

33. In the context of fluid (water) balance, the body is said to have "two compartments". What are they?
    A. The vascular and the interstitial compartments
    B. The intracellular and the interstitial compartments

C. The lymph and the vascular compartments
D. The extracellular and the intracellular compartments

Answer is D: Outside the cells and inside the cells include everything.

34. What is an insufficient concentration of potassium in the blood known as?

A. Hypokalemia
B. Hyponatremia
C. Hypopotassaemia
D. Hypocalcaemia

Answer is A: The Latin word for potassium is kalium. The prefix "-hypo" refers to too little. Arguably, choice C is not wrong, but this expression is not used.

35. If a solution is shown to be able to conduct electricity, then what is true?

A. The solution is free of impurities.
B. The solution contains a dissolved electrolyte.
C. The solution contains dissolved molecules.
D. The solution is an aqueous solution.

Answer is B: A dissolved electrolyte will ensure that ions are in solution, and the solution will conduct electricity as these ions are free to move through the solution.

36. How many grams of sodium chloride are there in a 1 L bag of 4% glucose and 0.18% sodium chloride solution?

A. 0.18
B. 1.8
C. 18
D. 41.8

Answer is B: 0.18% = 0.18 g/100 mL, so in 1000 mL there would be $10 \times 0.18 = 1.8$ g.

37. In the vascular compartment of the body, what is the solvent?

A. Blood
B. Plasma
C. Serum
D. Water

Answer is D: All these things are in the vascular compartment, but water is the solvent.

38. The extracellular fluid compartment consists of which of the following?

A. Vascular and transcellular
B. Interstitial, vascular and connective tissue fluid
C. Intracellular and transcellular
D. Transcellular, intracellular and connective tissue fluid

Answer is B: Vascular and interstitial liquids must be included.

39. Electrolyte balance is achieved largely by:
    A. The kidneys and aldosterone
    B. Drinking sufficient water
    C. Antidiuretic hormone and isotonic fluids
    D. Ensuring that daily water intake is the same as daily water output

Answer is A: The kidneys have the ability to reabsorb and secrete ions and water as required. Aldosterone causes the kidney tubules to reabsorb sodium ions while promoting the secretion of potassium ions.

40. What may hyponatremia be described as?
    A. Insufficient potassium in the blood
    B. Insufficient iron in the blood
    C. Insufficient sodium in the blood
    D. Excess sodium in the blood

Answer is C: The prefix "hypo-" refers to too little, and natrium is a Latin word that means sodium.

41. Which of the following ways of expressing a solution's concentration is written as a number without units?
    A. % concentration
    B. Molarity
    C. Osmotic pressure
    D. Specific gravity

Answer is D: Specific gravity of a solution is the ratio of the density of the solution to the density of water (which is 1.0). Consequently the density units "cancel out".

42. A solution of 5% glucose is infused over a period of 3 h. If 250 mL of solution was used, how many grams of glucose was infused?
    A. 5 g
    B. 12.5 g
    C. 15.0 g
    D. 50 g

Answer is B: 5% glucose means 5 g/100 mL of solution. Two fifty millilitre was used, therefore $2.5 \times 5 = 12.5$ g.

43. A solution is prepared by dissolving 10 g of glucose in 250 mL of water. What will be the concentration of this solution expressed as a percentage?
    A. 4
    B. 6
    C. 25
    D. 40

Answer is A: % concentration states how many grams of solute are in 100 mL of solution. 10 g/250 mL is the same as 10/2.5 per 100 mL. This is 4%.

44. What quantity is used to express the concentration of dissolved oxygen in blood?

A. Partial pressure
B. Osmotic pressure
C. Molarity
D. Osmolarity

Answer is A: Gases can be expanded and compressed as their pressure changes, and this makes determining their dissolved concentration problematical. For a gas to dissolve in water, it must be in contact with the water. The amount of gas that will dissolve in water is proportional to the pressure (or partial pressure) of the gas that is above and in contact with the water surface. It is conventional (and convenient) to use the partial pressure (in millimetre of mercury) of the gas in contact as a statement of the amount of dissolved gas.

45. If the label on a bottle of beer states that it contains 5.9% alcohol, how much alcohol does the bottle contain?

A. 5.9 g/100 mL
B. 0.9 g/100 mL
C. 5.9 g
D. 5.0 g/100 mL

Answer is A: 5.9% means that there is 5.9 g of alcohol per 100 mL of beer. "Percent" means "in a 100".

46. A healthy concentration of cholesterol in the blood is deemed to be less than 5.5 mmol/L. Which value below is greater than this value?

A. 0.0045 mol/L
B. 0.005 mol/L
C. 5000 μmol/L
D. 6000 μmol/L

Answer us D: 6000 μmol/L is the same as 6.0 mmol/L. All other values are less than 5.5 mmol/L.

47. Which of the following refers to the concentration of a solution?

A. Systolic pressure
B. Osmotic pressure of blood
C. Diastolic pressure
D. Partial pressure of O_2 in blood

Answer is B: Although "pressure" appears in this term, it refers to the solution concentration when expressed in units of pressure. Choice D also refers to a concentration, but of only one of the dissolved species.

48. Which of the following is NOT one of the body's fluid compartments?
    A. The interstitial compartment
    B. The intracellular compartment
    C. The gastro-intestinal compartment
    D. The extracellular compartment

Answer is C: The contents of the gut are not regarded as one of the body's fluid compartments.

49. Which fluid compartment holds the greatest volume of the body's aqueous solutions?
    A. Intracellular compartment
    B. Extracellular compartment
    C. Plasma compartment
    D. Interstitial compartment

Answer is A: The intracellular compartment contains about 27 of the body's 42 L or so of solution. The next largest is the extracellular with about 15 L. The other two are subsets of the extracellular compartment.

50. Calculate the "anion gap" for a patient whose electrolyte concentrations (in millimole per litre) are: sodium = 140, potassium = 3.9, chloride = 101, bicarbonate = 26.
    A. −9.1
    B. 9.1
    C. 175.1
    D. 61.1

Answer is B: The anion gap is calculated by subtracting the serum concentrations of chloride and bicarbonate (anions) from the concentrations of sodium and potassium (cations). Normal range is 7–17: $([Na^+] + [K^+]) - ([Cl^-] + [HCO^-_3]) = (140 + 3.9) - (101 + 26) = 136.1 - 127 = 9.1$

51. What mass of sodium chloride is in 100 mL of 0.3% Na^+Cl^- solution?
    A. 0.3 g
    B. 3.0 g
    C. 30 g
    D. 100 g

Answer is A: "0.3%" means 0.3/100. That is 0.3 g of the 100 g of solution is sodium chloride. One millilitre of water has a mass of 1 g. So 0.3 g/100 mL.

52. What mass of glucose is in 100 mL of 3.3% glucose solution?
    A. 0.3 g
    B. 3.0 g
    C. 3.3 g
    D. 100 g

Answer is C: "3.3%" means 3.3/100. That is 3.3 g of the 100 g of solution is glucose (the other 96.7 g is water). One millilitre of water has a mass of 1 g. So 3.3 g/100 mL.

53. What mass of potassium chloride is in 100 mL of 0.224% K^+Cl^- solution?
   A. 0.224 g
   B. 2.24 g
   C. 22.4 g
   D. 224 g

Answer is A: "0.224%" means 0.224/100. That is 0.224 g of the 100 g of solution is potassium chloride (the other 99.776 g is water). One millilitre of water has a mass of 1 g. So 0.224 g/100 mL. Choice D is nonsense as 224 g cannot exist in 100 mL.

54. What is extracellular fluid AND where is it found?
   A. Extracellular fluid is found within cells. It is all liquid in the body except blood, lymph and interstitial fluid.
   B. Extracellular fluid is that fluid outside of cells. It is synonymous with blood.
   C. Extracellular fluid is found within and between cells. It is any fluid that is NOT in blood, gut or lymph vessels.
   D. Extracellular fluid is that fluid outside of cells. It is in the interstitial space, in the blood vessels and lymph vessels.

Answer is D: Extracellular means outside of cells (intracellular means inside cells). Extracellular fluid includes not only interstitial fluid and blood mostly but also lymph, synovial fluid, CSF, saliva, vitreous and aqueous humour, endolymph, perilymph, etc.

4.3 Diffusion and Osmosis

Ions and molecules in a solution are continually bumping into each other and moving independently in random directions. This continual motion results in the solute particles and solvent molecules being evenly distributed. Because if there is a greater concentration of one type of molecule in one place compared to elsewhere, the random motion will result in more molecules moving away from that place than are moving towards the place. Diffusion is the name given to this random motion of molecules and ions. If water is the molecule that is moving and it is passing through a semi-permeable membrane from one solution to the solution on the other side, the movement is called "osmosis". If it is a solute molecule or ion that is passing through a semipermeable membrane, the movement is called "dialysis". The result of osmosis is that the more concentrated solution becomes more dilute. If two solutions that are separated by a semi-permeable membrane experience no net flow of water one way or the other, the two solutions are said to be isotonic.

Another way to describe the concentration of a solution is to state its "osmotic pressure" (in pressure units). The osmotic pressure of blood is about 7.3 atm

(740 kPa). Osmotic pressure is described as the tendency of water to move into a solution via osmosis. The higher the solution's concentration, the higher is its osmotic pressure. Osmotic pressure may be measured by an apparatus that determines the amount of pressure that must be applied to a solution to prevent water from crossing the membrane and entering the solution by osmosis.

1. Consider two aqueous solutions of different concentration separated by a semi-permeable membrane. In this situation, osmosis results in:
   A. Water molecules moving to the side where the solution concentration is lower
   B. The more concentrated solution becoming even more concentrated
   C. The more dilute solution becoming even more dilute
   D. The more concentrated solution becoming more dilute

Answer is D: The result of osmosis is that the more concentrated solution becomes more dilute.

2. Osmosis may be defined as which of the following?
   A. The diffusion of water molecules across a semi-permeable membrane from the solution with lower water concentration into the solution of higher water concentration
   B. The movement of water molecules across a semipermeable membrane from the solution of higher concentration into the solution of lower concentration
   C. The diffusion of solute particles across a semipermeable membrane from the solution of higher concentration into the solution of lower concentration
   D. The movement of water molecules across a semipermeable membrane from the solution of lower concentration into the solution of higher concentration

Answer is D: The definition should include "water", movement through an SP membrane, direction of water flow from more dilute solution into the solution of higher concentration.

3. What is the movement of water molecules across a plasma membrane from the side where the solution concentration is more dilute to the side where the solution is more concentrated called?
   A. Osmosis
   B. Reverse osmosis
   C. Diffusion
   D. Hydration

Answer is A: Osmosis is the diffusion of WATER molecules through a membrane, from where there is a higher concentration of water molecules to where there is a lower concentration of water molecules (i.e. into the more concentrated solution).

4. The movement of water molecules through a plasma membrane from the side where there is a higher concentration of water molecules to the side where there are fewer is best known as:
   A. Diffusion
   B. Osmosis
   C. Pinocytosis
   D. Hydrolysis

Answer is B: Osmosis is the diffusion of water molecules through a membrane, down their concentration gradient.

5. The diffusion of water molecules across a cell membrane from the side where the solution concentration is more dilute to the side where it is greater is known as which of the following?
   A. Osmosis
   B. Filtration
   C. Hydrolysis
   D. Buffer action

Answer is A: This is a definition of osmosis.

6. If a semipermeable membrane separates two aqueous solutions with different osmotic pressures, what will be the direction of water flow between solutions?
   A. From higher osmotic pressure to the solution of lower osmotic pressure
   B. From lower osmotic pressure to the solution of higher osmotic pressure
   C. From higher concentration to the solution of lower concentration
   D. From higher hydrostatic pressure to the solution of lower hydrostatic pressure

Answer is B: Lower osmotic pressure means a lower solution concentration (and a higher concentration of water molecules). Water moves from the dilute solution into the more concentrated one.

7. During dialysis, what moves across a semi-permeable membrane (and how)?
   A. Water molecules by diffusion from the region of high solute concentration to the region of low solute concentration
   B. Water molecules by filtration from the region of high hydrostatic pressure to the region of low hydrostatic pressure
   C. Solutes by diffusion from the region of high solute concentration to the region of low solute concentration
   D. Solutes by filtration from the region of low hydrostatic pressure to the region of high hydrostatic pressure

Answer is C: Dialysis refers to movement of solutes (not water). Choice D is wrong as filtration refers to movement due to hydrostatic pressure difference from the solution under higher pressure to low pressure.

8. A suitable definition of osmosis would be movement:
   A. Of solute particles through a plasma membrane from the side where their concentration is greatest to the side where it is lower
   B. Of water molecules through a plasma membrane from the side where their concentration is greatest to the side where it is lower
   C. Of a substance from a region where it is in high concentration to where its concentration is lower
   D. Caused by a hydrostatic pressure difference

Answer is B: Osmosis refers to movement of water molecules (not other molecules), by diffusion.

9. The difference between dialysis and diffusion is that
   A. Dialysis involves the movement of water molecules.
   B. Diffusion involves movement against the concentration gradient.
   C. Dialysis involves passive movement through a cell membrane.
   D. Diffusion is caused by a hydrostatic pressure difference.

Answer is C: Diffusion is passive and occurs in the direction of the concentration gradient. Diffusion can occur within a solution or across a membrane. Dialysis, on the other hand, requires a membrane and is a term applied to solutes, not water molecules.

10. Osmosis involves the movement of:
    A. Water molecules through a membrane from a region of higher concentration of water molecules to a region of lower water molecule concentration
    B. Solute particles from a region of higher solution concentration to a region of lower solution concentration
    C. Water molecules from a region of lower concentration of water to a region of higher water molecule concentration
    D. Solute particles through a membrane from a region of lower solute concentration to a region of higher solute concentration

Answer is A: Osmosis involves water molecules (not solutes) moving down their concentration gradient into a solution of lower concentration of water molecules.

11. What does "osmosis" refer to?
    A. The constant random motion of ions and molecules
    B. The movement of ions and molecules from regions of high concentration to regions of low concentration
    C. The movement of water molecules through a semi-permeable membrane
    D. The movement of water molecules through a semipermeable membrane from the side with higher water concentration to the side with lower water concentration

Answer is D: The definition must include water molecules; crossing a SP membrane; a correct direction of movement.

12. What is the difference between filtration and diffusion?
    A. Diffusion can occur through a biological membrane whereas filtration cannot.
    B. Filtration can occur through a biological membrane whereas diffusion cannot.
    C. Filtration is the movement of molecules caused by a pressure difference but diffusion does not involve a difference in pressure.
    D. Diffusion is the movement of molecules caused by a pressure difference but filtration does not involve a difference in pressure.

Answer is C: Filtration requires a pressure difference, diffusion does not. Both diffusion and filtration can occur through a membrane.

13. What is the difference between osmosis and dialysis?
    A. Dialysis involves the movement of solute molecules, whereas osmosis refers to water molecules.
    B. Osmosis involves the movement of solute molecules, whereas dialysis refers to water molecules.
    C. Osmosis involves movement of molecules across a membrane, but dialysis does not involve a membrane.
    D. Dialysis involves movement of molecules across a membrane, but osmosis does not involve a membrane.

Answer is A: Osmosis refers to the movement of water molecules (only) through a membrane. Only choice A is consistent with this.

14. Which one of the following processes that describe movement of the particles in a solution does **NOT** require passing through a membrane?
    A. Diffusion
    B. Filtration
    C. Dialysis
    D. Osmosis

Answer is A: While diffusion can occur through a membrane, its presence is not required in order to define diffusion.

15. Which statement about the osmotic pressure of an aqueous solution is correct? Osmotic pressure:
    A. Is an indication of the force with which pure water moves into that solution
    B. Is a measure of the tendency of water to move into the solution
    C. Is the drawing power of water and depends on the number of molecules in the solution
    D. Of a solution is called its osmolarity in milliosmole per kilogram

Answer is B: Solutions with high "osmotic pressure" are concentrated solutions and so water will diffuse into such solutions from solutions of lower concentration—a process known as osmosis. The other answers are nonsense.

16. Diffusion is the term given to the process where:
    A. Molecules move along their concentration gradient from high concentration to low concentration.
    B. Water moves along its concentration gradient from low concentration to high concentration.
    C. ATP is used to move ions along their concentration gradient.
    D. A membrane protein, by changing shape after binding to a molecule, moves the molecule across the plasma membrane.

Answer is A: This is the only correct definition of diffusion.

17. By what name is the movement of solute particles through a selectively permeable membrane in the direction of their concentration gradient known?
    A. Diffusion
    B. Dialysis
    C. Osmosis
    D. Filtration

Answer is B: While the solute particles are indeed diffusing through the membrane, the presence of a membrane makes dialysis the appropriate term to use. Diffusion is also applied to the movement of particles within a solution even when they do not cross a membrane.

18. Blood has a slightly higher osmotic pressure than the interstitial fluid that surrounds capillaries. What is the effect of this?
    A. Water will tend to move from the interstitial fluid into the capillaries.
    B. The solution concentration of blood is less than the solution concentration of interstitial fluid.
    C. Water will tend to move from the capillaries into the interstitial fluid.
    D. Capillaries will expand in diameter.

Answer is A: Water will move through a membrane into a solution of higher osmotic pressure (the blood) from a solution of lower osmotic pressure.

19. Consider a patient undergoing kidney dialysis, whose blood has bicarbonate at a concentration of 14 mmol/L and urea at 23 mmol/L. The dialysing liquid has bicarbonate at 32 mmol/L and urea at 0 mmol/L. In which direction will these substances flow?
    A. Bicarbonate will flow from patient's blood to dialysing liquid, urea will flow from patient's blood into dialysing liquid.
    B. Bicarbonate will flow from dialysing liquid to patient's blood, urea will flow from dialysing liquid into patient's blood.
    C. Bicarbonate will flow from dialysing liquid to patient's blood, urea will flow from patient's blood into dialysing liquid.
    D. Bicarbonate will flow from patient's blood to dialysing liquid, urea will flow from dialysing liquid to patient's blood.

Answer is C: Molecules will flow from areas of high concentration towards areas of low concentration. Hence bicarbonate will flow from the dialysing liquid at 32 mmol/L to blood at 14 mmol/L, while urea will flow from blood at 23 mmol/L to the dialysing liquid at 0 mmol/L.

20. Which is the best description for the osmotic pressure of a solution?
    A. The pressure that needs to be applied to the solution while it is separated from pure water by a membrane, to prevent a net flow of water through the membrane into the solution
    B. The force with which pure water moves through a membrane into that solution as a result of its solute concentration
    C. The movement of particles through a membrane, where the movement is caused by a hydrostatic pressure
    D. It is the force of attraction for water by undissolved particles in the solution.

Answer is A: The application of a hydrostatic pressure to a solution in order to oppose and prevent the osmotic flow of water into that liquid is the basis for assigning a value to that solution for its "osmotic pressure". Osmotic pressure is the value of this externally applied hydrostatic pressure. Choice C describes filtration. Choices B and D are nonsense.

21. In which of the following situations would the osmotic pressure of blood be the greatest?
    A. In a patient whose blood osmolarity is 290 mOsm/L
    B. In a patient whose blood osmolarity is 280 mOsm/L and whose urine specific gravity is 1.002
    C. In a patient with hyperthermia
    D. In a patient who is dehydrated

Answer is D: The healthy range for blood osmolarity is 280–300 mOsm/L. A dehydrated person would have a blood osmolarity approaching or exceeding 300 mOsm/L.

22. Osmosis involves movement of water from where the:
    A. Water concentration is lower to where it is higher
    B. Solute concentration is higher to where it is lower
    C. Solution is more concentrated to where it is less concentrated
    D. Water concentration is higher to where it is lower

Answer is D: In osmosis water molecules flow down their concentration gradient (and from weaker solutions to more concentrated ones).

23. What is "osmotic pressure"?
    A. The pressure exerted by a solution due to its concentration
    B. A measure of solution concentration expressed in the units of pressure
    C. The pressure exerted by the blood colloidal plasma proteins

D. The pressure that drives water movement out of the arterial end of capillaries

Answer is B: Osmotic pressure is a way of expressing solution concentration. The word pressure in the term "osmotic pressure" makes it tempting to erroneously think of the solution exerting some type of pressure due to its solutes.

24. If a red blood cell (RBC) is placed in a solution that has a greater concentration than that inside the RBC, what will happen?
    A. The RBC will crenate.
    B. The RBC will haemolyse.
    C. There will be a net movement of water out of the RBC into the solution.
    D. There will be no net movement of water out of the RBC.

Answer is C: Water will flow by osmosis from the RBC into the surrounding solution. If the difference in concentration is large enough, the outflow of water will be large, and the RBC will also shrivel (crenate) as a result of this outflow.

25. Which one of the following is an example of osmosis?
    A. Water moving from the glomerulus of a nephron into the Bowman's capsule
    B. Water leaving a blood capillary from close to its arteriole end, to enter the interstitial fluid
    C. Water entering a red blood cell that is in a 0.8% sodium chloride solution, by passing through its plasma membrane
    D. Water evaporating from perspiration on the skin

Answer is C: Osmosis refers to the movement of water through a membrane in the direction of its concentration gradient. A 0.8% solution is hypotonic to the contents of the RBC, so water would enter the cell. In both choices A and C, the water is moving due to a hydrostatic pressure difference.

26. The Na^+/K^+ ATPase pump in the plasma membrane moves Na out of the cell and K into the cell against their concentration gradient. Then Na re-enters the cell and K leaks out of the cell, along their concentration gradients through their membrane channels. What is the movement of Na and K along their concentration gradients called?
    A. Active transport
    B. Diffusion
    C. Facilitated diffusion
    D. Osmosis

Answer is B: The movement of these ions down their concentration gradient is diffusion. The concentration gradient is produced by the active transport of the Na^+/K^+ ATPase pump.

27. Which of the following statements is FALSE?
    A. Filtration is movement of water caused by a difference in hydrostatic pressure, while diffusion results from a difference in concentration.
    B. Both diffusion and filtration will tend to continue till there is an equal amount on both sides of the membrane.
    C. Any hypertonic solution has a concentration lower than that of blood while a hypotonic solution has a concentration greater than the blood.
    D. Water moves into a red blood cell resulting in haemolysis, and out of a cell by a process called plasmolysis.

Answer is C: Hypertonic solutions have a concentration **greater** than blood. Plasmolysis is the process in which cells lose water in a hypertonic solution.

28. Which of the following may happen "by osmosis"?
    A. The propagation of an action potential along an axon
    B. Sodium leaving a cell and potassium entering the cell
    C. Learning the names of skeletal muscles
    D. Water passing through a membrane

Answer is D: Osmosis is a word used to describe the movement of water through a membrane from the side where the concentration of water molecules is greater to the side where the water concentration is lower.

29. Given that deoxygenated blood in the pulmonary capillaries has an oxygen concentration of 40 mmHg and carbon dioxide concentration of 46 mmHg. Alveolar air has an oxygen concentration of 104 mmHg and carbon dioxide concentration of 40 mmHg. In which directions will oxygen and carbon dioxide diffuse?
    A. Oxygen will diffuse from capillary blood to alveolar air, carbon dioxide will diffuse from the blood into the alveoli.
    B. Oxygen will diffuse from alveolar air to capillary blood, carbon dioxide will diffuse from the alveoli into the blood.
    C. Oxygen will diffuse from alveolar air to capillary blood, carbon dioxide will diffuse from blood to the alveoli.
    D. Oxygen will diffuse from capillary blood to alveolar air, carbon dioxide will diffuse from blood to the alveoli.

Answer is C: Of course O_2 diffuses from alveoli to blood while CO_2 diffuses from blood to alveoli. Or by considering movement down the concentration gradients: O_2 from 104 to 40 mmHg and CO_2 from 46 to 40 mmHg.

30. Only one of the definitions of osmotic pressure below is correct. Which one?
    A. The hydrostatic pressure required to halt osmosis is called osmotic pressure.
    B. Water pressure that develops in a solution as a result of osmosis into that solution is called osmotic pressure.

C. Osmotic pressure is the pressure exerted by the movement of water across a semipermeable membrane due to the difference in solution concentration.
D. Osmotic pressure is the force required to prevent water from moving by osmosis across a selectively permeable membrane.

Answer is A: The other three choices, despite all being quotations from anatomy and physiology text books, are wrong. Osmotic pressure is not a force; water movement does not exert a pressure; osmosis between compartments in the body does not cause water pressure to develop.

31. Which form of transport through the plasma membrane requires the expenditure of energy by the cell?
   A. Facilitated diffusion
   B. Osmosis
   C. Active transport
   D. Diffusion

Answer is C: The term "active" implies using energy (in the form of ATP) to move a molecule against its concentration gradient, while the other processes are all passive.

32. What process does facilitated diffusion refer to?
   A. Movement along a concentration gradient assisted by protein carrier molecules
   B. Movement of ions and molecules along a concentration gradient
   C. Transport of molecules and ions against their concentration gradient
   D. Water movement through a semipermeable membrane

Answer is A: Facilitation is a role performed by protein carriers. The other choices refer to diffusion, active transport and osmosis, respectively.

33. Which statement distinguishes between osmosis and diffusion?
   A. Diffusion is movement of particles down their concentration gradient while osmosis refers to the movement of water down its concentration gradient.
   B. Diffusion is the movement of water while osmosis refers to the movement of water through a membrane.
   C. Diffusion occurs because of an applied pressure gradient while osmosis is a passive process.
   D. Diffusion is the movement of any particles while osmosis is the movement of water particles through a membrane.

Answer is D: Osmosis is a term reserved for use with water molecules that move through a semipermeable membrane from the side of the membrane with more water molecules to the side with fewer water molecules (that is, down the concentration gradient for water molecules). Diffusion does not involve a pressure gradient but filtration does and also involves a membrane.

4.4 Tonicity, Moles and Osmoles

The "mole" is the standard international unit for amount of substance (or number of particles). Just as dozen is the word for 12 things, the mole is the word for 6.02×10^{23} things. In the case of a mole, the things are atoms or molecules. The concentration of a solution can be expressed as the number of moles per litre. For example 5% glucose contains 50 g of glucose dissolved in a litre of solution. Fifty grams of glucose is 0.278 mol of glucose, so 5% glucose = 50 g/L = 0.278 mol/L = 278 mmol/L. A glucose solution with this concentration is said to be "isotonic" with blood. That is, it will not cause a net flow of water into or out of red blood cells. A hypotonic solution would cause net inflow, while a hypertonic solution would cause a net outflow from RBC.

The mass of glucose that contains 1 mol of glucose molecules (180 g) is determined from its formula ($C_6H_{12}O_6$) and the relative atomic masses of the atoms (12, 1 and 16, respectively). The same can be done for Na^+Cl^- so that 1 mol of sodium chloride has a mass of 58.5 g. However, sodium chloride is an ionic substance, so when it dissolves in water, the Na^+ ions and Cl^- ions separate and move independently. This means that 1 mol of solid sodium chloride produces 2 mol of particles (ions) when it dissolves—1 mol of Na^+ ions and 1 mol of Cl^- ions. We use the term "osmoles" to refer to the number of ions (particles) in solution when an ionic substance dissolves. One osmole = 6.02×10^{23} ions. Hence 1 mol of the ionic substance sodium chloride (6.02×10^{23} pairs of Na^+Cl^-) produces 2 Osm of dissolved ions (12.04×10^{23} ions).

1. What information does the molarity of a solution provide?
   - A. The density of a solution
   - B. The number of dissolved particles per litre of solution
   - C. The mass of a mole of the substance
   - D. The tonicity of the solution

Answer is B: If molarity is 1.2 mmol/L = $1.2 \times 10^{-3} \times (6.02 \times 10^{23})$ particles per litre.

2. What term is applied to an intravenous solution that would cause a net movement of water out of red blood cells?
   - A. Hypertonic
   - B. Supertonic
   - C. Epitonic
   - D. Hypotonic

Answer is A: "Hyper-" refers to greater (tonicity) than inside an RBC. Hence, water would move from the less concentrated solution (within the RBC) to the exterior.

3. What is a definition of an osmole? The amount of substance that:
   A. Must be dissolved to produce 6×10^{23} solute particles
   B. Must be dissolved to produce an osmotic pressure of 1.0 mmHg
   C. Must be dissolved to produce an isotonic solution
   D. Contains 6×10^{23} particles

Answer is A: Ionic substances (e.g. $Na^{+}Cl^{-}$) separate into ions when dissolved. Hence, a mole of NaCl would produce 2 Osm of ions (1 mol of Na ions and 1 mol of Cl ions). Hence, half of a mole of NaCl, when dissolved, would result in 1 Osm of dissolved particles (ions).

4. Choose the solution below that has the lowest concentration of dissolved particles.
   A. 0.9% NaCl
   B. Isotonic *powerade*
   C. 5% Glucose
   D. Hypotonic saline

Answer is D: Choices A, B and C are all isotonic solutions. "Hypo" refers to a solution concentration that is less than that of human blood.

5. What can be correctly said of an isotonic intravenous solution? An isotonic solution:
   A. Causes water to move out of red blood cells
   B. Causes no net movement of water into or out of red blood cells
   C. Has the same solutes in the same solution concentration as blood plasma
   D. Causes water to move into red blood cells

Answer is B: Isotonic means that the IV solution has the same concentration of particles (albeit the particles themselves may be different) as in RBC; hence, there is no net osmotic flow.

6. What would be the concentration of a solution that causes red blood cells placed in it to swell?
   A. Hypotonic
   B. Isotonic
   C. Hypertonic
   D. Iso-osmotic

Answer is A: If the RBC swell, that means water is entering the cells; hence, the surrounding liquid is at a lower concentration (i.e. hypotonic) to the liquid inside RBC.

7. The unit milliosmoles per litre (mOsm/L) refers to which of the following?
   A. The number of particles in solution, in multiples of 6×10^{20} per litre.
   B. 10^3 Times the number of moles of particles in a litre of solution.
   C. The number of molecules per litre of solution.
   D. The number of moles per millilitre of solution.

Answer is A: A milliosmole (mOsm) is $10^{-3} \times (6 \times 10^{23})$ particles which is 6×10^{20}. Choice B means 1000x the correct answer. Choice D means 1/1000x the correct answer.

8. A hypertonic solution is one which:
    A. Has an osmotic pressure that is different to that inside red blood cells
    B. Has an osmolarity less than that of red blood cells
    C. Causes no net movement of water through the membrane of red blood cells
    D. Has an osmolarity greater than that of red blood cells

Answer is D: "Hyper" means greater than the tonicity (or osmolarity) inside an RBC.

9. Given that 1 mol of Na^+Cl^- has a mass of 58.5 g, what would be the concentration of particles when 0.9 g of Na^+Cl^- is dissolved in 100 mL of water?
    A. 117 mmol/L
    B. 150 mmol/L
    C. 150 mOsm/L
    D. 300 mOsm/L

Answer is D: The number of moles in 0.9 g = 0.9 g/58.5 g/mol = 0.0154 mol of Na^+Cl^-. However, NaCl separates into ions when dissolved, so that there will be 0.0154 mol of Na and 0.0154 mol of Cl, hence 0.0308 Osm of particles = 30.8 mOsm/100 mL. In 1 L, the number of moles would be ten times greater, so 300 mOsm/L (is closest).

10. The unit millimoles per litre (mmol/L) refers to which of the following?
    A. The number of particles in solution, in multiples of 6×10^{20} per litre
    B. 10^3 Times the number of moles of particles in a litre of solution
    C. The number of molecules per litre of solution
    D. The number of moles per millilitre of solution

Answer is A: A mole = 6×10^{23} particles, a millimole (mmol) = 6×10^{20} particles. When dissolved, whether the particles are ions or molecules makes no difference.

11. An isotonic solution is one which:
    A. Has an osmotic pressure that is different to red blood cells
    B. Has an osmolarity less than that of red blood cells
    C. Causes no net movement of water between the solution and red blood cells
    D. Has an osmolarity greater than that of red blood cells

Answer is C: "Iso" means "the same". When the concentration of the surrounding liquid is the same as that inside a RBC, then no net water movement occurs.

12. A hypotonic solution may be characterised by which of the following?
    A. A solution whose osmolarity is greater than that of blood
    B. One that causes red blood cells to crenate

C. A solution within the range 280–300 mOsm/L
D. One that causes a net water movement into red blood cells

Answer is D: "Hypo" means less concentrated than the solution inside a RBC. This in turn means that the concentration of water molecules is greater outside the cell than inside. So there will be a net water flow by osmosis into the RBC.

13. Which of the following statements could be applied to a hypertonic solution?
    A. It causes red blood cells to shrink and crenate.
    B. It causes red blood cells to swell and perhaps lyse.
    C. It is a solution with an osmolarity less than that of blood.
    D. It causes movement of water into red blood cells.

Answer is A: RBC placed in a hypertonic solution would lose water to the solution, so their volume would decrease (they would shrink). This would cause their membrane to become wrinkled (to crenate).

14. What does the term "mole" refer to?
    A. The smallest particle of a molecular compound
    B. The amount of solute that must be dissolved in water to make an isotonic solution
    C. An amount of substance that contains 6.02×10^{23} particles
    D. A group of two or more atoms bonded together

Answer is C: Mole is the SI unit for "amount of substance". Choice C is the correct number.

15. What do solutions of 5% glucose, 9.5% sucrose and 0.9% sodium chloride have in common?
    A. They all have the same concentration.
    B. They are all hypotonic to plasma.
    C. They all contain the same number of dissolved particles per unit volume.
    D. They are all hypertonic to plasma.

Answer is C: The three solutions are isotonic to blood. Choice A is incorrect; they have different (weight/volume) concentrations.

16. An intravenous fluid that is **hypertonic** to blood would have what effect on the red blood cells?
    A. It would have no effect.
    B. The number of red blood cells would increase
    C. Red blood cells would lyse.
    D. It would cause red blood cells to crenate.

Answer is D: RBC in contact with a hypertonic solution would lose water to the solution and shrink. This would cause their plasma membrane to crenate (wrinkle).

17. An intravenous fluid that is **isotonic** to blood would have what effect on the red blood cells?
    A. It would have no effect.
    B. It would cause red blood cells to crenate.
    C. Red blood cells would lyse.
    D. The blood volume would increase.

Answer is A: RBC would be unaffected by an isotonic solution as there will be no net flow of water into or out of the cells.

18. What may be said of isotonic solutions? They have:
    A. Had added the same number of moles of solid substance per volume of solution
    B. The same number of grams of solute per volume of solution
    C. The same percent concentration
    D. The same number of dissolved particles per volume of solution

Answer is D: Choice A is incorrect as a mole of an ionic substance will result in 2 or 3 Osm of dissolved particles. (Choices B and C are incorrect.)

19. Fructose is a sugar with a molecular formula $C_6H_{12}O_6$. Given that the atomic weights are: C = 12, H = 1 and O = 16, how many moles of fructose are there in 36 g?
    A. 0.01 mol
    B. 0.10 mol
    C. 0.20 mol
    D. 0.50 mol

Answer is C: Mass of 1 mol of fructose = (6 × 12) + (12 × 1) + (6 × 16) = 72 + 12 + 96 = 180 g. Thirty-six gram is less than 1 mol. Number of moles = 36 g/180 g/mol = 0.2 mol.

20. Ribose is a sugar with molecular formula $C_5H_{10}O_5$. Given that the atomic weights are: C = 12, H = 1 and O = 16, how many moles of ribose are there in 3 g?
    A. 0.01 mol
    B. 0.02 mol
    C. 0.05 mol
    D. 0.10 mol

Answer is B: Mass of 1 mol of ribose = (5 × 12) + (10 × 1) + (5 × 16) = 60 + 10 + 80 = 150 g. Three grams is less than 1 mol. Number of moles = 3 g/150 g/mol = 0.02 mol

21. A solution is prepared containing 2% sodium chloride (Na^+Cl^-). Which of the following best describes the osmolarity of this solution? (Note: Na = 23, Cl = 35.5)

    A. 340 mOsm/L
    B. 680 mOsm/L
    C. 340 Osm/L
    D. 680 Osm/L

Answer is B: 2% = 2 g/100 mL = 20 g/L. Mass of 1 mol of Na^+Cl^- = 23 + 35.5 = 58.5 g. Number of moles of Na^+Cl^- required for 1 L of 2% solution = 20 g ÷ 58.5 g/mol = 0.34 mol. Sodium chloride is an ionic substance so dissociates into ions when dissolved. This would produce 0.34 mol of Na and 0.34 mol of Cl = 0.68 Osm of dissolved particles in the 1 L. 0.68 Osm/L = 680 mOsm/L.

22. The isotonic concentration for sodium chloride is 0.9%, and for glucose is 5%. Which of the following solutions is isotonic with the above solutions? A solution containing:

    A. 5% glucose and 0.25% sodium chloride
    B. 0.9% sodium chloride and 0.5% glucose
    C. 2.5% sodium chloride and 2.5% glucose
    D. 4% glucose and 0.18% sodium chloride

Answer is D: Choices A and B are both hypertonic. Four percent glucose contains 4/5 of the concentration of dissolved particles that are in (isotonic) 5% glucose. 0.18% sodium chloride contains 1/5 of the concentration of dissolved particles that are in 0.9% sodium chloride (0.18/0.9 = 0.2). One fifth plus four fifths add to five fifths of the number of particles that are required for an isotonic solution.

23. Given that 1 mol of glucose has a mass of 180 g, how many millimole of glucose does 40 g of glucose contain?

    A. 0.222
    B. 180
    C. 222
    D. 278

Answer is C: Number of mole of glucose = 40/180 = 0.222 mol. Number of millimole = 222.

24. Which of the following statements about an "osmole" is correct? An osmole is the:

    A. Same as a mole for ionic substances
    B. Formula weight of a substance expressed in grams
    C. Number of moles multiplied by the number of molecules in the chemical formula
    D. Amount of substance that must be dissolved in order to produce 1.6×10^{23} dissolved particles

Answer is D: An osmole refers to the number of solute particles (ions or molecules) that exist in solution, after a substance has dissolved.

25. **Glucose** is a sugar with a molecular formula $\mathbf{C_6H_{12}O_6}$. Given that the atomic weights are: C = 12, H = 1 and O = 16, what is the mass of 1 mol of glucose?
    A. 1.8 g
    B. 18 g
    C. 180 g
    D. 1800 g

Answer is C: Mass of 1 mol of glucose = (6 × 12) + (12 × 1) + (6 × 16) = 72 + 12 + 96 = 180 g.

26. **Lactose** is a sugar with a molecular formula $\mathbf{C_{12}H_{22}O_{11}}$. Given that the atomic weights are: C = 12, H = 1 and O = 16, what is the weight of 1 mol of lactose?
    A. 3.42 g
    B. 34.2 g
    C. 342 g
    D. 3420 g

Answer is C: Mass of 1 mol of lactose = (12 × 12) + (22 × 1) + (11 × 16) = 144 + 2 2 + 176 = 342 g.

27. Red blood cells are added to a hypotonic solution of glucose. Which of the following best describes what you would observe?
    A. The cells would most likely sink to the bottom unaffected.
    B. The cells would shrink due to water loss.
    C. The cells would coagulate.
    D. The cells would swell and burst due to intake of water.

Answer is D: A hypotonic solution would result in an inflow of water to RBC. This would stretch their plasma membrane, and eventually it would break (burst).

28. A solution that is said to be isotonic to blood has the same:
    A. Percent concentration as blood
    B. Number of moles of dissolved particles as blood
    C. Number of osmoles per litre of dissolved particles as blood
    D. Number of dissolved particles as blood

Answer is C: Isotonic means the same number of solute particles per litre. Choice B would also be correct if "moles" was replaced with "moles per litre".

29. If the following amounts of the given substances were dissolved in water, which would result in 4 Osm of dissolved particles?
    A. 2 mol of $C_6H_{12}O_6$
    B. 2 mol of Na^+Cl^-
    C. 2 mol of $C_{12}H_{22}O_{11}$
    D. 2 mol of $(Na^+)_2SO_4^{2-}$

Answer is B: Osmole is a term used when the solute particles came from an ionic substance. Osmoles = moles × number of ions in formula. There are two ions in the formula for sodium chloride, hence 4 Osm. There are three ions in the formula for sodium sulphate, hence 6 Osm.

30. What is the difference between "molarity" and "osmolarity"?
    A. Molarity applies only to covalent compounds while osmolarity applies only to ionic compounds.
    B. Osmolarity is molarity multiplied by 2.
    C. The molarity and osmolarity of a solution is the same for dissolved ionic compounds, but are different for dissolved covalent compounds.
    D. Osmolarity refers to the concentration of dissolved particles in a solution which may not be the same as the number of moles of substance that was dissolved per litre of solution.

Answer is D: Choice D is better than choice A as molarity can also be applied to ionic compounds.

31. Human blood has an osmolarity that lies within the range 280–300 mOsm/L. Which of the following statements is correct?
    A. An isotonic solution has osmolarity that is either less than 280 or greater than 300 mOsm/L.
    B. A hypotonic solution has osmolarity between 280 and 300 mOsm/L.
    C. A hypertonic solution has osmolarity between 280 and 300 mOsm/L.
    D. An isotonic solution has osmolarity between 280 and 300 mOsm/L.

Answer is D: An isotonic solution has the same osmolarity as blood.

32. What distinguishes an osmole from a mole?
    A. In a sample of substance, the number of osmoles is twice the number of moles.
    B. Both mole and osmole may be used in reference to ionic compounds, while covalent compounds are described by mole alone.
    C. The mass of a mole is the sum of the relative atomic masses (RAM) of the atoms in the formula stated as grams. An osmole is half of this mass.
    D. An osmole applies only to covalent molecular substances, whereas a mole is validly applied to both covalent and ionic substances.

Answer is B: Both mole and osmole may be used in reference to ionic compounds (because the atoms of ionic compounds separate into their constituent ions when dissolved), while the atoms within covalent compounds remain as molecules even when dissolved.

33. Which of the four IV solutions below is hypotonic?
    A. 0.18% sodium chloride and 4% glucose
    B. 0.3% sodium chloride and 3.3% glucose

C. 0.45% sodium chloride
D. 0.9% sodium chloride

Answer is C: As 0.9% NaCl is isotonic, and 0.45% is less than 0.9%, it must be hypotonic (and there is only one best correct answer!). A is "4% and a fifth" while B is "3.3% and a third".

34. Which of the following solutions is the most hypertonic?
   A. 0.3% sodium chloride and 3.3% glucose
   B. 0.45% sodium chloride and 2.5% glucose
   C. 0.45% sodium chloride
   D. 10% glucose

Answer is D: 5% glucose is isotonic, so 10% must be hypertonic. Choices A and B are isotonic, while choice C is hypotonic.

35. Given that a mole is 6×10^{23} particles, what is a millimole?
   A. 6×10^{20} particles
   B. 1000 mol
   C. 6×10^{26} particles
   D. 0.0001 mol

Answer is A: A millimole is one thousandth (or 0.001) of a mole. 1 mmol = $10^{-3} \times (6 \times 10^{23}) = 6 \times 10^{20}$ particles.

36. A solution that is isotonic to blood plasma is one which
   A. Contains 0.5% glucose
   B. Must contain the same solutes as blood and in the same concentration as in blood
   C. Has an osmolarity between 280 and 300 mOsm/L
   D. Causes no net movement of water into or out of cells

Answer is D: The concept of tonicity involves reference to a membrane (in this case the plasma membrane of a RBC). Osmolarity is a property of the solution (and of how much solute is dissolved in it) and exists whether a semipermeable membrane is present or not.

37. Which of the following quantities of substance would, when dissolved in 1 L of water, produce a solution with the highest osmolarity?
   A. 1 mol of glucose molecules ($C_6H_{12}O_6$)
   B. 1 mol of Na^+Cl^- (sodium chloride)
   C. 1 mol of haemoglobin molecules
   D. 1 Osm of K^+ (potassium ions)

Answer is B: Sodium chloride is an ionic substance, so when it dissolves, 1 mol of it would separate into 1 mol of individual Na^+ ions and 1 mol of Cl^- ions giving a total of 2 Osm of separate particles (ions) per litre.

38. Given the concentration of Na^+ inside the cell is about 10 mmol/L while it is 140 mmol/L in the extracellular fluid. The values for K^+ are 140 mmol/L inside the cell and 4 mmol/L extracellularly. In what directions would these ions diffuse?
    A. Na^+ would diffuse into the cell while K^+ would diffuse out.
    B. K^+ would diffuse into the cell while Na^+ would diffuse out.
    C. Both Na^+ and K^+ would diffuse out of the cell.
    D. Both Na^+ and K^+ would diffuse into of the cell.

Answer is A: Na would diffuse from 140 to 10 mmol/L—the direction of its concentration gradient which is from outside to inside. K would diffuse from 140 to 4 mmol/L, that is from inside to outside.

39. Which of the following statements best describes what all isotonic solutions have in common?
    A. They contain the same concentration of glucose molecules.
    B. They contain the same total concentration of particles.
    C. They contain the same concentration of sodium chloride.
    D. They have the same specific gravity.

Answer is B: While "iso-" refers to the same, it is the concentration of all the dissolved particles be they ions or molecules of any substance that determines tonicity.

40. Which statement is **INCORRECT**? Colloid osmotic pressure:
    A. Is the difference in solution concentration between plasma and interstitial fluid
    B. Refers to the greater solution concentration of plasma, compared to interstitial fluid, due to the plasma proteins
    C. Causes water to be drawn into blood from the interstitial fluid
    D. Is the pressure exerted on capillary walls due to the collision of the plasma proteins

Answer is D: Although colloid osmotic pressure has the word "pressure", this does not mean that a physical pressure is exerted on capillary walls due to the colloids (the plasma proteins).

41. Consider a volume of 100 mL of each of the four solutions below. Which would contain the greatest number of osmoles of dissolved particles? (FYI the atomic masses of H = 1, C = 12, O = 16, Na = 23, Cl = 35.5)
    A. 5% ethanol: C_2H_6O
    B. 5% glucose: $C_6H_{12}O_6$
    C. 5% sodium chloride: Na^+Cl^-
    D. 10% glucose: $C_6H_{12}O_6$

Answer is C: By adding the masses of the atoms in the formula, we can determine that 1 mol of each substance has a mass of: 46 g (for ethanol: 2 × 12 + 6 × 1 + 1 × 16 = 46); 180 g (glucose); and 58.5 g (NaCl). "5%" means 5 g of substance (much less than 1 mol of each substance) is dissolved in 100 mL of solution. An ethanol molecule is much less than half the mass of a glucose molecule, so there are **more** than twice as many ethanol molecules as glucose molecules in a 5% solution. Hence the ethanol solution has more osmoles than either glucose solution. While NaCl has a greater formula mass than ethanol, NaCl is an ionic substance, so separates into Na ions and Cl ions when dissolved. This doubling in the number of particles when NaCl dissolves means that choice C contains the greatest number of osmoles. Or by calculation: the number of osmoles in each 100 mL of solution is determined by dividing the mass of dissolved substance by the mass of 1 mol of that substance: for 5% ethanol: 5/46 = 0.109; 5% glucose: 5/180 = 0.0278; NaCl: (5/58.5) × 2 = 0.171; 10% glucose: 10/180 = 0.0556 Osm/100 mL.

42. A healthy concentration of sodium in the blood is typically 140 mmol/L. How many grams of sodium are there in a litre of blood? (atomic mass of Na = 23)

    A. 23/140 = 0.164 g
    B. 23/0.14 = 164.3 g
    C. 23 × 140 = 3220 g
    D. 23 × 0.14 = 3.22 g

Answer is D: The number of moles of a substance is determined by dividing the mass of substance by the mass of 1 mol of substance. One mole of sodium has a mass of 23 g: 140 mmol/L = 0.140 mol/L = mass of Na dissolved in 1 L (in gram)/23. Hence mass of Na = 23 × 0.14 = 3.22 g/L.

43. A healthy concentration of chloride in the blood is typically 105 mmol/L. How many grams of chloride are there in a litre of blood? (atomic mass of Cl = 35.5)

    A. 35.5 × 105 = 3727 g
    B. 35.5 × 0.105 = 3.73 g
    C. 35.5/0.105 = 338.1 g
    D. 35.5/105 = 0.3381 g

Answer is B: The number of moles of a substance in a solution is determined by dividing the mass of dissolved substance by the mass of 1 mol of the substance. One mole of chloride has a mass of 35.5 g: 105 mmol/L = 0.105 mol/L. So 0.105 mol/L = mass of Cl (in gram)/35.5. Rearrange this formula, hence mass of Cl = 35.5 × 0.105 = 3.73 g/L.

44. A blood concentration of glucose that is indicative of diabetes is 12 mmol/L. What % concentration of glucose is this equivalent to? (the formula for glucose is $C_6H_{12}O_6$; atomic mass of C = 12, H = 1, O = 16; mass of a mole = sum of masses of atoms in formula).

    A. 0.012%
    B. 0.216%

C. 5%
D. 12%

Answer is B: (by calculation) Percent concentration refers to how many grams are dissolved in 100 mL of solution.

One mole of glucose has a mass of 180 g. $(6 \times 12) + (12 \times 1) + (6 \times 16) = 180$

12 mmol/L = 0.012 mol/L

mol/L = (mass/L)/(mass of 1 mol): hence 0.012 mol/L = (mass/L)/180; so mass = $180 \times 0.012 = 2.16$ g/L

2.16 g/L = 0.216 g/100 mL = 0.216%

(Alternatively: The answer cannot be 5% as this is an isotonic glucose solution. Hence neither is it 12% as this is hypertonic. % concentration is not the same number as mole per litre, so the answer is not A.)

45. If 1 mol of magnesium chloride ($Mg^{2+}Cl^{-}{}_2$) was dissolved in a litre of water, how many osmoles would be present in the solution?
 A. 1 Osm
 B. 2 Osm
 C. 3 Osm
 D. 4 Osm

Answer is C: From the given formula, there are three ions in the solution, one magnesium and two chloride which would disassociate in solution. This would produce 3 Osm of ions.

46. A bag containing 1 L of prepared intravenous saline solution is printed with "Isotonic approximate osmolality 300 mOsm". What is meant by osmolality?
 A. It is the number of osmoles dissolved in kilogram of water.
 B. It is the number of osmoles dissolved per litre of saline solution.
 C. Osmolality is an alternate spelling for osmolarity, the number of osmole per litre of solution.
 D. Osmolality is used for covalent solutes, while osmolarity is used for ionic solutes.

Answer is A: As osmolality (osmol/kg) refers to the concentration within a mass (a kilogram) of solution, it is unaffected by the temperature of the solution. Osmolarity (with an "r") in osmoles per litre refers to the concentration within a volume of a litre of solution. For human physiological solutions which are relatively dilute, there is virtually no difference between the concentrations of solutions when expressed as osmolarity or osmolality.

47. Given that urea has the formula H_2NCONH_2 and the atomic weights: H = 1; C = 12; N = 14; O = 16, what is the mass of 1 mol of urea?
 A. 71 g
 B. 60 g

C. 45 g
D. 43 g

Answer is B: Multiply the number of atoms of each element (H, C, N, O) in the formula by the element's atomic weight to get mass of a mole. Hence: $(4 \times 1) + (1 \times 12) + (2 \times 14) + (1 \times 16) = 4 + 12 + 28 + 16 = 60$ g.

48. A human excretes approximately 450 mmol of urea in urine each day. Given the molar mass of urea is 60 g, what is the mass of urea excreted daily?
    A. 27 g
    B. 60 g
    C. 270 g
    D. 450 g

Answer is A: As the molar mass of urea is 60 g, this means that 1 mol of urea has a mass of 60 g. Now 450 mmol is LESS than 1 mol, so the number must be less than 60 g.

By numbers: 450 mmol × 60 g/mol = 0.45 × 60 = 27 g

49. Given that a solution of 0.9% sodium chloride is isotonic to blood, which of the following concentrations of sodium chloride solution would cause the greatest haemolysis?
    A. 1.05%
    B. 0.9%
    C. 0.75%
    D. 0.6%

Answer is D: Choices A and B are hypertonic and isotonic, respectively, so would not cause any haemolysis. A solution of 0.75%, being hypotonic, would cause some but not all of the red blood cells to lyse. 0.6% would cause the haemolysis of more red blood cells than would a solution of 0.7%.

50. How many millimole of sodium chloride is required to make a 1 L IV bag of normal saline solution (0.9% Na^+Cl^-)?
    A. 0.154 mmol
    B. 1.54 mmol
    C. 15.4 mmol
    D. 154 mmol

Answer is D: 0.9% = 0.9 g/100 mL. So in 1 L (1000 mL), there is 10 × 100 mL and 10 × 0.9 g = 9 g of NaCl. Na has atomic mass of 23 while Cl has an atomic mass of 35.5. Hence 1 mol of NaCl has mass of 23 + 35.5 = 58.5 g. Now 9 g is way less than 58.5 g, so we have less than 1 mol of NaCl. In fact we have 9/58.5 = 0.154 mol = 154 mmol.

51. If 1 mol of sodium chloride has a mass of 58.5 g. What mass of Na^+Cl^- would constitute 1 Osm?

    A. 29.25 g
    B. 58.5 g
    C. 87.3 g
    D. 117 g

Answer is A: Sodium chloride is an ionic compound. 29.25 g is 0.5 mol of sodium chloride. The number of osmoles = the number of moles × the number of ions = 0.5 × 2 = 1.0 Osm. When 29.25 g is dissolved in water, there will be half a mole of Na^+ and half a mole of Cl^- making 1 mol (1 Osm) of dissolved particles.

52. What does the term osmolarity refer to? (1 Osm = 6×10^{23} particles).

    A. It is the number of ions expressed as osmoles per litre of solution.
    B. It is the number of molecules expressed as osmoles per litre of solution.
    C. It is the number of particles (ions and molecules) expressed as osmoles per litre of solution.
    D. It is the number of particles expressed as osmoles per kilogram of solution.

Answer is C: OsmolaRity is the total number of particles (ions + molecules) expressed as osmoles per litre of solution. Choice D refers to OsmolaLity—spelled with "l" instead of "r".

4.5 Acids, Bases and Buffers

An acid is a substance that reacts with water to produce hydronium ions (H_3O^+) in solution. A base is a substance that reacts with water to produce hydroxide ions (OH^-) in solution. Water has equal concentrations of hydronium and of hydroxide ions so is said to be "neutral". That is, water is neither acid nor basic. The level of acidity of a solution is denoted by its pH value (puissance of Hydrogen). Acidic solutions have pH < 7.0, while basic solutions have pH > 7.0. pH is the (negative) logarithm of the hydrogen ion concentration which means that when the "pH value" changes by one unit, the concentration of hydronium ions changes by a factor of 10. Being negative means that as the acidity of a solution increases, the pH value decreases. Thus if the hydronium ion concentration of two solutions are 10^{-6} mol/L and 10^{-5} mol/L, their pH is 6 and 5, respectively. Note that the pH 5 solution has ten times the concentration of hydronium ions that are in the pH 6 solution (and a pH 4 solution would have 100 times the hydronium ion concentration of a pH 6 solution).

Normally the pH of a solution will fall to less than 2 when a strong acid is added to it, while if, instead, a strong base is added, its pH will rise to above 10. A buffered solution is one whose pH changes only very slightly if an acid or a base is added to it. This resistance to change of pH occurs because a buffered solution has two components in it (such as HPO_4^{2-} and $H_2PO_4^-$) one of which (HPO_4^{2-}) is able to destroy any acid that is added to the solution while the other component ($H_2PO_4^-$) is able to destroy any added base. The human body has three major buffer systems: the monohydrogen phosphate/dihydrogen phosphate buffer (above); the bicarbonate ion/carbonic acid buffer and the protein buffer. These act to maintain the blood's pH within the very narrow range of 7.35–7.45. Blood pH above this range is termed alkalosis, while blood pH below this range is termed acidosis.

An acid will react with a base to form their "salt" and water (this is a neutralisation reaction). Often the salt, being ionised, is more soluble in water than either of the acid or base were and this is medically useful. For example aspirin is an acid and when reacted with NaOH produces its soluble salt sodium acetylsalicylate (soluble aspirin). Morphine is a base. When reacted with sulphuric acid, it forms soluble morphine sulphate which can be given intravenously for rapid pain relief.

1. Which of the following statements is **CORRECT**?
   A. An acid is an electron donor and a base is an electron acceptor.
   B. An acidic solution has a pH less than 7, and a basic solution has a pH greater than 7.
   C. Neutralisation of a strong acid by a strong base gives only water.
   D. The pH of saliva is normally in the range of 8.5–9.5

Answer is B: All students should know that pH < 7 indicates acidity. Salivary pH is about 6.0–7.4.

2. Which statement about buffers below is most correct?
   A. A buffer is any acid and base which together control the concentration of pH in the blood.
   B. A buffer is the solution which allows CO_2 to be lost from the lungs in order to control the pH of the lungs.
   C. A buffer is a weak base and its acid salt or a weak acid and its basic salt. The ratio of the two components helps maintain blood pH levels.
   D. A buffer is a mixture of two acids which together help to maintain blood pH.

Answer is C: A buffer has two conjugate components, one dealing with acid while the other destroys base. Their concentration ratio determines the pH of the solution they are in.

3. Which of the following statements is **FALSE**?
   A. An acid is a proton donor and a base is a proton acceptor.
   B. An acidic solution has a pH greater than 7, and a basic solution has a pH less than 7.
   C. Neutralisation of an acid by a base gives a solution of salt in water.
   D. The pH of the stomach is normally in the range of 1.6–1.8.

Answer is B: The opposite of this statement is the truth.

4. Neutralisation of an acid by a base can be represented by the following equation:

$$H_3O^+ + OH^- \rightarrow 2H_2O$$

This means that the pH of the solution after the neutralisation is approximately:
   A. 6
   B. 7
   C. 8
   D. 9

Answer is B: The solution resulting from a neutralisation reactions should have pH = 7 (should be neutral)—provided that it is properly titrated.

5. A buffer solution consisting of citric acid and citrate in a ratio of 5:1 maintains the pH at 7.4. What would be the ratio of the two components if the pH were to become 6.4?
   A. 6:1
   B. 10:1
   C. 50:1
   D. 1:10

Answer is C: As the ratio of the two components changes, so does the pH of the solution. As pH is a logarithmic scale, a pH value change by 1.0 requires a change of one of the components by a factor of 10. This is achieved if the ratio changes from 5 to 1, to 50–1. Furthermore, as the citric acid component has been increased, the pH decreases to be more acidic.

6. Which of the following statements is closest to a correct definition of an acid?
   A. A substance that ionises in a solution to produce hydronium and hydroxide ions
   B. A substance that reacts with water to produce hydroxide ions
   C. A substance that dissociates in water to produce a solution with pH greater than 7.0
   D. A substance that reacts partly with water to produce a low concentration of hydronium ions

Answer is D: An acid produces hydronium ions (H_3O^+) when added to water. If the concentration of hydronium ions produced is low, then the acid is a weak acid.

7. What does the pH of a buffered solution depend on?
   A. The ratio of the components of the buffer solution
   B. The amount of acid added to the buffer solution
   C. The amount of base added to the buffer solution
   D. The amount of acid and of base added to the solution

Answer is A: The pH of the buffered solution will change very little when either acid or base is added.

8. What is the long term acid–base balance in the body controlled by?
   A. The phosphate and carbonic acid/bicarbonate buffers in the blood
   B. The kidneys and the lungs
   C. The phosphate, carbonic acid/bicarbonate and protein buffers in the blood and cells
   D. The kidneys

Answer is B: Buffers temporarily manage the pH in the body's liquids. It is from both the lungs and kidneys that acid is excreted (or not) from the body.

9. You have a bottle of a strong acid and you add 100 mL of this to 1 L of water. What would the pH of this final solution most likely be?
   A. pH = 1
   B. pH = 5
   C. pH = 7
   D. pH = 11

Answer is A: Strong acid produces a high concentration of hydronium ions in aqueous solution, so the pH would be low.

10. What is the approximate range of pH of gastric juice in the stomach?
    A. 1.6–1.8
    B. 6.2–7.4
    C. 7.3–7.5
    D. 7.8–8.6

Answer is A: The stomach has the lowest pH in the body due to the secretion of hydrochloric acid (a strong acid) into it.

11. A buffer solution consisting of acetic acid and acetate ions (base) in a ratio of acid to base of 1:20 maintains the pH at 7.4. What is the ratio of the two components if the pH were to become 8.4 after addition of more of the basic component?
    A. 1:200
    B. 1:21
    C. 1:30
    D. 1:40

Answer is A: As pH is a logarithmic scale, a rise in pH value by 1.0 (from 7.4 to 8.4) requires a change in the concentration of the basic component by a factor of 10. This is achieved if the ratio becomes 1–200.

12. Acidity is stated as a pH value. If the pH of urine sample "A" is 6 and the pH of urine sample "B" is 7, then which of the following is true?
    A. The most acidic sample is sample B.
    B. Sample A has ten times the hydroxide ion concentration of sample B.
    C. Sample B has ten times the hydrogen ion concentration of sample A.
    D. Sample A has ten times the hydrogen ion concentration of sample B.

Answer is D: The solution with higher hydrogen ion (=hydronium ion) concentration has the lower pH.

13. One form of acid–base imbalance in the body is called acidosis. In this situation, which of the following is true?
    A. The blood is less alkaline than it should be.
    B. The blood's pH is less than 7.0.
    C. The blood is less acidic than it should be.
    D. The concentration of hydrogen ions in the blood is less than it should be.

Answer is A: Acidosis = blood pH < 7.35. Blood pH less than 7.0 is extremely rare and presages imminent death.

14. One of the buffer systems in the blood is the carbonic acid/bicarbonate buffer. It helps to maintain the body's acid–base balance by destroying any excess:
    A. Hydrogen ions in the blood
    B. Acid or base in the blood
    C. Hydroxide ions formed in the blood
    D. Bicarbonate ions formed in the blood

Answer is B: A buffer has the ability to destroy either acid or base.

15. The major buffer system in the extracellular compartment is the:
    A. Protein buffer
    B. Carbonic acid/bicarbonate buffer
    C. Ammonia buffer
    D. Phosphate buffer

Answer is B: "Extracellular compartment" is almost synonymous with blood. The protein buffer is important within cells. The ammonia buffer is important in the kidney's filtrate.

16. If a patient was suffering from "acidosis", what would this mean?
    A. Blood pH is not sufficiently alkaline.
    B. Blood pH is acidic.
    C. There is too little hydronium ion in the plasma.
    D. Blood pH is too acidic.

Answer is A: Healthy blood is slightly alkaline. Acidosis is applied to blood with pH below the normal range of 7.35–7.45. If blood has pH < 7.35, it is not sufficiently alkaline.

17. What happens when an acid is added to a buffered solution?
    A. The solution becomes acidic.
    B. The pH of the solution decreases significantly.
    C. The pH of the solution decreases very slightly.
    D. The pH of the solution increases slightly.

Answer is C: A buffer will resist change to its pH. But pH does alter very slightly as the basic component destroys the added acid and in the process becomes its cognate partner. This changes the ratio of the two components of the buffer slightly, and hence the pH.

18. Which of the statements below is correct?
    A. The dihydrogen phosphate component of the phosphate buffer releases hydrogen ions into the lungs for excretion, and in the process, reverts to monohydrogen phosphate.
    B. Haemoglobin as it passes through the lungs releases hydronium ions, which are breathed out.
    C. Carbon dioxide that is dissolved in blood diffuses into the alveoli and is breathed out.
    D. Hydronium ions react with bicarbonate ions to form carbonic acid which moves into the lungs for exhalation.

Answer is C: Carbon dioxide that we breathe out enters the alveoli after diffusing there from the blood plasma. No other gas is excreted.

19. When aspirin (an acid) is in the stomach (an acidic environment), what may be said of the aspirin molecules?
    A. Most molecules will be un-ionised and therefore able to be pass through the stomach mucosa.
    B. Most molecules will be ionised and therefore able to be pass through the stomach mucosa.
    C. Most molecules will be un-ionised and therefore NOT able to be pass through the stomach mucosa.
    D. Aspirin will be in the form of its salt and therefore able to be pass through the stomach mucosa.

Answer is A: Aspirin in an acidic environment will be in its molecular form (and un-ionised). This makes it soluble in non-polar media like the plasma membrane (which is lipid). Being soluble in lipid, molecular aspirin is able to be absorbed from the stomach.

20. Which statement below best describes an acid solution? The pH is less than:
    A. 5
    B. 6
    C. 7
    D. 8

Answer is C: An acidic solution has a pH of less than 7.0.

21. If a patient has blood pH that is 7.3 (which is below the healthy range of blood pH values), which of the following is a correct statement?
    A. The patient has alkalosis.
    B. The patient has excessive alkali.

C. The patient has acidosis.
D. The patient has insufficient acid.

Answer is C: A pH lower than the healthy blood pH range is called acidosis.

22. Which of the following mechanisms results in acid being excreted from the body?
    A. Breathing out $H_2PO_4^-$ (dihydrogen-phosphate) from the lungs
    B. Excreting HCO_3^- in the urine
    C. Excreting NH_3 (ammonia) in the urine
    D. Breathing out CO_2 from the lungs

Answer is D: As CO_2 exhaled from the lungs, carbonic acid in the blood converts into carbon dioxide and water. So as carbon dioxide is exhaled, the amount of carbonic acid within the body decreases.

23. If the pH of a patient's blood is 7.4, then it can be said that the patient has:
    A. Acidic blood
    B. Alkaline blood
    C. Neutral blood
    D. Alkalosis

Answer is B: pH above 7.0 indicates an alkaline solution. However, 7.4 is within the healthy range, so there is no alkalosis.

24. Which of the following would happen when hydrochloric acid is added to water?
    A. The acid would react with water to form a low concentration of hydronium ions.
    B. The acid would react with water to form a high concentration of hydroxide ions.
    C. The acid would react with water to form a high concentration of hydronium ions.
    D. The acid would react with water to form a high concentration of bicarbonate ions.

Answer is C: HCl is a strong acid. Hence a high concentration of hydronium ions would be produced.

25. If the pH of the stomach contents changed from 3 to 2, which of the following is true?
    A. At pH 2, the concentration of hydronium ions is two thirds of the concentration at pH 3.
    B. At pH 3, the concentration of hydronium ions is 50% more than the concentration at pH 2.
    C. At pH 2, the concentration of hydronium ions is ten times the concentration at pH 3.

D. At pH 3, the concentration of hydronium ions is ten times the concentration at pH 2.

Answer is C: When pH decreases from 3 to 2, this indicates an increase in hydronium ion concentration by a factor of 10. Remember pH is a logarithmic scale.

26. The carbonic acid and bicarbonate buffer system is one if the buffers that help to maintain the blood's pH within the healthy range by doing which of the following?
    A. Carbonic acid destroys excess base in the blood while bicarbonate destroys excess acid.
    B. Carbonic acid destroys excess acid in the blood while bicarbonate destroys excess base.
    C. Carbonic acid and bicarbonate destroy excess acid.
    D. Carbonic acid and bicarbonate destroy excess base.

Answer is A: A buffer has two components. One is capable of destroying acid while the other is capable of destroying base.

27. In a solution like blood that is buffered to minimise changes in pH, what determines the pH of the solution?
    A. The concentration of the hydronium ions
    B. The logarithm of the concentration of the hydronium ions
    C. The ratio of the concentration of hydronium ions to hydroxide ions
    D. The ratio of the concentrations of the two components of the buffer

Answer is D: Any hydronium and hydroxide ions present will react with one of the two components of the buffer. So the amount of buffer component (viz. their ratio) governs the concentration of these ions and hence the pH of the solution.

28. When in solution, an acidic drug also exists as its salt. What feature does the salt of a drug have that may not be shared by the acid form of the drug?
    A. The salt of the drug does not carry an electrical charge but the acid form does.
    B. The salt of the drug may be more soluble in lipid than the acid form of the drug.
    C. The acid form of the drug is ionised, whereas the salt is not.
    D. The salt of the drug may be more soluble in water than the acid form of the drug.

Answer is D: The salt of the drug, being ionised, is more soluble in water (a polar liquid) than is the drug molecule.

29. Which one of the following best defines an acid?
    A. An aqueous solution with a pH of less than 7.0
    B. A substance that reacts with water to produce hydrogen ions
    C. A substance that reacts with water to produce hydroxide ions

D. A substance that reacts with water to produce hydronium ions

Answer is D: A "hydrogen ion" H^+ is a bare proton which, if it freely exists, would attach to a water molecule to form the hydronium ion (H_3O^+). So choice D is a better answer than choice B.

30. Which one of the following acids is not normally found in the body?
    A. Hydrochloric acid
    B. Carbonic acid
    C. Nucleic acid
    D. Acetylsalicylic acid

Answer is D: Acetylsalicylic acid is the medicine known as aspirin.

31. If a sample of blood has a pH of 7.25, which of the following is true?
    A. It is in the healthy range.
    B. It is neutral.
    C. It is acidic.
    D. It is basic.

Answer is D: pH above 7.0 indicates basic. 7.25 is below the healthy range.

32. If a blood sample has a pH of 7.25, which of the following is **NOT** true?
    A. The blood is acidic.
    B. The person may be suffering from metabolic acidosis.
    C. The person may be suffering from respiratory acidosis.
    D. The blood is alkaline.

Answer is A: pH must be less than 7.0 for acidic to be true. The condition certainly is an acidosis.

33. Why does drinking acidic solutions such as orange juice **NOT** make the blood acidic?
    A. Blood is alkaline and alkaline solutions react with ingested acids to neutralise them.
    B. The volume of water in the body is great enough that the effect of ingested acidic food and drink on pH is negligible.
    C. The blood has buffers whose components are able to resist changes to blood pH.
    D. The kidney is able to rapidly filter acidic components from blood to avoid pH changes.

Answer is C: The blood has the ability to buffer normal intakes of acidic or basic substances. So excursions of blood pH outside the healthy range are prevented.

34. Which statement about the phosphate buffer is correct?
    A. $H_2PO_4^{2-}$ destroys acid and HPO_4^- destroys base.
    B. $H_2PO_4^{2-}$ destroys base and HPO_4^- destroys acid.

C. $H_2PO_4^{2-}$ is basic and HPO_4^- is acidic.
D. The ratio of $H_2PO_4^{2-}$:HPO_4^- decreases following the addition of acid.

Answer is B: Dihydrogen phosphate is able to destroy base by donating one of its hydrogen atoms (and in the process it becomes mono-hydrogen phosphate). Choice D is wrong as addition of acid would decrease the proportion of monohydrogen phosphate while increasing dihydrogen phosphate. This would make the ratio increase, that is the "fraction" would be larger.

35. If the pH of a sample of blood was 7.4, which term below would be applied to it?
    A. Acidotic
    B. Acidic
    C. Alkalotic
    D. Alkaline

Answer is D: 7.4 is in the healthy range for blood. Being greater than 7.0 indicates alkaline.

36. Three of the solutions below have the SAME level of acidity (or alkalinity), the other is different. Which is the different one?
    A. Blood
    B. A neutral solution
    C. A solution with pH = 7
    D. A solution with a hydrogen ion concentration of 10^{-7} mol/L

Answer is A: Blood has pH in the range 7.35–7.45. The other choices all have pH = 7.0

37. Which of the following statements is a definition of a buffer solution?
    A. One where the number of osmoles of solute equals the number of moles
    B. One where the concentration of hydronium ions equals the concentration of hydroxide ions
    C. A solution that resists change to its pH
    D. A solution that reacts partly with water to produce a relatively low concentration of hydronium ions

Answer is C: This is a definition of a buffer solution. Choice B indicates a neutral solution, while choice D alludes to a weak acid.

38. Which of the following solutions is the most acidic?
    A. Blood plasma
    B. Gastric juice
    C. A solution with pH = 7
    D. A solution with hydronium ions at a concentration of 10^{-5} mol/L

Answer is B: Gastric juice has a pH of about 1.5. The solution in choice D has a pH of 5 (=negative log of 10^{-5}).

39. Pure water contains small but equal amounts of hydronium (H_3O^+) ions and hydroxide (OH^-) ions. This means that water is:
    A. Acidic
    B. Basic
    C. Acidic and basic
    D. Neither acidic nor basic

Answer is D: When the concentrations of hydronium and hydroxide ions is the same, their effects are cancelled and the solution is neutral.

40. What do we call a substance that, when added to water, reacts with water to produce hydronium ions?
    A. An ionic compound
    B. A base
    C. An electrolyte
    D. An acid

Answer is D: This is a definition of an acid.

41. How does the blood's phosphate buffer behave when a hydronium ion is encountered?
    A. HPO_4^{2-} destroys the hydronium ion and becomes $H_2PO_4^-$ and blood pH decreases very slightly,
    B. $H_2PO_4^-$ destroys the hydronium ion and becomes HPO_4^{2-} and blood pH decreases very slightly,
    C. HPO_4^{2-} destroys the hydronium ion and becomes $H_2PO_4^-$ and blood pH increases very slightly,
    D. $H_2PO_4^-$ destroys the hydronium ion and becomes HPO_4^{2-} and blood pH increases very slightly,

Answer is A: Monohydrogen phosphate is the species that destroys acid (by accepting a hydrogen ion from the acid). Blood pH would decrease very slightly as the concentration of monohydrogen phosphate decreases slightly and the concentration of dihydrogen phosphate increases slightly.

42. Which one of the statements below about buffers is correct?
    A. A buffer consists of an acid and a base.
    B. A buffer will react with water to produce hydronium ions.
    C. A solution that is neither acidic nor basic is called a buffer.
    D. The components of a buffer are conjugate pairs.

Answer is D: A buffer has two components. Each component converts to the other as it destroys acid (or base). Choice A is incorrect as a buffer is a weak acid and its slightly basic salt (or vice versa).

43. An acid may be defined as a substance that:

   A. Reacts completely with water producing a high concentration of hydronium ions
   B. Reacts with water to produce hydroxide ions
   C. Produces a pH of more than seven when in solution
   D. Reacts with water to produce hydronium ions

Answer is D: An acid produces hydronium ions. Choice A defines a strong acid (but not a weak one).

44. Which one of the statements below about buffers is **WRONG**?

   A. A buffer consists of a weak acid and its slightly basic salt.
   B. A buffer will react with water to produce hydronium ions.
   C. A solution that resists change to its pH is called a buffer.
   D. The components of a buffer are conjugate pairs.

Answer is B: The ability to produce hydronium ions with water defines an acid, not a buffer.

45. An acid may be defined by which one of the following statements?

   A. A substance with equal concentrations of hydronium and hydroxide ions
   B. A substance that reacts partly with water to produce hydronium ions
   C. A substance that maintains the pH of the solution it is in
   D. A substance that reacts with water producing hydroxide ions

Answer is B: This is a definition of an acid.

46. Which of the following defines an acid?

   A. When in solution, resists change to pH.
   B. Reacts with water to produce hydronium ions.
   C. Reacts with a buffer to produce water.
   D. Has a pH of more than 7.

Answer is B: This is a definition of an acid. All other choices are wrong statements.

47. Which of the following is **NOT** an acid?

   A. A solution having a pH of 3
   B. A substance which reacts with water to produce hydronium ions
   C. Carbonic acid
   D. A solution that resists change to its pH

Answer is D: Such a solution is a buffer.

48. What effect would a **base** have when added to the following solutions?

   A. Tap water—it would react to produce hydronium ions.
   B. Tap water—it would react to produce hydroxide ions.

C. Buffered water—it would raise the pH of the solution.
D. Un-buffered water—it would lower the pH of the solution.

Answer is B: A base has the ability to produce hydroxide ions when added to water.

49. What effect would an **acid** have when added to the following solutions?

A. Tap water—it would react to produce hydronium ions.
B. Tap water—it would react to produce hydroxide ions.
C. Buffered water—it would lower the pH of the solution.
D. Un-buffered water—it would raise the pH of the solution.

Answer is A: An acid has the ability to produce hydronium ions when added to water.

50. A fresh sample of blood (a buffered solution) is withdrawn from the body and its pH measured at 7.4. A small quantity of strong acid (as an isotonic solution) is added to it. Which of the following values is most likely to be the pH of the resulting blood solution?

A. 3.0
B. 7.3
C. 7.4
D. 7.5

Answer is C: Blood is buffered so very little pH change will result from the addition of acid. Certainly not enough to change the pH by 0.1.

51. Blood in the body is different to a buffered solution held in a beaker. This is because hydronium (hydrogen) ions can be removed from solution in blood almost immediately by:

A. Filtration in the kidney from where it is eliminated from the body in urine
B. Combining with hydroxide ions to form water which is eliminated as urine
C. Combining with bicarbonate to form carbonic acid which disassociates to water and carbon dioxide which is breathed out
D. Reacting with monohydrogen phosphate to form dihydrogen phosphate

Answer is C: This is the best answer as it describes the buffer action and the continuous nature of exhalation means that there is no delay in excreting carbon dioxide. Choice A is also correct, but the elimination of acid via kidneys is not immediate.

52. The definition of an acid is a substance that reacts:

A. Completely with water to produce a large concentration of hydronium ions
B. With water to produce hydroxide ions
C. With a base to produce a neutral salt
D. With water to produce hydronium ions

Answer is D: This is the best answer. Choice A is true of a strong acid only. Choice C is incorrect as the salt produced may also be basic or acidic.

53. Choose the **INCORRECT** statement about buffer solutions.
    A. The pH of a buffer does not change on addition of acid or base.
    B. A buffer has two components which may be a weak acid and its salt.
    C. A buffer has two components which may be a weak base and its salt.
    D. pH of a buffer depends only on the ratio of its components.

Answer is A: The pH of a buffer does change, albeit very slightly, on addition of acid or base.

54. In the condition known as acidosis, the blood pH would be:
    A. Greater than 7.45
    B. Less than 7.45
    C. Less than 7.35
    D. Less than 7.00

Answer is C: The healthy pH range for blood is 7.35–7.45. A pH value below this, while still alkaline, indicates insufficient alkalinity (or too much acidity) and is termed acidosis.

55. As the blood becomes more acidic, hydrogen (hydronium) ions combine with bicarbonate ions to form carbonic acid which in turn dissociates into carbon dioxide (to be breathed out by the lungs) and water. Which is the equation describing this reaction?
    A. $CO_2 + H_2O \leftarrow H_2CO_3 \leftarrow H^+ + HCO_3^-$
    B. $CO_2 + H_2O \rightarrow H_2CO_3 \rightarrow H^+ + HCO_3^-$
    C. $CO_2 + H_2O \leftarrow H^+ + HCO_3^- \leftarrow H_2CO_3$
    D. $H^+ + HCO_3^- \leftarrow CO_2 + H_2O \leftarrow H_2CO_3$

Answer is A: Choice B is going in the wrong direction, while choices C and D are scrambled.

56. A buffer solution consisting of $H_2PO_4^-$ (acid) and HPO_4^{2-} (base), in a ratio of 1:4, maintains the pH at 7.4. What is the ratio of these two components if the pH were to become 8.4 after the addition of more of the basic component of the buffer?
    A. 1:5
    B. 2:4
    C. 1:8
    D. 1:40

Answer is D: The pH of a buffer is determined by the ratio of its two components. pH is a logarithmic scale of the hydronium ion concentration of a solution. Hence the ratio must change by a factor of 10 in order for the pH to change by one (i.e. change from 7.4 to 8.4).

57. Which of the following pH values would you expect a dilute solution of a weak base to have?
    A. 2.4
    B. 6.2
    C. 8.2
    D. 12.4

Answer is C: A base has a pH greater than 7.0. A weak base will have a value that is not much greater than 7.0

58. A buffer solution having a pH of 8.5 has a small amount of strong acid added to it. What is its pH value likely to be closest to?
    A. 5.0
    B. 8.4
    C. 8.6
    D. 12.4

Answer is B: A buffer solution resists change to its pH. This means that its pH will change very little, but it will change slightly. As acid has been added, its pH will drop by a small amount.

59. Which of the following statements about the role of the kidneys or lungs in the maintenance of acid–base balance within the body is **FALSE**?
    A. The kidneys are able to affect several buffer systems but the lungs affect only one.
    B. The lungs exert their influence on the carbonic acid/bicarbonate buffer system.
    C. Neither the lungs nor the kidneys have much influence over the protein buffer.
    D. The kidneys are able to respond more rapidly than the lungs to changes in acid–base balance.

Answer is D: The lungs with each breath are able to excrete acid, by exhaling CO_2. So the lungs are able to respond more rapidly than the kidneys.

60. A buffered solution is able to resist change to its pH when an acid is added to the solution by doing which of the following?
    A. Removing hydrogen ions from the solution
    B. Removing hydroxide ions from the solution
    C. Removing hydrogen peroxide from the solution
    D. Removing hydronium ions from the solution

Answer is D: A buffered solution contains two entities, one of which can remove hydronium ions from solution, turning them into water. Sometimes "hydrogen ions" is used instead of the more correct hydronium ions. Hydrogen peroxide is not produced by adding acid to a solution.

61. The healthy range for blood pH is from 7.35 to 7.45. To what blood pH does the term "acidosis" refer?

    A. pH greater than 7.45
    B. pH less than 7.35
    C. pH below 7.00
    D. pH above 7.00

Answer is B: Acidosis refers to any blood pH below the lower limit of normal which is 7.35.

62. What blood pH does the term "alkalosis" refer to?

    A. pH greater than 7.45
    B. pH less than 7.35
    C. pH below 7.00
    D. pH above 7.00

Answer is A: Alkalosis refers to any blood pH above the upper limit of normal which is 7.45.

63. Which body fluid is at the lowest pH?

    A. Urine
    B. The contents of the duodenum
    C. Blood
    D. The contents of the stomach

Answer is D: Acids have low pH (less than 7). The stomach contents contain hydrochloric acid secreted by the parietal cells of the gastric glands so typically has a pH of less than 2.0. Urine, while usually acidic, rarely falls below pH of 5.

64. Which of the following reactions demonstrates the action of the ammonia (NH_3) buffer?

    A. $H_3O^+ + NH_4^+ \rightarrow H_2O + NH_3$
    B. $H_3O^+ + NH_3 \rightarrow H_2O + NH_4^+$
    C. $H_2O + NH_3 \rightarrow H_3O^+ + NH_4^+$
    D. $H_2O + NH_4^+ \rightarrow H_3O^+ + NH_3$

Answer is B: A buffer must destroy acid, so choices C and D, in which acid is "created" are wrong. A buffer destroys acid by accepting a hydrogen atom from the hydronium molecule so converting it to water: choice B describes this and has equal numbers of hydrogen atoms (6) on each side of the arrow, while choice A has an imbalance of H atoms and of charge.

65. Which reaction below demonstrates the action of the bicarbonate buffer?

    A. $H_3O^+ + HCO_3^- \rightarrow H_2O + H_2CO_3$
    B. $H_3O^+ + H_2CO_3 \rightarrow H_2O + HCO_3^-$
    C. $H_2O + HCO_3^- \rightarrow H_3O^+ + H_2CO_3$
    D. $H_2O + H_2CO_3 \rightarrow H_3O^+ + HCO_3^-$

Answer is A: A buffer must destroy acid, so choices C and D, in which acid is "created" are wrong. A chemical reaction must have equal numbers of H atoms on each side of the arrow, but choice B has 5H on one side and 3H on the other.

66. Which reaction below demonstrates the action of the phosphate buffer?
    A. $H_2O + HPO_4^{2-} \rightarrow H_3O^+ + H_2PO_4^-$
    B. $H_3O^+ + H_2PO_4^- \rightarrow H_2O + HPO_4^{2-}$
    C. $H_3O^+ + HPO_4^{2-} \rightarrow H_2O + H_2PO_4^-$
    D. $H_2O + H_2PO_4^- \rightarrow H_3O^+ + HPO_4^{2-}$

Answer is C: A buffer must destroy acid, by accepting a hydrogen atom from a hydronium ion and thus converting it to water. This means that mono-hydrogen-phosphate must appear on the left of the arrow and dihydrogen phosphate on the right. So choices B and D are wrong. There must be equal numbers of H atoms (4) on each side of the arrow, but choice B has 5H on one side and 3H on the other.

67. If humans venture to above 5000 m in altitude, hyperventilation (in response to low atmospheric pressure) results in a very low concentration of dissolved CO_2 in the blood. What is the consequence of this?
    A. Alkalosis with pH less than 7.3
    B. Acidosis with pH less than 7.3
    C. Alkalosis with pH greater than 7.5
    D. Acidosis with pH greater than 7.5

Answer is C: A deficiency of CO_2 dissolved in blood will mean that the amount of carbonic acid produced is too low. This will cause blood pH to rise—a condition known as alkalosis.

68. Which of the following is a substance known as a salt?
    A. Potassium chloride
    B. Sodium
    C. Bicarbonate
    D. Lactate

Answer is A: A salt is produced when an acid reacts with a base. Hence it has two words in its name: the first comes from the base (potassium hydroxide) while the second word is derived from the name of the acid (hydrochloric acid).

69. "Quick-eze" is an oral antacid tablet that contains calcium carbonate as the active ingredient. It is intended to relieve "indigestion" by neutralising stomach acid. Which equation below represents the neutralisation reaction?
    A. $H_2O + Ca^{2+}\,CO_3^{2-} \rightarrow Ca^{2+}\,Cl_2^- + 2HCl + CO_2$
    B. $2HCl + Ca^{2+}\,CO_3^{2-} \rightarrow Ca^{2+}\,Cl_2^- + H_2O + CO_2$
    C. $Ca^{2+}\,Cl_2^- + Ca^{2+}\,CO_3^{2-} \rightarrow 2HCl + H_2O + CO_2$
    D. $CO_2 + 2HCl + Ca^{2+}\,Cl_2^- \rightarrow Ca^{2+}\,CO_3^{2-} + H_2O$

Answer is B: A neutralisation reaction involve stomach acid (hydrochloric acid) reacting with a base (calcium carbonate is a weak base) to produce a salt (calcium chloride) and water. Choice D is wrong as calcium carbonate does not appear as a reagent.

70. Pure water contains hydronium ions at a concentration of 10^{-7} mol/L. Why is water not acidic?
    A. Water also contains a buffer which destroys the hydronium ions.
    B. Water also contains hydroxide ions at the concentration 10^{-7} mol/L.
    C. This concentration of hydronium ions is equivalent to a pH of 7, which indicates a neutral solution
    D. By definition, pure water is neither acidic nor alkaline.

Answer is B: Water self ionises. That is two water molecules can collide to produce the ions $H_2O + H_2O \rightarrow H_3O^+ + OH^-$. These ions are produced in equal amounts so there is never an excess of hydronium (which would make water acidic) or of hydroxide (which would make water alkaline). Choice C is true, but does not offer an explanation.

71. If the pH a urine sample falls from 6.5 to 5.5, what may be said of its acidity?
    A. The urine is acidic, and acidity has increased.
    B. The urine is acidic, and acidity has decreased.
    C. The urine is alkaline, and alkalinity has decreased.
    D. The urine is alkaline, and alkalinity has increased.

Answer is A: The urine is acidic as it pH is below 7. As the pH then decreases, the solution has become more acidic.

72. If the pH difference between the two urine samples is 1. What is the difference in **H^+ concentration** (stated in mole per litre) between the two samples?
    A. 100 mol/L
    B. 10 mol/L
    C. 1 mol/L
    D. 0.1 mol/L

Answer is B: pH is the negative logarithm (in base 10) of the hydronium ion concentration (!) Log of 10 is 1. So if pH differs by 1, the concentration of hydronium ions differs by a factor of 10.

73. Given the pH values for aqueous solutions below, which is the most highly acidic?
    A. 1.2
    B. 1.8
    C. 3.2
    D. 7.35

Answer is A: The lower is the pH value, the greater is the concentration of hydronium ions, that is, the more acidic is the solution.

74. Aspirin has the formula $C_9H_8O_4$: hence what type of substance is it?
    A. Ionic, molecular
    B. Covalent, molecular
    C. Ionic, non-molecular
    D. Covalent, non-molecular

Answer is B: The three elements that comprise aspirin: carbon, hydrogen and oxygen are all non-metals. Hence aspirin is a covalent compound. All covalent substances are composed of molecules.

75. Acetyl salicylic acid (aspirin) is insoluble in water. However when aspirin is added to sodium hydroxide solution, it dissolves. What is happening?
    A. Aspirin is an ionic compound and so dissolves in sodium hydroxide solution.
    B. Aspirin is a covalent compound and so dissolves in sodium hydroxide solution.
    C. Aspirin reacts with sodium hydroxide to produce its salt and water. The salt is an ion and so dissolves in water.
    D. Aspirin reacts with sodium hydroxide to produce its salt and water. The salt is covalent and so dissolves in water.

Answer is C: Aspirin is an acid and so reacts with a base (sodium hydroxide) to produce its salt (sodium acetyl salicylate) and water. A salt is composed of ions (Na^+ and $C_9H_7O_4^-$). Ions, being particles with an electrical charge, dissolve in polar solvents such as water. Covalent compounds are usually insoluble in water.

76. Morphine ($C_{17}H_{19}NO_3$) is a weak base but is usually administered medically as its salt: morphine sulphate ($C_{17}H_{21}NO_7S$). Why is this?
    A. Morphine sulphate is less toxic than morphine.
    B. Morphine sulphate is less addictive than morphine.
    C. Morphine sulphate is more soluble in water than morphine.
    D. Morphine sulphate may be taken orally, but morphine must be administered intravenously.

Answer is C: Morphine sulphate is the salt of morphine, formed when morphine (a base) is neutralised by sulphuric acid. The salt is soluble in water and may be taken orally, or intravenously for more rapid pain relief. Morphine as a base cannot be given IV.

77. When water is mixed with pure (glacial) acetic acid, the resulting solution is called dilute acetic acid. Which form, pure acetic acid or dilute, has the greater electrical conductivity?
    A. Pure acetic acid contains more ions than dilute acetic acid and so is more able to conduct electricity.
    B. Dilute acetic acid contains more water and ions than pure acetic acid and so conducts electricity less readily.

C. Pure acetic acid is not diluted with water and so is a better electrical conductor than dilute acetic acid.
D. Dilute acetic acid is water mixed with acetic acid. Water reacts with acetic acid to produce ions which allow the solution to conduct electricity. Pure acetic acid contains no ions.

Answer is D: The definition of an acid is "a substance that produces hydronium ions when added to water". Ions are necessary for a solution to conduct electricity. Pure acetic acid has no water in it, so also has zero ions. Without ions to conduct the electric current, no current can flow through pure acetic acid.

78. In the carbonic acid/bicarbonate ion buffer system, which species destroys acid and which destroys base?
    A. Carbonic acid destroys acids while bicarbonate ions destroy base.
    B. Carbonic acid destroys base while bicarbonate ions destroy acids.
    C. Carbonate ions destroys acids while bicarbonate ions destroy base.
    D. Carbonate ion destroys base while bicarbonate ions destroy acid.

Answer is B: Carbonic acid reacts with hydroxide ions (base) to form bicarbonate ions and water. Bicarbonate ions react with hydronium ions (acid) to form carbonic acid and water.

4.6 Organic Chemistry and Macromolecules

In the context of anatomy and physiology, I refer to organic chemistry (or carbon chemistry) as the chemistry of carbohydrates, proteins and lipids. These are very large molecules, so are called macromolecules, are polymers and contain the elements: C, H, O and (in the case of proteins) N. Carbohydrates, which are many monosaccharide units linked together, may have between 24 atoms (for glucose) up to 500,000 atoms for glycogen. Proteins consist of amino acid units linked together. They can have between several thousand and several million atoms in their molecule. Lipids may have 73 atoms (for cholesterol) to about 150 for a triglyceride. Triglycerides consist of a glycerol molecule, to which are attached three fatty acid molecules.

The carbon atom always forms four chemical bonds and can bond to itself to form long chains. If a fatty acid has four single bonds surrounding each of its carbon atoms, it is called a saturated fat. If a fatty acid has a double bond between adjacent C atoms, it is termed "unsaturated"—or "polyunsaturated" if there is more than one double bond. Two fatty acids with 18 C in their molecular chain (linoleic and alpha-linolenic) are termed "essential" as they cannot be made in the body. Hence they must be included in the diet. Nine of the 20 amino acids found in humans are essential—the others can be manufactured within the body. The order in which particular amino acids are placed in "the line" is referred to as the primary structure of the protein. If the protein molecule has a repeating pattern of some amino acids in some sections of the protein chain, this is referred to as the secondary structure. The large protein molecule can fold and bend as different parts of the molecule interact with each other to give it its characteristic shape—this is the tertiary structure. If several polypeptide chains (each with their own primary, secondary and tertiary structure) interact to form the protein—as is the case for haemoglobin—this is called their quaternary structure. Carbohydrates (such as starch, glycogen, cellulose) are composed of monosaccharides (such as glucose, fructose, galactose, ribose). Two monosaccharides can join to form a 12 carbon atom disaccharide such as sucrose, lactose or maltose.

1. Denaturation of proteins involves disruption of the:
   A. Primary structure
   B. Secondary structure
   C. Tertiary structure
   D. Covalent bonding

Answer is C: Protein "tertiary structure" is held together by the relatively weak attraction between different parts of the protein molecule known as "hydrogen bonding". This structure is the most easily disrupted.

2. Which one of the following is **NOT** a function of lipids in the body?
   A. Function as co-enzymes
   B. Used in phospholipids
   C. Used to make prostaglandins
   D. Steroids are produced from them

Answer is A: Co-enzymes are proteins (not lipids).

3. Amino acids are the building blocks of all proteins. As more amino acids are added the protein chain tends to fold upon itself in a characteristic way. This characteristic of a protein is called the:
   A. Primary structure
   B. Secondary structure
   C. Tertiary structure
   D. Quaternary structure

Answer is C: The shape produced by folding is the tertiary structure. Primary structure is the sequence in which the amino acids are joined. Secondary structure is any repetition of the primary structure. Quaternary structure does not exist.

4. Which of the following is the correct combination for the nucleic acid DNA?
   A. Phosphate, ribose, uracil
   B. Phosphate, deoxyribose, uracil
   C. Phosphate, ribose, adenine
   D. Phosphate, deoxyribose, adenine

Answer is D: "Deoxyribose" supplies the "D" in DNA. The base uracil occurs in RNA (not DNA) where it substitutes for the thymine that occurs in DNA.

5. To which of the following class of biological compounds do all enzymes belong?
   A. Hormones
   B. Proteins
   C. Carbohydrates
   D. Lipids

Answer is B: All enzymes are proteins.

6. The twisting of a polypeptide chain into a characteristic shape such as a helix is an example of which of the following?
   A. Primary structure
   B. Secondary structure
   C. Tertiary structure
   D. Quaternary structure

Answer is C: Shape of the protein molecule is the tertiary structure of the protein.

7. Which one of the following is **NOT** a steroid compound?
   A. Stearic acid
   B. Oestrogen
   C. Cholesterol
   D. Testosterone

Answer is A: Stearic acid is a fatty acid. The sex hormones oestrogen and testosterone are derived from cholesterol.

8. Which of the following carbohydrates are able to pass through the plasma membrane?
   A. Disaccharides
   B. Sucrose
   C. Glycogen
   D. Monosaccharides

Answer is D: For example glucose maybe absorbed from the gut by passing thought the plasma membrane of the cells lining the gut. Glucose is also absorbed by cells for respiration inside mitochondria.

9. Fructose is a simple sugar or carbohydrate. What is it an example of?
   A. Monosaccharide
   B. Disaccharide
   C. Polysaccharide
   D. Oligosaccharide

Answer is A: Fructose is a simple sugar or monosaccharide—it cannot be converted to a simpler sugar by hydrolysis.

10. Cells use glucose as an energy source, but store it as glycogen. When needed the glycogen is broken down by a process called:
    A. Glycolysis
    B. Glycogenesis
    C. Gluconeogenesis
    D. Glycogenolysis

Answer is D: Glycogen-o-lysis refers to the lysis (splitting) of glycogen into smaller glucose units.

11. A lipid such as oleic acid contains a number of double bonds in the carbon chain. Because of this, what term is applied to it?
    A. Monounsaturated
    B. Diunsaturated
    C. Polyunsaturated
    D. Saturated

Answer is C: "Unsaturated" because addition hydrogen atoms may be attached to the carbon bonds if the double bond was replaced by a single bond. "Poly-" because there is more than one instance of a double bond between adjacent carbon atoms.

12. What are the structural components of proteins?
    A. Amino acids
    B. Fatty acids
    C. Peptides
    D. Monosaccharides

Answer is A: Peptides is used for several amino acids joined together.

13. Which of the following chemical formulae represents a typical carbohydrate?
    A. $Ca_{10}(PO_4)_6(OH)_2$
    B. $N_2C_5H_{12}O_4$
    C. $C_{120}H_{240}O_{120}$
    D. $C_{57}H_{110}O_6$

Answer is C: A carbohydrate is composed of carbon, hydrogen and oxygen atoms. Usually there are twice as many H atoms as C atoms. Choiçe D, having relatively few O atoms, is typical of a lipid.

14. Proteins perform a wide range of functions in the body including which one of the following?
    A. They are our major source of energy
    B. They act as enzymes
    C. They are used to make sex hormones
    D. They are necessary for the absorption of vitamins A, D, E and K

Answer is B: All enzymes are proteins. Carbohydrates should be the major energy source. Choice D lists the fat-soluble vitamins.

15. Which of the following refers to a carbohydrate?
    A. Diglycerol
    B. Adenosine diphosphate
    C. Disaccharide
    D. Dipeptide

Answer is C: A disaccharide is a sugar (hence a carbohydrate) that can be simplified into simpler sugars.

16. What are the bonds that maintain the secondary and tertiary structure of proteins?
    A. Hydrogen bonds
    B. Covalent bonds
    C. Ionic bonds
    D. Peptide bonds

Answer is A: Hydrogen bonds acting between N and H. Covalent bonds act between the atoms of an amino acid, while peptide bonds (=covalent bond) is the term used for the bond between one amino acid to the adjacent one.

17. Which of the following formulae would be most likely to represent a lipid?
    A. $C_{2936}H_{4624}N_{786}O_{889}S_{41}$
    B. $C_{57}H_{110}O_6$
    C. $C_6H_{12}O_6$
    D. $Ca_{10}(PO_4)_6(OH)_2$

Answer is B: Lipids consist of carbon, hydrogen and oxygen, with relatively few O compared to C. Choice C is a monosaccharide, choice A is a protein (note the nitrogen). No one should have chosen D (=calcium hydroxyapatite).

18. What are the four most common elements found in proteins?
    A. C, H, N, Ca
    B. C, O, N, Fe
    C. C, H, N, O
    D. N, C, H, Na

Answer is C: Nitrogen must be present. Choice B is wrong as H is absent.

19. To what type of fatty acids is the term "saturated" applied?
    A. Those with four single bonds around each carbon atom
    B. Those with at least one double bond between carbon atoms
    C. Those that are not implicated in coronary heart disease
    D. Those which are essential in our diet

Answer is A: Saturated refers to the number of single bonds around carbon atoms. The maximum is four if there are no double bonds.

20. Which of the following is a polysaccharide?
    A. Glucuronidase
    B. Glucagon
    C. Glucose
    D. Glycogen

Answer is D: Choice A is an enzyme ("-ase"), B is a hormone, C is a monosaccharide.

21. What holds the primary structure of a protein together?
    A. Hydrogen bonds
    B. Covalent bonds
    C. Peptide bonds
    D. Ionic bonds

Answer is C: Protein primary structure refers to the sequence of the linked amino acids. The link between adjacent amino acids is called a "peptide bond"—a special term reserved for the covalent bond between the C of one amino acid and the N of the adjacent amino acid.

22. Which of the following are **NOT** proteins?
    A. Glycolipids
    B. Enzymes
    C. Haemoglobin
    D. Albumin

Answer is A: Glycolipids are lipids not proteins.

23. What is a "saturated fat"?
    A. One that contains cholesterol
    B. A triglyceride that has three fatty acids
    C. One where the carbon atoms that are connected by single bonds
    D. One that must be included in our diet

Answer is C: Saturated means each carbon atom is directly bonded to four other atoms.

24. What is a function of carbohydrates in the body?
    A. To act as enzymes
    B. To provide energy
    C. To function as local hormones
    D. To provide the building blocks for proteins

Answer is B: Dietary carbohydrates are converted to simple sugars which are disassembled in mitochondria to produce ATP.

25. Which of the following is a typical formula for a polypeptide?
    A. $C_{57}H_{110}O_6$
    B. $C_{18}H_{29}O_2$
    C. $C_{21}H_{41}O_5N_8$
    D. $C_6H_{12}O_6$

Answer is C: A polypeptide is several amino acids joined together. Hence the element nitrogen must be present.

26. In what form is most of the lipid component of our food in?
    A. Triglycerols
    B. Polysaccharides
    C. Complex carbohydrates
    D. Polypeptides

Answer is A: Our dietary lipids are in the form of triglycerols (a glycerol molecule and three fatty acids).

27. In organic compounds, how many bonds do carbon atoms always form?
    A. Four single bonds
    B. Four covalent bonds
    C. Two double bonds
    D. Four ionic bonds

Answer is B: Carbon has four valence electrons and "would like" a share of four more to fill its outer shell. This is achieved by four covalent bonds, where a double bond counts as two covalent bonds.

28. A molecule of an organic compound can be thought of as having two portions: a radical with an attached functional group. What is the correct definition of a functional group?
    A. It consists of carbon atoms and hydrogen atoms.
    B. It is responsible for most of the properties of the molecule.
    C. It has a modest influence on the properties of the molecule.
    D. It consists of a ring of carbon atoms.

Answer is B: As organic molecules all contain C, H and O atoms in similar bonding situations, and the differences in their chemical behaviour (function) is largely governed by the differences in structure of small groups of atoms that are bonded differently or contain atoms other than C, H and O. These groups of atoms are known as the functional group.

29. The one medicine may have many names. What is the "generic name" of a medicine?
    A. The name that describes the effect of the drug on the body.
    B. The systematic name of the chemical involved.

C. The name given to the product by a manufacturer of the drug.
D. The name proposed by the inventor and approved by a government agency.

Answer is D: A medicine has only one "generic" name, but each manufacturer will give it their own "trade" name. Choice A refers to the naming of a class of drugs each of which may be a different molecule.

30. When a protein is denatured, which aspect of its structure is affected the **LEAST**?

A. Primary structure
B. Secondary structure
C. Tertiary structure
D. Quaternary structure

Answer is A: Denaturation does not affect primary structure. That is, the sequence of amino acids within the polypeptide chain.

31. What chemical symbol for the element/atom occurs in an amino acid but not in fats and carbohydrates?

A. N
B. H
C. O
D. C

Answer is A: Amino acids contain an "amine" group—a nitrogen atom bonded to a hydrogen atom. Nitrogen atoms do not occur in fats or carbohydrates.

32. The element carbon is said to be tetravalent. This means it always has the following number of covalent bonds in neutral compounds.

A. One
B. Two
C. Three
D. Four

Answer is D: Carbon forms four covalent bonds. "Tetra" refers to four.

33. Which of the following structures represents a molecule belonging to the **alcohol** family?

A. CH_3–COOH
B. CH_3–OH
C. CH_3–O–CH_3
D. CH_3–CO–CH_3

Answer is B: An "alcohol" has an oxygen bonded to a hydrogen (-OH) atom as a functional group. Choice A has a carboxylic acid (-COOH) functional group.

34. Which statement describing the chemical reactivity of organic molecules is **FALSE**?
    A. Reactivity depends on the nature of the functional group.
    B. Reactivity is independent of the number and nature of the carbon rings.
    C. Reactivity is dependent on the nature and size of the radical group.
    D. Reactivity in the body may depend on the internal body factors such as temperature and pH.

Answer is B: Reactivity is in fact influenced by the number and nature of the carbon rings. So the statement B is false.

35. Which of the following structures represents molecules belonging to the **amide** family?
    A. CH_3–SH
    B. CH_3–NH_2
    C. CH_3–CO–NH_2
    D. CH_3–OH

Answer is C: The "amide" functional group consists of a carbon, oxygen and a nitrogen atom. The carbon atom is double bonded to an oxygen atom while also being single bonded to a nitrogen atom.

36. A hydrocarbon is a type of compound that contains:
    A. Only carbon atoms
    B. Carbon and Hydrogen atoms
    C. Carbon, Hydrogen and Oxygen atoms
    D. Water and carbon atoms

Answer is B: Hydrocarbon means hydrogen combine with carbon.

37. A hydrocarbon molecule which contains at least one double bond could be an example of which type of compound:
    A. Alkane
    B. Alkene
    C. Alkyne
    D. Alcapone

Answer is B: Alkenes contain at least one double bond between carbon atoms. Alkanes have only single bonds, while alkynes have at least one triple bond between carbon atoms.

38. Which of the following structures represents a molecule belonging to the **ketone** family?
    A. CH_3–CO–NH_2
    B. CH_3–NH_2
    C. CH_3–O–CH_3
    D. CH_3–CO–CH_3

Answer is D: A ketone has a "carbonyl" functional group (C double bonded to O) while the carbon atom of the carbonyl group is also bonded to two other carbon atoms (that is, is attached to two radicals). Choice C is incorrect as the O is single bonded to the carbon(s). Choice A is incorrect as the NH group is not a radical.

39. Only one of the following patterns of bond is ever displayed by carbon atoms in organic compounds. (Each line represents a bond) Which is it?

A. $-\mathrm{C}\equiv$

B. $\overset{\|}{-\mathrm{C}-}$ with a single bond below: $\underset{|}{\overset{\|}{-\mathrm{C}-}}$

C. $\overset{|}{-\mathrm{C}=}$

D. $\underset{|}{-\mathrm{C}\equiv}$

Answer is C: Carbon always forms four bonds, never more or less. A double bond = 2 bonds; a triple bone = 3 bonds.

40. Certain components of food are often referred to as being saturated. This term refers to the bonding of carbon atoms present in molecules of the food. Which of the following bonding arrangements is a saturated environment?

A. $-\mathrm{C}\equiv\mathrm{C}-$

B. $=\mathrm{C}=\overset{|}{\mathrm{C}}-$

C. $\underset{|}{\overset{|}{-\mathrm{C}}}-\underset{|}{\overset{|}{\mathrm{C}-}}$

D. $\underset{|}{-\mathrm{C}}=\underset{|}{\mathrm{C}-}$

Answer is C: In this case each carbon atom has four single bonds. A double bond would indicate an unsaturated compound.

41. What is the sequence of amino acids in a protein known as?

A. Primary structure
B. Secondary structure
C. Tertiary structure
D. Quaternary structure

Answer is A: Primary structure refers to the order in which amino acids are attached to each other.

42. Which one of the following structures represents a saturated organic compound?

A. 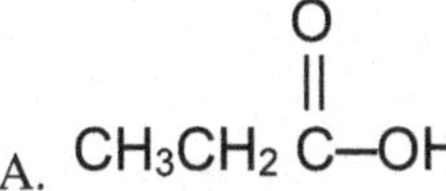
B. CH_3CH_2OH
C. $CH_3CHCHCH_3$
D. CH…CH

Answer is B: A saturated compound must have each carbon atom directly bonded to four other atoms. Only the arrangement in choice B can satisfy this requirement. In choices C and D, the arrangement CH means that there must be a double bond between two C atoms.

43. Which of the following is NOT classed as a lipid?

A. Fatty acids
B. Oils
C. Steroids
D. Keratin

Answer is D: Examples of lipids are waxes, fats, oils, steroids, cholesterol and phospholipids. Keratin is a protein (the "-in" ending to keratin denotes that).

44. Oleic acid has the formula: $CH_3(CH_2)_7CH{=}CH(CH_2)_7COOH$. (The "=" indicates a double bond) How many atoms of carbon are there in a molecule of oleic acid?

A. 6
B. 10
C. 18
D. 20

Answer is C: Counting from left to right of the formula: 1 C in CH_3; The 7 outside the brackets means 7 lots of everything inside $(CH_2)_7$ so 7 more C; one in each CH, so 2 more C; Then 7 more C in the brackets; then one C before the 2 oxygen atoms. Total = 1 + 7 + 2 + 7 + 1 = 18.

45. What feature of oleic acid $[C_8H_{17}CH{=}CH(CH_2)_7COOH]$ indicates that it is a mono-unsaturated fatty acid?

A. The double bond between two atoms.
B. The double bond between two carbon atoms.
C. The lone double bond between two carbon atoms.
D. The double bond between two carbon atoms and between carbon and oxygen.

Answer is C: There must be a double bond between two carbon atoms and to be mono-unsaturated, and there must be just the one double bond. If there were more, the molecule would be poly-unsaturated. The double bond between C and O is not relevant for saturation.

Chapter 5
Integumentary System

The skin covers the body with the superficial epidermis, made of epithelial tissue, and the deeper dermis, containing much connective tissue. Under the dermis is the superficial fascia (also called the hypodermis despite NOT being part of the skin). The most superficial layer of the epidermis is the waterproof stratum corneum. It is not comprised of living cells, almost impervious to water soluble materials and hence affords the body great protection against dehydration and from environmental bacteria in the same way that the peel of (say) an apple protects its contents. The deepest layer of the epidermis is the stratum germinativum. Here keratinocytes are actively dividing (and producing keratin and previtamin D_3). Melanocytes also inhabit this layer and produce melanin. Dendrocytes migrate to the skin from the bone marrow and become macrophages there. Merkel cells are one of the several types of sensory receptors found in the skin.

The dermis contains blood capillaries, hair roots and several types of sensory receptors which may be free nerve endings or encapsulated. The dermis also contains sudoriferous (sweat) and sebaceous (oil) glands—these are "exocrine" glands because they secrete their product through a duct. The papillary dermal layer overlies the deeper reticular dermal layer. The former rises up to push papillae into the epidermis and forms troughs to receive epidermal ridges, hence has a wavy boundary with the epidermis. Both layers of the dermis are connective tissue, the deeper layer containing collagen and elastic fibres.

5.1 Layers of the skin

1. What is the name given to the most superficial layer of the integument?

   A. Stratum corneum
   B. Papillary dermal layer
   C. Stratum lucidum

M. Caon, *Examination Questions and Answers in Basic Anatomy and Physiology*, https://doi.org/10.1007/978-3-030-47314-3_5

D. Superficial fascia

Answer is A: The stratum corneum is the horny (i.e. keratinised), dead covering to the epidermis. Choice D is another name for the hypodermis.

2. Which layer of the integument contains rapidly dividing keratinocytes?
   A. Stratum lucidum
   B. Papillary dermal layer
   C. Stratum germinativum
   D. Reticular dermal layer

Answer is C: Stratum germinativum (=stratum basale) implies that germination is happening. In fact new cells are formed here by mitosis to replace those lost from the stratum corneum.

3. Which layer of the integument is the most superficial layer?
   A. Hypodermis
   B. Stratum granulosum
   C. Stratum corneum
   D. Reticular dermal layer

Answer is C: Stratum corneum is the outermost dead layer of the epidermis.

4. Which is the most superficial layer of the integument that also has capillaries, lymphatics and sensory neurons?
   A. Reticular dermal layer
   B. Papillary dermal layer
   C. Stratum granulosum
   D. Stratum lucidum

Answer is B: The papillary layer is superficial to the reticular layer. Both are part of the dermis which is where the most superficial capillaries are.

5. Which of the following lists layers of the integument in the order from most superficial first, to deep?
   A. Epidermis, hypodermis, dermis
   B. Epidermis, papillary dermal layer, reticular dermal layer
   C. Dermis, stratum germinativum, stratum corneum
   D. Stratum corneum, stratum germinativum, epidermis

Answer is B: The integument consists of the epidermis and dermis. The dermis consists of the papillary dermal layer and the reticular dermal layer, which is deeper.

6. Which skin layer is the most superficial?
   A. Stratum lucidum
   B. Stratum corneum

C. Papillary dermal layer
D. Reticular dermal layer

Answer is B: Stratum corneum is the outermost corny or horny layer made largely of keratin.

7. The integument consists of which of the following layers?

   A. Epidermis and dermis
   B. Epidermis, dermis and hypodermis
   C. Stratum germinativum, stratum spinosum, stratum granulosum, stratum lucidum and stratum corneum
   D. Stratum corneum, dermis and reticular dermal layer

Answer is A: Hypodermis is not part of the integument. Choice C excludes the dermis.

8. Which layer of the skin is the most superficial?

   A. Epidermis
   B. Dermis
   C. Papillary dermal layer
   D. Stratum germinativum

Answer is A: Epidermis refers to the outermost layer of the skin. While stratum germinativum is part of the epidermis, it is the deepest part.

9. The "stratum corneum" is part of the skin that:

   A. Contains the youngest, rapidly dividing cells
   B. Anchors the skin to muscle while allowing it to slide over muscle
   C. Contains collagen, elastin and reticular fibres
   D. Protects the body against heat, chemicals and bacteria

Answer is D: The stratum corneum is "not alive" so forms the protective barrier between the human internal environment and our surroundings.

10. The hypodermis is which of the following?

   A. The outer layer of the skin
   B. The inner layer of the skin
   C. The superficial fascia and not regarded as part of the skin
   D. Not vascularised, getting its nutrients by diffusion

Answer is C: Hypodermis is "below" the dermis. Although it has "dermis" in its name, it is not part of the integument.

11. Which is the outermost layer of the skin?

   A. Dermis
   B. Epidermis
   C. Stratum lucidum

D. Reticular dermal layer

Answer is B: The epidermis is "on top of" (superficial to) the dermis.

12. Which one of the following is **NOT** part of the integumentary system?
    A. Hypodermis
    B. Sebaceous glands
    C. Finger nails
    D. Stratum corneum

Answer is A: Hypodermis is the superficial fascia, also known as the subcutaneous fat layer.

13. Which list of structures is **NOT** all part of the integumentary system?
    A. Sebaceous glands, hair, nails, mammary glands
    B. Meissner's corpuscles, hypodermis, eccrine sweat glands, oil glands
    C. Apocrine sweat glands, sebaceous glands, Merkel disc, hair follicles
    D. Melanocytes, keratinocytes, Merkel cells, dendrocytes

Answer is B: The hypodermis is not part of the integumentary system.

14. What is one difference between the dermis and the epidermis?
    A. Epidermis is composed of fibrous connective tissue while the dermis is composed of epithelial cells.
    B. Dermis is the most exterior layer.
    C. Dermis is not part of the skin while the epidermis is.
    D. Dermis is vascularised while the epidermis is not.

Answer is D: The stratum germinativum of the epidermis and the more superficial layers, being without blood capillaries receives their nutrition via diffusion from capillaries in the dermis.

15. Which stratum of the epidermis protects the body against water loss and abrasion?
    A. Stratum germinativum
    B. Stratum basale
    C. Stratum lucidum
    D. Stratum corneum

Answer is D: The stratum corneum is the outermost layer of the epidermis and is composed of dead cells filled with keratin. This layer is harmlessly rubbed off (abrades) as the body comes into contact with objects outside the body.

16. On which part of the integument is most of the body's normal flora located?
    A. Dermis
    B. Epidermis
    C. Microdermis

D. Hypodermis

Answer is B: The normal flora (bacteria) is outside the body, residing on the surface of the epidermis.

17. Which is the deepest layer of the integument?
    A. Epidermis
    B. Dermis
    C. Stratum corneum
    D. Papillary dermal layer

Answer is B: The dermis includes the papillary layer and the deeper reticular layer.

18. Which one of the following statements about the stratum corneum is correct?
    A. Cells in this layer undergo cell division to replace the skin.
    B. It consists of dead cells.
    C. It contains collagen, elastin and reticular fibres.
    D. The layer has sensory receptors known as Merkel discs, Meissner's and Pacinian corpuscles.

Answer is B: Stratum corneum is the outermost layer of the epidermis and is subject to physical abrasion. The now flattened, keratinised cells are not living.

19. What is the "superficial fascia"?
    A. It is the skin.
    B. It is the epidermis.
    C. It is the hypodermis.
    D. It is the subcutaneous fat.

Answer is C: Hypodermis is not part of the skin but is synonymous with superficial fascia. Choice D is incorrect as while subcutaneous fat may be located between the skin and the muscle layer, it is not superficial fascia.

20. What distinguishes the stratum corneum of the epidermis from the other layers?
    A. The stratum corneum is not vascularised, while the other layers are.
    B. The stratum corneum produces new cells, while the other layers do not.
    C. The stratum corneum is only present in the epidermis of the hands and feet.
    D. The cells of this stratum do not possess a nucleus, while the cells of the other layers do.

Answer is D: The cells of the stratum corneum are dead, having lost all of their organelles and being filled with keratin.

21. Where must the ink in a skin tattoo be placed for the tattoo to be permanent?
    A. In the hypodermis
    B. In the dermis

C. In the epidermis
D. Under the stratum corneum

Answer is B: The epidermis is continually replaced as new cells from the stratum basale move the over lying cells more superficially until eventually the stratum corneum flakes off. If the ink is placed below the stratum basale of the epidermis into the dermis, it remains there and can be seen through the translucent epidermis.

22. Which layer(s) of the skin are penetrated by the lance when a finger is pricked to obtain a drop of blood?
    A. Epidermis
    B. Epidermis and dermis
    C. Epidermis, dermis and hypodermis
    D. Stratum corneum

Answer is A: The epidermis does not have a blood supply, but the dermis does. To obtain a blood sample, the lance must penetrate the epidermis in order to cut the capillaries in the dermis. The stratum corneum is the outer dead layer of the epidermis.

23. Which of the following describes the structure of one the two strata in the dermis?
    A. Single-cell thick, containing the youngest rapidly dividing keratinocytes
    B. Most superficial layer containing dead, flat cells filled with keratin
    C. Contains capillaries and lymphatics that supply skin as well as sensory neurons
    D. Composed of adipose and some areolar connective tissue

Answer is C: This describes the papillary dermal layer. Choice A describes the stratum germinativum; B describes the stratum corneum of the epidermis. D describes the hypodermis.

24. Which of the following does NOT describe one of the layers/strata of the epidermis?
    A. Cells contain keratin filaments that span cells, attach to desmosomes and so hold cells together.
    B. Composed of adipose and some areolar connective tissue.
    C. Most superficial layer containing dead, flat cells filled with keratin.
    D. Contain granules which aggregate keratin and contain waterproofing glycolipid.

Answer is B: This describes the hypodermis (or subcutaneous) layer. A refers to the stratum spinosum; C refers to the stratum corneum; D refers to the stratum granulosum.

5.2 Receptors and Glands of the Skin

1. Where are sebaceous glands found?
   A. In the digestive system
   B. In the hypodermis
   C. In the dermis
   D. In the stratum corneum

Answer is C: Sebaceous glands produce sebum (oil) and are located in the dermis.

2. Which of the following is NOT a sensory receptor of the skin?
   A. A Meissner corpuscle
   B. An apocrine gland
   C. A root hair plexus
   D. A nociceptor

Answer is B: This is a type of sudoriferous (sweat) gland.

3. Which of the following statements is **INCORRECT**?
   A. Sudoriferous glands secrete sebum.
   B. Sebaceous glands secrete oil.
   C. Apocrine glands secrete sweat,
   D. Ceruminous glands secrete cerumen.

Answer is A: Sudoriferous glands secrete sweat. Sebum (or oil) is produced by sebaceous glands.

4. What does the term "nociceptors" in the skin refer to?
   A. Sensors that detect movement of hair follicles
   B. Any lamellated sensory corpuscle
   C. The sensory receptor that is associated with the Merkel cells of the epidermis
   D. Free nerve endings with large receptive fields that detect pain

Answer is D: The prefix "noci" is derived from the Latin for hurt.

5. What do sudoriferous glands do?
   A. Secrete sebum into a hair follicle
   B. Secrete sweat through a duct to the skin
   C. Secrete sweat through a duct to the skin or into a hair follicle
   D. Secrete cerumen through a duct to the skin or into a hair follicle

Answer is C: Eccrine sweat glands secrete through a duct, apocrine sweat glands secrete into a hair follicle.

6. Which glands secrete "oil" into a hair follicle?
   A. Apocrine
   B. Eccrine
   C. Ceruminous
   D. Sebaceous

Answer is D: Eccrine sudoriferous glands secrete to skin surface via a duct.

7. Which one of the following is **NOT** a type of sweat gland?
   A. Eccrine gland
   B. Merocrine gland
   C. Endocrine gland
   D. Apocrine gland

Answer is C: Endocrine is a general term for glands that produce hormones. Merocrine is another name for eccrine.

8. What do the apocrine glands of the skin secrete?
   A. Apocrin
   B. Cerumin
   C. Milk
   D. Sweat

Answer is D: Apocrine glands are a type of sweat gland.

9. Another name for oil glands in the skin is:
   A. Sebaceous glands
   B. Eccrine glands
   C. Merocrine glands
   D. Apocrine glands

Answer is A: The other three choices are all sweat glands.

10. What is another name for sweat glands?
    A. Ceruminous glands
    B. Sebaceous glands
    C. Sudoriferous glands
    D. Apocrine glands

Answer is C: Choice D does list a sweat gland, but not all of them.

11. Sudoriferous glands are also known as:
    A. Eccrine or apocrine glands
    B. Sebaceous or oil glands
    C. Ceruminous or apocrine glands
    D. Mammary or eccrine glands

Answer is A: Sudoriferous gland are sweat glands, of which there are two types.

12. One of the following is **NOT** a gland found in the integument. Which one?
    A. Sudoriferous gland
    B. Mammary gland
    C. Pineal gland
    D. Sebaceous gland

Answer is C: The pineal gland is in the brain.

13. All but one of the following are sensory receptors in the skin. Which one is NOT?
    A. Reticular dermal layer
    B. Merkel discs
    C. Nociceptors
    D. Pacinian corpuscles

Answer is A: The reticular dermal layer is not a receptor, it is the deepest layer of the dermis.

14. What is the secretion produced by sudoriferous glands?
    A. Sweat
    B. Sebum
    C. Cerumin
    D. Merocrin

Answer is A: Sudoriferous glands are sweat glands.

15. Which of the following is an example of an exocrine gland that is located in the dermis?
    A. Pancreas
    B. Sebaceous glands
    C. Lacrimal glands
    D. Meissner's corpuscle

Answer is B: Sebaceous (oil) glands are located in the skin. Choices A and C are not, while Meissner's corpuscle is a sensory structure.

16. Which two types of glands are found in the skin?
    A. Eccrine and merocrine glands
    B. Sudoriferous glands and sweat glands
    C. Oil glands and sebaceous glands
    D. Sebaceous glands and sudoriferous glands

Answer is D: Sebaceous glands and sudoriferous glands are also known as oil glands and sweat glands, respectively. Eccrine glands and merocrine glands are the same thing as each other. Sudoriferous glands and sweat glands are the one thing. Oil glands and sebaceous glands are the same thing.

5.3 Cells, Functions and Burns of the Skin

1. Which statement below is correct?
   A. Dendrites are produced by dendrocytes.
   B. Glycolipids are produced by lipocytes.
   C. Keratin is produced by keratinocytes.
   D. Melanin is produced by Merkel cells.

Answer is C: Keratin is the horny protein that gives the epidermis its protective qualities and is produced by keratinocytes.

2. When a medicine is delivered via a patch attached to the skin, it is said to be delivered:
   A. Transdermally
   B. Subcutaneously
   C. Topically
   D. Intramuscularly

Answer is A: Some medicines can be absorbed through the skin to enter the bloodstream. Choice C applies to ointments that act on the skin they are applied to. Choices B and D are delivered by a penetrating injection to locations deep to the skin.

3. Full-thickness burns to more than 20% of the skin surface is a life-threatening situation. Why is this?
   A. The synthesis of vitamin D (calcitriol) is severely compromised.
   B. The loss of skin sensation prevents access to environmental information.
   C. The body is not able to prevent water loss from the burnt area.
   D. Soft internal tissues are abraded by the external environment.

Answer is C: The body is not able to prevent water loss from the exposed tissues. (Also, but not part of the provided choices, the loss of skin means that the body is unprotected from infectious agents.)

4. Which is a notable feature of the stratum corneum layer of the integument?
   A. It is where melanocytes and keratinocytes are rapidly dividing.
   B. It is richly supplied with blood capillaries.
   C. It consists of keratin-filled cells with glycolipid in between cells.
   D. It has protruding epidermal ridges that push the overlying epidermis into "fingerprints".

Answer is C: This property makes the skin a barrier between the internal body and the environment.

5. When a pharmaceutical is administered hypodermically, it is:
   A. Wiped onto the skin
   B. Applied via a patch that adheres to the skin

C. Injected into the dermis
D. Injected into subcutaneous fat

Answer is D: "Hypo" means below, in this case, the dermis. Subcutaneous fat (superficial fascia) lies beneath the dermis.

6. What is the protein that fills the outermost dead cells of the epidermis?
   A. Granstein
   B. Dermin
   C. Melanin
   D. Keratin

Answer is D: Choice C is also a protein, but is deep to the outermost layer, lying just superficial to the stratum germinativum.

7. Choose the incorrect statement below.
   A. Keratinocytes produce keratin.
   B. Merkel cells are associated with a sensory nerve ending.
   C. Melanocytes produce melanin.
   D. Dendrocytes produce dendrocidin.

Answer is D: Dendrocytes are macrophages, "dendrocidin" is not produced.

8. Which of the following situations could produce life-threatening fluid loss and infection?
   A. Stomach ulcers
   B. Full-thickness skin burn
   C. Severe muscle tear
   D. Displaced bone fracture

Answer is B: The waterproof outer layer of the epidermis is lost if skin is burnt off, so evaporation from the wet tissues beneath proceeds rapidly.

9. What do sudoriferous glands do?
   A. Secrete sebum into a hair follicle
   B. Secrete sweat through a duct to the skin
   C. Secrete sweat through a duct to the skin or into a hair follicle
   D. Secrete cerumen through a duct to the skin or into a hair follicle

Answer is C: Eccrine sweat glands secrete through a duct, apocrine sweat glands secrete into a hair follicle.

10. Which cell type produces a pigment that affords the skin some protection against ultraviolet radiation?
    A. Keratinocytes
    B. Melanocytes
    C. Dendrocytes
    D. Merkel cells

Answer is B: Melanin causes skin to darken (tan) which absorbs some of the incoming UV radiation.

11. A drug that is applied to the skin and exerts its therapeutic effect systemically is said to be administered:
    A. Transdermally
    B. Topically
    C. Intradermally
    D. Subcutaneously

Answer is A: The prefix "trans" implies crossing the skin to be absorbed into the bloodstream, rather than affecting the skin onto which it is applied. "Topically" is reserved for the latter type of administration.

12. What must be the case for a drug to be administered transdermally?
    A. It must be water soluble.
    B. It must be lipid soluble.
    C. It must be injected subcutaneously.
    D. It must not irritate the skin's chemoreceptors.

Answer is B: Integument is virtually waterproof. Lipid-soluble drugs may penetrate between the keratinised cells.

13. The skin participates in the production of vitamin D when which of the following occurs? When:
    A. Calcium is present
    B. Signalled to by the hormone PTH
    C. Exposed to ultraviolet radiation
    D. The precursor molecule is produced by the liver

Answer is C: UV light converts cholecalciferol (=provitamin D_3) into vitamin D precursor. The liver then adds a hydroxyl group.

14. Three functions of the skin are to:
    A. Store fat, produce sweat and receive stimuli
    B. Synthesise vitamin D, excrete bile and provide a barrier to the entry of bacteria
    C. Produce keratin, assist in the immune response and produce lymphocytes
    D. Produce melanin, secrete sebum and minimise water loss

Answer is D: Choice A is incorrect as fat is not stored. Choice B is incorrect as bile is not excreted. Choice C is incorrect as lymphocytes are not produced.

15. Which of the following is **NOT** a cell?
    A. Macrophage
    B. Chondroblast
    C. Lysosome

D. Melanocyte

Answer is C: A lysosome is an organelle within a cell. -Phage, -blast and -cyte are all suffixes indicating a type of cell.

16. What purpose is vitamin D (calcitriol) used for?

    A. Required for several stages of haemostasis
    B. Required for uptake of calcium from the gut
    C. Required for erythropoiesis
    D. Required for uptake of intrinsic factor

Answer is B: Dietary calcium requires vitamin D for absorption.

17. The skin begins the production of vitamin D in which of the following situations? When:

    A. Exposed to ultraviolet radiation
    B. Signalled to by the hormone PTH
    C. Calcium is present
    D. It adds a hydroxyl group to a cholesterol molecule

Answer is A: Choice D happens in the liver and in the kidney. PTH signals the kidney to act on the modified precursor, whereupon a hydroxyl group is added.

18. What are the three functions of the skin?

    A. To store fat, produce sebum and trap a layer of air beneath hair to insulate against heat loss
    B. To synthesise vitamin D, excrete bile and protect against abrasion
    C. To produce melanin, regulate body temperature and minimise water loss
    D. To produce keratin, assist in the immune response and produce dendrocytes.

Answer is C: Fat is not stored in the skin. Bile is not excreted. Dendrocytes are not produced.

19. Which statement about vitamin D is **INCORRECT**?

    A. It is required for uptake of calcium from the gut.
    B. It is made in the skin, liver and kidneys.
    C. It is an essential part of our diet.
    D. It is a hormone.

Answer is C: As our body can make "vitamin D", it is not an essential part of our diet. As it circulates in the bloodstream, it is regarded as a type of steroid hormone.

20. If a drug is administered "transdermally", which of the following applies?

    A. It is absorbed through the skin and acts locally.
    B. It is injected into the dermis.

C. It is absorbed through the skin and acts systemically.
D. It is injected into the subcutaneous fat layer.

Answer is C: Transdermally refers to the drug having to pass through the skin, rather than bypass the skin as in hypodermic administration (choice D), or intradermal injection (choice B).

21. A drug that is administered "transdermally" is one that:
    A. Acts at (or close to) the skin area to which it is administered
    B. Is injected into the muscle
    C. Is applied to the epidermis
    D. Is inserted into the anus

Answer is C: Transdermal administration is applied to those drugs that can be absorbed from an adhesive "patch" (say), through the skin and enter the blood stream.

22. People with full-thickness burns to more than 20% of their body are in a life-threatening situation. This is due to which of the following?
    A. The body's inability to thermoregulate
    B. The loss of the ability to produce vitamin D
    C. The build-up of urea and uric acid which would otherwise have been excreted by the skin
    D. The body's inability to prevent water loss

Answer is D: Without a waterproof outer layer, the body would lose water faster than it could be replaced.

23. Which one of the following cell types is responsible for forming the skin's ability to tan on exposure to sunlight?
    A. Melanocytes
    B. Keratinocytes
    C. Dendrocytes
    D. Lymphocytes

Answer is A: Melanocytes produce melanin, the pigment responsible for tanning and for protecting the cell nucleus from UV radiation.

24. Which list below contains functions that are NOT performed by the integumentary system?
    A. Protection, secretion of sebum, role in immunity
    B. Body temperature regulation, excretion, synthesis of vitamin D
    C. Perception of stimuli, production of sweat, protection
    D. Body temperature regulation, synthesis of vitamin E, social function

Answer is D: The skin does not produce vitamin E.

25. Full-thickness burns to more than 20% of the body surface is life-threatening because of the:

    A. Fluid loss and inability to produce vitamin D
    B. Loss of ability to thermoregulate and infection
    C. Inability to excrete lactic acid, urea and uric acid, loss of thermoregulation
    D. Fluid loss and loss of the barrier against infection

Answer is D: Removing the skin barrier means that bacteria can enter the body and water can leave the body. This can be fatal.

26. Which one of the following cell types is responsible for forming the skin's stratum corneum?

    A. Melanocytes
    B. Keratinocytes
    C. Dendrocytes
    D. Lymphocytes

Answer is B: Keratinocytes secrete keratin which fills the dead cells of the stratum corneum of the epidermis.

27. Which of the following is **NOT** a function of the skin?

    A. Fat storage
    B. Waterproofing the body
    C. Production of vitamin D
    D. Immunity

Answer is A: Fat is stored in the hypodermis, which is also known as the subcutaneous layer and the superficial fascia.

28. What is the most common protein in the epidermis?

    A. Elastin
    B. Keratin
    C. Melanin
    D. Cholesterol

Answer is B: Most epidermal cells are keratinocytes that produce the fibrous protein keratin.

29. What is the fibrous protein in our skin that protects it from abrasion?

    A. Melanin
    B. Keratin
    C. Sebum
    D. Elastin

Answer is B: The outermost layer of the epidermis (stratum corneum) is composed of dead cells filled with keratin. This layer is harmlessly rubbed off (abrades) as the body comes into contact with objects outside the body.

30. Which one of the following cell types is a macrophage?
    A. Keratinocytes
    B. Melanocytes
    C. Dendrocytes
    D. Merkel cells

Answer is C: Dendrocytes are able to engulf bacteria and debris. The melanocytes and keratinocytes of the skin produce melanin and keratin, respectively. While Merkel cells are a sensory structure.

31. Which list contains only the types of cells found in the epidermis?
    A. Keratinocytes, melanocytes, dendrocytes, Meissner's corpuscles
    B. Basal cells, keratinocytes, melanocytes, dendrocytes
    C. Meissner's corpuscles, Pacinian corpuscles, red blood cells, Merkel cells
    D. Melanocytes, dendrocytes, Pacinian corpuscles, Merkel cells

Answer is B: Meissner's corpuscles and Pacinian corpuscles are not cells, but are types of sensory receptor found in the dermis, not the epidermis.

32. Which of the following is NOT one of the principal functions of the skin?
    A. Storage of fat
    B. Synthesis of vitamin D
    C. Body temperature regulation
    D. Perception of stimuli

Answer is A: Fat is stored in the subcutaneous layer, also known as the hypodermis, but it is not part of the skin. Skin is the epidermis and dermis.

Chapter 6
Homeostasis

Homeostasis is the body's automatic tendency to maintain a relatively constant internal environment in terms of temperature, cardiac output, ion concentrations, blood pH, hydration, dissolved CO_2 concentration in blood, blood glucose concentration, concentrations of wastes, etc. This constancy of internal environment is maintained despite energy and molecules continuously entering and leaving the body. The values of these (and other) variables oscillate within a narrow range. The body is able to monitor these variables and uses negative feedback (almost exclusively) to raise the values if they get too low and to lower the values if they get too high. "Negative" feedback means that the body's response opposes the stress. Thus the body is in a dynamic state of equilibrium because its internal conditions change and vary (oscillate) within relatively narrow limits.

Receptors monitor changes in these physiological variables, that is, they receive a stimulus. This stimulus is transmitted via an afferent pathway to an integrating centre (e.g. the brain or a gland). The integrating centre compares the stimulus to the normal level of the variable—the "set point". If a response is required, a message is sent via an efferent pathway to the effector organ. The effector produces a response that moves the value of the variable back towards the set point. The responses include altering the breathing or heart rate or blood pressure; vasoconstricting or vasodilating; eating, drinking and secreting.

1. Choose one answer below that completes the sentence so that it makes a true statement. Positive feedback:

   A. Is the way the body maintains homeostasis
   B. Is a response that opposes a stimulus
   C. Is a response that maintains a dynamic state of equilibrium
   D. Is a response that enhances a stimulus

   Answer is D: "Positive" feedback will reinforce the stimulus, making it greater or more insistent.

M. Caon, *Examination Questions and Answers in Basic Anatomy and Physiology*, https://doi.org/10.1007/978-3-030-47314-3_6

2. The hypothalamus and core thermoreceptors notice if body temperature is too low. The hypothalamus signals smooth muscle in blood vessels to vasoconstrict and skeletal muscle to shiver. This activity causes body temperature to rise, which is again noticed by hypothalamus and core thermoreceptors, so the hypothalamus turns off these heat gain mechanisms. In this scenario, which of the following statements is correct?
   A. The scenario describes positive feedback in action.
   B. The core thermoreceptors are the effector organs.
   C. The hypothalamus is the control centre.
   D. Skeletal muscle is the efferent pathway to the effector organ.

Answer is C: The hypothalamus interprets the sensory input and determines the response. Choice A is wrong as this is a negative feedback. The effector organs are the skeletal muscles (which are not a "pathway").

3. The human body's ability to maintain a relatively constant internal temperature is an example of what?
   A. Respiratory heat loss
   B. Homeostasis
   C. Vasodilation and evaporative heat loss
   D. Positive feedback

Answer is B: Homeostasis is derived from words that mean remaining similar and standing still and refers to physiological conditions remaining more or less the same.

4. Some of the body's homeostatic responses rely on "negative feedback". Which of the following happens in negative feedback?
   A. The body ignores changes in a physiological variable that are directed towards the set point for that variable.
   B. The body ignores changes in a physiological variable that are directed away from the set point for that variable
   C. The body's response acts to oppose the change in the physiological variable.
   D. The body's response acts to enhance the change in the physiological variable

Answer is C: "Negative" refers to the response being in the opposite direction to the stimulus. Thus if a variable is increasing, then the body's response is to produce a change that causes the variable to decrease. The body does not "ignore" stimuli.

5. In homeostasis, what is it that produces the response that moves the physiological variable back towards the middle of its healthy range?
   A. The effector
   B. The receptor
   C. The integrating centre

D. The efferent pathway

Answer is A: The effector produces the effect that it is directed to by the integrating centre. That direction is passed along the efferent (or outgoing) pathway.

6. The “afferent pathway” in the description of a feedback loop in homeostasis, refers to the:

   A. Circulating blood
   B. Pathway from the integrating centre to the effector
   C. Outgoing signal
   D. Path taken by the signal produced by a stimulus

Answer is D: The afferent signal is incoming from the receptor to the integrating centre. It may be via the blood, but may be via a nerve.

7. Homeostasis usually returns the body to a healthy state after stressful stimuli by:

   A. Negative feedback
   B. Positive feedback
   C. Means of the immune system
   D. Means of the nervous system

Answer is A: Negative feedback is far more common in maintaining homeostasis than is positive feedback.

8. What does the word “homeostasis” refer to?

   A. The steps leading to repair of a blood vessel and the coagulation of blood
   B. The maintenance of internal body conditions within narrow limits
   C. The controlled response that opposes the influence that caused it
   D. The production of blood cells in active bone marrow

Answer is B: This is the best definition. Choice C refers to negative feedback only.

9. How does homeostasis return the body to a healthy state after stressful stimuli? By producing a:

   A. Responses that oppose the stress
   B. Learned behaviour
   C. Reflex action
   D. Buffering mechanism

Answer is A: Refers to negative feedback. Positive feedback is also possible (but rare), nevertheless choice A is the best answer.

10. What causes the body to maintain a relatively constant internal environment?

    A. Positive feedback
    B. Homeostasis
    C. Reflexes

D. pH buffers

Answer is B: Homeostasis matches the definition in the question.

11. Homeostasis relies on feedback to achieve its aims. "Negative" feedback refers to which situation below? The body's response:
    A. Travels from the effector to the integrating centre via the afferent pathway
    B. Opposes the stressful stimulus
    C. Is to decrease the set point
    D. Enhances the stressful stimulus

Answer is B: "Negative" refers to the opposing nature of the response.

12. The term "homeostasis" is described by which one of the following statements? The body's ability to:
    A. Respond to a stimulus or stress in such a way as to enhance the stress
    B. Maintain a relatively constant internal temperature
    C. Respond to a stimulus or stress in such a way as to reduce the stress
    D. Maintain a relatively constant internal environment

Answer is D: Choice A refers to positive feedback. Choice C refers to negative feedback. Choice B is an example of homeostasis.

13. Synthesis and release of most hormones are regulated by negative feedback. Negative feedback means:
    A. A rise in hormone levels affects the target organ which acts to inhibit further hormone release.
    B. A rise in hormone levels affects the target organ which acts to stimulate further hormone release.
    C. The effect of hormones on target cells does not control further release of hormone.
    D. Neural stimuli result in the release of oxytocin and antidiuretic hormone from the hypothalamus.

Answer is A: When the response opposes a change, it is negative feedback.

14. Which of the following is a correct definition of "positive feedback"?
    A. The process by which the body maintains homeostasis
    B. A mechanism in which the body's response to a stimulus opposes the stimulus
    C. A mechanism whereby the body responds to a stimulus by acting to enhance the stimulus
    D. The dynamic equilibrium maintained by an integrating centre which causes an effector to respond to the stimulus received by the receptor

Answer is C: When the stimulus is enhanced, the feedback is termed "positive".

15. Which of the following is a correct definition of "negative feedback"?
    A. The process by which the body maintains homeostasis
    B. A mechanism in which the body's response opposes the stimulus
    C. A mechanism whereby the body responds to a stimulus by acting to enhance the stimulus
    D. The dynamic equilibrium maintained by an integrating centre which causes an effector to respond to the stimulus received by the receptor

Answer is B: When an action is taken to cause the stimulus to be reduced, the feedback is termed "negative".

16. Homeostasis refers to maintaining:
    A. A constant internal body environment through negative feedback
    B. Body conditions relatively constant within a narrow range through negative and positive feedback
    C. Adequate concentrations of respiratory gases
    D. Blood glucose level within the healthy range

Answer is B: The maintenance of body conditions through feedback which is usually negative but is positive in some rare situations.

17. What does the term "homeostasis" refer to?
    A. The chemical processes that take place in the organelles of the cells
    B. The body's tendency to maintain a relatively constant internal environment
    C. The body's use of energy to produce chemical substances and parts for growth
    D. Any body response that opposes the stimulus that initiated the response

Answer is B: This is the best definition of homeostasis. Choice D refers to negative feedback.

18. The body returns to a healthy state after stressful stimuli thanks to which of the following?
    A. Positive feedback
    B. Metabolism
    C. Anabolism
    D. Negative feedback

Answer is D: Negative feedback is more common than positive feedback.

19. Which of the following would be a negative feedback response by the body to hyperthermia?
    A. Shivering
    B. Sweating
    C. Vasoconstriction of blood vessels in the dermis
    D. An increase in metabolic rate

Answer is B: Hyperthermia is an increase on body temperature above 38 °C. Negative feedback would be a response that opposed the temperature rise. Allowing perspiration to evaporate from the skin would cool the skin and lead to a decrease in body temperature. The effect of the other choices would be to increase body temperature.

20. How would the hypothalamus respond if its osmoreceptors noticed an increase in plasma osmotic pressure? It would send a message to:
    A. The posterior pituitary to release more ADH
    B. The posterior pituitary to release less ADH
    C. The adrenal cortex to release less aldosterone
    D. The atria to release more ANP

Answer is A: The hypothalamus would respond in a way that opposed the rise in osmotic pressure. Releasing more ADH would make the distal convoluted tubules of the nephrons to become more permeable to water. This would allow more water to be reclaimed from the filtrate which would act against the osmotic pressure of the plasma from rising further. It would also stimulate thirst.

21. During the delivery of a baby, the baby's head is pushing against the cervix causing the cervix wall to stretch. This stretching causes nerve impulses to be sent to the hypothalamus which directs the posterior pituitary to release oxytocin in the blood. Oxytocin stimulates the uterus to contract which pushes the baby's head deeper into the cervix, stretching it further. This situation is a description of which of the following?
    A. Negative feedback
    B. Positive feedback
    C. Homeostasis
    D. An afferent pathway to an integrating centre

Answer is B: The stimulus (cervical stretching) causes the integrating centre (the hypothalamus) to respond by enhancing the stimulus—releasing oxytocin so that the uterus contracts. Enhancing the stimulus is an example of positive feedback. The feedback stops when the cervix is no longer being stretched, that is when the baby is delivered.

22. Which of the following statements about positive feedback and homeostasis is correct?
    A. They are regulation mechanisms that control most fluctuations in the internal environment of the body.
    B. The response to the stimulus serves to exaggerate the feedback effect.
    C. This type of feedback response only involves an effector not a specific stimulus receptor site.
    D. This feedback mechanism involves adjustments at the organ level but not at the cellular level.

Answer is B: Positive feedback causes the stimulus to increase in magnitude which in turn elicits a more exaggerated response.

23. Which of the following is an example of negative feedback?
    A. The uterine contractions during childbirth push the baby's head against the cervix which stimulates the uterus to contract.
    B. The release by platelets that aggregate at the site of a blood vessel injury, of compounds that promote platelet aggregation.
    C. As resting membrane potential rises, voltage-gated ion channels open which allow ions to enter the axon which causes further rise in the membrane potential and more ion channels to open.
    D. As blood sugar level rises, insulin is released which promotes the absorption of glucose from the blood by liver, muscle and fat cells.

Answer is D: In this choice, the response to rising glucose is the release of insulin, which promotes a decrease in blood glucose by causing it to be absorbed from the blood. That is, the response is to oppose the change that initiated the response. This is termed "negative" feedback. The other three choices are examples of positive feedback.

24. Which of the following is the efferent pathway for the control of body temperature?
    A. The blood stream
    B. The hypothalamus
    C. The nerve(s) that connect the brain to the adrenal glands
    D. The core body temperature receptors

Answer is C: Only the blood stream and nerves are "pathways". Efferent refers to a signal that is leaving the brain and travelling towards an effector. The signalling by the brain to the adrenal glands to alter the release of epinephrine is one way that body temperature is maintained.

25. As blood glucose concentration rises, what homeostatic response would the body produce?
    A. Insulin would be released to cause blood glucose to fall.
    B. Glucagon would be released to cause blood glucose to fall.
    C. Insulin would be released to cause blood glucose to rise.
    D. Insulin and glucagon would be released to cause blood glucose to rise.

Answer is A: Homeostasis tends to result in a response that opposes the detected change. So something would be done to arrest the rise in glucose concentration. Insulin is the hormone that causes cells to absorb glucose from the blood.

26. As the osmotic concentration of blood increases, what homeostatic response would the body produce?
    A. The glomerular filtration rate would increase.
    B. The feeling of thirst would increase, so that a liquid with low osmolarity is ingested.
    C. The rate of release of ADH would decrease.

D. The feeling of satiety would increase, so that a low osmolarity liquid is less likely to be ingested.

Answer is B: Homeostasis tends to result in a response that opposes the detected change. So something would be done to reverse the rise in osmotic concentration of blood. Drinking water would do this. The other three choices would not oppose the rise in osmotic concentration.

27. As body temperature increases, what homeostatic response would the body produce?
    A. Thyroxine would be released to increase the metabolic rate.
    B. Vasoconstriction would occur, so that blood is withdrawn from close to the skin.
    C. Skeletal muscles may shiver to increase muscular heat production.
    D. Sweating would increase, so that evaporation would transfer heat to the air.

Answer is D: Homeostasis tends to result in a response that opposes the detected change. So something would happen to promote the loss of heat from the body. The evaporation of sweat from the skin will produce a cooling effect on the superficial tissues. The other choices would generate more heat or decrease heat loss.

28. As acidic chyme enters the duodenum from the stomach, what homeostatic response would the digestive system produce?
    A. Intestinal gastrin would be released to stimulate gastric secretion and move more chyme into the duodenum.
    B. The pancreas would decrease its secretion of alkaline pancreatic juice into the duodenum.
    C. The enterogastric reflex would tighten the pyloric sphincter and decrease gastric secretion.
    D. The duodenal brush border would secrete digestive enzymes that function in an acidic environment.

Answer is C: Homeostasis tends to respond to a detected change with "negative feedback". The duodenum requires an alkaline environment, so the arrival of acidic chyme would result in a response that causes duodenal pH to rise. By closing the pyloric sphincter, the alkaline pancreatic juice and alkaline brush border secretions will have time to restore duodenal pH to the required level of about 7.4.

29. When blood sodium ion concentration moves close to the lower limit of normal (137 mmol/L), what would the body do to maintain homeostasis?
    A. More aldosterone would be released to cause more reabsorption of Na^+ from the filtrate.
    B. Additional aquaporins would be inserted in the nephron tubule to allow more water to pass into the filtrate.
    C. More aldosterone would be released to cause more reabsorption of K^+ from the filtrate.

D. Water would move into the intracellular compartment to increase the osmotic concentration of blood.

Answer is A: Homeostasis almost always results in a response that opposes the detected change. Aldosterone promotes the reabsorption of Na^+ from the filtrate, this would slow the decline in the blood's sodium concentration. The other three choices are wrong.

30. If blood pressure is detected to be lower than required to optimally perfuse the tissues, what homeostatic response would occur?
    A. The vasomotor centre would be inhibited and so heart rate would decrease.
    B. Vasoconstriction would occur hence total peripheral resistance would increase.
    C. The strength of myocardial contraction would increase and so stroke volume would decrease.
    D. Smooth muscle in the walls of veins would relax resulting in vasodilation.

Answer is B: The homeostatic response to decreased blood pressure is to undertake action to increase blood pressure. Vasoconstriction would achieve this by increasing the peripheral resistance. Choices A and D are the opposite of what would happen.

31. When peripheral chemoreceptors detect an increase in the concentration of dissolved carbon dioxide in arterial blood above 45 mmHg (the upper limit of normal), which of the following homeostatic responses would occur?
    A. Respiratory rate increases, so that more carbon dioxide may be exhaled.
    B. Respiratory rate decreases, so that less carbon dioxide may be exhaled.
    C. Respiratory rate increases, so that less carbon dioxide may be exhaled.
    D. Respiratory rate decreases, so that more carbon dioxide may be exhaled.

Answers is A: The homeostatic response opposes the stimulus, so if blood pCO_2 is rising, the response must produce a decrease in pCO_2. If respiratory rate increases, the increase in lung ventilation will blow off more CO_2 (not less). This will lead to a decrease in the gas dissolved in arterial blood and a restoration of the healthy dissolved carbon dioxide concentration.

32. At the start of each ovarian cycle, FSH and LH are released. These hormones stimulate the granulosa and thecal cells (respectively) of the follicle that surrounds the ovum. Consequently a few ovarian follicles are stimulated to mature. Which of the listed subsequent developments is an example of **positive** feedback?
    A. The developing follicle produces oestrogens at a low level which act on the anterior pituitary to inhibit the release of FSH and LH.
    B. Granulosa cells secrete inhibin which inhibits the release of FSH.
    C. Rapidly increasing plasma concentration of oestrogens stimulates a surge in LH release which stimulates ovulation.

D. High plasma concentrations of progesterone and oestrogen act on the hypothalamus to inhibit the release of GnRH.

Answer is C: The rapidly increasing production of oestrogens by the developing follicle stimulates a surge in the release of LH by the anterior pituitary (rather than a decrease). In turn the follicle is stimulated to ovulate which means that the granulosa and thecal cells cease to secrete oestrogens. Instead they become the corpus luteum and secrete progesterone.

33. Select the example below that describes an example of positive feedback.
    A. The hypothalamus, noticing that body temperature is falling, initiates the muscular contractions we know as shivering.
    B. The macula densa of the nephron notices an increase in the sodium ion concentration in the filtrate and signals the afferent arteriole to constrict.
    C. The contractions of the uterus force the head of the foetus against the cervix which causes the uterus to contract more forcefully.
    D. The stretch receptors of the carotid sinus notice that the wall of the carotid artery is more stretched. This information is sent to the brain which then causes the heart rate to decrease.

Answer is C: Examples of positive feedback in human physiology are rare. The contractions of childbirth is one. The other three describe negative feedback.

34. What is meant by acid–base balance?
    A. The ability to maintain body solutions as neither acidic nor basic.
    B. The body's ability to maintain the solutions in the three main compartments at pH 7.4.
    C. The situation where the concentration of hydronium ions equals the concentration of hydroxide ions.
    D. Any excess acid entering or being produced by the body is removed by buffer systems.

Answer is B: The buffers in the body operate to maintain the pH of its aqueous solutions close to 7.4 (not pH 7.0 as implied by choices A and C). Excess acid is excreted via the lungs and kidneys.

35. What is meant by electrolyte balance?
    A. The maintenance of the dissolved ion concentrations in solution within their narrow healthy ranges.
    B. Ensuring that the solution concentrations of sodium and chloride ions are appropriate.
    C. The function of the kidneys in reabsorbing required electrolytes from the urine filtrate.
    D. Matching electrolyte excretion in sweat and faeces with intake via foodstuffs.

Answer is A: Electrolytes such as sodium, potassium, chloride, bicarbonate, calcium, phosphate, and magnesium ions have an upper and lower limit to their healthy concentrations in the body's solutions. Electrolyte balance are the metabolic processes that ensure that these ions are within their range. Electrolyte balance involves more than just the two ions in choice B; more than just preventing their loss in urine as in choice C; but does include urinary losses, which are missing from choice D.

36. What is meant by fluid balance in the body?
    A. Avoiding dehydration
    B. Avoiding over-hydration
    C. Replacing the losses via sweating during exercise with balanced electrolyte drinks
    D. Ensuring that water inputs are equal to water losses

Answer is D: Unless water intake via food, drink, metabolic production and intravenous infusion exactly matches losses via sweat, respiration, faeces and urine, etc., the body will either dehydrate or "swell up" with water retention.

37. Which of the following could be used to define what is meant by homeostasis?
    A. The maintenance of a healthy blood pressure at all levels of physical activity
    B. The prevention of blood loss by rapidly plugging holes in damaged blood vessels
    C. An alternative medicine based on the theory of "like cures like" and dilution increases potency
    D. Noticing the values of all physiological parameters and keeping them within their healthy ranges

Answer is D: The body is sensitive to changes in any of its physiological variables and is able to initiate actions to restore them to their healthy range if they stay outside acceptable limits. Choice B refers to haemostasis; C refers to homeopathy.

38. Which of the following is **wrong** when used to explain how negative feedback maintains homeostasis?
    A. A negative feedback loop is one that tends to decrease the effect of a process.
    B. The body is in a dynamic state of equilibrium, internal conditions change and oscillate within relatively narrow limits.
    C. A process that can continue to amplify your body's response to a stimulus until the stimulus is removed.
    D. Feedback that produces a response that moves the variable value back towards the set point.

Answer is C: This describes positive feedback. In the case of childbirth, the stimulus is removed when the baby is born. In the case of haemostasis, the stimulus is removed when the platelet plug seal the leak in the blood vessel.

Chapter 7
Skeletal System

There are 206 individual bones in the adult skeleton. Each one is an organ, and you should know the names of most of them. They are classified as part of either the axial skeleton or the appendicular skeleton. Bone may be termed compact (dense) or spongy (cancellous). Compact bone is found in the shaft (the diaphysis) of long bones and is dense. Microscopically it can be seen to be composed of cylindrical structures called osteons. These are lamellar structures with osteocytes within lacunae that surround a central canal that contains blood vessels. The central canal can exchange material with the lacunae via little channels called canaliculi. Spongy bone is found in the ends (the epiphyses) of long bones and is composed of bony trabeculae separated by spaces. Marrow is found in the shaft of long bones and between the trabeculae of spongy bone. Active marrow produces red and white blood cells by haemopoiesis. Yellow marrow is inactive. Bone is a storage place for calcium and is continually being remodelled by osteoclasts (which remove bone) and osteoblasts (which deposit bone). In the process, Ca is released or stored.

The surface of bones is marked by features that may rise above the surface (projections or roughenings) or be indented below the surface. The features mark attachment points for tendons and ligaments, places where a bone articulates with another, grooves where tendons may lie or an opening for nerves and blood vessels to pass through. These features are given names such as tuberosity, condyle, foramen and tubercle.

A bone is connected to an adjacent bone at an articulation (a joint), and they are bound to each other by ligaments. All the joints in the appendicular skeleton are freely moveable (synovial) joints and stabilised by tendons and intracapsular menisci. Their free movement is produced when the muscles attached to them contract and is ensured by the smooth hyaline cartilage that covers the articulating bone surfaces and the lubricating synovial fluid within the joint capsule.

M. Caon, *Examination Questions and Answers in Basic Anatomy and Physiology*, https://doi.org/10.1007/978-3-030-47314-3_7

7.1 Bones

1. Which of the following is **NOT** a bone of the axial skeleton?
   A. Deltoid
   B. Ethmoid
   C. Sphenoid
   D. Hyoid

Answer is A: The "deltoid" bone does not exist. Choice D, the hyoid does not articulate with any bone, but being in the neck is considered part of the axial skeleton.

2. Which of the following is a function of the skeletal system?
   A. Haemopoiesis
   B. Haemostasis
   C. Peristalsis
   D. Glycogenolysis

Answer is A: New red and white blood cells are made in the bone marrow (haemopoiesis). Haemostasis refers to stopping bleeding.

3. In which of the following bone structures do osteocytes live?
   A. Osteons
   B. Canaliculi
   C. Lacunae
   D. Lamellae

Answer is C: Lacunae are the spaces within the lamellae of an osteon which enclose a bone cell (osteocyte). These cells extend their processes into the canaliculi.

4. Which bone is most superior?
   A. Manubrium
   B. Occipital bone
   C. Cervical vertebra #3
   D. Patella

Answer is B: The occipital bone forms the base of the skull so is above (superior) to all the others when the body is in the anatomical position.

5. What is a "trochanter"?
   A. Part of a femur
   B. A feature of the pelvis
   C. A projection that forms part of an articulation
   D. A groove in which lies a tendon

Answer is A: The “greater” trochanter is a bone marking (a bump) that lies on the lateral surface of the proximal femur, while the “lesser” trochanter lies on the medial surface of the proximal femur.

6. One of the functions of bones is to make red blood cells. What is this process known as?
   A. Haemolysis
   B. Haemopoiesis
   C. Haematuria
   D. Haemostasis

Answer is B: Haemopoiesis (or haematopoiesis) refers to making all of cellular components of blood. The other (wrong) choices are, respectively, the rupturing (lysis) of RBC, blood in the urine, the process by which we cause bleeding to stop.

7. Where do osteocytes reside?
   A. In lamellae
   B. In endosteum
   C. In trabeculae
   D. In lacunae

Answer is D: Lacunae are the spaces within the lamellae of an osteon which are occupied by a bone cell (osteocyte).

8. Which of the following describes the movements known as pronation and supination?
   A. The flexing of the arm with respect to the forearm around the elbow
   B. The swelling of the foot to the medial and lateral directions
   C. The twisting of the wrist while the elbow is held motionless
   D. The rotation at the shoulder that causes the arm to describe a cone shape

Answer is C: Pronation occurs when the radius is moved from a position parallel to the ulna to one where it crosses over the ulna, and supination is the return motion (may be thought of as the action we perform when turning off a wall = mounted tap).

9. Which of the following is **NOT** a “long” bone?
   A. The humerus
   B. The tibia
   C. A carpal
   D. A metacarpal

Answer is C: A “long bone” has a length that is significantly longer than its width. A carpal (a bone of the wrist) is a short bone.

10. Which one of the following is a bone that is embedded within a tendon?
    A. Sphenoid
    B. Hyoid
    C. Ethmoid
    D. Sesamoid

Answer is D: Sesamoid refers to like a sesame seed. Many are normal parts of anatomy that ossify within tendons (patella, fabella, hallux sesamoid).

11. In which one of the following structures do osteocytes reside?
    A. Haversian canals
    B. Lacunae
    C. Trabeculae
    D. Endosteum

Answer is B: Lacuna refers to a hole, pit, “lake” or cavity within a bone.

12. What are the bones of the fingers known as?
    A. Short bones
    B. Metacarpals
    C. Carpals
    D. Phalanges

Answer is D: Phalanges (singular phalanx) are the bones of the fingers or toes. Metacarpals are the bones of the hand.

13. Which of the following comprise seven bones?
    A. Cervical vertebrae
    B. Carpals
    C. Cranial bones
    D. Lumbar vertebrae

Answer is A: C1–C7. There are eight carpals in each wrist. There are eight cranial bones. There are five lumbar vertebrae.

14. Which term below refers to a depression in a bone?
    A. Tuberosity
    B. Fossa
    C. Tubercle
    D. Condyle

Answer is B: Fossa is a depression reminiscent of a (small) dinner table plate. A tuberosity is a roughening on a bone surface. The other two choices are projections above a bone surface.

15. What body part is able to perform pronation and supination?
    A. Forearm
    B. Foot

C. Thigh
D. Wrist

Answer is A: It is the radius and ulna bones whose movement relative to each other produces pronation and supination.

16. Where are blood vessels in compact bone found?
    A. In the canaliculi
    B. In the periosteum
    C. In the lacunae
    D. In the central canal

Answer is D: The central canal (or haversian canal) of an osteon houses blood vessels.

17. Which of the following is **NOT** a depression or cavity on a bone?
    A. Tuberosity
    B. Facet
    C. Meatus
    D. Sinus

Answer is A: A tuberosity is a rough area on the surface of a bone to which a muscle's tendon attaches.

18. One of the following lists contains only bones in the appendicular skeleton. Which one?
    A. Patella, ethmoid, femur, coccyx, tibia
    B. Clavicle, fibula, metatarsal, phalange, radius
    C. Humerus, scapula, occipital, metacarpal, sternum
    D. Ulna, radius, phalange, mandible, coxal

Answer is B: The coccyx, occipital, sternum, mandible are parts of the axial skeleton.

19. What is contained within the medullary canal of a long bone?
    A. Trabeculae
    B. Lamellae
    C. Marrow
    D. Osteoblasts and osteoclasts

Answer is C: Red (and/or yellow) marrow fills the medullary canal (or cavity).

20. Where in the skeleton is the scapula located?
    A. In the axial skeleton
    B. In the appendicular skeleton
    C. In the carpal region
    D. In the shoulder girdle

Answer is D: The scapula (shoulder blade) is part of the shoulder girdle. It is also part of the appendicular skeleton, but choice D is the more specific answer.

21. Which of the following bone markings forms part of an articulation?
    A. The deltoid tuberosity of the humerus
    B. The lateral condyle of the femur
    C. The greater trochanter of the femur
    D. The greater tubercule of the humerus

Answer is B: The lateral condyle of the femur articulates with the lateral facet of the superior articular surface of the tibia, in the knee joint.

22. Where is the epiphyseal plate of a long bone located?
    A. In the diaphysis
    B. Between the diaphysis and the epiphysis
    C. In the epiphysis
    D. In the medullary canal

Answer is B: The "plate" is the hyaline cartilage between the shaft (diaphysis) and end (epiphysis) of a long bone in children and adolescents. In adults it is replaced by an ossified "epiphyseal line".

23. In which structure are osteoclasts and osteoblasts found?
    A. In the periosteum
    B. In the haversian canals
    C. In the lacunae of osteons
    D. In the trabeculae of osteons

Answer is A: The membrane around bones houses the bone-forming cells (osteoblasts) and the bone-reabsorbing cells (osteoclasts).

24. Which of the following is a **NOT** a projection from a bone surface?
    A. Trochanter
    B. Tubercle
    C. Trabeculum
    D. Tuberosity

Answer is C: A trabeculum (plural trabeculae) is the internal bony structure of spongy (cancellous) bone.

25. Which of the listed bones is superior to the rest?
    A. Manubrium
    B. Xiphoid process
    C. Coccyx
    D. Femur

Answer is A: The manubrium is the upper part of the sternum. The xiphoid process is the lower part.

26. Choose the correct sentence. Compact bone contains:
    A. Lamellae and osteocytes but no osteons
    B. Trabeculae, canaliculi and osteons
    C. Haversian systems and canaliculi but no osteons
    D. Osteons and lamellae but no trabeculae

Answer is D: Compact bone does contain osteons, but not trabeculae (which occur in cancellous bone).

27. Which of the following bone markings is **NOT** a projection for muscle attachment?
    A. Fossa
    B. Tuberosity
    C. Tubercle
    D. Trochanter

Answer is A: A fossa is a depression into which fits a projecting part of another bone when the two bones form a joint.

28. Which of the list below is a cell that reabsorbs bone?
    A. Osteon
    B. Osteoblast
    C. Osteocyte
    D. Osteoclast

Answer is D: The suffix "-blast" refers to an immature cell. In this case, it develops into an osteocyte after it has secreted bone matrix around itself.

29. The formula for the inorganic salts in bone is:
    A. NH_6C_3COOH
    B. $C_6H_{12}O_6$
    C. $Ca_{10}(PO_4)_6(OH)_2$
    D. $CH_3(CH_2)_7CH{=}CH(CH_2)_7COOH$

Answer is C: This is hydroxyapatite. It is the only one with calcium and phosphorus. All the others are organic molecules.

30. Which of the following is a projection from a bone surface?
    A. Fossa
    B. Fissure
    C. Foramen
    D. Facet

Answer is D: Facet is an articular surface that projects from a bone surface. A fossa is a shallow basin-like depression in a bone, often serving as articular surface.

31. Which of the listed bones is the most inferior?
    A. Ethmoid
    B. Sphenoid
    C. Femoid
    D. Hyoid

Answer is D: Hyoid is in the neck. Choices A and B are in the skull. Femoid is not a bone.

32. Choose the correct sentence. Cancellous bone contains:
    A. Lamellae and osteocytes but no trabeculae
    B. Trabeculae, canaliculi and osteons
    C. Haversian systems and canaliculi but no osteons
    D. Trabeculae and lamellae but no osteons

Answer is D: Cancellous (spongy) bone does not contain osteons, but does contain the little beams known as trabeculae.

33. Which of the following bones is part of the cranium?
    A. Occipital
    B. Mandible
    C. Hyoid
    D. Carpal

Answer is A: The occipital bone forms the floor of the cranium.

34. The appendicular skeleton includes all of the following **EXCEPT** one. Which one?
    A. The pectoral girdle
    B. The thoracic cage
    C. The phalanges
    D. The lower limbs

Answer is B: The thoracic cage (the ribs) are part of the axial skeleton.

35. What is the name given to the central tunnel of an osteon that contains blood vessels?
    A. Canaliculus
    B. Endosteum
    C. Haversian canal
    D. Medullary canal

Answer is C: Also known as the central canal. The medullary canal is a macroscopic structure within the diaphysis (shaft) of a long bone.

36. Which list contains the bones of the pelvic and pectoral girdles?
    A. Coxal, scapulae, manubrium, ilium
    B. Clavicles, cervical, coccyx, innominate

C. Clavicles, scapulae, coxal
D. Clavicles, scapulae, sacrum, coxal

Answer is C: The manubrium, coccyx, sacrum are within the axial skeleton.

37. The manubrium and the xiphoid process are located on which part of the skeleton?

A. The lower jaw
B. The sternum
C. The pelvis
D. The hand

Answer is B: The manubrium and xiphoid process (and the body) are all parts of the sternum.

38. Carpals refers to:

A. The points of attachment of ribs to vertebrae
B. Bones of the wrist
C. Bones that are embedded within a tendon
D. The thumbs

Answer is B: Wrist bones are carpals, hand bones are metacarpals (ankle bones are tarsals).

39. Haemopoiesis refers to:

A. Blood cell formation in bone marrow
B. The process of blood clotting
C. The crenation of red blood cells in a hypotonic solution
D. An excessively large proportion of red blood cells to plasma

Answer is A: While the prefix "haemo-" refers to blood, the "-poiesis" part means making (or forming).

40. On which bone is the greater trochanter found?

A. Pelvic
B. Femur
C. Radius
D. Humerus

Answer is B: Only the femur has a structure named "trochanter".

41. To which bones does the word phalanges apply? Those in the:

A. Fingers and toes
B. Wrist and ankle
C. Ankle and foot
D. Fingers and hand

Answer is A: Choice B refers to carpals and tarsals. Choices C and D include metacarpals and metatarsals.

42. The axial skeleton groups together which sets of bones?
    A. The arms and hands, the legs and feet, shoulder girdle and pelvic girdle
    B. The head, shoulder girdle, arms and hands
    C. The thoracic cage, vertebral column, shoulder girdle, the pelvic girdle, the skull and facial bones
    D. Bones of the skull and face, thoracic cage and vertebral column

Answer is D: Choice A describes the appendicular skeleton. Choices A and D are complementary.

43. What are the cells that are found in the lacunae of compact bone called?
    A. Osteocytes
    B. Osteons
    C. Osteoblasts
    D. Osteoclasts

Answer is A: Osteocytes develop from osteoblasts.

44. The appendicular skeleton groups together which sets of bones?
    A. The arms and hands, the legs and feet, shoulder girdle and pelvic girdle
    B. The head, shoulder girdle, arms and hands
    C. The thoracic cage, vertebral column, shoulder girdle, the pelvic girdle, the skull and facial bones
    D. Bones of the skull and face, thoracic cage and vertebral column

Answer is A: The appendages (arms and legs) and the girdles form the appendicular skeleton.

45. What does the term "haversian canal" refer to in bone?
    A. The larger examples of foramina
    B. A groove that receives a muscle's tendon
    C. The centre of an osteon that contains blood capillaries
    D. The space within a long bone that contains marrow

Answer is C: Named after the English physician Clopton Havers. Also known as the central canal.

46. What is the structure that attaches one bone to another?
    A. Ligament
    B. Cartilage
    C. Tendon
    D. Diaphysis

Answer is A: A ligament is the fibrous connective tissue that ties articulating bones together.

47. Which of the following describes what an "epiphysis" is?
    A. The shaft of a long bone
    B. The line that separates the shaft from the end of a long bone
    C. The membrane that surrounds a bone
    D. The end of a long bone

Answer is D: The similar terms "epiphyseal plate" refer to the region between the diaphysis and the epiphysis where growth of a long bone takes place. The "epiphyseal line" replaces the plate in adults, when the long bone has ceased to grow.

48. To what does the term "osteon" refer in bone?
    A. The bone cells (osteocytes) in lacunae
    B. A small rounded projection on a bone
    C. Concentric cylinders of calcified bone matrix
    D. The membrane covering the outside of a bone

Answer is C: These concentric cylinders surround a central canal (haversian canal) and occur in compact bone.

49. Blood cell formation (haemopoiesis) occurs in which of the following structures?
    A. Red marrow
    B. Yellow marrow
    C. Medullary cavity
    D. Epiphyseal plate

Answer is A: Red marrow is "active" marrow, in that blood cells are being formed there. Red marrow occurs in the medullary cavity as well as in the cancellous bone of the epiphyses.

50. Compact bone differs from spongy (cancellous) bone because compact bone:
    A. Does not contain osteons
    B. Is used to form short bones
    C. Contains marrow
    D. Has haversian canals

Answer is D: Osteons are the structural units of compact bone, and each one surrounds a channel, called a haversian canal, which contains blood vessels.

51. What is the role of hyaline cartilage in the body?
    A. It attaches muscle to bone.
    B. It reinforces joints by tying one bone to another.
    C. It covers articulating bone surfaces.
    D. It produces synovial fluid.

Answer is C: Hyaline cartilage is very smooth, and by covering the articulating surfaces of bones, ensures that the joint moves without noticeable friction.

52. What are bone-forming cells called?
    A. Osteons
    B. Osteocytes
    C. Osteoclasts
    D. Osteoblasts

Answer is D: Osteoblasts secrete bone around themselves, eventually isolating themselves within lacunae, whereupon they mature into osteocytes.

53. Where does the increase in the length of a long bone take place?
    A. Diaphysis ossification centres
    B. Epiphyseal plates
    C. Cartilaginous plates
    D. Medullary canal

Answer is B: Bone is deposited on the side of the plate that is distal to the epiphysis while bone is removed from the proximal side. In this way, the diaphysis increases in length (and is remodelled).

54. The human skeleton consists of which of the following?
    A. Pectoral girdle, the hip girdle and the bones of the limbs
    B. Axial skeleton and the appendicular skeleton
    C. Cranial bones, the thoracic cage, the two girdles and the limb bones
    D. Appendicular skeleton, skull bones and the vertebral skeleton

Answer is B: Choice A omits the axial skeleton. Choice C omits the facial bones and lumbar vertebrae. Choice D omits the ribs.

55. The functions of bones may be stated as:
    A. Fat storage, movement, mineral storage, protection, blood cell formation
    B. Mineral storage, haemopoiesis, movement, leverage, protection
    C. Blood cell formation, hormone production, movement, support, protection
    D. Support, storage, movement, haemopoiesis, protection

Answer is D: Each choice includes protection, movement and blood cell formation (haemopoiesis), so these terms cannot be used to choose the answer. Bones do not produce hormones, so choice C is eliminated. Bones do provide support to the body, hence only choice D is a complete list of functions.

56. The tibia articulates distally with which one of the following?
    A. Tarsals
    B. Metatarsals
    C. Phalanges
    D. Femur

Answer is A: Distal refers to the end that is further away from the trunk. Hence, the ankle bones (tarsals) is the correct choice.

57. What is the term “osteon” used for?

    A. A bone cell
    B. An haversian system in compact bone
    C. The bony structure in spongy bone
    D. The space in a bone where a bone cell lives

Answer is B: Osteon and haversian system are synonyms. A haversian canal (central canal) is at the centre of the haversian system.

58. Hypochondriac refers to which of the following?

    A. Someone who complains chronically of ill health
    B. The abdominal region inferior to your ribs
    C. An abnormally low level of chondria in the body
    D. That part of your head surrounding your chin

Answer is B: “Hypo-” refers to below (or inferior to), while “chondral” refers to cartilage—specifically the costal cartilages that join ribs 7–10 to the sternum. Hypochondriac refers to the abdomen immediately inferior to the rib cage. Choices C and D are nonsense. While choice A is true in layman’s terms, we are dealing with anatomical terminology here.

59. Where are the bones known as the humerus and radius located?

    A. In the axial skeleton
    B. In the arm
    C. In the leg
    D. In the arm and leg, respectively

Answer is B: This is a better answer than the also true choice A.

60. Where does blood cell formation occur?

    A. Blood
    B. Endosteum
    C. Haversian canal
    D. Red marrow

Answer is D: Red marrow is actively forming blood cells. Yellow marrow is inactive.

61. On what bone does the acetabulum occur?

    A. Occipital
    B. Humerus
    C. Pelvis
    D. Tibia

Answer is C: The acetabulum is the “socket” into which fits the ball of the femur.

62. Where is the xiphoid process?
    A. On the sternum
    B. On the humerus
    C. On the temporal bone
    D. On the tibia

Answer is A: At the inferior end of the body of the sternum. The temporal bone has a mastoid and a styloid process.

63. What is the metaphysis?
    A. The shaft of a long bone
    B. The region that separates the narrow shaft of a long bone from its end
    C. The end of a long bone
    D. The canal inside a long bone that contains marrow

Answer is B: The metaphysis is the bony remnant of the growth plate. The cartilaginous component of the growth plate is the epiphyseal plate.

64. What term is applied to moving the thigh laterally away from the midline of the body?
    A. Extension
    B. Adduction
    C. Abduction
    D. Flexion

Answer is C: "Abduct" is to take away (from the midline).

65. Where is the hyoid bone?
    A. In the sternum
    B. In the wrist
    C. In the knee
    D. In the neck

Answer is D: It is just superior to the cricoid cartilage, inferior to the mandible and superficial to the pharynx. Also known as the lingual (tongue) bone.

66. What is the occipital bone?
    A. It is one of the carpals.
    B. It is a bone of the cranium.
    C. It is one of the vertebrae.
    D. It is a facial bone.

Answer is B: The occipital bone forms the base of the cranium.

67. What is a "foramen"?
    A. A basin-like depression serving as an articular surface
    B. A raised roughening which is a site for muscle attachment

C. A hole through a bone for a nerve or blood vessel
D. A sharp slender projection of bone

Answer is C: Nerves and blood vessels penetrate through bone via these spaces.

68. Which statement correctly defines an osteon?

A. The arrangement of trabeculae and osteocytes that make up spongy bone
B. The membrane that lines the medullary cavity
C. Concentric cylinders of calcified bone matrix
D. The distal or proximal end of a long bone

Answer is C: In two dimensions, osteons appear as circles or ovals under the microscope, while in three dimensions, they are cylindrical.

69. What is the significance of the cribriform plate of the ethmoid bone?

A. It supports the optic nerve.
B. It contains foramina through which pass the olfactory neurons.
C. The external auditory meatus passes through it.
D. It is an attachment point for the tongue.

Answer is B: The plate is a sieve-like structure, due to the many foramina which allow neurons from the olfactory bulb to contact the olfactory membrane in the superior nasal concha. Thus inhaled air can stimulate our sense of smell.

70. What is the function of the sella turcica of the sphenoid bone?

A. It supports the pituitary gland.
B. It protects the mammillary body.
C. It surrounds the pineal gland.
D. It supports the optic chiasma.

Answer is A: The sella turcica (Latin for Turkish saddle) is a small depression of the sphenoid bone into which fits the pituitary gland.

71. What is the purpose of the foramen magnum of the occipital bone?

A. It allows passage for the optic radiation to reach the occipital lobe.
B. It supports the cerebellum.
C. The spinal cord passes through it to connect to the brain.
D. It allows the brainstem to connect with the spinal cord.

Answer is C: Foramen refers to a hole in a bone, while magnum means large. This large opening in the occipital bone contains the superior part of the spinal cord.

72. Which bone(s) is/are part of the appendicular skeleton?

A. Manubrium
B. Metacarpals
C. Maxilla
D. Mandible

Answer is B: The appendicular skeleton includes the four limbs and shoulder and pectoral girdles. The metacarpals are the bones of the hands.

73. Which of the following bones is classified as a "flat" bone?
    A. Femur
    B. Fibula
    C. Frontal
    D. Fabella

Answer is C: The frontal bone, and all cranial bone are "flat". Choices A and B are long bones, while the fabella is a sesamoid bone in the tendon of the gastrocnemius muscle.

74. Which of the following surface features of bones serves as part of an articulation?
    A. Facet
    B. Foramen
    C. Fissure
    D. Phalange

Answer is A: A facet is a shallow dish-like depression in a bone that receives a projection such as a condyle, from another bone. Foramina are holes in bone through which blood vessels or nerve pass. A phalange is a finger or toe bone.

75. What feature listed below ties two articulating bones together?
    A. Retinaculum
    B. Cartilage
    C. Tendon
    D. Ligament

Answer is D: A ligament ties adjoining bones together. A retinaculum is a band of connective tissue that holds a group of tendons in place, e.g. at the wrist and ankle.

76. The term "canaliculi" is applied to structures that occur in which of the following anatomical features?
    A. Kidney nephrons and pancreatic islets
    B. Liver lobules and bone osteons
    C. Neuromuscular junctions and the juxtaglomerular apparatus
    D. Lung alveoli and thyroid follicles

Answer is B: Canaliculi (little canals) are tubes that collect the bile produced by liver hepatocytes. In compact bone, the term is applied to the minute channels that connect lacunae (which contain bone osteocytes) with the haversian canal at the centre of an osteon.

77. Where can red marrow be found in the adult?
    A. In the pelvis, sternum and ribs
    B. At the epiphyseal ends of long bones such as the radius, ulna and tibia
    C. In the pelvis, sternum, ribs and the proximal ends of the humerus and femur
    D. In the medullary cavity of long bones

Answer is C: All bone marrow in children is red (actively forming blood cells). It gradually changes to "yellow" with age except for the sites in C. The medullary cavities contains yellow marrow. Red marrow is found mainly in the flat bones, such as the pelvis, sternum, cranium, ribs, vertebrae and scapulae, and in the cancellous ("spongy") material at the proximal epiphyseal ends of long bones such as the femur and humerus.

78. How many phalanges are there in the body?
   A. 20
   B. 48
   C. 56
   D. 60

Answer is C: Phalanges are the bones of the fingers and toes. There are 14 in each hand or foot. Two in the thumb/big toe and three phalanx (phalanges) in each of the four other fingers/toes.

79. How many bones in total, that are proximal to the ankles, are counted as being in the lower limbs?
   A. 4
   B. 6
   C. 8
   D. 12

Answer is C: Being proximal to the ankles excludes the bones of the ankle and feet. There are four bones in each leg (femur, patella, tibia, fibula): 4 + 4 = 8.

80. Which list has the correct numbers of vertebrae for an adult?
   A. 7 cervical, 5 lumbar, 12 thoracic, 1 coccyx, 1 sacrum
   B. 5 cervical, 7 lumbar, 12 thoracic, 4 coccyx, 5 sacrum
   C. 5 cervical, 7 lumbar, 10 thoracic, 1 coccyx, 5 sacrum
   D. 7 cervical, 5 lumbar, 10 thoracic, 4 coccyx, 1 sacrum

Answer is A: There are 26 vertebrae: 7C, 12T, 5L, 1S, 1C in that order. The sacrum and coccyx are single fused bones in adults.

81. What is the significance of the lumbar curve of the spine?
   A. It allows the torso to twist about a vertical axis.
   B. It places the body's centre of gravity above the feet to enable a standing posture.
   C. It places the head's centre of gravity above the atlas vertebrae, so that the head may be held level.
   D. It allows the torso to bend forwards to pick up objects from the ground.

Answer is B: Without the lumbar curve directing some body mass anteriorly, the centre of gravity could not be held over the base of support (the area bounded by the feet when placed on the ground). Choice C refers to the cervical curve.

82. How do the long bones of a child differ from those of an adult?
    A. Children still have osteons in their bones.
    B. The long bones of children have a medullary cavity while those of adults do not.
    C. In children, the majority of their marrow is inactive yellow marrow.
    D. In children, the epiphyseal cartilage separates the epiphysis from the diaphysis.

Answer is D: The long bones of children have not finished growing. Hence, the epiphyseal cartilage is still present (and can be seen in X-ray images)—in adults it has ossified.

83. Which of the following is not stored in bone?
    A. Calcium
    B. Iron
    C. Phosphorus
    D. Lipid

Answer is B: Iron is stored bound to ferritin in intestinal epithelial cells. Bone stores about 99% of our calcium and about 85% of our phosphorus lipid (fat) is stored in yellow bone marrow.

84. What are the two major divisions of the skeletal system?
    A. Axial; appendicular
    B. Vertebrae, pelvis, lower limbs; head, ribcage, upper limbs
    C. Appendages and girdles; head, ribcage, vertebrae
    D. Pelvis and lower limbs; vertebrae and superior bones

Answer is A: Axial (head ribcage and vertebrae) and appendicular (limbs and shoulder and pelvic girdles).

85. How does compact bone differ from cancellous bone?
    A. Compact bone contains osteocytes while cancellous bone does not.
    B. Compact bone is muscular tissue while cancellous bone is connective tissue.
    C. Compact bone consists of osteons while cancellous bone consists of marrow-filled spaces surrounded by trabeculae.
    D. Compact bone has haversian canals while cancellous bone had Volkmann's canals.

Answer is C: Closely packed osteons make compact bone dense, while marrow-filled spaces give cancellous bone a spongey appearance. Both types have osteocytes and are connective tissue.

86. What is the function of red marrow?
    A. To produce blood cells
    B. To produce red blood cells

C. To act as a fat store
D. To act as a store of bone minerals

Answer is A: Red marrow is "active" in the sense that it is forming red and white blood cells. Yellow marrow is inactive but acts as a fat store.

87. Which of the following structures does NOT appear in an osteon?
    A. Lamellae
    B. Lacunae
    C. Canaliculi
    D. Periosteum

Answer is D: The periosteum is the membrane on the outer surface of a bone.

88. Which is NOT a difference between a foetal skull and an adult skull?
    A. The adult has one frontal bone, the foetus frontal bone is in two halves.
    B. The foetal skull bones are not joined by sutures.
    C. The foetal skull bones have different names to those in the adult skull.
    D. The relative size of the foetal facial bones compared to the cranial bones are smaller in the foetal skull than for the adult.

Answer is C: The skull bones have the same names whether in the foetal or adult skull.

89. What feature distinguishes thoracic vertebrae from other vertebrae?
    A. Thoracic vertebrae are attached to a pair of ribs.
    B. Thoracic vertebrae are located in the neck.
    C. Thoracic vertebrae do not have a vertebral foramen.
    D. Thoracic vertebrae are classified as "flat" bones, while other vertebrae are "irregular".

Answer is A: Thoracic vertebrae are attached to a pair of ribs consequently also have an articulating facet at the point of attachment.

90. At what stage during human development does the lumbar curve of the spine develop and for what purpose?
    A. It develops as baby begins to crawl so that its head can be lifted to see where it is going.
    B. It develops as the toddler learns to run. The curve allows the legs to move freely to produce the running gait.
    C. It develops prior to puberty to allow the pregnant female to carry the foetus in utero.
    D. It develops as a baby begins walking (it allows the body's centre of gravity to be positioned over the base of support).

Answer is D: The lumbar curve is concave to the dorsal side. This curve allows the body's centre of gravity to be shifted forward, so that it is above where the feet are placed. This allows the baby to balance while standing. Choice A refers to the cervical curve, while choices B and C are nonsense.

91. On which bone would you find a feature called a “trochanter”?
    A. Fibula
    B. Femur
    C. Patella
    D. Humerus

Answer is B: Only the femur (thigh bone) has the greater trochanter and the lesser trochanter.

92. What is the effect of soaking a bone in an acid solution for several days?
    A. The organic component of bone is dissolved, leaving the residual bone hard and rigid.
    B. The calcium and phosphorus salts are dissolved, leaving the residual bone hard and rigid.
    C. The inorganic part of the bone is dissolved, leaving the residual soft and flexible.
    D. The acid breaks the chemical bonds in the bone structure, leaving a sludge-like residue.

Answer is C: Bone structure has an organic component and an inorganic component. Soaking in nitric acid removes the inorganic component (Ca and P) leaving the residual organic component (collagen, cells and glycoproteins) which still retains the original shape of the bone but is now soft and flexible.

93. In compact bone tissue, what are the structures called “haversian canals”?
    A. Passages that contain blood vessels
    B. Small passages that contain cellular processes from osteocytes
    C. The spaces between lamellae that contain osteocytes
    D. The functional units of bone consisting of concentric lamellae

Answer is A: Haversian (or central) canals are at the centre of an osteon and contain small blood vessels. Choice B refers to canaliculi, C to lacunae and D to osteons.

7.2 Joints

1. Which bone of the head has a synovial joint?
    A. The sphenoid
    B. The maxilla
    C. The mandible
    D. The hyoid

Answer is C: The ramus of the mandible articulates with the temporal bone at the mandibular fossa. The joint is freely moveable, allowing us to chew and speak.

2. A synovial joint is also known as one of the following, which one?
   A. Synarthrosis
   B. Immovable joint
   C. Slightly moveable joint
   D. Freely moveable joint

Answer is D: Synovial fluid is the lubricant that allows friction-free joint movement. A synarthrosis is an immovable joint.

3. Freely moveable joints are also known as:
   A. Fibrous joints
   B. Cartilaginous joints
   C. Amphiarthroses
   D. Synovial joints

Answer is D: Synovial joints have synovial fluid between the articulating bones. Fibrous joints have fibre between the articulating bones. Cartilaginous joints have cartilage between the bones. Amphiarthroses are "slightly moveable" joints.

4. Which of the following is an example of a synovial joint? The joint between the:
   A. Tibia and fibula
   B. Sternum and rib 1
   C. Thoracic vertebrae 4 and 5
   D. Proximal ends of the radius and ulna

Answer is D: The radius and ulna are freely moving at the elbow joint.

5. Synovial joints have all of the following features **EXCEPT** one. Which one?
   A. Are surrounded by an articular capsule
   B. Have synovial fluid filling the space between articulating bones
   C. Have synovial membrane covering the articulating bone surfaces
   D. Are supported by reinforcing ligaments

Answer is C: Articulating bone surfaces are covered by hyaline cartilage. Synovial membranes cover all other internal joint surfaces.

6. Articulating bones are joined by:
   A. Aponeuroses
   B. Tendons
   C. Fasciculi
   D. Ligaments

Answer is D: Ligaments join bone to bone. Tendons (and aponeuroses) join muscle to bone.

7. What does "articulation" refer to?
   A. The joining of a ligament to a bone
   B. The contact made between a tendon and a bone
   C. The contact between two bones
   D. The connection between a muscle and a bone

Answer is C: Articulation means joint between bones.

8. How do synovial joints differ from the other types of bone articulation?
   A. They have a joint cavity.
   B. The bones are joined by fibrous tissue.
   C. The articulating bones are joined by cartilage.
   D. The articulating bone surfaces are covered by tendons.

Answer is A: The cavity is filled with synovial fluid. Choice D is nonsense.

9. Synovial joints differ from the other types of joint between bones in the body because:
   A. They are immovable joints.
   B. They are slightly moveable.
   C. The bones are joined by cartilage.
   D. The ends of the articulating bones are covered by hyaline cartilage.

Answer is D: Hyaline cartilage is very smooth which (along with synovial fluid) makes the movement of the articulating bones almost friction free.

10. What is true of synovial joints? They:
    A. Are also known as amphiarthroses
    B. All have an articular disc to aid shock absorption
    C. Have a fluid-filled space between the articulating bones
    D. Have articulating bones held together by cartilage

Answer is C: The fluid is synovial fluid. Amphiarthroses are "slightly moveable" joints, whereas synovial joints are freely moveable.

11. Which of the following is an amphiarthrotic joint?
    A. Symphysis pubis
    B. Suture in the skull
    C. Elbow
    D. Shoulder

    Answer is A: Choice B is an immovable joint, while choices C and D are freely moveable synovial (diarthrotic) joints.

12. In a long bone, which of the following parts are involved in an articulation?
    A. Epiphysis
    B. Metaphysis
    C. Diaphysis

D. Symphysis

Answer is A: Epiphysis refers to the end of a long bone. The ends articulate with adjacent bones.

13. What is a distinguishing feature of synovial joints?
    A. There is fluid between the articulating bones.
    B. They are immovable joints.
    C. The articulating bones are held together by tendons.
    D. They involve a "ball and socket" articulation.

Answer is A: The fluid is synovial fluid. Ball and socket joints are synovial, but are only one type of synovial joint.

14. Which term refers to the joint between the two pelvic bones?
    A. Pubic epiphysis
    B. Pubic symphysis
    C. Pubic diaphysis
    D. Pubic metaphysis

Answer is B: The pubic symphysis joins the two pelvic bones anteriorly. It is an immovable joint.

15. Which of the following joints is a synovial joint?
    A. Intervertebral joints
    B. The tibiofibular joint
    C. Skull sutures
    D. Carpometacarpal joints

Answer is D: The joints between the wrist bones (carpals) and bones of the hand (metacarpals) are synovial and freely movable joints.

16. Which two bones are involved in the movement known as "pronation"?
    A. Radius and ulna
    B. Tibia and fibula
    C. Carpals and metacarpals
    D. Mandible and temporal bone

Answer is A: The bones of the forearm (radius and ulna) are able to pronate. That is, cross over when the forearm twists.

17. What is the physical function(s) performed by synovial fluid?
    A. It forms a film that keeps the articulating surfaces of bone from touching.
    B. It provides nutrients to the cells in cartilage and removes their waste.
    C. It removes the heat generated by friction when the joint is in continuous use.
    D. It lubricates the movement of articulating bones and distributes the pressure evenly across the joint.

Answer is D: The lubricant action ensures a friction-free movement, while the relative incompressibility of the fluid allows (for example) the body weight to be distributed over the surface area of the knee joint (think about Pascal's principle). Choice B is true but is a biological function.

18. What is the biological function(s) performed by synovial fluid?
    A. It forms a film that keeps the articulating surfaces of bone from touching.
    B. It provides nutrients to the cells in cartilage and removes their waste.
    C. It removes the heat generated by friction when the joint is in continuous use.
    D. It lubricates the movement of articulating bones and distributes the pressure evenly across the joint.

Answer is B: Cartilage is avascular, so the chondrocytes get their nutrients and dispose of their wastes via the synovial fluid that bathes them.

19. Which of the following refers to an articulation between bones?
    A. Sinus
    B. Sulcus
    C. Symphysis
    D. Synovium

Answer is C: A symphysis is an immovable joint between two bones (e.g. the pubic symphysis or the mandibular symphysis). Choices A and B refer to structures within a bone. Synovium refers to the membrane that produces synovial fluid.

20. What type of joint has the space between bones filled with cartilage?
    A. Cartilaginous joints
    B. Synovial joints
    C. Fibrous joints
    D. Diarthroses

Answer is A: Cartilaginous joints have cartilage filling the space between bones. Choices B and D both refer to freely moveable joints.

21. By what name is something that attaches a bone to another bone known?
    A. Aponeurosis
    B. Sarcomere
    C. Ligament
    D. Tendon

Answer is C: Ligament is the "ligature" that join articulating bones. A tendon (or an aponeurosis) attaches a muscle to a bone.

22. What type of joint is a "ball and socket" joint?
    A. Synovial
    B. Cartilaginous
    C. Fibrous
    D. Bony

Answer is A: The hips and shoulders are ball and socket joints and are freely moveable. All freely moveable joints are synovial—that is, have synovial fluid between the articulating bones.

23. Which two pairs of terms that describe joints do NOT go together?
    A. Immovable and synarthrosis
    B. Slightly moveable and amphiarthrosis
    C. Freely moveable and diarthrosis
    D. Unmoveable and syndesmosis

Answer is D: A syndesmosis (e.g. the distal joint between fibula and tibia) is slightly moveable.

24. What type of joint is that between two adjacent vertebral bodies?
    A. Cartilaginous
    B. Freely moveable
    C. Synovial
    D. Fibrous

Answer is A: Adjacent vertebral bodies are separated by a fibro-cartilaginous intervertebral disc; hence, they are cartilaginous joints. They are not freely moveable (or synovial).

25. What type of joint is that between the facet of the superior articular process of one vertebra and the facet of the inferior articular process of the vertebra immediately above?
    A. Bony
    B. Fibrous
    C. Synovial
    D. Cartilaginous

Answer is C: These facet joints have hyaline cartilage on the facets and synovial fluid between them. They function to permit but limit the flexion and rotation of the spine.

26. Why does the knee joint have such a complicated arrangement of ligaments?
    A. The knee joint is the articulation of three bones and bears the majority of the body weight.
    B. The knee joint is the articulation of three bones and is the largest joint in the body.
    C. The knee joint bears the majority of the body weight and is the largest joint in the body, and the ligaments limit the extent of allowable knee movement.
    D. The knee joint bears the majority of the body weight and is the largest joint in the body, and the ligaments allow for supination of the leg.

Answer is C: Choices A and B are also correct, but choice C is better. The leg does not supinate.

27. Which of the following explains how supination of the forearm is achieved?
    A. Proximal end of the radius spins beneath the lateral epicondyle of humerus while distal end moves from medial to lateral side of ulna.
    B. Distal end of the radius spins beneath the lateral epicondyle of humerus while the proximal end moves from medial to lateral side of ulna.
    C. Proximal end of the ulna spins beneath the lateral epicondyle of humerus while distal end moves from medial to lateral side of radius.
    D. Distal end of the radius moves from medial to lateral side of ulna while the proximal end spins beneath the lateral epicondyle of humerus.

Answer is A: Describes the movements in supination. Choice D describes pronation, while choices B and C describe incorrect bone positions.

28. What is the PHYSICAL role played by the fluid secreted by the synovial membrane of diarthrotic joints?
    A. It supplies oxygen and nutrients to the chondrocytes in the avascular articular cartilage.
    B. The incompressible fluid spreads the pressure produced by the weight of body evenly to the whole joint surface and joint capsule.
    C. It removes wastes from the chondrocytes in living in the articular cartilage.
    D. The fluid contains phagocytic cells that remove microbes, and the detritus that results from wear and tear in the joint.

Answer is B: Diarthrotic joints are also known as synovial joints and freely moveable joints. The physical role is to transmit the pressure evenly to all joint surfaces. The other choices are also roles performed by synovial fluid, but are biological roles.

29. Where is a gomphosis found in the body?
    A. In the wrist
    B. In the knee
    C. Between the radius and ulna
    D. In the mandible

Answer is D: A gomphosis is the name given to a synarthrosis (non-moveable joint) that binds a tooth to the jaw.

30. What is the name for an immovable joint?
    A. Amphiarthrosis
    B. Diarthrosis
    C. Synarthrosis
    D. Monoarthrosis

Answer is C: Synarthroses are immoveable bone joints, for example brain sutures, between first rib and the manubrium, between teeth and the jaw. A monoarthrosis does not exist!

31. Both the knee and elbow joints are "hinge" joints allowing the limbs to flex. What is the major difference between the structure of the knee and that of the elbow?

    A. The elbow involves three bones while the knee involves only two.
    B. The knee involves three bones while the elbow involves only two.
    C. There are three articulating bones in the elbow, but four in the knee.
    D. Both joints involve three bones, but in addition the knee has a sesamoid bone.

Answer is D: The knee has the patella (a large sesamoid bone) in addition to the three articulating long bones that are common to both joints. Hence choice C is correct, but D is the better answer.

32. Which of the following is a "ball and socket joint"?

    A. The sternoclavicular joint
    B. The temporomandibular joint
    C. The wrist joint
    D. The shoulder joint

Answer is D: The shoulder joint is the ball and socket joint, albeit with a shallow socket. The other ball and socket joints are the hip joints.

33. Which of the following features on a bone form part of an articulation?

    A. Spine
    B. Tubercle
    C. Tuberosity
    D. Facet

Answer is D: A facet is a shallow depression in a bone that is shaped to accept the articular process, such as a condyle, from an articulating bone.

34. How many ribs does each of the thoracic vertebrae T2–T10 articulate with?

    A. One rib
    B. Two ribs
    C. Three ribs
    D. Four ribs

Answer is D: Each of the thoracic vertebrae T2–T10 articulate with four ribs. For example T5 articulates with the head of its own rib (rib 5) and at the transverse process. In addition, T5 also articulates with the superior costal facet of the head of rib 6. Hence, T5 articulates with left and right rib 5 and left and right rib 6.

35. The tendons of skeletal muscles of the arms and legs often lie over a joint and insert on the bone that is moved when the muscle contracts. While the muscle shortens by a small amount, the bone to which it is attached moves over a large angle. To maximise the range of movement of the bone, where should the muscle insertion be?

A. Quite far from the joint
B. About midway along the long bone
C. Quite close to the joint
D. The distance from the joint does not matter, the range of bone movement depends on how much the muscle contracts

Answer is C: An insertion that is close to the joint allows a muscle contraction that takes place over a short distance to cause the bone to move through a large angle. Consequently, the end of the bone that is distal to the insertion moves through a relatively large distance.

36. Which bones rotate during the movement known as "supination"?

A. The ulna rotates laterally over the radius.
B. The ulna rotates medially over the radius.
C. The radius rotates laterally over the ulna.
D. The radius medially over the ulna.

Answer is D: The head of the radius rotates against the radial notch of the ulna, so that it changes from being parallel with the ulna, to crossing over it. Choice C describes pronation of a forearm that had been previously supinated.

Chapter 8
Muscular System

Skeletal muscle is attached to the skeleton, is voluntarily controlled and has cells that are very long "fibres" that are multinucleate and striated. Each skeletal muscle is an organ and is named. The names may seem strange, but they contain information about the muscle. The name may indicate the location of the muscle in the body (femoris, brachii), the shape of the muscle (deltoid, trapezius), its action (adductor, pronator) or its relative size (maximus, longus) of the muscle. The names may indicate the origin and insertion points (sternocleidomastoid) of the muscle, or the number of origins (biceps, quadriceps), or the direction (rectus, transversus) in which the muscle fibres lie. Or they may be named whimsically (sartorius).

Skeletal muscle cells are distinctly different from cardiac and smooth muscle cells. They are also distinct from other body cells. The cells are called fibres, their plasma membrane is called sarcolemma, the cytoplasm is called sarcoplasm, the endoplasmic reticulum is called sarcoplasmic reticulum, and the contracting structures within a myofibril of a muscle cell are called sarcomeres. Within a sarcomere are bundles of thick and thin myofilaments. The sarcolemma bundles many myofibrils together into a cell somewhat like a packet of spaghetti. Sarcomeres are joined end to end to form a long strand called a myofibril. The thick myofilaments are composed of the protein myosin. The thin myofilaments are composed of the protein actin, while the proteins, troponin and tropomyosin along with calcium ions and ATP participate in the physiology of contraction of a sarcomere. Each cell/fibre is surrounded by a membrane called the endomysium which overlies the sarcolemma. The endomysium contains nerve axons and blood capillaries. Nerve axons make contact with muscle fibres via a structure called the neuromuscular junction. A bundle of muscle fibres is called a fascicle and is surrounded by a connective tissue membrane called the perimysium. A muscle is a bundle of fascicles, and the membrane surrounding a muscle is called the epimysium.

M. Caon, *Examination Questions and Answers in Basic Anatomy and Physiology*, https://doi.org/10.1007/978-3-030-47314-3_8

8.1 Muscle Naming

1. Which muscle naming criteria are used to name the quadriceps femoris?
   A. Muscle action and location
   B. The origin and insertion
   C. Location and direction of muscle fibres
   D. Location and number of origins

Answer is D: Femoris refers to location on the femur, quadriceps refers to four origins.

2. Which of the following muscles is named according to its origin and insertion?
   A. Transversus abdominus
   B. Semimembranosus
   C. Sternocleidomastoid
   D. Deltoid

Answer is C: The origin is on the sternum and clavicle (sternocleido-), while the insertion (on the "moving bone") is to the mastoid process of the temporal bone.

3. Which of the muscles listed below is named according to its action?
   A. Adductor longus
   B. Temporalis
   C. Sternocleidomastoid
   D. Peroneus longus

Answer is A: Adduction is the action of bringing an abducted bone back towards the body's midline.

4. Where are the semimembranosus and semitendinosus muscles located?
   A. Seminal vesicle
   B. Thigh
   C. Forearm
   D. Back

Answer is B: They are two of the three "hamstring" muscles on the dorsal thigh, the other being the biceps femoris.

5. What information is contained in the muscle name "biceps brachii"?
   A. The muscle location and the number of origins
   B. The number of origins and the muscle action
   C. The muscle size and location in the body
   D. The muscle's shape and its action

Answer is A: "Brachii" indicates location on the brachus (arm), while "biceps" refers to the two origins (attachments) of the muscle.

6. What is the gluteus maximus named for? Its:
   A. Size
   B. Shape
   C. Action
   D. Origin and insertion

Answer is A: The gluteus maximus is larger than either the gluteus medius or gluteus minimus.

7. Which of the following muscles is **NOT** named after its location in the body?
   A. Deltoid
   B. Extensor carpi ulnaris
   C. Rectus abdominus
   D. Biceps femoris

Answer is A: The deltoid is named after its shape, which is likened to the capital Greek letter delta (a triangle).

8. Which of the following muscles is **NOT** named after its location in the body?
   A. Biceps brachii
   B. Sternocleidomastoid
   C. Rectus abdominus
   D. Flexor carpi radialis

Answer is B: The sternocleidomastoid is named after its origins (on the sternum and clavicle) and insertion (the mastoid process).

9. Which of the following muscles is **NOT** named after its location in the body?
   A. Latissimus dorsi
   B. Adductor longus
   C. Rectus femoris
   D. Biceps brachii

Answer is B: "Adductor" refers to the action of adduction, while longus refers to its size. Latissimus dorsi refers to the dorsal surface of the body.

10. Which muscles extend the leg?
    A. Quadriceps
    B. Hamstrings
    C. Gluteus muscles
    D. Soleus, gastrocnemius and tibialis anterior

Answer is A: Anatomically "leg" refers the limb between the knee and ankle. Extension of the leg is achieved by contacting the quadriceps (on the front of the thigh).

11. Which of the following muscles is named after its location in the body?
    A. Sartorius
    B. Triceps brachii
    C. Soleus
    D. Trapezius

Answer is B: "Brachii" refers to the location on the arm (brachus).

12. The muscles involved in mastication include which of the following?
    A. Sternocleidomastoid, scalene
    B. Sartorius, gracilis, soleus
    C. Temporalis, masseter, buccinator
    D. Orbicularis oculi, mentalis

Answer is C: The muscles in choice A move the neck and head, those in choice B are in the thigh and leg, while the orbicularis oculi is a facial muscle that encircles the eye.

13. What action does the flexor carpi ulnaris perform?
    A. It flexes the lower arm.
    B. The same as the extensor carpi ulnaris.
    C. It flexes the fingers.
    D. The same as the flexor carpi radialis.

Answer is D: The flexor carpi ulnaris and the flexor carpi radialis perform the same action—flexing the wrist—the former lies superficial to the ulna bone, while the latter is adjacent and superficial to the radius bone.

14. Which of the following muscles is named using the criterion of its size?
    A. Sternocleidomastoid
    B. Gluteus medius
    C. Flexor digitorum profundus
    D. Trapezius

Answer is B: "Medius" implies intermediate, in this case between the gluteus maximus and the gluteus minimus.

15. Which of the following muscles causes the wrist to bend?
    A. Extensor digitorum
    B. Extensor carpi ulnaris
    C. Flexor digitorum profundus
    D. Abductor pollicis longus

Answer is B: The term "carpi" refers to the carpal bone (the eight bones of the wrist), so this muscle extends the wrist.

16. Which of the following muscles increases the angle between the bones of the fingers and hand?

    A. Extensor digitorum
    B. Extensor carpi ulnaris
    C. Flexor digitorum profundus
    D. Abductor pollicis longus

Answer is A: "Extensor" refers to the action of extension (increasing the angle between the hand and fingers), while "digitorum" refers to the digits (fingers).

17. What are the muscles known as triceps brachii, biceps femoris and quadriceps femoris named according to? Their:

    A. Relative size and location of muscle's origin
    B. Number of origins and location in the body
    C. Shape of muscle and direction of muscle fibres
    D. Number of insertions and location in the body

Answer is B: Bi-, tri- and quad- refer to the number of origins, while the second word refers to location (arm or femur).

18. The space between the ribs is filled with:

    A. Intercostal muscle
    B. Costal cartilage
    C. Intercostal space
    D. Pleura

Answer is A: "Costa" means rib. Intercostal means between the ribs.

19. What are the muscles known as gluteus maximus, gluteus medius and gluteus minimus named according to? Their:

    A. Size
    B. Shape
    C. Whimsy
    D. Direction of their muscle fibres

Answer is A: maximus > medius > minimus.

20. Which muscle and bone listed below do **NOT** work together in combination?

    A. Humerus and biceps femoris
    B. Quadriceps and tibia
    C. Femur and gluteal muscles
    D. Radius and biceps brachii

Answer is A: The biceps femoris lies adjacent to the femur, not the humerus.

21. Which of the following muscles is named after its origin and insertion points?
    A. Tibialis anterior
    B. Extensor digitorum longus
    C. Rectus femoris
    D. Sternocleidomastoid

Answer is D: The sternocleidomastoid has its origin on the sternum and clavicle, while its insertion is on the mastoid process of the temporal bone.

22. The muscle known as the "transversus abdominus" is named according to:
    A. Its size and number of origins
    B. The direction of its muscle fibres and its action
    C. Its action and its location in the body
    D. Location in the body and direction of muscle fibres

Answer is D: Its location = abdominus, its muscle fibre direction = transversus.

23. Which muscle is one of the "hamstrings" group?
    A. Biceps brachii
    B. Biceps femoris
    C. Triceps brachii
    D. Quadriceps femoris

Answer is B: The hamstrings are in the thigh. "Femoris" refers to the femur bone. Biceps femoris is one of the three hamstring muscles. Choices A and C are on the arm (from "brachii").

24. Which one of the following muscle names refers to a superficial muscle that lies between the ribs?
    A. Serratus anterior
    B. Rectus femoris
    C. Extensor carpi radialis brevis
    D. External intercostals

Answer is D: "Costa" refers to rib. There are internal intercostal muscles, so the external intercostals are more superficial. The muscles in choices B and C lie on the femur and radius bones, respectively.

25. Which of the following muscles is **NOT** a flexor of the knee?
    A. Semimembranosus
    B. Semitendinosus
    C. Rectus femoris
    D. Biceps femoris

Answer is C: The rectus femoris is on the anterior of the thigh and extends the knee. The other three together constitute the "hamstrings" and flex the knee.

26. Which muscle is located on the posterior part of the forearm?
    A. Flexor digitorum superficialis
    B. Flexor digitorum profundus
    C. Extensor digitorum
    D. Triceps brachii

Answer is C: This question is testing the knowledge that muscles on the posterior part of the forearm are extensors (extend the fingers), while muscles on the anterior part are flexors. It also relies on knowing that these muscles are named according to their action. Hence the answer is C.

27. Which of the listed muscles is named according to its action?
    A. Rectus abdominus
    B. Peroneus longis
    C. Pronator teres
    D. Latissimus dorsi

Answer is C: The pronator teres pronates the forearm (turns the radius so that it crosses over the ulna).

28. Which muscle is located on the front of the anatomical leg?
    A. Tibialis anterior
    B. Rectus femoris
    C. Sartorius
    D. Flexor carpi radialis

Answer is A: The "leg" is between the knee and ankle—it contains the tibia and fibula. Anterior refers to the "front". Hence tibialis anterior. The other muscles are not on the leg. Rectus femoris is on the front of the "thigh".

29. Which muscle named below includes its origin and insertion?
    A. Pubococcygeus
    B. Sartorius
    C. External abdominal obliques
    D. Serratus anterior

Answer is A: This muscle (part of the pelvic floor) has two parts to its name: origin is on the pubis (pubic bone), insertion is to the coccyx. Choice B is named "whimsically", choice C named according to location and direction of muscle fibres, and choice D is named according to location and appearance.

30. One muscle below is NOT named after its shape. Which one?
    A. Deltoid
    B. Rhomboid
    C. Masseter
    D. Trapezius

Answer is C: The masseter is named after its action which allows mastication (chewing) of food.

31. On what part of the body is the serratus anterior muscle located?
    A. On the front of the torso
    B. On the lower back
    C. At the front of the thigh
    D. In the serratus region

Answer is A: "Anterior" refers to front, so choice B is wrong. There is no "serratus" region.

32. Which list below has the superficial muscles of the superior medial thigh in the correct order from medial to lateral?
    A. Gracilis, adductor longus, pectineus
    B. Gracilis, pectineus, adductor longus
    C. Adductor longus, gracilis, pectineus
    D. Pectineus, adductor longus, gracilis

Answer is A: Gracilis is most medial while pectineus is most lateral.

33. Which list has the muscles of the hamstrings in order from most lateral to medial?
    A. Semitendinosus, biceps femoris, semimembranosus, gracilis
    B. Rectus femoris, semitendinosus, semimembranosus
    C. Vastus lateralis, semitendinosus, semimembranosus
    D. Biceps femoris, semitendinosus, semimembranosus

Answer is D: There are three (not four) muscles in the hamstrings. The rectus femoris and vastus lateralis are in the quadriceps, not the hamstrings.

34. Which two muscles are on the dorsal side of the body?
    A. Rectus abdominus, pectoralis major
    B. Latissimus dorsi, trapezius
    C. External obliques, erector spinae
    D. Infraspinatus, vastus medialis

Answer is B: While the erector spinae and infraspinatus are on the dorsal side, their partners are not. Choice A muscles are on the ventral surface.

35. Which muscle is named according to its action?
    A. Semitendinosus
    B. Rhomboid major
    C. Triceps brachii
    D. Supinator

Answer is D: The supinator (origin on distal humerus and ulna, insertion on radius) performs "supination" of the forearm.

36. Which muscle is named according to "whimsy" rather than scientific rationale?
    A. Pronator
    B. Sartorius
    C. Trapezius
    D. Popliteus

Answer is B: Sartorius is named after a sartor (or tailor) who may have sat cross-legged. Pronator is named after its action; Trapezius after its shape; Popliteus after its location.

37. Which muscle is named after its points of origin and insertion?
    A. Brachialis
    B. Extensor digitorum
    C. Pubococcygeus
    D. Semimembranosus

Answer is C: One of the muscles of the "pelvic floor", its origin is on the pubis and insertion is on the coccyx and sacrum.

38. Which muscle below is descriptively named according to the direction of its muscle fibres?
    A. Transversus abdominis
    B. Serratus anterior
    C. Pectoralis major
    D. External intercostals

Answer is A: "Transversus" refers to muscle fibres that are horizontal when we are in a standing position (fibres lie in the transverse plane).

39. Which muscle below is descriptively named according to its location in the body?
    A. Gracilis
    B. Adductor magnus
    C. Anconeus
    D. Brachioradialis

Answer is D: "Brachio-" refers to the muscle origin on the arm (brachus) and "-radialis" to the insertion point on the radius bone. Hence also named for its origin and insertion ☺.

40. Which muscle below is descriptively named according to its relative size?
    A. Biceps brachii
    B. Adductor magnus
    C. Gastrocnemius
    D. Masseter

Answer is B: “Magnus” in Latin means great (as in “not small”). There are also the adductor minimus, adductor longus and adductor brevis muscles in the upper medial thigh.

41. Which muscle below is descriptively named according to its number of origins?
    A. Temporalis
    B. Rectus abdominus
    C. Triceps brachii
    D. Brachialis

Answer is C: “Triceps” refers to the three origins of this muscle: one on the scapula and two on the humerus.

42. Which muscle below is descriptively named according to its shape?
    A. Supinator
    B. Trapezius
    C. Latissimus dorsi
    D. Erector spinae

Answer is B: Trapezius is named after its resemblance to a trapezoidal shape: like a triangle with its top point cut off.

43. Which muscle below is descriptively named according to its origin and insertion?
    A. Sternocleidomastoid
    B. Platysma
    C. Orbicularis oris
    D. Supraspinatus

Answer is A: The sternocleidomastoid originates on the manubrium of the sternum (sterno-) and the medial end of the clavicle (-cleido-) and inserts on the mastoid process of the temporal bone (-mastoid).

44. Which muscle below is descriptively named according to its muscular action?
    A. Gluteus maximus
    B. Pronator teres
    C. Soleus
    D. Vastus lateralis

Answer is B: Pronation is the action of crossing the radius over the ulna, so that your hands are ready to type at a keyboard.

45. Which muscle below is descriptively named according to whimsy?
    A. Tibialis anterior
    B. Vastus medialis
    C. Semimembranosus
    D. Sartorius

Answer is D: A "sartor" is a tailor (a maker of men's clothes). Perhaps the name was chosen in reference to the cross-legged position in which tailors once sat. Also known as the honeymoon muscle. That is whimsy for you.

46. In which direction (relative to the spine) do the external oblique muscle fibres lie? External oblique muscles lie:
    A. Parallel to the spine
    B. Horizontal to the spine
    C. At an angle to vertical direction of the spine
    D. Perpendicular to the spine

Answer is C: External oblique muscle fibres run at an angle to vertical direction/spine.

47. In which directions (considering spine of the standing person) do the rectus abdominis muscle fibres lie? Rectus muscle fibres:
    A. Are vertical, hence are "erect"
    B. Are horizontal when contracted
    C. Horizontally, vertically and at an angle to the spine
    D. Lie at an angle to spine

Answer is A: Rectus fibres are vertical, hence are "erect".

48. What is the relative size of the gluteus medius?
    A. It is the biggest of the three gluteus muscles.
    B. It is the smallest of the three gluteus muscles.
    C. It is located in the medium position.
    D. It is smaller than gluteus maximus, but bigger than gluteus minimus.

Answer is D: The title gluteus medius implies that the muscle is the smaller than gluteus maximus, but bigger than gluteus minimus.

49. Select the correct pair of actions performed by the following muscles:
    A. Pronator quadratus pronates the forearm; extensor digitorum extends the fingers.
    B. Flexor carpi ulnaris extends the wrist; extensor digitorum flexes the fingers.
    C. Extensor carpi ulnaris extends the wrist; extensor digitorum pronates the fingers.
    D. Pronator quadratus extends the forearm; flexor carpi ulnaris flexes the wrist.

Answer is A: Pronator quadratus pronates the forearm; extensor digitorum extends the fingers (the digits).

50. Over which bones do the extensor carpi ulnaris (ECU) and the extensor carpi radialis brevis (ECRB) lie?
    A. The ECU is over the ulna; the ECRB is over the radius.
    B. The ECU is over the radius; the ECRB is over the ulna.

C. The ECU is over the humerus; the ECRB is over the ulna.
D. The ECU is over the scapula; the ECRB is over the manubrium.

Answer is A: The ECU is over the ulna ("ulnaris" is in the name); the ECRB is over the radius ("radialis" is in the name).

51. What do the two words in the name of the biceps femoris muscle refer to?
    A. "Femoris" refers to the femur bone on the thigh, "biceps" to the two origins of the muscle.
    B. "Femoris" refers to female musculature, "biceps" refers to its location on the arm.
    C. "Biceps" refers to its location on the two ceps regions, "femoris" refers to female musculature.
    D. "Biceps" refers to location on each arm and thigh, "femoris" refers to femur.

Answer is A: Biceps refers to the two origins (or two bellies of the muscle), while femoris refers to the femur in the thigh. Not to be confused with the biceps brachii muscles which are on the anterior surface of the humerus.

52. What broad general shapes do the (a) rhomboid major muscle and (b) deltoid muscle resemble?
    A. (a) Rhombus; (b) an inverted Greek capital letter delta
    B. (a) Square; (b) an inverted Greek capital letter delta
    C. (a) Rectangle; (b) the Greek letter delta
    D. (a) Rhombus; (b) the Roman numeral for ten

Answer is A: Rhombus is a "tilted" square, while capital delta looks like an equilateral triangle standing in its point.

53. Select the correct list of the muscles that comprise the quadriceps group:
    A. Rectus femoris, semimembranosus, vastus intermedius, iliocostalis
    B. Rectus femoris, vastus intermedius, vastus lateralis, vastus medialis
    C. Longissimus, vastus lateralis, vastus medialis, sternocleidomastoid
    D. Vastus intermedius, vastus lateralis, vastus medialis, gluteus minimus

Answer is B: The quadriceps, comprised of four muscles, lies on the ventral surface of the femur. Rectus femoris, vastus intermedius, vastus lateralis, vastus medialis.

54. Select the correct list of the muscles that comprise the hamstrings group.
    A. Biceps brachii (upper), semimembranosus (medial-ish), semitendinosus (medial)
    B. Gluteus medius (upper), semimembranosus (medial-ish), triceps femoris (posterior)
    C. Biceps femoris (lateral), anterior radialis (upper), semitendinosus (medial)
    D. Biceps femoris (lateral); semimembranosus (medial-ish), semitendinosus (medial)

Answer is D: Hamstrings lie on the dorsal surface of the femur and include biceps femoris (lateral); semimembranosus (medial-ish), semitendinosus (medial).

55. Where the triceps brachii is located?
    A. On the anterior surface of the arm
    B. On the posterior surface of the arm
    C. On the sacrum
    D. On the femur

Answer is B. Triceps brachii is on the posterior surface of the arm. The "forearm" lies between the elbow and wrist.

56. Where is the location of the triceps surae?
    A. Triceps surae is on the posterior surface of the leg.
    B. Triceps surae is on the anterior surface of the leg.
    C. Triceps surae is located on the posterior surface of the femur.
    D. Triceps surae is located on humerus.

Answer is A: Triceps surae is on the posterior surface of the leg. This region is known as the calf and also as the sural region. The "leg" lies between the knee and ankle. The thigh lies between the pelvis and knee.

8.2 Anatomy and Physiology

1. By what name is the plasma membrane of a muscle cell known?
    A. Sarcoplasm
    B. Sarcomere
    C. Sarcoplasmic reticulum
    D. Sarcolemma

Answer is D: "Sarco-" refers to flesh (muscle), "lemma-" refers to sheath (membrane) around the cell.

2. Smooth muscle is different from skeletal muscle because smooth muscle:
    A. Is found in the walls of arteries
    B. Can be voluntarily contracted
    C. Has many nuclei in a cell
    D. Has intercalated discs between cells

Answer is A: Smooth muscle occurs in the walls of tubes, whereas skeletal muscle does not.

3. All of the following structures are part of a muscle cell except one. Which one?
    A. Sarcoma
    B. Sarcolemma
    C. Sarcoplasm
    D. Sarcoplasmic reticulum

Answer is A: Sarcoma refers to a malignant tumour (a cancer) of connective or other non-epithelial tissue (bones, muscles, tendons, cartilage, nerves, fat, and blood vessels).

4. Which of the following muscle cell structures is the longest?
   A. A myofilament
   B. A myofibril
   C. A sarcomere
   D. A troponin molecule

Answer is B: A muscle cell is a bundle of myofibrils. Myofibrils contain many sarcomeres joined end to end. Within sarcomeres are found (the shorter) thick and thin myofilaments.

5. Which of the following is the smallest structure within a muscle fibre?
   A. Myosin
   B. Myofilament
   C. Myofibril
   D. Sarcomere

Answer is A: Myosin is a molecule that makes up a thick myofilament. Many thick and thin myofilaments make up a sarcomere. Many sarcomeres joined end to end form a myofibril.

6. What is a "sarcomere"?
   A. A cancer of connective tissue
   B. The cytoplasm of a muscle cell
   C. A section of a myofilament.
   D. The plasma membrane of a muscle cell

Answer is C: A myofilament is a long line of sarcomeres joined end to end. So one section of a myofilament is a sarcomere.

7. Smooth muscle cells may be described by which of the following?
   A. Striated, voluntary, multinucleate
   B. Not striated, voluntary, multinucleate
   C. Striated, involuntary, uninucleate
   D. Not striated, involuntary, uninucleate

Answer is D: Smooth muscle is not-striated, and it is involuntary and has one nucleus.

8. By which term is a muscle that opposes or reverses a particular movement called?
   A. Agonist
   B. Synergist
   C. Antagonist

D. Fixator

Answer is C: An agonist muscle performs the action, while the antagonist must relax (be stretched) while the action is being performed (and can reverse the action of the agonist).

9. Which term is given to the unit of a myofibril that contracts?
   A. Sarcoplasm
   B. Sarcomere
   C. Sarcolemma
   D. Sarcoplasmic reticulum

Answer is B: The contraction of a myofibril is due to the shortening of its component sarcomeres.

10. Which is the largest of the structures in a muscle fibre?
   A. Myofibril
   B. Myofilament
   C. Myosin
   D. Myopic

Answer is A: A myofibril extends the length of a muscle cell. A myofilament is shorter than a sarcomere, while myosin is a molecule in a thick myofilament.

11. Which protein(s) are found in thin myofilaments?
   A. Actin
   B. Actin and tropomyosin
   C. Actin, tropomyosin and troponin
   D. Actin, myosin, tropomyosin and troponin

Answer is C: Actin is the major component of a thin filament. Tropomyosin covers the biding site, while troponin provides the mechanism for removing tropomyosin from the binding site.

12. What characteristic of a smooth muscle cell distinguishes it from cardiac and from skeletal muscle?
   A. Being branched
   B. Being under involuntary control
   C. Lack of striations
   D. Being uninucleate

Answer is C: Both cardiac and skeletal muscles show striations when viewed under the microscope, but smooth muscle does not.

13. Which of the following muscles is a common intramuscular injection site?
   A. Deltoid
   B. Gluteus maximus

C. Vastus medialis
D. Latissimus dorsi

Answer is A: The upper arm at the shoulder is the location of the deltoid. It is the gluteus medius (rather than the gluteus maximus) that is used for IM injections to avoid the sciatic nerve.

14. What is the source of the majority of the energy needed by muscles for physical activity that continues for longer than 30 or 40 min?

A. ATP stored in muscle fibres
B. Glycolysis of glucose in the cell cytoplasm
C. ATP produced from creatine phosphate stored in muscle fibres
D. Aerobic respiration of pyruvic acid in mitochondria

Answer is D: Choice A lasts for a few seconds only. Choice B can provide energy for a couple of minutes. Choice C can provide energy for vigorous activity lasting about 15 s.

15. Which one of the following is **NOT** a characteristic of skeletal muscle?

A. Excitability
B. Autonomic innervation
C. Contractility
D. Extensibility

Answer is B: Skeletal muscle is voluntary so is innervated by the somatic nervous system, not the autonomic system.

16. Skeletal muscle cells have all of the following characteristics **EXCEPT** one. Which one?

A. A neuromuscular junction crossed by ACh (acetyl choline).
B. Invaginations of sarcolemma called "T tubules".
C. They are branched.
D. They are striated.

Answer is C: Skeletal muscle cells are not branched (but cardiac muscle cells are).

17. Which of the following muscle structures is the largest?

A. Sarcomere
B. Fascicle
C. Myofibril
D. Muscle fibre

Answer is B: A muscle fascicle is a bundle of muscle fibres (cells). Choices A and B are smaller than a cell.

18. Which feature is shared by cardiac muscle cells and skeletal muscle cells?

A. Striations
B. Intercalated discs

C. Branching
D. Involuntary nature

Answer is A: Both types of muscle cells are striated. Choices B, C and D are characteristics of cardiac muscle cells only.

19. What structures attach a muscle to a bone?
    A. A tendon
    B. A fasciculus
    C. A sarcomere
    D. An internal intercostal

Answer is A: Tendons (thin rope-like structures) attach muscle to bone.

20. Which of the following muscle structures is the smallest?
    A. Sarcomere
    B. Fasciculus
    C. Myofibril
    D. Muscle fibre

Answer is A: A sarcomere is a section of a myofibril. A muscle fibre (cell) is a bundle of myofibrils, while a fasciculus is a bundle of muscle fibres.

21. A feature of skeletal muscle that is **NOT** shared with cardiac or smooth muscle is:
    A. Striations
    B. Branched cells
    C. Intercalated discs
    D. Many nuclei

Answer is D: A skeletal muscle cell is a "syncytium" being derived from many cells, so retains their many nuclei.

22. What is the role of acetylcholine in muscle cell contraction?
    A. It is a neurotransmitter.
    B. It binds to troponin causing it to change shape.
    C. It supplies the energy for contraction.
    D. It engages with the binding site on actin.

Answer is A: ACh crosses the synaptic cleft to transmit a neural impulse to the muscle sarcolemma.

23. What is a sarcomere?
    A. It is the plasma membrane of a muscle cell.
    B. It is the cytoplasm of a muscle cell.
    C. It is a section of myofibril.
    D. It is a bundle of thick and thin myofilaments.

Answer is C: While there are thick and thin myofibrils within a sarcomere, choice C is the better answer.

24. Which of the following is the smallest unit in a muscle?
    A. Muscle fibre
    B. Myosin
    C. Fasciculus
    D. Myofibril

Answer is B: Myosin is a molecule, many of which combine to make a thick myofibril.

25. Skeletal muscle cells can be characterised as:
    A. Unstriated, involuntary, multinucleate
    B. Unstriated, voluntary, multinucleate
    C. Striated, voluntary, uninucleate
    D. Striated, voluntary, multinucleate

Answer is D: Being multinucleate (and voluntary) distinguishes skeletal muscle cells from smooth, being voluntary distinguishes skeletal muscle from cardiac muscle cells.

26. Which one of the following is not made of skeletal muscle?
    A. Diaphragm
    B. Pyloric sphincter
    C. Vastus lateralis
    D. Tongue

Answer is B: We are able to exercise conscious control over the other three muscles so they are skeletal muscle.

27. What is the cytoplasm of a skeletal muscle cell called?
    A. Sarcolemma
    B. Sarcomere
    C. Sarcoplasm
    D. Fasciculus

Answer is C: The prefix "sarco" is commonly used to describe microscopic muscle cell structures.

28. What does the term "origin" refer to in the musculoskeletal system?
    A. The point of attachment of a muscle to the "moveable" bone
    B. The line that separates the shaft from the end of a long bone
    C. The point of attachment of a muscle to the "stationary" bone
    D. The end of a long bone

Answer is C: Muscles are said to "originate" on the bone that does not move when that muscle contracts, and to insert on the bone that does flex or extend when the muscle contracts.

29. What is the protein of thick myofilaments in a skeletal muscle cell?
    A. Tropomyosin
    B. Myosin
    C. Actin
    D. Acetylcholine

Answer is B: Choices A and C are both proteins of thin myofilaments. ACh is a neurotransmitter.

30. A skeletal muscle fibre (cell) consists of many sections (units) which contract. What is the name given to one of the units that contract?
    A. Sarcomere
    B. Sarcolemma
    C. Sarcoplasm
    D. Fasciculus

Answer is A: A muscle fibre consists of many sections (sarcomeres) connected end to end that shorten (contract) as the thick and thin filaments slide past each other.

31. The neurotransmitter that causes an action potential to occur in a muscle cell membrane is called:
    A. Inorganic phosphate (HPO_4^{2-})
    B. Adenosine diphosphate (ADP)
    C. Calcium (Ca^{2+})
    D. Acetylcholine (ACh)

Answer is D: Acetylcholine is the neurotransmitter. All somatic motor neurons release ACh at their synapses with skeletal muscle fibres.

32. Which of the following describes skeletal muscle?
    A. Striated, voluntary, multinucleate, individually named
    B. Striated, branched, uninucleate, involuntary
    C. Not striated, uninucleate, voluntary, individually named
    D. Not striated, multinucleate, involuntary, with intercalated discs

Answer is A: Skeletal muscle is striated and is not branched (cardiac muscle cells are branched).

33. Which list is in the correct order of **DECREASING** size?
    A. Muscle fibre, sarcomere, myofilament, myofibril
    B. Muscle, fasciculus, muscle fibre, myofibril
    C. Sarcomere, fasciculus, myofibril, myofilament
    D. Muscle, muscle fibre, myosin, myofibril

Answer is B: A myofibril consists of sarcomeres joined end to end; A myofibril contains many myofilaments; A myofibril is larger than a molecule of myosin.

34. Select the one **INCORRECT** statement about skeletal muscles:
    A. An "agonist" opposes or reverses a particular movement.
    B. A muscle's attachment point to a stationary bone is called its "origin".
    C. A skeletal muscle cell is a "syncytium".
    D. Muscles that immobilise a bone are called "fixators".

Answer is A: An antagonist opposes a movement. All other statements are correct.

35. Which of the following groupings of muscle type and their characteristics is **INCORRECT**?
    A. Skeletal, striated, voluntary
    B. Smooth, visceral, involuntary
    C. Cardiac, striated, voluntary
    D. Skeletal, striated, syncytium

Answer is C: Cardiac muscle is not voluntary.

36. Microscopically, muscle fibres contain parallel myofibrils. What are the units joined end to end within a myofibril called?
    A. Myofilament
    B. Motor unit
    C. Myosin
    D. Sarcomere

Answer is D: Sarcomeres joined end to end, like train carriages, form a myofibril.

37. The part of a skeletal muscle cell that is able to contract is called:
    A. Sarcoplasm
    B. Sarcolemma
    C. Sarcomere
    D. Sarcoplasmic reticulum

Answer is C: Sarcomeres contain myofilaments that slide past each other as the sarcomere contracts.

38. The energy for muscle contraction is derived from the mechanisms below **EXCEPT** for one. Which one is **NOT** a method of producing ATP?
    A. Anaerobic glycolysis
    B. Aerobic respiration
    C. Direct phosphorylation of ADP by creatinine phosphate
    D. Anaerobic digestion of lactic acid

Answer is D: ATP is not produced from lactic acid—lactic acid is produced from pyruvic acid after glucose is anaerobically lysed into ATP and pyruvic acid.

39. With respect to the flexion of the forearm, which of the following statements is correct?
    A. The origin of the biceps brachii is on the radius, and its insertion is on the scapula.

B. The origin of the biceps brachii is on the ulna, and its insertion is on the scapula.
C. The agonist muscle is the biceps brachii, and the antagonist is the triceps brachii.
D. The agonist muscle is the biceps brachii, and the antagonist is the brachialis.

Answer is C: Antagonist for flexion is the triceps brachii. (The origin, not the insertion of the biceps brachii is on the scapula.)

40. Patients confined to bed and those with plaster casts immobilising a bone fracture suffer muscle wasting. What is the term used for this condition?
    A. Disuse atrophy
    B. Denervation atrophy
    C. Muscle dystrophy
    D. Muscle hypertrophy

Answer is A: Immobilising a bone will mean the attached muscles are not used. Their nerve supply is intact, albeit nervous stimulation is reduced.

41. What is the source of the ATP used by muscles for vigorous activity that may last for 10–15 s?
    A. Glycolysis of glucose in the cell cytoplasm forms ATP.
    B. The ATP that is stored in muscle cells as ATP.
    C. Aerobic respiration in the mitochondria produces the ATP.
    D. Creatinine phosphate in muscle and ADP react to form the required ATP.

Answer is D: Creatinine phosphate is used to "recharge" the ADP that forms from ATP when the cross-bridge is energised.

42. Which of the following is **NOT** a common intramuscular injection site?
    A. Gluteus medius
    B. Deltoid
    C. Gluteus maximus
    D. Vastus lateralis

Answer is C: The gluteus maximus is avoided, so that the sciatic nerve is not pierced by the needle.

43. What structure attaches a muscle to a bone?
    A. A meniscus
    B. A ligament
    C. A cartilage
    D. A tendon

Answer is D: A tendon is part of a muscle. Ligaments attach bones to each other, while cartilage covers the articulating surfaces of bone.

44. When a muscle contracts, exactly what structure gets shorter?
    A. The fascicles of a muscle
    B. The myosin molecules of a myofilament
    C. The actin molecules of a myofilament
    D. The sarcomeres of a myofibril

Answer is D: The length of the sarcomeres decreases.

45. What is the neurotransmitter that crosses the neuromuscular junction?
    A. Acetylcholine (ACh)
    B. Adrenalin (epinephrine)
    C. Noradrenalin (norepinephrine)
    D. Ca^{2+}

Answer is A: All somatic motor neurons release ACh at their synapses with skeletal muscle fibres, that is, at the neuromuscular junction.

46. What does aerobic respiration refer to?
    A. Glycolysis in the cytoplasm in the absence of oxygen
    B. Oxidative phosphorylation in the mitochondria in the presence of oxygen
    C. Glycolysis in the liver in the presence of oxygen
    D. Gluconeogenesis in the liver in the absence of oxygen

Answer is B: Respiration occurs in the mitochondria of a cell. The term aerobic refers to the presence of oxygen.

47. What is the name given to the membrane that surrounds a muscle fascicle?
    A. Pericardium
    B. Peritoneum
    C. Periosteum
    D. Perimysium

Answer is D: The syllable "mys-" refers to muscle. The other choices refer to the heart, abdomen, and bone, respectively.

48. What does the term "endomysium" refer to?
    A. The protein of thick myofilaments
    B. The tissue that surrounds a muscle fibre
    C. The gap between the axon terminal of a motor nerve cell and a motor end plate of a muscle cell
    D. The vessel from which Ca^{2+} ions are released prior to a muscle cell contraction

Answer is B: Endomysium is the connective tissue that surrounds a muscle cell (a fibre). Perimysium surrounds a muscle fascicle, while epimysium surrounds the whole muscle. The other choices are wrong.

49. Why does anaerobic respiration occur during vigorous exercise?
    A. There may be insufficient oxygen supplied to the muscle cells.
    B. Carbon dioxide builds up in the respiring muscle cells.
    C. Anaerobic respiration releases more energy from glucose than aerobic respiration.
    D. During anaerobic respiration enzymes make more ATP than during aerobic exercise.

Answer is A: Vigorous exercise may require more oxygen than can be supplied by the circulation. In this case, ATP must be produced anaerobically which is less efficient than when sufficient oxygen is present. Hence choices C and D are wrong.

50. Which is the correct equation for anaerobic respiration in humans?
    A. Glucose + lactic acid → carbon dioxide + water
    B. Glucose → lactic acid + carbon dioxide
    C. Glucose → lactic acid + (little energy)
    D. Glucose → lactic acid

Answer is C: Anaerobic respiration produces lactic acid and (of course) energy.

51. Which is the correct equation for aerobic respiration in humans?
    A. Glucose + oxygen → carbon dioxide + water
    B. Glucose + oxygen → carbon dioxide
    C. Glucose → carbon dioxide + water + energy
    D. Glucose + oxygen → carbon dioxide + water + energy

Answer is D: Aerobic means that oxygen must be an input. The result must have energy being produced. Hence D.

52. What is true about the energy released from glucose respiration?
    A. Anaerobic respiration releases more than aerobic respiration.
    B. Aerobic respiration releases more than anaerobic respiration.
    C. Aerobic and anaerobic respiration release about the same amount.
    D. Anaerobic respiration release no energy.

Answer is B: Aerobic respiration produces 36 ATP molecules, while anaerobic respiration produces just two molecules of ATP.

53. Which pair of terms below refers to a contraction and relaxation?
    A. Flexor and extensor
    B. Origin and insertion
    C. Brachialis and radialis
    D. Agonist and antagonist

Answer is D: An agonist muscle causes a movement to occur (by contracting) while an antagonist muscle cooperates in the movement by relaxing (allowing itself to lengthen).

54. Which muscles extend the spine?
    A. Extensor indicis proprius
    B. Erector spinae
    C. Rectus abdominis
    D. Rectus femoris

Answer is B. Spine is extended (brought erect from the flexed position) by the erector spinae, a group of muscles composed of the three groups iliocostalis, longissimus, spinalis.

55. Which muscles extend the forearm?
    A. Biceps brachii
    B. Deltoid
    C. Triceps brachii
    D. Extensor digitorum

Answer is C. Forearm is extended by triceps brachii. The biceps brachii flexes the forearm.

56. Select the correct origins of the triceps brachii:
    A. Infraglenoid tubercule of humerus, posterior surface of radius, posterior ulnar shaft
    B. Infraglenoid tubercule of radius, posterior surface of ulna, anterior humeral shaft
    C. Infraglenoid tubercule of scapula, posterior surface of humerus, superior lateral margin of humerus
    D. Infraglenoid tubercule of clavicle, anterior surface of humerus, posterior radial margin

Answer is C: The triceps brachii lies on the dorsal surface of the humerus and extends the forearm. Muscle origins lie on the "non-moving" bone. Infraglenoid tubercule of scapula, posterior surface of humerus, posterior humeral shaft distal to radial groove.

57. Where is the insertion of the sternocleidomastoid muscle?
    A. Manubrium of the sternum and clavicle
    B. Mastoid process of temporal bone
    C. Mastoid process of mandible
    D. Mastoid process of manubrium

Answer is B: Insertion is on the "moving" bone = mastoid process of temporal bone. Origins are on the superior margin of the manubrium of the sternum and medial end of clavicle.

58. What is the correct action that is performed by the masseter?
    A. Elevates mandible to "masticate" food
    B. Flexes maxillae to "masticate" food

C. Pronates mandible to "masticate" food
D. Supinates mandible to "masticate" food

Answer is A: Elevates mandible to "masticate" food.

59. Under which gluteal muscle does the sciatic nerve lie?
    A. Gluteus maximus
    B. Gluteus minimus
    C. Gluteus medius
    D. All of the above

Answer is A: Sciatic nerve is mostly under gluteus maximus.

60. Into which gluteal muscle do you insert the needle for IM injection?
    A. Gluteus medius
    B. Gluteus medius and gluteus maximus
    C. Gluteus minimus
    D. Gluteus minimus and gluteus medius

Answer is B: Gluteus medius and gluteus maximus at a point superior and lateral to the middle of the gluteus maximus.

61. Select the most appropriate answer for why the deltoid, vastus lateralis and gluteus medius muscles are chosen as the sites for intramuscular injection.
    A. All three muscles have extensive fascia so have a large surface area for drug absorption and are well supplied with blood vessels.
    B. All three muscles are located close to major nerves and arteries—so drug will be quickly absorbed.
    C. All three are traditional areas for injection chosen according to the practical considerations.
    D. All three muscles are easily accessible, can gradually absorb large quantities of drug and are located at a safe distance from major nerves and arteries.

Answer is D: All three muscles are easily accessible, can gradually absorb larger quantities of drug, located on a safe distance from major nerves and arteries. Choice A is factually correct, but omits the safety aspect of avoiding major arteries and nerves.

62. What muscle attaches to the deltoid tuberosity?
    A. The biceps brachii
    B. The deltoid muscle
    C. The tuberosity muscle
    D. The triceps brachii

Answer is B: "Tuberosity" is a name given to a bone surface marking that is a site for a muscle's tendon to attach. The deltoid muscle tendon attaches to the deltoid tuberosity.

8.3 Sliding Filament Theory

1. Of the events that lead to myofilaments sliding over each other, which of the following happens first?
   A. The myosin head engages with the binding site on actin.
   B. Troponin changes shape and pulls on tropomyosin.
   C. Calcium ions enter the cell cytoplasm.
   D. ATP is hydrolysed to ADP and inorganic phosphate.

Answer is C: Calcium must first enter the cytoplasm in order to bind with troponin. Once the binding site is exposed, the myosin head may engage the site. Prior to engagement, ATP must be hydrolysed to ADP.

2. During muscle cell contraction, what happens because of Ca^{2+} binding to troponin?
   A. The binding site on actin is uncovered.
   B. Acetylcholine (ACh) is released.
   C. The cross-bridge disengages from the thin filament.
   D. ATP hydrolyses to ADP.

Answer is A: Troponin causes tropomyosin (which covers the binding site of actin) to be shifted away.

3. Which of the events below is the FIRST to occur prior to a muscle cell contracting?
   A. ATP binds to myosin.
   B. ADP detaches from myosin.
   C. The active site on actin is exposed.
   D. Ca^{2+} is released from the sarcoplasmic reticulum.

Answer is D: The release of Ca from where it is stored (in the SR) is necessary before the active binding site of actin can be exposed.

4. Which statement below best describes the role of Ca^{2+} in muscle contraction?
   A. Ca^{2+} binds to troponin, thereby changing its shape to expose the binding site.
   B. Ca^{2+} causes ADP and inorganic phosphate to detach from the myosin cross-bridge.
   C. Ca^{2+} attaches to the myosin head, causing it to disengage from its binding site.
   D. Ca^{2+} crosses the sarcolemma from the axon terminal which allows the action potential to propagate along the sarcolemma.

Answer is A: Ca attaches to troponin causing a shape change which shifts the attached tropomyosin away from binding sites of actin.

5. What causes the myosin binding site of an actin molecule to be exposed?
   A. ATP attaching to the myosin cross-bridge
   B. A nerve impulse reaching the motor end plate of a motor nerve.
   C. Calcium ions attaching to troponin
   D. Acetylcholine crossing the neuromuscular junction

Answer is C: Ca causes a change of shape to the troponin molecule when they attach. This causes troponin to wrench tropomyosin away from its resting position covering the binding site.

6. What is the role of Ca^{2+} in muscle contraction?
   A. Ca causes an action potential to travel along the sarcolemma.
   B. Ca binds to troponin changing its shape.
   C. Ca attaches to the binding site of myosin, energising it.
   D. Ca engages with the binding site of actin causing the power stroke.

Answer is B: The myosin head cannot attach to actin until its binding site is exposed. Ca causes this to happen.

7. What is the substance that binds to troponin in order to cause muscle contraction?
   A. PO_4^{2-}
   B. H_3O^+
   C. Ca^{2+}
   D. Fe^{2+}

Answer is C: Calcium ions do the job.

8. Which of these events is necessary for the contraction of a muscle cell?
   A. The shortening of myofilaments
   B. The conversion of ADP and HPO_4^{2-} to ATP in the cross-bridge
   C. Ca^{2+} binding to troponin causing it to change shape
   D. The movement of ACh from the sarcolemma to the axon terminal

Answer is C: The binding of calcium is essential. (Choice A is wrong as myofilaments do not shorten—they slide past each other. Choices B and D are the reverse of what actually happens).

9. Which statement about thick or thin myofilaments is CORRECT?
   A. Thick myofilaments contain the three proteins myosin, tropomyosin and troponin.
   B. Thin myofilaments contain the three proteins actin, tropomyosin and troponin.
   C. Thick myofilaments contain about 300 myosin molecules each of which has a binding site for a cross-bridge.
   D. Thin myofilaments contain about 300 myosin molecules each of which has a cross-bridge.

Answer is B: Thin myofilaments contain these three proteins. Choices C and D are wrong as myosin has a cross-bridge rather than a binding site.

10. What is the role of calcium ions in muscle contraction?
    A. Bind to troponin, thus changing its shape and pulling it away from the actin molecule
    B. Cause the myosin cross-bridge to detach from its binding site
    C. Cause the action potential to propagate along the sarcolemma
    D. Bind with ADP during aerobic respiration to produce ATP to provide energy

Answer is A: Calcium binds to troponin. All other choices are wrong statements.

11. Which of these events is necessary for the contraction of a muscle cell?
    A. The shortening of myosin molecules
    B. The hydrolysis of ATP to ADP and HPO_4^{2-} in the myosin cross-bridge
    C. Ca^{2+} binding to tropomyosin causing it to change shape
    D. The movement of Ca^{2+} from the sarcoplasm into the sarcoplasmic reticulum

Answer is B: The hydrolysis of ATP energises the cross-bridge. (Myosin molecules do not get shorter; calcium binds to troponin, not tropomyosin; calcium returns to the SR after contraction.)

12. What is the role of Ca^{2+} in the contraction of a muscle cell?
    A. Ca^{2+} binds to troponin to change its shape which reveals actin's binding site.
    B. Ca^{2+} attaches to the binding site of actin.
    C. Ca^{2+} detaches from ATP as it forms ADP.
    D. Ca^{2+} causes the myosin head to detach from the binding site of actin.

Answer is A: Actin's binding site is covered until calcium causes it to be exposed.

13. What binds to troponin causing it to expose the binding site on actin to enable muscle cell contraction?
    A. Ca^{2+} (calcium)
    B. ACh (acetylcholine)
    C. PO_4^{2-} (phosphate)
    D. ADP (adenosine diphosphate)

Answer is A: Troponin has a binding site for calcium.

14. Which molecule in a muscle cell has a cross-bridge that attaches to a binding site to effect the shortening of the cell?
    A. Tropomyosin
    B. Myosin
    C. Actin
    D. Troponin

Answer is B: Myosin (the thick filament) has a cross-bridge that attaches to a binding site on actin (the thin filament) when it is exposed.

15. Which event causes the thick filament to slide over the thin filament (i.e. the "power stroke") during muscle contraction?
    A. The attachment of Ca ions to troponin
    B. The hydrolysis of ATP to ADP, inorganic phosphate and energy
    C. The engagement of the head of the myosin cross-bridge with its binding site on actin
    D. The attachment of a molecule of ATP to the head of myosin's cross-bridge

Answer is C: The attachment of the already energised cross-bridge to its binding site stimulates the head to swivel and so cause the filaments to slide over each other. Choice A causes the binding site to be exposed. Choice B energises the cross-bridge. Choice D causes detachment of the cross-bridge.

16. On which molecule is the energised cross-bridge which produces muscle cell contraction located?
    A. Fibrin
    B. Troponin
    C. Actin
    D. Myosin

Answer is D: The myosin molecule has a "cross-bridge". The actin molecule has the binding site for the cross-bridge which is covered by tropomyosin. Ca binds to troponin in order to move tropomyosin away from the binding site.

17. Which of the following describes the relationship between the three proteins, actin, troponin and tropomyosin in thin myofilaments?
    A. Tropomyosin moves troponin away from actin's binding site.
    B. Actin moves tropomyosin away from troponin's binding site.
    C. Troponin moves tropomyosin away from actin's binding site.
    D. Actin moves troponin away from tropomyosin's binding site.

Answer is C: The actin molecule has the binding site for myosin's cross-bridge. The site is covered by tropomyosin. Ca binds to troponin in order to expose the binding site by moving tropomyosin away from the binding site.

18. What is the relationship between the three proteins, actin, troponin and tropomyosin in thin myofilaments?
    A. Troponin changes shape on binding with calcium ions and pulls tropomyosin away from the binding sites on the actin molecule.
    B. Troponin changes shape on binding with calcium ions and pulls actin away from the binding sites on the tropomyosin molecule.
    C. Tropomyosin changes shape on binding with calcium ions and pulls actin away from the binding sites on the troponin molecule.
    D. Actin changes shape on binding with calcium ions and pulls tropomyosin away from the binding sites on the troponin molecule.

Answer is A: Actin is the protein of thin filaments and has a binding site that is covered by tropomyosin. It is troponin that binds with Ca.

19. What structure are the "thin filaments" within a sarcomere attached to?
    - A. The thick filaments
    - B. The M line
    - C. The H band
    - D. The Z line

Answer is D: Z lines are the boundary between one sarcomere and the next. When thin filaments slide over the thick filaments, the Z lines are pulled closer together and the sarcomere contracts.

20. Which protein attaches the thick filament of a sarcomere to the Z line?
    - A. Nebulin
    - B. Tropomyosin
    - C. Troponin
    - D. Titin

Answer is D: Titin extends along the length of the thick filament and beyond the end of each thick filament to attach to the Z line. Titin is elastic so that when the sarcomere is relaxed, titin is longer than when the sarcomere is contracted.

Chapter 9
Gastrointestinal System

The contents of the gut are outside the body and are potentially dangerous to the body if they cross the gut wall. The contents include hydrochloric acid, protein-digesting enzymes and bacteria. We are protected from the gut contents by its mucosal lining and the cells (macrophages, dendritic cells, B-lymphocytes and T-lymphocytes) in the lymphoid tissue known as Peyer's patches. The gut also contains ingested food molecules of protein, carbohydrate and triglyceride. Digestion (hydrolysis) is the disassembling of large food molecules and is necessary to reduce the very large polymerised molecules in food to particles small enough to pass into the cells lining the gut. The products of hydrolysis of proteins are amino acids and di- and tripeptides. Carbohydrates are hydrolysed into monosaccharides (glucose, fructose, galactose), while triglycerides are hydrolysed into free fatty acids and monoglycerides. Nucleic acids (DNA and RNA) are hydrolysed by pancreatic nucleases into nucleotides and then hydrolysed by brush border enzymes (nucleosidases and phosphatases) into their free bases, pentose sugars and phosphate ions.

Ingested food passes through the oesophagus, stomach, small diameter intestine (duodenum, jejunum, ileum) and large diameter intestine (caecum, ascending colon, transverse colon, descending colon, sigmoid colon rectum), before the residue is eliminated. The movement is produced by peristalsis and is unidirectional, being controlled by deglutition, the gastro-oesophageal and pyloric sphincters, the ileocecal valve and the anal sphincters.

Hydrolysis is achieved by enzymes in saliva and from glands in the stomach and duodenal mucosae and from the pancreas. The products of carbohydrate and protein digestion are absorbed by active transport (for glucose and amino acids), enter the blood capillaries and are transported to the liver by the hepatic portal vein. The products of triglyceride digestion diffuse through the plasma membranes of the cells lining the small intestine where they are reconstituted into triglycerides. They then combine with phospholipids and cholesterol and are coated with protein to form water-soluble chylomicrons. Chylomicrons enter the lymphatic vessels (via lacteals) and eventually the blood stream via the thoracic duct.

M. Caon, *Examination Questions and Answers in Basic Anatomy and Physiology*, https://doi.org/10.1007/978-3-030-47314-3_9

Several accessory organs contribute to the digestive process by adding their secretion to the contents of the gut. These are the salivary glands, the pancreas and the liver.

9.1 Anatomy and Function

1. Which type of cell produces hydrochloric acid?
   A. Zymogenic cells
   B. Parietal cells
   C. Chief cells
   D. Enteroendocrine cells

Answer is B: The parietal cells of gastric glands produce hydrochloric acid. (Zymogenic cells are the same as chief cells, and they produce pepsinogen.)

2. What is the role of gastrin in the digestive system?
   A. To stimulate the release of bile and pancreatic juice
   B. To stimulate gastric secretion
   C. To activate pepsinogen
   D. To hydrolyse proteins to polypeptides

Answer is B: Gastrin is a hormone that stimulates gastric secretion (from the stomach wall).

3. What feature of the small intestine enhances its ability to absorb digested food?
   A. Its large surface area
   B. The gaps between adjacent epithelial cells
   C. Secretion of the hormone absorptin
   D. Its longer length compared to the large intestine

Answer is A: The large surface area allows the products of digestion ample opportunity to make contact with the absorbing surface. "Absorptin" is not a hormone.

4. Which of the following gut structures are listed in the correct order that food would pass through them, from input to exit?
   A. Pyloric sphincter, ileum, jejunum, transverse colon
   B. Pancreas, jejunum, ascending colon, sigmoid colon
   C. Ileum, duodenum, descending colon, ascending colon
   D. Duodenum, ileum, caecum, transverse colon

Answer is D: Jejunum is before the ileum; food does not pass through the pancreas; duodenum is before the ileum.

5. Which statement about the layers of the alimentary canal is correct?
   A. The serosa absorbs the products of digestion.
   B. The mucosa protects against self-digestion.

C. The submucosa is involved in segmentation and peristalsis
D. The muscularis externa is dense connective tissue.

Answer is B: The mucosa (not serosa) absorbs the products of digestion; the muscularis externa causes segmentation and peristalsis; the submucosa (not muscularis) is dense connective tissue.

6. Which of the following pairs of substances are **NOT** secreted by the stomach as part of “gastric juice”?

   A. Hydrochloric acid and pepsinogen
   B. Hormones and intrinsic factor
   C. Nuclease and amylase
   D. Mucus and gastrin

Answer is C: Nuclease and amylase are enzymes secreted by the pancreas.

7. From which of the gut structures below is most digested food absorbed?

   A. Duodenum
   B. Stomach
   C. Ileum
   D. Ascending colon

Answer is C: Ileum is the part of the small intestine distal to the duodenum which has a structure suited for absorption.

8. What is the term applied to the production of glucose from non-carbohydrate molecules?

   A. Deamination
   B. Transamination
   C. Glycogenolysis
   D. Gluconeogenesis

Answer is D: Gluco refers to glucose; neo = new; genesis = producing. (Glycogenolysis is the production of glucose form glycogen—by lysis.)

9. What is the purpose of “intrinsic factor” in gastric juice?

   A. To activate pepsinogen
   B. To assist with the absorption of vitamin B_{12}
   C. To protect the stomach lining against hydrochloric acid
   D. To stimulate the release of gastrin

Answer is B: Vitamin B_{12} is a large molecule which cannot be absorbed without forming a complex with intrinsic factor.

10. Which of the following does **NOT** contribute to increasing the surface area of the small intestine?

    A. The brush border
    B. Plicae circulars

C. Intestinal crypts
D. Villi

Answer is C: The crypts produce the secretion known as intestinal juice.

11. Which layer of the gastrointestinal tract is in contact with the contents of the gut?
    A. Muscularis externa
    B. Mucosa
    C. Serosa
    D. Submucosa

Answer is B: It secretes mucus and absorbs the products of digestion.

12. What is the name given to the process of moving the gut contents along the tract in the right direction?
    A. Peristalsis
    B. Emesis
    C. Segmentation
    D. Deglutition

Answer is A: Segmentation is “to and fro” movement about the one spot.

13. Which of the following substances is **NOT** produced by the cells of the gastric glands?
    A. Mucus
    B. Hydrochloric acid
    C. Gastrin
    D. Pepsin

Answer is A: Mucus cells are in the epithelium, not in the gastric glands, and produce mucus.

14. Which hormone stimulates the release of bile and pancreatic juice?
    A. Cholecystokinin
    B. Secretin
    C. Intestinal gastrin
    D. Pepsin

Answer is A: Duodenum produces and releases CCK when protein and fats enter the duodenum.

15. Where is the gastro-oesophageal sphincter?
    A. Between the stomach and the duodenum
    B. Between the stomach and the caecum
    C. At the entrance to the stomach
    D. Before the external anal sphincter

Answer is C: It is where the oesophagus enters the stomach. It prevents the stomach contents refluxing into the oesophagus.

16. Which sections of the gut perform the majority of the digestion of food and absorption of the digested products?
    A. Stomach and duodenum
    B. Jejunum and ileum
    C. Ascending colon and transverse colon
    D. Duodenum and jejunum

Answer is D: While a small amount of digestion and absorption occurs in the stomach, most occurs in the duodenum (which receives pancreatic enzymes and bile) and the jejunum (which is adapted for absorption).

17. What is the function of the oesophagus in digestion?
    A. It is a site of mechanical digestion.
    B. It transfers food from the mouth to the stomach.
    C. The oesophagus secretes amylase to begin carbohydrate digestion.
    D. The oesophagus secretes hydrochloric acid.

Answer is B: It is merely a conduit to transfer food from the mouth to the stomach while bypassing the thoracic structures.

18. What is the purpose of the mucosal barrier between the cells of the stomach wall and the stomach contents?
    A. It prevents the enzymes in the stomach contents from digesting the stomach.
    B. It converts pepsinogen to its active form.
    C. It prevents bacteria in the stomach from invading the stomach wall.
    D. It prevents undigested food molecules from being absorbed by the stomach lining.

Answer is A: The stomach contents include the enzyme pepsin and hydrochloric acid, both of which would damage the epithelial cells of the stomach if in contact with them.

19. Which parts of the alimentary canal prepare food for chemical digestion?
    A. The mouth, oesophagus and stomach
    B. The mouth, stomach and small intestine
    C. The mouth, stomach and duodenum
    D. The teeth, stomach and pancreas

Answer is C: The mouth (mastication and mixing with saliva); the stomach (churning); the duodenum (segmentation). Choice C is better than choice B as digestion has occurred before food reaches the distal parts of the SI.

20. Which three sections does the small intestine consists of?

    A. Ileum duodenum, caecum
    B. Antrum, jejunum, duodenum
    C. Rectum, ileum, duodenum
    D. Ileum, duodenum, jejunum

Answer is D: Caecum is part of the LI; antrum is part of the stomach; rectum is the last part of the colon.

21. What name is given to the movement of food material through the gastrointestinal tract?

    A. Peristalsis
    B. Segmentation
    C. Deglutition
    D. Bowel movement

Answer is A: While segmentation is also movement, it is a to and fro movement rather than unidirectional, so peristalsis is the best answer.

22. Emulsification is the name of the process carried out by:

    A. Lipase
    B. Bile
    C. Micelles
    D. Lacteals

Answer is B: Bile contains "surfactants" which congregate on the interface between lipid and water and so prevent small droplets of fat from aggregating into larger ones (=emulsification).

23. The pH of the stomach and the pH of the small intestine are BEST described (respectively) as:

    A. Acidic and alkaline/basic
    B. Strongly acidic and weakly alkaline/basic
    C. Acidic and weakly alkaline/basic
    D. Strongly acidic and strongly alkaline/basic

Answer is B: The pH of the stomach contents is about 2 (strongly acidic), while the pH of small intestine is about 8 (weakly alkaline). Thus, B is the **best** answer.

24. Which layer of the alimentary canal is responsible for absorbing the products of digestion?

    A. Muscularis interna
    B. Mucosa
    C. Serosa
    D. Submucosa

Answer is B: The mucosal layer is in closest proximity to the contents of the canal, and the digested material must pass through it.

25. Which of the following is **NOT** a part of the gastrointestinal tract?
    A. Ileum
    B. Pancreas
    C. Rectum
    D. Caecum

Answer is B: Pancreås is an accessory gland outside of the GI tract, but delivering its secretion via a duct.

26. Which of the following presents the structures through which chyme travels in the correct sequence?
    A. Oesophagus, ileum, duodenum, ileocecal valve, transverse colon, rectum
    B. Stomach, duodenum, ileum, transverse colon, ileocecal valve, rectum
    C. Stomach, duodenum, transverse colon, ileum, rectum, ileocecal valve
    D. Ileocecal valve, stomach, duodenum, transverse colon, ileum, rectum

Answer is B: Food is not referred to as "chyme" until after it is in the stomach (excludes A). Rectum is the last structure (excludes C). Stomach should be before ileocecal valve (excludes D).

27. What is the function of gastrin?
    A. To facilitate the absorption of vitamin B_{12} from the gut
    B. To inhibit gastric secretion
    C. To stimulate gastric secretion
    D. To stimulate pancreatic secretion

Answer is C: Gastrin is a hormone that stimulates the cells lining the stomach gastric pits to secrete.

28. Which list of sections of the intestine has them in correct order from nearest to furthest from the mouth?
    A. Duodenum caecum, jejunum, ileum
    B. Caecum, sigmoid colon, transverse colon, rectum
    C. Duodenum, ileum, rectum, jejunum
    D. Jejunum, ileum, caecum, ascending colon

Answer is D: Caecum is after the ileum; sigmoid colon is after the transverse colon; jejunum is before the ileum.

29. Forward movement of food material through the gastrointestinal tract is achieved by which process?
    A. Peristalsis
    B. Emesis
    C. Deglutition
    D. Hydrolysis

Answer is A: Peristalsis is the name given to the smooth muscle movements that propel the gut contents.

30. The lowest pH is found in which of the listed body sites?
    A. Pancreas
    B. Stomach
    C. Duodenum
    D. Blood

Answer is B: The stomach has the lowest pH (is the most acidic), pH < 2.

31. What is true about the "muscularis externa"?
    A. It is the muscle used for peristalsis.
    B. It secretes mucus.
    C. It refers to superficial skeletal muscles.
    D. It is composed of connective tissue.

Answer is A: Muscularis externa is a layer (a tunic) of the gut wall.

32. The surface area available for absorption in the small intestine is increased by all of the following structures **EXCEPT** one. Which one?
    A. Villi
    B. Haustra
    C. Plicae circularis
    D. Microvilli

Answer is B: Haustra are pouches (or sacculations) in the large intestine that do NOT occur in the small intestine.

33. Emulsification is the process where:
    A. Procarboxypeptidase and chymotrypsinogen become active enzymes.
    B. Chyme is moved backwards and forwards across the surface of the small intestine.
    C. Fat droplets are dispersed into smaller droplets.
    D. Dietary fat is digested by lipase.

Answer is C: Producing smaller droplets of fat increases the surface area available for lipase and allows digestion to proceed faster.

34. Which one of the following is **NOT** a function of the large intestine?
    A. Absorption of electrolytes
    B. Synthesis of some vitamins
    C. Absorption of water
    D. Digestion of fats

Answer is D: Digestion occurs and is completed in the small intestine.

35. How are the pH of the small intestine and the pH of the stomach (respectively) best described?
    A. Alkaline (basic) and acidic
    B. Weakly alkaline (basic) and strongly acidic

C. pH of 4 and pH of 8
D. pH of 7 and pH of 1.5

Answer is B: The pH of the stomach contents is about 2 (strongly acidic), while the pH of small intestine is about 8 (weakly alkaline). The numbers in choices C and D are wrong. Thus B is the best answer.

36. Choose the list which has the selected structures of the alimentary canal in the same order that chyme would pass through them.
    A. Larynx, jejunum, ileum, descending colon, transverse colon, sigmoid colon
    B. Mouth, pharynx, oesophagus, large intestine, small intestine, anus
    C. Oesophagus, stomach, duodenum, ascending colon, sigmoid colon, rectum
    D. Stomach, duodenum, ileum, descending colon, transverse colon, ascending colon

Answer is C: Descending colon is after the transverse colon; small intestine is before the large intestine.

37. Which fluid within the body is likely to have the LOWEST pH?
    A. The chyme in the ileum
    B. Saliva
    C. The blood
    D. The chyme in the stomach

Answer is D: The stomach contents are the most acidic, that is, have the lowest pH.

38. What is a function of the stomach?
    A. Absorb the products of digestion
    B. Participate in deglutition
    C. Participate in mechanical digestion
    D. Release cholecystokinin

Answer is C: The churning of chyme is part of mechanical digestion. While some simple molecules (water, aspirin, glucose) can be absorbed from the stomach, most absorption occurs after digestion in the small intestine.

39. Which of the following is **NOT** a function of the mucosa of the small intestine?
    A. Protection against infectious disease
    B. Secretion of digestive enzymes
    C. The absorption of end products of digestion
    D. Segmentation

Answer is D: Segmentation is caused by the muscularis externa, not the mucosa.

40. Which fluid within the body is likely to have the HIGHEST pH?
    A. The contents of the ileum
    B. The contents of the start of the duodenum

C. The blood
D. The contents of the stomach

Answer is A: Highest pH means the most alkaline. Blood pH is about 7.4, which is less than that in the intestines. The proximal part of the duodenum will receive acidic chyme from the stomach, so its pH will be lowered by that. A is the best choice.

41. A function of the LARGE intestine is to:

A. Absorb the products of digestion
B. Absorb water
C. Participate in mechanical digestion
D. Release intestinal gastrin

Answer is B: The products of digestion have been absorbed before the remaining contents reach the LI. It absorbs additional water to make faeces a paste.

42. Which fluid within the body is likely to have the **LOWEST** pH?

A. The contents of the ilium
B. The contents of the start of the duodenum
C. Urine in the bladder
D. The contents of the stomach

Answer is D: Lowest pH means most acidic. The stomach contents is most acidic.

43. Which one of the following is **NOT** a function of the stomach?

A. Digestion of fats
B. Digestion of proteins
C. Mechanical digestion
D. Storage of food

Answer is A: The stomach produces pepsinogen which activates to pepsin which begins the digestion of protein. Fat is not significantly digested in the stomach (gastric lipase initiates the digestion of milk protein).

44. What is the name of the hormone that stimulates the stomach to secrete hydrochloric acid?

A. Gastrin
B. Intestinal gastrin
C. Secretin
D. Cholecystokinin

Answer is A: Gastrin is the best answer—where it comes from is not pertinent.

45. Fatty acids are transported around the body by the blood in structures known as:

A. Micelles
B. Chylomicrons

C. Triglycerols
D. Low-density lipoproteins

Answer is B: Chylomicrons transport fats via the lymph system to the blood. Micelles exist in the lumen of the intestine.

46. The wall of the alimentary canal is made up of the four layers (not in any order): the muscularis externa, serosa, mucosa and submucosa. Which layer absorbs the end products of digestion?

    A. Submucosa
    B. Muscularis externa
    C. Serosa
    D. Mucosa

Answer is D: The mucosa is the innermost layer so digested food must pass through it to be absorbed.

47. The functions of the stomach are:

    A. Storage of meal, digestion of protein, mechanical digestion
    B. Mechanical digestion, digestion of carbohydrate, food storage
    C. Churning of chyme, digestion of fats, mechanical digestion
    D. Digestion of protein, absorption of glucose, storage, mechanical digestion

Answer is D: Choice A is not wrong, but D is better as it includes glucose, which being a small molecule, can be absorbed through the stomach wall.

48. What is the pH of the duodenum?

    A. Highly acidic (pH between 1.5 and 3)
    B. Highly alkaline (pH of 10–12)
    C. Alkaline in the pH range 7.35–7.45
    D. Alkaline (pH in the range 7.1–8.2)

Answer is D: The duodenum is slightly alkaline.

49. In the stomach which cells secrete pepsinogen?

    A. Parietal cells
    B. Zymogenic cells
    C. Kupffer cells
    D. Enteroendocrine cells

Answer is B: "Zymogenic" means producing an enzyme—also known as Chief cells.

50. Which part of the gastrointestinal tract contains three distinct layers of smooth muscle in its walls?

    A. Rectum
    B. Small intestine

C. Stomach
D. Oesophagus

Answer is C: The stomach uses its muscular wall to churn (mix) the contents as part of mechanical digestion.

51. Which hormone is responsible for the contraction of the gall bladder?
    A. Secretin
    B. Gastrin
    C. Cholecystokinin
    D. Intrinsic factor

Answer is C: One of the functions of CCK is to cause the gall bladder to contract.

52. During which phase are our gastric secretions stimulated by the sight and smell of food?
    A. Gastric
    B. Digestive
    C. Cephalic
    D. Intestinal

Answer is C: The cephalic phase refers to the thought and sight of food stimulating our secretions in preparation for food entering our stomach.

53. Which of the following is a function of the normal flora of the large intestine?
    A. To hydrolyse cellulose
    B. To synthesise blood clotting proteins
    C. To synthesise B vitamins and vitamin K
    D. To secrete intrinsic factor

Answer is C: Humans are unable to hydrolyse cellulose; blood clotting proteins are synthesised in the liver; intrinsic factor is secreted by the stomach.

54. Why is insulin not given as an oral drug?
    A. It is too irritating to the gastrointestinal mucosa.
    B. It is altered by passing through the liver.
    C. It is too big a molecule to be absorbed through the plasma membrane.
    D. It would be digested by enzymes in the stomach.

Answer is D: Insulin is a protein so would be hydrolysed by protein digesting enzymes in the stomach (and duodenum).

55. Which one of the listed molecule types are absorbed from the gut?
    A. Starch
    B. Monosaccharides
    C. Cellulose
    D. Polypeptides

Answer is B: All the other molecules are too large to be absorbed through the gut wall.

56. What is a function of the SMALL intestine?
    A. Temporarily store ingested food
    B. Absorb the products of digestion
    C. Participate in mechanical digestion
    D. Secrete hydrochloric acid

Answer is B: The small intestine has a very large surface area which makes it well suited to absorb the products of digestion.

57. What is the name of the hormone that inhibits the stomach from secreting gastric juice?
    A. Gastrin
    B. Pepsin
    C. Enterogastrin
    D. Cholecystokinin

Answer is D: One of the effects of CCK is to inhibit gastric secretion

58. Fatty acids are absorbed from the gut into structures known as:
    A. Triglycerides
    B. Sinusoids
    C. Capillaries
    D. Lacteals

Answer is D: Lacteals contain lymph in which digested fats are transported to rejoin the blood stream in the vena cava via the thoracic duct.

59. What does the hormone gastrin do?
    A. Stimulates the secretion of saliva
    B. Inhibits the secretion of gastric juice
    C. Stimulates the secretion of hydrochloric acid
    D. Stimulates the secretion of cholecystokinin

Answer is C: Gastrin stimulates the G cells of the gastric glands in the stomach wall to secrete hydrochloric acid.

60. The alimentary canal has four layers. Which list below has them in correct order starting from the lumen?
    A. Serosa, submucosa, mucosa, muscularis externa
    B. Submucosa, mucosa, muscularis externa, serosa
    C. Mucosa, submucosa, muscularis externa, serosa
    D. Muscularis externa, mucosa, submucosa, serosa

Answer is C: The mucosal layer is in contact with the canal contents. The serosa is continuous with the visceral peritoneum.

61. “Emulsification” is used to describe which of the following processes?
    A. The synthesis of cholesterol from acetyl CoA
    B. The separation of a large fat globule into small droplets
    C. The making of insoluble glycogen from individual glucose molecules
    D. The activation of prolipase to lipase

Answer is B: Emulsification surrounds small droplets of fat with a coating that prevents the droplets from amalgamating into larger droplets. This provides more surface area for lipases to work on than would be the case if only larger droplets existed.

62. When do the nutrients in food enter the body?
    A. When swallowed via the mouth or inserted intravenously
    B. When food is converted to chyme in the stomach
    C. When hydrolysis products are absorbed into the epithelial lining of the gut
    D. When monosaccharides, amino acids and fatty acids are transformed in the liver

Answer is C: Gut contents are outside of the body until absorbed through the mucosal lining.

63. When is it considered that food material is outside of the body?
    A. When in extracellular fluid
    B. When outside of a body cavity
    C. When urinated or defecated out
    D. When part of the contents of the gut

Answer is D: Food in the gut has not “entered the body” until it has been absorbed into the epithelial lining. This happens after being hydrolysed into molecules that are small enough to be absorbed. The material that is urinated or defecated out is indeed outside the body, but is not considered to be food.

64. Why are the contents of the stomach and intestines considered to be potentially dangerous to health?
    A. Gut contains acid, bacteria and digestive enzymes.
    B. Raw ingested food contains bacteria on its surface.
    C. Food potentially contains toxins that have not yet been cleansed by passing through the liver.
    D. The gut contains the poisonous products of metabolism such as urea waiting for excretion.

Answer is A: The stomach contains hydrochloric acid, the duodenum contains enzymes that will hydrolyse proteins, fats and carbohydrates, the large intestine contains bacteria (normal flora).

65. What is absorbed from the gut in the duodenum?
    A. Iron and water
    B. Virtually all foodstuffs, electrolytes, water

C. Water, free fatty acids and monoglycerols
D. Monosaccharides, amino acids and small peptides

Answer is A: Iron is absorbed by intestinal mucosa cells of the duodenum and attaches to transferrin. Most foodstuffs are absorbed from the jejunum. The duodenum is where enzymes are mixed with chyme. It is also much shorter than the jejunum so that most chyme passes through before it is ready to be absorbed.

66. What is absorbed from the gut in the ileum?
    A. Virtually all foodstuffs, electrolytes, water
    B. Iron, calcium and water
    C. Water, electrolytes, vitamin B_{12}, bile
    D. Free fatty acids and monoglycerols

Answer is C: Vitamin B_{12} is absorbed by endocytosis and bile is absorbed for recycling. Most other foodstuffs and water have been absorbed in the jejunum.

67. What is absorbed from the gut in the jejunum?
    A. Iron, calcium and water
    B. Virtually all foodstuffs, electrolytes, water
    C. Free fatty acids and monoglycerols
    D. Monosaccharides, amino acids and small peptides

Answer is B: Jejunum is the part of the small intestine where most of the products of digestion are absorbed.

68. What is the function of the mucosa of the wall of the alimentary canal?
    A. Secretes mucus, digestive enzymes and hormones and absorbs end products of digestion
    B. Contains blood vessels and lymphatics and transport the products of digestion
    C. Is responsible for gut motility known as peristalsis and segmentation
    D. Suspends and attaches the gut's abdominal organs to the body wall

Answer is A: The mucosa lines the cavity of the gut so is in contact with the contents. Absorption occurs into the mucosal cells.

69. What is the function of the submucosa of the wall of the alimentary canal?
    A. Contains blood vessels and lymphatics and transport the products of digestion
    B. Secretes mucus, digestive enzymes and hormones and absorbs end products of digestion
    C. Is responsible for gut motility known as peristalsis and segmentation
    D. Suspends and attaches the gut's abdominal organs to the body wall

Answer is A: The submucosa receives the absorbed foodstuffs which enter the blood capillaries and lymphatics in this layer.

70. What is the function of the muscularis externa of the wall of the alimentary canal?
    A. Suspends and attaches the gut's abdominal organs to the body wall
    B. Secretes mucus, digestive enzymes and hormones and absorbs end products of digestion
    C. Is responsible for gut motility known as peristalsis and segmentation
    D. Contains blood vessels and lymphatics and transport the products of digestion

Answer is C: Muscularis externa is a layer of smooth muscle that propels chyme along the gut.

71. What is the function of the serosa of the wall of the alimentary canal?
    A. Is responsible for gut motility known as peristalsis and segmentation
    B. Secretes mucus, digestive enzymes and hormones and absorbs end products of digestion
    C. Suspends and attaches the gut's abdominal organs to the body wall
    D. Contains blood vessels and lymphatics and transport the products of digestion

Answer is C: The serosa is the serous membrane that surrounds the gut and is continuous with the greater and lesser omentum and mesentery. Because of this, the gut is attached to the dorsal body wall.

72. In describing the hormonal regulation of gastric secretion, what is the "cephalic phase"?
    A. A conditioned reflex triggered by the thought, sight, smell or taste of food that increases gastric secretion.
    B. The release of intestinal gastrin when food enters the duodenum, which stimulates gastric secretion.
    C. The distension of the stomach that occurs when food enters and the presence of peptides stimulate release of gastrin.
    D. The distension of the duodenum and the presence of acidic chyme and partially digested food produce a tightening of the pyloric sphincter and decrease in gastric secretion.

Answer is A: "Cephalic" refers to the head. This phase occurs before food enters the stomach. Choices B and D refer to the intestinal phase; choice C to the gastric phase.

73. In describing the hormonal regulation of gastric secretion, what is the "gastric phase"?
    A. The release of intestinal gastrin when food enters the duodenum, which stimulates gastric secretion.
    B. A conditioned reflex triggered by the thought, sight, smell or taste of food that increases gastric secretion.

C. The distension of the stomach that occurs when food enters and the presence of peptides stimulate release of gastrin.
D. The distension of the duodenum and the presence of acidic chyme and partially digested food produce a tightening of the pyloric sphincter and decrease in gastric secretion.

Answer is C: "Gastric" refers to the stomach. This phase occurs when stomach pH increases as food enters the stomach. Choices A and D refer to the intestinal phase; choice B to the cephalic phase.

74. In describing the hormonal regulation of gastric secretion, what is the "intestinal phase"?
    A. A conditioned reflex triggered by the thought, sight, smell or taste of food that increases gastric secretion.
    B. The release of gastric juice and HCl stimulated by a rise in stomach pH level.
    C. The distension of the stomach that occurs when food enters and the presence of peptides stimulate release of gastrin.
    D. The distension of the duodenum and the presence of acidic chyme and partially digested food produce a tightening of the pyloric sphincter and decrease in gastric secretion.

Answer is D: "Intestinal" refers to gut distal to the stomach. This "enterogastric reflex" is initially stimulatory (i.e. increases gastric secretion) and then inhibitory (by decreasing gastric secretion). Choices A refers to the cephalic phase, choices B and C to the gastric phase.

9.2 Digestion and Enzymes

1. Which of the following statements is **WRONG**? The end products of:
    A. Protein digestion are transported to the liver via the hepatic portal vein
    B. Triglyceride digestion are transported to the liver via the lymphatic system
    C. Carbohydrate digestion are transported to the liver in the blood
    D. Triglyceride digestion are transported via the lymphatic system

Answer is B: The products of fat digestion are transported by the lymphatic system but do not pass through the liver.

2. Which one of the following processes is **NOT** part of mechanical digestion?
    A. Hydrolysis
    B. Peristalsis
    C. Segmentation
    D. Mastication

Answer is A: Hydrolysis is a chemical process that coverts large food molecules into smaller ones (i.e. performs digestion).

3. What are the end products of carbohydrate digestion?
   A. Chylomicrons
   B. Amino acids
   C. Free fatty acids
   D. Monosaccharides

Answer is D: Monosaccharides or "simple" sugars.

4. What are some products of lipid digestion?
   A. Free bases and pentose sugars
   B. Fructose and glucose
   C. Amino acids and small peptides
   D. Free fatty acids and monoglycerols

Answer is D: Lipids are digested into free fatty acids and monoglycerols.

5. Which of the following is an active enzyme?
   A. Procarboxypeptidase
   B. Pepsin
   C. Telophase
   D. Trypsinogen

Answer is B: The prefix "pro-" and suffix "-ogen" refer to inactive enzymes. Telophase is a stage of cell division (not an enzyme).

6. Which of the following terms is used to describe the changing of large food molecules into smaller molecules?
   A. Mechanical digestion
   B. Deglutition
   C. Segmentation
   D. Hydrolysis

Answer is D: Hydrolysis refers to the splitting (lysis) of large molecules into smaller ones using water (hydro).

7. What are the end products of carbohydrate digestion?
   A. Monosaccharides
   B. Disaccharides
   C. Glucose
   D. Trisaccharides

Answer is A: Glucose is one example of a monosaccharide, but there are others.

8. What molecules are the products of protein hydrolysis?
   A. Monoglycerols and free fatty acids
   B. Monosaccharides and disaccharides
   C. Amino acids

D. Amino acids and small peptides

Answer is D: Amino acids and small peptides (consisting of two or three amino acids) are hydrolysis products of proteins.

9. What happens to the products of lipid digestion in the gut?
   A. They are actively transported into the epithelial cells lining the gut.
   B. They diffuse into epithelial cells and are reconstituted into triglycerides.
   C. They are transported to the liver by the hepatic portal vein.
   D. They diffuse through the plasma membrane of epithelial cells, then diffuse into blood capillaries

Answer is B: Fats, being lipid soluble, diffuse into epithelial cells to be reconstituted into triglycerides.

10. Digestion of food molecules is necessary, so that:
    A. Indigestible food molecules are separated from digestible food molecules.
    B. Essential amino acids and fatty acids may be absorbed by the body.
    C. Excretion of waste products can occur via the bowel.
    D. Food may be converted into particles small enough to pass into the cells of the gut wall.

Answer is D: The production of molecules that are small enough to be absorbed into epithelial cells is the purpose of digestion.

11. What are the end products resulting from the digestion of carbohydrates?
    A. Monosaccharides
    B. Monoglycerols
    C. Pentose sugars
    D. Amino acids

Answer is A: Digestion results in carbohydrates (which are polysaccharides) being converted to monosaccharides (which are small molecules).

12. Which enzyme below digests proteins?
    A. Nuclease
    B. Maltase
    C. Carboxypeptidase
    D. Transaminase

Answer is C: The syllable "-peptid-" refers to proteins (which are polypeptides).

13. Which of the following could **NOT** be used to describe pepsinogen?
    A. It is a protein
    B. It is a hormone
    C. It is related to an enzyme
    D. It is inactive

Answer is B: Pepsinogen is an inactive protein enzyme. But it is not a hormone.

14. What food is digested by lipase?
    A. Nucleic acids
    B. Carbohydrates
    C. Polypeptides
    D. Triglycerides

Answer is D: Lipase digests lipids. Our dietary lipids are triglycerides.

15. What food is digested into monoglycerols?
    A. Protein
    B. Lipid
    C. Nucleic acid
    D. Starch

Answer is B: Dietary lipids are triglycerols which are digested into free fatty acids and monoglycerols.

16. What happens to the products of digestion of lipids? They are absorbed into a:
    A. Capillary and transported by the blood to the liver
    B. Capillary and transported by the blood to the heart
    C. Lacteal and transported by the lymph to the heart
    D. Lacteal and transported by the lymph to the liver

Answer is C: Digested fats absorbed from the gut are transported in lymph. They re-enter the blood stream via the thoracic duct then travel to the heart.

17. Which is an enzyme secreted by the gastric glands?
    A. Pepsin
    B. Gastrin
    C. Cholecystokinin
    D. Intrinsic factor

Answer is A. Pepsin (strictly pepsinogen) is the only enzyme on the list.

18. What are the products of protein digestion?
    A. Monoglycerols and fatty acids
    B. Dipeptides, tripeptides and amino acids
    C. Bases, pentose sugars and nitrate ions
    D. Monosaccharides and disaccharides

Answer is B: Peptides and amino acids are produced by the hydrolysis of protein.

19. Correctly complete the sentence. Pepsinogen is:
    A. Converted to pepsin by hydrochloric acid
    B. Converted to pepsin by intrinsic factor
    C. Secreted by the pancreas
    D. Involved in production of carbohydrate digesting enzymes

Answer is A: Pepsinogen "gen"erates pepsin in the presence of hydrochloric acid.

20. What feature do procarboxypeptidase, pepsinogen, fibrinogen and chymotrypsinogen have in common?

    A. They are all enzymes.
    B. They are all produced by the pancreas.
    C. They are all inactive.
    D. They all digest proteins.

Answer is C: The prefix and suffix "pro-" and "-ogen" refer to these molecules needing modification to become enzymes. In their present form, they are inactive.

21. Protein is digested to polypeptides by which of the following?

    A. Pepsinogen
    B. Intrinsic factor
    C. Hydrochloric acid
    D. Pepsin

Answer is D: Pepsinogen is inactive (otherwise it would digest the proteins with the zymogenic/chief cells) until converted to pepsin by HCl.

22. What are the products of carbohydrate digestion?

    A. Monosaccharides
    B. Amino acids
    C. Monoglycerides
    D. Monogylcerols

Answer is A: Carbohydrates are digested into simple sugars known as monosaccharides (e.g. glucose, fructose, ribose).

23. What does bile do?

    A. Bile stimulates the release of lipase.
    B. Bile emulsifies fat.
    C. Bile digests fat.
    D. Bile hydrolyses fat.

Answer is B: Emulsification of fat is the role played by bile.

24. What is the function of the hepatic portal vein?

    A. To return blood from the liver to the heart
    B. To transport blood rich in amino acids, monosaccharides and free fatty acids to the liver
    C. To transport the products of protein and carbohydrate digestion from gut to the liver
    D. To allow the nutrients absorbed from the gut to bypass the liver

Answer is C: The digestion products of protein and carbohydrate (but not of triglycerides) are trans**port**ed to the liver by the portal vein.

25. Which of the following are the end products of protein hydrolysis?
    A. Monoglycerols
    B. Keto acids and non-essential amino acids
    C. Polypeptides
    D. Amino acids and di- and tripeptides

Answer is D: Amino acids are the smallest molecules resulting from protein hydrolysis. Some two amino acid units (dipeptides) and three amino acid units (tripeptides) can also be absorbed from the gut.

26. What converts pepsinogen to pepsin in the stomach?
    A. Hydrochloric acid
    B. Gastrin
    C. Intrinsic factor
    D. Pepsinase

Answer is A: HCl cleaves off 44 amino acids from pepsinogen to form pepsin.

27. What the products of hydrolysis of lipids?
    A. Monosaccharides
    B. Chylomicrons
    C. Amino acids and small peptides
    D. Free fatty acids and monoglycerols

Answer is D: A triglyceride is hydrolysed into two free fatty acids plus a monoglycerol. Chylomicrons are the structure within which, fats are transported in the lymph and blood.

28. Monosaccharides are the product of digestion of what substance?
    A. Proteins
    B. Carbohydrates
    C. Triglycerides
    D. Dipeptides

Answer is B: Carbohydrates are composed of many monosaccharide units.

29. Which of the following is an enzyme?
    A. Amylase
    B. Gastrin
    C. Intrinsic factor
    D. Pepsinogen

Answer is A: Amylase is the only enzyme. Pepsinogen is not an enzyme until it is changed to pepsin.

30. What feature do procarboxypeptidase, pepsinogen, trypsinogen and chymotrypsinogen have in common?
    A. They are all enzymes.
    B. They are all produced by the pancreas.

C. They all digest proteins.
D. They are all inactive.

Answer is D: They are inactive so are not yet enzymes; when activated they would digest proteins, but not in their present form.

31. The products of fat digestion are absorbed into the epithelial cells of the intestinal wall differently from the way products of protein and carbohydrate digestion are. The reason is:

A. The products of protein and carbohydrate digestion are smaller.
B. The products of fat digestion are actively transported across the plasma membrane.
C. The products of fat digestion are smaller.
D. Monoglycerides are soluble in the plasma membrane.

Answer is D: The products of fat digestion are fat (lipid) soluble so can penetrate the plasma membrane which is made of lipid.

32. What pancreatic enzyme digests lipids to free fatty acids and monoglycerides?

A. Lipase
B. Bile
C. Cholecystokinin
D. Lingual lipase

Answer is A: The "lip-" part of lipase indicates that it digests lipids, while the "-ase" part is indicative of an enzyme. Lingual lipase is produced by the salivary glands, not the pancreas.

33. What is the product of carbohydrate digestion?

A. Glucose
B. Monosaccharides
C. Glycogen
D. ATP

Answer is B: Glucose is a monosaccharide, but there are others, so choice B is the better answer.

34. What stomach enzyme digests protein to polypeptides?

A. Pepsin
B. Hydrochloric acid
C. Gastrin
D. Intrinsic factor

Answer is A: Pepsin is the only enzyme in the list.

35. What are the products of protein digestion?

A. Polypeptides
B. Monosaccharides

C. Amino acids
D. Free fatty acids and monoglycerides

Answer is C: Polypeptides have not yet completed the digestion process.

36. What is the process in the digestion of food molecules that produces their monomers called?

A. Polymerisation
B. Hydrolysis
C. Isomerisation
D. Deamination

Answer is B: Digestion is the process of making smaller molecules from polymers by the addition of H or OH from water. It is called hydrolysis.

37. In the digestion of food molecules, the process known as "hydrolysis" involves an enzyme and what else?

A. Splitting a molecule into two smaller molecules using a water molecule
B. Splitting a molecule into two smaller molecules using hydrogen
C. The metabolism of glucose to produce water and energy
D. Splitting of triglycerols into molecules that are soluble in water

Answer is A: Hydro- refers to water; -lysis refers to splitting of a molecule.

38. The process of hydrolysis may be represented by which of the following?

A. $H^+Cl^- + Na^+OH^- \rightarrow Na^+ + Cl^- + H_2O$
B. $C_{12}H_{22}O_{11} + H_2O \rightarrow 2C_6H_{12}O_6$
C. Alanine + glycine $\rightarrow$ alanylglycine + H_2O
D. Glucose $\rightarrow$ glycogenesis $\rightarrow$ glycogen

Answer is B: Hydrolysis refers to using a water molecule to split a larger molecule into smaller ones. Here a disaccharide is being split into two monosaccharide molecules.

39. Which digestive enzyme in pancreatic juice digests proteins?

A. Trypsin
B. Lipase
C. Pepsin
D. Amylase

Answer is A: While pepsin also digests protein, it is produced in the stomach, not the pancreas.

40. Which digestive enzyme in saliva breaks down starch?

A. Trypsin
B. Lipase
C. Pepsin

D. Amylase

Answer is D: Amylase is the salivary enzyme that begins digesting starch.

41. What are the products of carbohydrate digestion?
    A. Free bases and pentose sugars
    B. Monosaccharides
    C. Cellulose and disaccharides
    D. Free fatty acids and monoglycerides

Answer is B: Carbohydrates are polymers of small units called monosaccharides. Digestion (hydrolysis) frees the monosaccharides.

42. Consider a laboratory experiment that investigates the ability of an enzyme to hydrolyse protein at different pH levels. The solutions in the different test tubes are all at 37 °C and may contain distilled water, protein, enzyme and be at different pH. Which test tubes are likely to be the controls?
    A. The ones containing distilled water
    B. The test tubes with pepsin added
    C. The test tube with a pH of 7.0
    D. The ones without any added enzyme

Answer is D: Such an investigation should have one tube that contains the enzyme (e.g. pepsin) and another lacking the enzyme, with the tubes being identical in all other respects (i.e. both containing protein dissolved in distilled water at the same pH). In this case, the one(s) without the enzyme is the control.

43. In terms of the physiology of digestion, to what does the term "hydrolysis" refer?
    A. Use of hydrochloric acid to digest food molecules
    B. Effect that hydronium ions have on the pH of a solution
    C. Generation of hydrogen gas flatus in the large intestine
    D. Production of two molecules from one by a reaction with water

Answer is D: The "hydro-" part of hydrolysis refers to water, while the "-lysis" part refers to the splitting of a large molecule into two smaller parts. The water molecule also splits into H^+ and OH^- which are attached to the two smaller fragments.

44. Which of the following is one of the processes of mechanical digestion?
    A. Stomach churning
    B. Hydrolysis
    C. Secretion of gastric juice
    D. Enzyme activation

Answer is A: In the stomach, peristaltic waves (~3 per minute) mix the contents with gastric juice which facilitates chemical digestion. Secretion, while being mechanical, is not digestion.

45. What does lipase digest?
    A. Lipids
    B. Protein
    C. Nucleic acids
    D. Carbohydrates

Answer is A: The prefix "lip-" should tell you that lipids is the molecule that is digested by lipase.

46. What does pepsin hydrolyse?
    A. Polypeptides to dipeptides
    B. Carbohydrates to monosaccharides
    C. Protein to polypeptides
    D. Polypeptides to amino acids

Answer is C: Pepsin is secreted into the stomach to act on ingested proteins to hydrolyse them into polypeptides before they move into the duodenum.

47. What molecule does lactase hydrolyse?
    A. Maltose
    B. Glucose
    C. Lactose
    D. Fructose

Answer is C: Enzymes are specific and act only on their intended substrate. The prefix "lact-" indicates that the molecule acted on is lactose.

48. What molecule does maltase hydrolyse?
    A. Sucrose
    B. Galactose
    C. Glucose
    D. Maltose

Answer is D: Enzymes are specific and act only on their intended substrate. The prefix "malt-" indicates that the molecule acted on is maltose.

49. What does the enzyme glucoamylase hydrolyse?
    A. Nucleic acids
    B. Carbohydrates
    C. Protein
    D. Triglycerols

Answer is B: The terms "gluco-" and "amyl-" refer to carbohydrate, so carbohydrates are the target of enzymes (e.g. amylase) with these letters in their name.

50. What are the most favourable temperature and pH for optimal amylase functioning?
    A. 0 °C and pH = 4
    B. 37 °C and pH from 4 to 7

C. 37 °C and pH between 7 and 9
D. 0 °C and pH from 4 to 7

Answer is C: 37 °C, which is healthy human body temperature and pH = between 7 and 9 (from neutral to alkaline pH) because amylase is released into the mouth and small intestine where the pH is alkaline.

51. What will result from an experiment where amylase is added to one test tube containing a maltose solution and maltase to another?

    A. Amylase digests maltose.
    B. Maltase digests maltose.
    C. Amylase digests all carbohydrates.
    D. Maltase digests only glucose.

Answer is B: Maltose is digested by maltase—but not amylase. Maltase and amylase are enzymes. Enzymes are specific to their substrate, and they can only digest the molecule that they are designed for.

52. What enzyme digests carbohydrates in the mouth?

    A. Salivary amylase, which acts on the starch and hydrolyses it to maltose
    B. Salivary maltase, which acts on sugar and hydrolyses it to glucose
    C. Salivary hydrolase, which acts on carbohydrate and hydrolyses it to carbonates.
    D. Salivary lactase, which acts on lactose and hydrolyses it to glucose.

Answer is A: Salivary amylase, which acts on the starch in food and hydrolyses it to maltose. Pancreatic amylase and maltase continue to digest carbohydrates in the small intestine.

53. What is the effect of bile on safflower oil? **AND** what enzyme digests lipids? Select the correct answer:

    A. Bile changes pH of oil; then amylase digests lipids.
    B. Bile dissolves the oil; then maltase digests lipids.
    C. Bile emulsifies the oil; then lipase digests lipids.
    D. Bile digests the oil; then lipase digests bile.

Answer is C: Bile emulsifies the oil—i.e. bile molecules, having a hydrophilic end and a hydrophobic end, surround small droplets of oil and prevents them coalescing with other small droplets to form a big blob. Emulsification increases the surface area of the oil, so enzyme lipase can digest the lipid.

54. Where is pepsin produced? **AND** what is the effect of pepsin on food molecules?

    A. Pepsin is produced in the pancreas. Pepsin is an enzyme that digests (hydrolyses) lipids into smaller oil pieces.
    B. Pepsin is produced by the oesophagus. Pepsin is an enzyme that digests (hydrolyses) carbohydrates into smaller polypeptide pieces.

C. Pepsin is produced in liver. Pepsin is an enzyme that digests (hydrolyses) bile acids into smaller polycarbonate pieces.
D. Pepsin is produced by the stomach. Pepsin is an enzyme that digests (hydrolyses) proteins into smaller polypeptide molecules.

Answer is D: Pepsin is produced in cells of the stomach wall (as pepsinogen!). Pepsin is an enzyme that hydrolyses protein into smaller polypeptide molecules.

55. In which solution would protein be hydrolysed the most? (AND why is this)?
    A. One with neutral pH and containing pepsin and distilled water. This is due to pepsin working best in the neutral environment of the stomach. Water is present in the stomach.
    B. One with acidic pH and containing pepsin and HCl. This is due to pepsin working best in the acidic environment of the stomach. HCl is produced in the stomach.
    C. One with basic pH and containing water and sodium carbonate. This is due to pepsin working best in the basic environment of the stomach. Sodium is present in the stomach.
    D. One with basic pH and containing pepsin and sodium carbonate. This is due to pepsin working best in basic environment of the stomach. Sodium is present in the stomach.

Answer is B: The stomach is an acidic environment due to containing HCl. Pepsinogen which is produced by the "chief" cells in the stomach wall is activated to pepsin when it encounters the acidic environment within the stomach. HCl is produced by the parietal cells in the stomach wall.

56. Which of the following is NOT true?
    A. Most enzymes work best at a particular pH.
    B. Some enzymes can be inhibited irreversibly by specific molecules.
    C. Most enzymes can increase the rate of many different chemical reactions.
    D. Some enzymes require co-factors to function.

Answer is C: Enzymes are specific to one substrate, so they will increase the rate of one chemical reaction only.

9.3 Liver and Accessory Organs

1. Which of the following glands are accessory organs of the digestive system?
    A. Adrenal glands
    B. Pancreatic islets
    C. Gastric glands
    D. Salivary glands

Answer is D: Salivary glands produce a secretion (saliva) that empties into the digestive tract via a tube. (Pancreatic islets are in the pancreas, but are themselves not a gland.)

2. Which liver cells produce bile?
   A. Kupffer cells
   B. Sinusoids
   C. Hepatocytes
   D. The acini

Answer is C: Hepatocytes are liver cells. Kupffer cells are macrophages, while sinusoids are blood capillaries.

3. Which of the following is a function of the liver?
   A. Recycling of non-viable red blood cells
   B. Conversion of pyruvic acid to lactic acid
   C. Synthesis of plasma proteins
   D. Production of renin

Answer is C: The liver produces many proteins. (Spleen recycles RBC; liver converts lactic acid back into pyruvic acid; kidney produces renin.)

4. Name the major cell type in a liver lobule.
   A. Kupffer cells
   B. Hepatocytes
   C. Sinusoids
   D. Epithelial cells

Answer is B: Hepato- = liver. Kupffer cells also occur in the liver, but are macrophages.

5. What does the term "gluconeogenesis" refer to?
   A. The conversion of glycogen to glucose
   B. The removal of an amine group from an amino acid
   C. The production of glucose from non-carbohydrate molecules
   D. The conversion of disaccharides to monosaccharides

Answer is C: "-neogenesis" refers to making glucose from something new (that is not a carbohydrate).

6. Which of the following is TRUE of bile?
   A. It converts inactive pancreatic enzymes to active form.
   B. Needed in the small intestine for the digestion of fats.
   C. Synthesised by the gall bladder.
   D. Needed in the small intestine for the emulsification of fats.

Answer is D: Bile emulsifies (rather than digests) fat. It is stored (but not synthesised) in the gall bladder.

7. Why are the blood capillaries in the liver lobules so permeable?
   A. To allow the products of digestion to leave the blood for processing in the liver
   B. To allow fatty acids to leave the liver cells to enter the blood
   C. To allow plasma proteins that are synthesised in the liver to enter the blood
   D. To allow red blood cells at the end of their life to leave the blood to be recycled in the liver

Answer is C: Plasma proteins are large molecules that otherwise would not be able to enter (or leave) blood capillaries.

8. Which one of these processes is **NOT** part of carbohydrate metabolism in the liver?
   A. Production of ATP from glucose
   B. Production of glucose from glycogen
   C. Production of glucose from amino acids
   D. Production of glycogen from glucose

Answer is A: ATP production occurs in the mitochondria of cells and is called cellular respiration (rather than carbohydrate metabolism).

9. The liver contains "leaky capillaries" known as sinusoids. This enables what liver product to enter the blood stream?
   A. Angiotensinogen
   B. Kupffer cells
   C. Plasma proteins
   D. Cholesterol

Answer is C: Plasma proteins are large molecules that otherwise would not be able to enter (or leave) blood capillaries.

10. What is the function of bile salts?
    A. To assist the absorption of digested lipids
    B. To emulsify lipids
    C. To hydrolyse lipids
    D. To digest lipids

Answer is B: Emulsification produces small droplets of fat which increases the surface area of fat that is available to lipases which may then hydrolyse the fat.

11. What does the term "gluconeogenesis" refer to?
    A. The conversion of non-carbohydrate molecules to glucose
    B. The formation of non-essential amino acids from a keto-acid
    C. The removal of an amine group from a molecule
    D. The release of glucose from stored glycogen

Answer is A: "-neogenesis" refers to making something (glucose) from new (a molecule that is not a carbohydrate).

12. A lobule of the liver contains several blood vessels. Which one carries nutrient-rich blood from the small intestine?
    A. Hepatic artery proper
    B. Hepatic portal vein
    C. Central vein
    D. Bile ductule

Answer is B: "Portal" refers to a vein that transports blood from one capillary bed to another (rather than returning it to the heart).

13. What is the function of bile?
    A. Bile hydrolyses polypeptides.
    B. Bile emulsifies fats and oils.
    C. Bile activates procarboxypeptidase.
    D. Bile stimulates the pancreas to secrete pancreatic juice.

Answer is B: Emulsification of fats is the function of bile salts.

14. The liver is able to deaminate amino acids forming ammonia in the process. What happens to the ammonia?
    A. It is phagocytosed by Kupffer cells.
    B. It is used in transamination to form non-essential amino acids.
    C. It is converted to bile to be excreted via the gut.
    D. It is converted to urea for excretion by the kidneys.

Answer is D: Urea is the molecule and vehicle for excretion of human nitrogenous waste.

15. If blood glucose is high, what does the liver do about it?
    A. The liver converts glucose to glycogen or triglycerides.
    B. The liver performs glycogenolysis.
    C. The liver performs gluconeogenesis.
    D. The liver transaminates glucose to produce amino acids.

Answer is A: This process removes glucose from circulation. Choices B and C would increase blood glucose. Transamination is done to amino acids to produce different amino acids.

16. Which one of the following is **NOT** a function of the liver?
    A. Recycling of red blood cells
    B. Storage of fat soluble vitamins
    C. Removal and recycling of lactic acid
    D. Activation of vitamin D

Answer is A: Red blood cells are recycled in the spleen.

17. Why are sinusoids the type of capillaries found within a liver lobule?
    A. To allow for mixing of blood from the hepatic artery and the hepatic portal vein.
    B. So that liver synthesised plasma proteins may enter the blood
    C. To allow worn out red blood cells to leave the blood stream.
    D. In order for the products of digestion to be removed from the blood.

Answer is B: The epithelial cells that make up sinusoids have gaps between them large enough for plasma proteins to fit through and to enter the blood.

18. What is the function of bile salts?
    A. To digest dietary fats through hydrolysis
    B. To excrete the products haeme breakdown
    C. To emulsify ingested fats and oils
    D. To activate trypsinogen and chymotrypsinogen

Answer is C: Emulsification is the function of bile salts (not hydrolysis). Some materials are excreted via the gut by incorporating the waste into bile, but not as bile salts.

19. Which of the functions below is **NOT** performed by the liver?
    A. Production of glucagon
    B. Synthesis of lipoproteins to transport fatty acids
    C. Deamination of amino acids to form keto acids
    D. Conversion of non-carbohydrate molecules to glucose

Answer is A: Glucagon is a hormone that is produced by the pancreas.

20. What are the cells in the pancreas that secrete "pancreatic juice" called?
    A. Hepatocytes
    B. Peyer's patches
    C. The acini
    D. Islets of Langerhans

Answer is C: Hepatocytes are in the liver; Peyer's patches are lymphoid tissue in the gut wall; the islets produce insulin and glucagon.

21. Which of the following is a function of bile?
    A. To attach to vitamin B_{12} to allow it to be absorbed
    B. To activate trypsinogen, chymotrypsinogen and procarboxypeptidase
    C. To digest fats
    D. To disperse large lipid globules into smaller droplets

Answer is D: This is known as emulsification.

22. What is the process that splits carbohydrates in the gut into smaller molecules called?
    A. Catalysis
    B. Hydrolysis

C. Catabolism
D. Glycogenolysis

Answer is B: Glycogenolysis also splits a carbohydrate (glycogen) into smaller molecules (glucose), but this happens in the liver.

23. Kupffer cells are macrophages. Where are they found?
   A. In the lymphatics of the submucosa and devour bacteria that escape the gut
   B. In the lumen of the large intestine and feed on our normal flora to produce vitamin K
   C. They are in the stomach wall as part of the mucosal barrier
   D. They occur in liver sinusoids and engulf bacteria in blood coming from the gut

Answer is D: They are in the liver sinusoids (capillaries).

24. Which one of the following statements is **UNTRUE**?
   A. Glucose is produced during the manufacture of ATP.
   B. Glycogenolysis is the process of releasing glucose from glycogen.
   C. Gluconeogenesis is the conversion of amino acids to glucose.
   D. The monosaccharides galactose and fructose can be converted to glucose.

Answer is A: In fact the reverse is true. Glucose is consumed during the production of ATP.

25. Which of the following structures produce bile?
   A. The gall bladder
   B. The liver
   C. The pancreas
   D. The duodenum

Answer is B: Bile is a liver product that is stored in the gall bladder.

26. Which one of the following is a function of the liver?
   A. Lipase
   B. Digestive enzymes
   C. Insulin
   D. Plasma proteins

Answer is D: Plasma proteins (albumins, globulins and fibrinogen) are produced in the liver.

27. Which of the following organs is an accessory organ rather than an organ of the gastrointestinal tract?
   A. Duodenum
   B. Rectum
   C. Caecum

D. Liver

Answer is D: The liver is not part of the digestive tract (chyme does not pass through it), rather it contributes bile to the gut contents via the common bile duct.

28. Which of the following pancreatic juice enzyme aids in the digestion of proteins?
    A. Amylase
    B. Lipase
    C. Nuclease
    D. Trypsin

Answer is D: Amylase digests carbohydrate; lipase digests lipids; nuclease digests nucleic acids.

29. What role do the Kupffer cells of the liver perform?
    A. They are sinusoids.
    B. They are hepatocytes.
    C. They are macrophages.
    D. They de-aminate amino acids.

Answer is C: Kupffer cells engulf any bacteria that have travelled from the gut in the blood.

30. Which one of the following is **NOT** secreted in pancreatic juice?
    A. Amylase
    B. Trypsinogen
    C. Pepsinogen
    D. Lipase

Answer is C: Pepsinogen is secreted by the Chief cells (zymogenic cells) in the gastric pits of the stomach.

31. All of the following statements regarding the liver are true except one, which one?
    A. It can convert amino acids to glucose during periods of fasting.
    B. Blood from the hepatic artery and portal vein travels away from the central vein of each lobule.
    C. It contains special phagocytic cells which remove worn-out blood cells from the circulation.
    D. It converts ammonia to urea.

Answer is B: It is not true. In fact blood from the hepatic artery and the hepatic portal vein mix and flow towards the central vein of a lobule.

32. Which statement best describes the process of glycogenesis?
    A. The digestion of glycogen in the diet
    B. The conversion of fat into glycogen in muscle tissue

C. The conversion of glucose into glycogen in the liver
D. The conversion of glycogen into glucose in muscle tissue

Answer is C: The term glycogenesis refers to the making (-genesis) of glycogen (glyco-) from glucose. This removes glucose from the blood circulation.

33. One of the functions of the liver is to produce:
    A. Blood cells
    B. Digestive enzymes
    C. Insulin and glucagon
    D. Glycogen from glucose

Answer is D: Blood cells are produced in bone marrow; digestive enzymes are produced in the pancreas and gut wall; the hormones insulin and glucagon are produced by the pancreas.

34. Which of the following organs is an accessory organ of the gastrointestinal tract?
    A. Jejunum
    B. Appendix
    C. Caecum
    D. Pancreas

Answer is D: The other three are all part of the digestive tract itself.

35. What role do the Kupffer cells of the liver perform?
    A. They transport plasma proteins.
    B. They perform gluconeogenesis.
    C. They are macrophages.
    D. They produce bile.

Answer is C: Kupffer cells are macrophages that engulf any bacteria in the sinusoids of the liver lobules transported there by the hepatic portal blood flow that may have originated from the gut.

36. Where is angiotensinogen produced?
    A. Adrenal glands
    B. Liver
    C. Kidney
    D. Endothelial cells

Answer is B: The liver produces many proteins including this precursor to angiotensin II.

37. What is unusual about the blood supply to the liver?
    A. The pressure in the liver capillaries is higher (about 55 mmHg) than is usual for other capillaries.
    B. Arterial and venous blood mix in the sinusoid capillaries.

C. Liver capillary beds are drained by an arteriole not a venule.
D. The osmolarity of blood in the liver lobules can reach 1200 mOsm/L.

Answer is B: Blood from the hepatic artery and the hepatic portal vein mix before returning to the heart via the hepatic vein. Choice A is true of the glomerular capillaries of the kidney while choice D occurs in the vasa recta of the kidney medulla.

38. Which of these processes occurs in the liver?
    A. Deamination: the removal of an amine group from a molecule
    B. The production of pepsinogen
    C. The production and storage of insulin
    D. The production and release of secretin, cholecystokinin and vasoactive peptide

Answer is A: Deamination, and making the non-essential amino acids, occurs in the liver. Pepsinogen is produced in the stomach lining; insulin in the pancreas; choice D in the intestine.

39. Which of the following describes the basic anatomy of the liver?
    A. Contains clusters of secretory cells called acini and others call islets of Langerhans. Both secrete into a common duct.
    B. The functional unit is the hexagonal lobule consisting of hepatocytes that are arranged in lines which radiate from a central vein. Between lines of hepatocytes are blood-filled sinusoids.
    C. Made up of four layers (tunics): the deepest is the mucosa, then the submucosa, the muscularis externa and the serosa.
    D. Consists of different cells called parietal cells (which secrete $H^{+}Cl^{-}$ and intrinsic factor), zymogenic cells which secrete pepsinogen, enteroendocrine G cells (secrete hormones) and hepatocytes which secrete bile.

Answer is B: Liver lobules are six-sided structures of hepatocytes with a "triad" of vessels at each corner. The hepatocytes are supplied with blood by "leaky" capillaries called sinusoids. Choice A describes the pancreas; C the wall of the gut and D the glands in the wall of the stomach.

40. Which of the following describes the basic anatomy of the pancreas?
    A. Consists of different cells called parietal cells (which secrete $H^{+}Cl^{-}$ and intrinsic factor), zymogenic cells which secrete pepsinogen, enteroendocrine G cells (secrete hormones), and hepatocytes which secrete bile.
    B. The functional unit is the hexagonal lobule consisting of hepatocytes that are arranged in lines which radiate from a central vein. Between lines of hepatocytes are blood-filled sinusoids.
    C. Made up of four layers (tunics): the deepest is the mucosa; then the submucosa, the muscularis externa and the serosa.
    D. Contains clusters of secretory cells called acini and others call islets of Langerhans. Both secrete into a common duct.

Answer is D: The pancreas is both exocrine (the acini secrete enzymes) and endocrine (the islets secrete insulin and glucagon). Choice A describes the glands in the wall of the stomach; B the liver; C the wall of the gut.

41. Which of the following are the major salivary glands?
    A. The sublingual glands; the subtonsal glands; the submandibular glands
    B. The parotid glands; the sublingual glands; the palatine glands
    C. The submandibular glands; the parotid glands
    D. The submandibular glands; the parotid glands; the sublingual glands

Answer is D: There are three pairs of glands (total six). The "palatine glands" and "subtonsal glands" do not exist.

9.4 Nutrition

1. In what form are the majority of dietary lipids ingested?
    A. Trisaccharides
    B. Tripeptides
    C. Cholesterol
    D. Triglycerides

Answer is D: Triglycerides (or triacylglycerols) are the major component of human dietary lipids.

2. Given the data in the table below for healthy % body fat range, what is the desirable (normal or healthy) % of body fat for average woman of age 22 years and an average man of 45 years?

Age	Female	Male
20–39	22–32%	9–20%
40–59	24–34%	12–22%
60–79	26–37%	14–24%

    A. Between 22% and 32% for the woman and 12–22% for the man
    B. Between 22% and 32% for the woman and 9–20% for the man
    C. Between 24% and 34% for the woman and 12–22% for the man
    D. Between 22% and 26% for the woman and 14–24% for the man

Answer is A: The 22-year-old woman falls into the 20–39 year age category whose normal % of body fat is between 22% and 32%. For the 45-year-old male, his healthy fat % is 12–22%. Women with over 35% body fat and men with over 25% body fat are considered obese.

3. Consider two men: one of 70 kg and 15% body fat (BF) and the other of 100 kg and 12% BF. Which one is leaner? AND which one has the greater mass of body fat?
   A. The 70 kg chap is leaner. The 100 kg chap has a less mass of fat.
   B. The 70 kg chap is leaner. The 100 kg chap has a greater mass of fat.
   C. The 100 kg chap is leaner. The 100 kg chap has a greater mass of fat.
   D. The 100 kg chap is leaner. The 100 kg chap has a less mass of fat.

Answer is C: The 100 kg chap is leaner (has a lower % body fat). The 100 kg chap has a greater mass of fat (12% of 100 kg = 12 kg) than the little bloke (who has 15% of 70 kg = 10.5 kg of fat).

4. If the body is unable to catabolise carbohydrates, what is it most likely to catabolise?
   A. Proteins
   B. Nucleotides
   C. Lipids
   D. Nucleic acids

Answer is C: If carbohydrates (glucose) are not available, then fat (lipids) will be catabolised. After that, protein is used for energy.

5. What happens during glycolysis?
   A. A molecule of glucose is converted into two molecules of pyruvic acid.
   B. Eighteen molecules of ATP are produced.
   C. Carbon dioxide is produced.
   D. More energy is consumed than is released.

Answer is A: Glycolysis involves the cleaving of a glucose molecule (with six C atoms) into two pyruvic acid molecules (pyruvate) each with three C atoms. This produces a net gain of two ATP molecules, but does not produce carbon dioxide.

6. In what form is most of the fat in the human body?
   A. Steroids
   B. Phospholipids
   C. Triglycerides
   D. Prostaglandins

Answer is C: All four choices are forms of fat. Triglycerides are eaten, digested and absorbed as fatty acids and monoglycerols. After being absorbed from the gut, they are reconstituted into triglycerides for transport via the blood stream.

7. Which of the following foods when catabolised in the body produces the highest yield of energy?
   A. 100 g of fat
   B. 100 g of chocolate
   C. 100 g of carbohydrate
   D. 100 g of protein

Answer is A: Fat is higher in energy (kilojoules) than either protein or carbohydrate. Chocolate has sugar (carbohydrate) as well as fat as an ingredient.

8. Consider a person whose diet provides a daily energy intake of 10,500 kJ and whose daily activities result in their body consuming 9500 kJ of energy daily. If this situation continues for several months, the possible outcome would be:
   A. A gradual rise in basal metabolic rate.
   B. The person increases the mass by several kilograms of fat tissue.
   C. The energy requirements of daily activities will rise to consume 10,500 kJ.
   D. An increase in radiation from the body to dissipate excess energy.

Answer is B: If energy intake is greater than energy expenditure, the person will store the excess energy as fat, so will get fatter.

9. What does it mean when the label on a packaged food item states that 100 g contains 650 kJ of energy? It means that:
   A. The human body is able to extract 650 kJ of heat and useful work by digesting the food.
   B. 650 kJ of energy was consumed in growing or producing or making the food.
   C. 650 kJ of heat energy is released when 100 g of the food is burned in an atmosphere of pure oxygen.
   D. Ultimately your body will be able to perform 650 kJ of work for every 100 g of the food that you eat.

Answer is C: Food energy values are determined by burning a sample of the food in a calorimeter and measuring the heat produced. If the body was 100% efficient, choice A would be true.

10. What is the term applied to malnutrition caused by a lack of protein in a child's diet?
    A. Scurvy
    B. Marasmus
    C. Proteinuria
    D. Kwashiorkor

Answer is D: Kwashiorkor is a disease caused by the lack of protein in a child's diet. Marasmus results from an inadequate intake of protein and energy. Scurvy is caused by a lack of vitamin C. Proteinuria refers to protein being present in urine.

11. Which "B vitamin" may be deficient in vegan diet?
    A. Thiamine (B_1)
    B. Riboflavin (B_2)
    C. Niacin (B_3)
    D. B_{12}

Answer is D: Vitamin B_{12} is found only in animal products, so a vegan diet that is not artificially fortified with B_{12} will be deficient. A lack of B_{12} leads to megaloblastic anaemia.

Chapter 10
Endocrine System

The organs of the endocrine system produce hormones. A hormone is a messenger molecule, made in small quantities by endocrine cells/glands and released into the blood circulation to coordinate cellular activities in distant tissues. Hormone molecules elicit a response in the target cells by attaching to a receptor protein on the cell's plasma membrane (or inside the cell) that is specific for that molecule.

The hypothalamus is a region of the brain that controls the endocrine system and integrates the activities of the nervous and endocrine systems in three ways. (1) The hypothalamus produces and secretes regulatory hormones that control the secretion of six or seven hormones from the endocrine cells in the anterior pituitary gland. In turn, the hormones released by the anterior pituitary stimulate their target endocrine organs to produce and release hormones which go on to influence the behaviour of the body cells which have receptors for these hormones. (2) The hypothalamus produces ADH and oxytocin which are transported to the posterior pituitary to be stored and released into the blood following appropriate stimuli. (3) The hypothalamus contains "autonomic centres" that exert neural control over endocrine cells of the medullae of the adrenal glands. When activated, the adrenal medullae release hormones into the bloodstream.

Many organs participate in the production of hormones and you should know (or know of) most of them. The hypothalamus and the anterior pituitary have already been mentioned. The thyroid gland produces thyroxine and calcitonin. The latter decreases the concentration of Ca ions in the blood. The four parathyroid glands produce parathyroid hormone (PTH) which increases the concentration of Ca ions in the blood. The adrenal glands produce many hormones. More than 24 corticosteroid hormones are produced from cholesterol by the adrenal cortex. These include "mineralocorticoids" (e.g. aldosterone), "glucocorticoids" (e.g. cortisol, corticosterone) and "gonadocorticoids" (e.g. estrodiol and the weak androgens androstenedione, dehydroepiandrosterone). The adrenal medullae produce epinephrine and norepinephrine. The pancreas contains regions of cells (called islets of Langerhans) that produce insulin (the beta cells) and glucagon (the alpha cells). The testes produce testosterone and inhibin while the ovaries produce oestrogens, inhibin and

M. Caon, *Examination Questions and Answers in Basic Anatomy and Physiology*, https://doi.org/10.1007/978-3-030-47314-3_10

progesterone. The pineal gland produces melatonin. The kidneys produce erythropoietin, calcitriol ("vitamin D_3") and the enzyme renin. The liver (not regarded as an endocrine organ) produces the prohormone angiotensinogen which is converted by renin and then by "angiotensin converting enzyme" to the hormone angiotensin II. The atria of the heart produce atrial natriuretic hormone (ANP) which blocks the secretion of renin.

10.1 Endocrine System and Hormones in General

1. Which statement below about hormones is true?
   A. Hormones are enzymes that catalyse reactions.
   B. Hormones are released into the blood circulation.
   C. Hormones affect all cells of the body.
   D. Hormones are released by neurones at synapses.

Answer is B: Hormones are circulating messengers that are transported in blood. A particular hormone does not necessarily affect every cell in the body. Some (but not all) hormones are released by neurones, for example in the hypothalamus.

2. Which hormones are soluble in blood?
   A. Steroid hormones
   B. Hormones produced by the adrenal cortex
   C. The sex hormones
   D. Those released by the pituitary gland

Answer is D: The pituitary gland releases peptide hormones which are soluble in blood. The other choices refer to steroid hormones which are insoluble in blood. They require transport via plasma proteins.

3. Which one of the following is **NOT** part of the endocrine system?
   A. The islets of Langerhans (pancreatic islets)
   B. The thyroid gland
   C. The acini cells of the pancreas
   D. The parathyroid glands

Answer is C: The acini cells produce digestive enzymes.

4. What is the difference between an exocrine gland and an endocrine gland?
   A. An endocrine gland secretes neurotransmitters (an exocrine gland does not).
   B. An endocrine gland secretes via a tube to the destination (an exocrine gland does not).
   C. An exocrine gland secretes into the blood (an endocrine gland does not).
   D. An endocrine gland secretes into the blood (an exocrine gland does not).

Answer is D: Endocrine glands secrete "circulating" hormones that secrete into the blood.

5. By what term are hormones derived from tyrosine also known?
   A. Amino acid derivatives
   B. Peptide hormones
   C. Steroid hormones
   D. Corticosteroids

Answer is A: Catecholamines (adrenaline, noradrenaline), dopamine and thyroid hormone are all derived from tyrosine.

6. Which hormones have their receptors inside their target cell?
   A. Amino acid-based hormones
   B. Hormones with a membrane carrier mechanism or that are lipid soluble
   C. Steroid hormones and peptide hormones of less than 50 amino acids
   D. Lipid-soluble hormones

Answer is B: Choice D is also correct but B is the better answer as thyroid hormones cross membrane by a carrier mechanism.

7. What effect does aldosterone have?
   A. It causes glucose to be absorbed from the blood.
   B. It causes Na^+ to be absorbed in the kidneys.
   C. It causes Ca^{2+} to be absorbed from the gut.
   D. It causes K^+ to be absorbed from the filtrate.

Answer is B: Aldosterone causes reclamation of sodium ions from the filtrate and potassium to be secreted in exchange.

8. Which of the following is a part of the endocrine system?
   A. The thalamus
   B. The pancreatic islets (islets of Langerhans)
   C. The renal glands
   D. The salivary glands

Answer is B: The pancreatic islets secrete the hormones insulin and glucagon.

9. Which of the following is an amino acid derivative hormone?
   A. Epinephrine
   B. Tyrosine
   C. Testosterone
   D. Prostaglandin

Answer is A: Epinephrine is derived from the amino acid tyrosine (so tyrosine is a wrong answer).

10. Which statement below is true of steroid hormones?
    A. They do not have a specific receptor to bind with.
    B. They are not lipid soluble so bind to receptor proteins on the cell membrane.
    C. They are lipid soluble so diffuse through the cell membrane.
    D. They cross the cell membrane via a carrier mechanism.

Answer is C: Steroids are lipid soluble so diffuse through the plasma membrane to bind to a receptor within the cell.

11. What effect does aldosterone have? It causes:
    A. Angiotensin to be formed from angiotensinogen
    B. Na^+ to be absorbed from the filtrate
    C. Na^+ and Ca^{2+} to be absorbed from the filtrate and K^+ to be secreted into the filtrate
    D. Na^+ to be absorbed from the filtrate and K^+ to be secreted into the filtrate

Answer is D: Aldosterone promotes the absorption of sodium from the filtrate while potassium (also a positively charged ion) is secreted to maintain electrical neutrality.

12. Which of the following is **NOT** part of the endocrine system?
    A. The thymus
    B. The pineal gland
    C. The acini cells of the pancreas
    D. The posterior pituitary gland

Answer is C: The acini cells produce digestive enzymes.

13. Which statement applies to steroid hormones?
    A. They are transported and dissolved in blood.
    B. They bind to receptor proteins on the outside of the plasma membrane.
    C. They cross the plasma membrane by using a protein carrier mechanism.
    D. They bind to receptors in the cell cytoplasm or nucleus.

Answer is D: As steroid hormones are lipid soluble, their receptors are inside the cell as they can cross the plasma membrane.

14. What is one mechanism of hormone action?
    A. They act as second messengers in the cytoplasm.
    B. They act as enzymes for reactions.
    C. They act as receptor proteins.
    D. They activate genes in the nucleus.

Answer is D: Some hormones (corticosteroids) determine which genes are transcribed in the nucleus.

15. What is the effect of ADH (antidiuretic hormone)?
    A. Allows walls of collecting duct to become permeable to water
    B. Inhibits the reabsorption of Na^+
    C. Causes an increase in the volume of urine produced
    D. Promotes diuresis

Answer is A: ADH causes more aquaporins to be inserted into the collecting duct walls which allow water molecules to pass through along the osmotic gradient.

16. Which one of the following is **NOT** true of peptide hormones?
    A. They are water soluble.
    B. They are derived from amino acids.
    C. Their receptors are located in the cell cytoplasm.
    D. They are transported and dissolved in blood.

Answer is C: The receptors for peptide hormones are located on the plasma membrane as they are unable to penetrate the plasma membrane.

17. How do steroid hormones differ from amino acid-based hormones?
    A. Steroid hormones are water soluble, whereas amino acid-based hormones are not.
    B. The receptors for steroid hormones are only in the cytoplasm or the nucleus (and amino acid-based hormones are not).
    C. Steroid hormones are made only in the adrenal glands, while amino acid-based hormones are produced by a variety of glands.
    D. Steroid hormones activate a G-protein and exert their effect via "second messengers", but the action of amino acid-based hormones results directly on binding to their receptor.

Answer is B: Steroid hormones are lipid soluble so can enter the cell to bind to receptors located inside the cell. Amino acid-based hormones are not lipid soluble, so their receptors are located outside the cell on the plasma membrane.

18. Which of the following organ(s) are **NOT** endocrine organs?
    A. Renal
    B. Adrenal
    C. Thyroid
    D. Parathyroid

Answer is A: Renal gland is another name for the kidney.

19. To which group of hormones does aldosterone belong?
    A. Catecholamines
    B. Glucocorticoids
    C. Mineralocorticoids
    D. Gonadocorticoids

Answer is C: As aldosterone is concerned with the absorption of sodium ions, a "mineral" and it is produced by the adrenal **cort**ex, it is termed a mineralocorticoid.

20. Where are the receptors for almost all of the amino acid-derived hormones located?
    A. On the mitochondria
    B. In the nucleus
    C. On the outside of the plasma membrane
    D. On the inside of the plasma membrane

Answer is C: Amino acid-derived hormones cannot get through the plasma membrane, so they attach to receptors located on the outside of the membrane.

21. Which of the following hormones **CANNOT** cross the plasma membrane?
    A. Sex hormones
    B. Amino acid-based hormones
    C. Thyroid hormones
    D. Steroids

Answer is B: Amino acid-based hormones are water soluble but not lipid soluble, so cannot cross the plasma membrane.

22. Which of the following is a substantial difference between amino acid-based hormones and steroid hormones?
    A. Endocrine glands release steroid hormones while amino acid hormones are released from exocrine glands.
    B. Amino acid hormones are circulating hormones while steroid hormones are local hormones.
    C. Amino acid hormones are fat soluble while steroid hormones are not.
    D. Steroid hormones can pass through the plasma membrane while amino acid hormones cannot.

Answer is D: Steroid hormones are fat soluble, AA-based hormones are water soluble (and not fat soluble).

23. What type of molecule is cAMP, or what role does it play?
    A. A second messenger
    B. An amino acid-based hormone
    C. A catecholamine
    D. A steroid hormone

Answer is A: Cyclic AMP is a second messenger. That is, it is released from the inside of the plasma membrane when an AA-based hormone binds to its receptor.

24. Which one is **NOT** a mode of action of hormones on their target cell?
    A. A hormone may stimulate the synthesis of an enzyme in the target cell.
    B. A hormone may activate an enzyme by altering its shape.

C. A hormone may deactivate an enzyme by altering its structure.
D. Some hormones are enzymes that promote a chemical reaction in a cell.

Answer is D: Hormones are not enzymes (but may cause enzymes to be formed).

25. Which of the following could be a definition of a hormone?
    A. Chemicals released to communicate between adjacent cells in contact.
    B. A chemical messenger released into blood to coordinate activities in distant tissues.
    C. A chemical messenger released by a neurone at a synapse.
    D. A chemical messenger in the extracellular fluid between cells of a single tissue.

Answer is B: Hormones are also known as circulating hormones, meaning they are transported in the blood.

26. Which one of the following groups of hormones has their receptor inside the cell?
    A. Steroid hormones
    B. Catecholamines
    C. Adrenalin and noradrenalin
    D. Peptide hormones

Answer is A: Steroids are lipids, hence are lipid soluble so can enter the cell.

27. Which of the following statements about endocrine hormones is always true?
    A. They are secreted by neurones.
    B. They are derived from amino acids.
    C. They are produced by exocrine glands.
    D. They are released into the bloodstream.

Answer is D: Endocrine hormones are also known as circulating (in blood) hormones.

28. Which of the following statements about corticosteroids is true?
    A. They may also act as neurotransmitters.
    B. They are transported dissolved in blood.
    C. They are produced by the adrenal gland.
    D. They are amino acid derivatives.

Answer is C: They are produced by the adrenal cortex.

29. Finish the sentence so that it is correct: A hormone may be defined as a chemical messenger that:
    A. Is released into circulation in small quantities
    B. Facilitates communication between adjacent cells
    C. Moves through extracellular fluid between cells of a single tissue

D. Crosses the synaptic cleft and binds to a receptor

Answer is A: Hormones travel through the blood to all parts of the body, and only small quantities are required.

30. Endocrine communication involves hormones of two types:
    A. First messengers and second messengers
    B. Steroid hormones and amino acid-based hormones
    C. Amino acid derivatives and peptide hormones
    D. Peptide hormones and corticosteroids

Answer is B: Hormones are of two structural types, and this determines whether they are lipid soluble (can pass through the plasma membrane) or not.

31. Which of the following statements is **FALSE**?
    A. Peptide hormones are not able to penetrate the cell membrane.
    B. Thyroid hormone can cross the membrane.
    C. Steroid hormones bind to receptors on the outside of the cell membrane.
    D. Catecholamines are not lipid soluble.

Answer is C: Steroid hormones are lipid soluble, so their receptors are inside the cell as they can pass through the plasma membrane.

32. An amino acid-based hormone binds to its receptor. This has the effect of:
    A. Activating an enzyme to produce cAMP
    B. Causing it to diffuse through the cell to trigger a cascade of reactions
    C. Activate a G-protein
    D. Allowing it to move along the membrane

Answer is C: After a hormone binds to its receptor, a G (for guanine) protein is activated as the first effect. The G-protein then moves along the membrane to activate an enzyme that produces—for example—cAMP, which is a "second messenger".

33. Why are the receptors for amino acid-based hormones on the outside surface of the cell membrane?
    A. Cells can respond faster when the hormone does not need to enter the cell.
    B. Amino acid-based hormones cannot penetrate cell membranes.
    C. Lysosomes in the intracellular fluid digest amino acid-based hormones.
    D. Amino acid-based hormones have no role in activating genes in the nucleus.

Answer is B: Amino acid-based hormones are not lipid soluble.

34. What is the difference between endocrine glands and exocrine glands?
    A. Endocrine glands produce hormones, whereas exocrine glands do not.
    B. Exocrine glands secrete into the blood stream, whereas endocrine glands do not.

C. Endocrine glands are controlled by the autonomic nervous system, whereas exocrine glands are not.
D. Exocrine glands secrete steroid hormones, whereas endocrine glands secrete amino acid-based hormones.

Answer is A: By definition endocrine glands are part of the endocrine system and they produce hormones.

35. Which statement is **NOT CORRECT** about hormones:

A. They are chemical substances that alter cell activity.
B. They regulate metabolic function of other cells in the body.
C. Steroid hormones are amino acid-based and are synthesised from cholesterol.
D. They are produced in glands and transported via the blood stream.

Answer is C: Steroid hormones are not based on amino acids. They are lipids.

36. The hormones known as "catecholamines" (adrenaline, noradrenaline and dopamine) are not lipid soluble. Therefore their receptor sites are:

A. On the inside of the plasma membrane
B. On the outside of the plasma membrane
C. In the cell cytoplasm
D. In the cell nucleus

Answer is B: With the receptors on the outside of the membrane, catecholamines do not need to be able to penetrate the membrane.

37. Which of the following is **NOT** a mechanism of hormone action?

A. A hormone may activate an enzyme.
B. A hormone may bind to a protein to alter the permeability of the plasma membrane.
C. A hormone may bind to troponin to expose the binding site the myosin cross-bridge.
D. A hormone may enter the cell to stimulate cell mitosis.

Answer is C: It is Ca ions (not a hormone) that bind to troponin to initiate a muscle cell contraction. The other three choices are true.

38. Hormones may produce their action by several mechanisms. Which of the following is one such mechanism?

A. By stimulating the synthesis of a structural protein in the cell cytoplasm
B. By detecting the concentration level of a substance in the blood
C. By causing a nerve impulse to be sent to the brain
D. By stimulating a reflex muscle contraction

Answer is A: Steroid hormones may diffuse into the cell to bind to receptors that activate a gene to synthesise a specific protein.

39. Which of the following statements is true?
    A. Luteinising hormone is produced by the luteal gland.
    B. Pituitary hormone is produced by the pituitary gland.
    C. Adrenal hormone is produced by the adrenal gland.
    D. Parathyroid hormone is produced by the parathyroid gland.

Answer is D: There is no luteal gland. The other three are all endocrine glands and produce several hormones. Only the parathyroid produces a single hormone that is named after the gland.

40. Which type of chemical does the definition: "a messenger made in small quantities and released into circulation" refer to?
    A. A plasma protein
    B. A neurotransmitter
    C. An enzyme
    D. A hormone

Answer is D: A hormone is a chemical messenger that is transported in blood (the circulation).

41. What is a collective name for hormones that are derived from the amino acid tyrosine?
    A. Corticoids
    B. Catecholamines
    C. Tropic hormones
    D. Androgens

Answer is D: The "catechol" group is a benzene ring with two hydroxyl groups attached. The amine is NH_2. They are made from tyrosine by removing the carboxylic acid group and adding a second hydroxyl group.

42. Which one of the following statements is correct?
    A. Catecholamines have their receptors inside the cell.
    B. Thyroid hormones have their receptors on the outside of the plasma membrane.
    C. Steroid hormones have their receptors on the outside of the plasma membrane.
    D. Peptide hormones have their receptors on the outside of the plasma membrane.

Answer is D: Peptide hormones are not soluble in lipid so cannot diffuse through the plasma membrane, hence their receptors are necessarily on its outside. Similarly, catecholamines are not soluble in lipid, hence their receptors are on the outside the plasma membrane.

43. Which organ produces the most testosterone?
    A. Ovaries
    B. Testes

C. Adrenal glands
D. Pituitary gland

Answer is B: The testes produce testosterone, but the ovaries produce a small amount too.

44. Which organ produces the most oestrogens?
    A. The placenta
    B. The testes
    C. The corpus luteum
    D. The ovaries

Answer is D: The ovaries, specifically, the cells surrounding the ovum produce oestrogens. The corpus luteum produces a smaller amount.

45. What are hormones?
    A. Steroid messenger molecules that bind to receptors inside the target cell.
    B. Chemicals released from a pre-synaptic membrane that bind to a receptor on the receptor cell.
    C. Messenger molecules produced and released from a gland that bind to a receptor on a cell.
    D. Blood-borne chemicals that transport insoluble molecules in blood.

Answer is C: Hormones are chemical messenger molecules released from endocrine glands when the appropriate stimulus is present. Choice A describes one type of hormone (the others are amino acid based); Choice B describes a neurotransmitter; D describes plasma proteins.

46. What terms can be used to classify the structure of hormones? Either:
    A. Nucleic acids or phospholipids
    B. Steroids or amino acid derivatives
    C. Proteins or carboxylic acids
    D. Carbohydrates or fatty acids

Answer is B: Hormones are either steroids (and have their receptors inside cells) or amino acid derivatives and, except for thyroid hormone, have their receptors on the cell membrane.

47. If the receptor for a particular hormone is located on the plasma membrane of the cell, what may be said of the hormone?
    A. The hormone is a steroid hormone.
    B. The hormone is a peptide hormone.
    C. The hormone is unable to cross the cell membrane.
    D. The hormone is an amino acid derived hormone.

Answer is C: Steroid hormones, being fat-soluble, are able to penetrate the plasma membrane and so have their receptors inside the target cell. Choice B is wrong as most AA-derived hormones also have their receptors on the plasma membrane. Choice D is wrong as one AA-derived hormone (thyroid hormone) can cross the cell membrane.

48. If a particular hormone is not lipid soluble, what else may be said of the hormone?
    A. It cannot enter the cell unless they have a cross-membrane transport mechanism.
    B. It works by binding to receptors located on the plasma membrane.
    C. It works by binding to receptors within the cell nucleus or mitochondria.
    D. It works by binding to receptors in the cell cytoplasm or nucleus.

Answer is A: If a hormone is not lipid soluble, it cannot cross the membrane without a specific carrier mechanism—thyroid hormone is not lipid soluble, but has just such a mechanism. Choice B is not necessarily correct as thyroid hormone can cross the membrane. Choices C and D are wrong as the membrane must be crossed for this to happen.

49. If the receptor for a particular hormone is located on the plasma membrane of the cell, what may be said of the mode of action of the hormone?
    A. It works by altering plasma membrane permeability state by opening or closing membrane channels.
    B. It works by activating appropriate genes in cell nucleus to stimulate the synthesis of an enzyme.
    C. It stimulates mitosis.
    D. It works by activating a "second messenger" with the cell.

Answer is D: The first messenger is the hormone that binds to a receptor on the plasma membrane. This binding activates a "G-protein" inside the cell which then moves along the membrane to activate an enzyme that produces a molecule that is the second messenger. The second messenger then stimulates a cascade of reactions within the cell.

50. What functions are performed by the endocrine system?
    A. It allows the ovarian cycle, conception and childbirth to occur.
    B. It controls the body with chemicals called hormones.
    C. It protects the body by destroying bacteria and tumour cells.
    D. It directs body functions with chemicals called neurotransmitters.

Answer is B: The endocrine system works with hormones. Choice A is almost correct but conception is not an endocrine process.

51. Which hormone is a mineralocorticoid?
    A. Aldosterone
    B. Cortisol
    C. Epinephrine
    D. Corticosterone

Answer is A: Aldosterone is the major mineralocorticoid.

10.2 Hypothalamus and Pituitary

1. Which statement about the hypothalamus is correct?
   A. The hypothalamus is connected to the brain by the infundibulum.
   B. The hypothalamus is composed of glandular epithelial tissue.
   C. The hypothalamus secretes "releasing hormones".
   D. The hypothalamus secretes epinephrine and norepinephrine.

Answer is C: The hypothalamus is part of the brain so is composed of neural tissue. Epinephrine and norepinephrine are released from the adrenal medulla.

2. Where in the body is the hypothalamus located?
   A. On the inferior surface of the brain
   B. In the cortex of the adrenal gland
   C. In the anterior pituitary gland
   D. On the dorsal surface of the thyroid gland

Answer is A: The hypothalamus is below the thalamus, more or less on the "floor" of the brain, but above the pituitary gland.

3. From where are antidiuretic hormone and oxytocin released?
   A. The anterior pituitary
   B. The posterior pituitary
   C. The adrenal cortex
   D. The adrenal medulla

Answer is B: ADH and OT are produced in the hypothalamus and transported to the posterior pituitary.

4. Which structure controls the endocrine system and integrates the activities of the nervous and endocrine systems?
   A. The infundibulum
   B. The pituitary gland
   C. The thalamus
   D. The hypothalamus

Answer is D: Hypothalamus secretes regulatory hormones that control endocrine cells in the anterior pituitary Gland, produces ADH and oxytocin, contains "autonomic centres" that exert neural control over endocrine cells of the adrenal medullae.

5. Which one of the statements below is true?
   A. The anterior pituitary produces testosterone from cholesterol and releases it when releasing hormones arrive from the hypothalamus.
   B. The hypothalamus produces ADH and oxytocin which are stored in the posterior pituitary.

C. The posterior pituitary contains autonomic centres that exert neural control over the adrenal glands.
D. The thalamus produces ADH and oxytocin which are stored in the anterior pituitary

Answer is B: Testosterone is not produced in the pituitary, nor does the pituitary contain autonomic centres.

6. Which endocrine organ produces "releasing hormones" and "inhibitory hormones"?

A. Thyroid
B. Anterior pituitary
C. Hypothalamus
D. Thalamus

Answer is C: The hypothalamus controls the secretion of the anterior pituitary by the use of "releasing hormones" and "inhibitory hormones".

7. Which structure integrates the activities of the endocrine system and the nervous system?

A. The hypothalamus
B. The thalamus
C. The posterior pituitary
D. The anterior pituitary

Answer is A: The hypothalamus is located in the brain and also produces regulatory hormones.

8. Which structure produces ADH and oxytocin?

A. The thalamus
B. The anterior pituitary
C. The hypothalamus
D. The posterior pituitary

Answer is C: ADH and OT are then transported to the posterior pituitary for storage and release.

9. Which structure is composed of glandular epithelial tissue?

A. The thalamus
B. The anterior pituitary
C. The posterior pituitary
D. The hypothalamus

Answer is B: The anterior pituitary produces and releases six hormones.

10. Which structure produces the hormones ADH and oxytocin?

A. The posterior pituitary
B. The anterior pituitary

C. The thalamus
D. The hypothalamus

Answer is D: ADH and OT are produced in the hypothalamus and are then transported to the posterior pituitary for release.

11. The hypothalamus produces "releasing hormones". What do these releasing hormones do?
    A. They direct the posterior pituitary to release hormones.
    B. They direct the anterior pituitary to release hormones.
    C. They direct the gonads to release hormones.
    D. They act as "second messengers" when hormones bind to their receptor site.

Answer is B: The anterior pituitary produces hormones that stimulate other endocrine organs, but the anterior pituitary does not release hormones until directed to the hypothalamus.

12. Which of the following "controls" the endocrine system?
    A. The posterior pituitary
    B. The thalamus
    C. The anterior pituitary
    D. The hypothalamus

Answer is D: The hypothalamus secretes releasing hormones and inhibitory hormones that stimulate the anterior pituitary gland to secrete hormones which in turn control the activities of endocrine cells in the thyroid, cortex of adrenal glands and reproductive organs. The hypothalamus produces ADH and oxytocin which are stored and released into blood from posterior pituitary. The hypothalamus exerts neural control over the adrenal medullae.

13. Which part of the pituitary gland is comprised of neural tissue?
    A. The posterior pituitary
    B. The pars intermedia
    C. The adenohypophysis
    D. The anterior pituitary

Answer is A: The posterior pituitary is a part of the brain.

14. What may be correctly said about oxytocin and antidiuretic hormone?
    A. They are made and released from the posterior pituitary.
    B. They are made in the hypothalamus and stored and released from the posterior pituitary.
    C. They are made in the hypothalamus and stored and released from the anterior pituitary.
    D. They are made and released from the anterior pituitary.

Answer is B: The hypothalamus produces ADH and OT. The posterior pituitary releases them.

15. What are the two parts of the pituitary gland known as?
    A. The thalamus and the hypothalamus
    B. Anterior and posterior
    C. Alpha and beta cells
    D. Cortex and medulla

Answer is B: Anterior pituitary (epithelial tissue) and posterior pituitary (neural tissue).

16. Which two hormones are stored in the posterior pituitary prior to their release?
    A. Luteinising hormone and follicle-stimulating hormone
    B. Adrenalin and noradrenalin
    C. Calcitonin and calcitriol
    D. Oxytocin and antidiuretic hormone

Answer is D: OT and ADH are produced in the hypothalamus before being transported to the posterior pituitary.

17. The posterior pituitary does which one of the following?
    A. Produces growth hormone, prolactin and tropic hormones
    B. Secretes regulatory hormones that control endocrine cells in the anterior pituitary
    C. Exerts neural control over other endocrine glands
    D. Stores and releases oxytocin and antidiuretic hormone

Answer is D: OT and ADH are made in hypothalamus and passed along neurones to the posterior pituitary.

18. The structure that secretes regulatory hormones that control the pituitary gland is known as the:
    A. Hypothalamus
    B. Hypophysis
    C. Hypothyroid
    D. Hypothymus

Answer is A: Hypothalamus is a structure. Hypophysis is another name for the pituitary gland.

19. The pituitary gland has an anterior portion and a posterior portion. One difference between the two is:
    A. The anterior portion releases hormones following a nervous stimulus while the posterior portion releases hormones following a hormonal stimulus.
    B. The posterior portion is able to produce hormones while the anterior portion merely stores hormones made elsewhere.
    C. The anterior portion releases antidiuretic hormone (ADH) while the posterior portion does not.

D. The posterior portion is neural tissue while the anterior portion is glandular tissue.

Answer is D: Choices A, B and C are the reverse of the true situation.

20. What does the anterior lobe of the pituitary gland synthesise and release?
    A. Growth hormone-releasing hormone
    B. Corticotropin-releasing hormone
    C. Thyroid-stimulating hormone
    D. Gonadotropin-releasing hormone

Answer is C: All the releasing hormones are produced in the hypothalamus.

21. Complete the following sentence correctly: The hypothalamus:
    A. Is the major link between the nervous and the endocrine systems
    B. Is situated in the brain superior to the thalamus
    C. Produces a hormone that stimulates the thyroid gland
    D. Does not produce antidiuretic hormone (ADH)

Answer is A: The hypothalamus integrates the nervous and endocrine systems of body control.

22. Which of the following secretes growth hormone?
    A. The adrenal glands
    B. The thyroid gland
    C. The posterior lobe of the pituitary gland
    D. The anterior lobe of the pituitary gland

Answer is D: The anterior lobe produces and releases GH.

23. Which statement is **NOT CORRECT**?
    A. Males and females produce testosterone and oestrogen, respectively, in their gonads.
    B. Both male and females produce follicle-stimulating hormone and luteinising hormone.
    C. Oxytocin is produced and released from the anterior lobe of the pituitary gland.
    D. Production of testosterone and oestrogen inhibits the release of gonadotrophin-releasing hormone from the hypothalamus.

Answer is C: Oxytocin is produced in the hypothalamus and released from the posterior lobe of the pituitary gland.

24. Which organ produces luteinising hormone?
    A. The anterior pituitary
    B. The posterior pituitary
    C. The corpus luteum

D. The adrenal cortex

Answer is A: The anterior pituitary releases LH along with FSH.

25. How may the anatomical relationship between the hypothalamus and the pituitary gland be described?
    A. Both are located in the brainstem, with the hypothalamus being inferior to the pituitary.
    B. The hypothalamus wraps around the trachea and the pituitary glands are on its posterior surface.
    C. The pituitary is part of the pre-central gyrus, while the hypothalamus is part of the post-central gyrus.
    D. The pituitary lies inferior to the hypothalamus, connected to it by the infundibulum.

Answer is D: The pituitary sits in the hypophyseal fossa of the sella turcica of the sphenoid bone, inferior to the hypothalamus. The pituitary releases its hormones when directed to by the hypothalamus. Choice B describes the thyroid gland.

26. Which of the following is NOT a way that the hypothalamus controls the endocrine system?
    A. Hypothalamus produces and releases oestrogens, progesterone and human chorionic gonadotropin.
    B. Hypothalamus contains "autonomic centres" that exert neural control over endocrine cells of the adrenal glands.
    C. Hypothalamus produces ADH and oxytocin which are stored and released from posterior pituitary gland.
    D. Hypothalamus secretes regulatory hormones that control endocrine cells in the anterior pituitary gland.

Answer is A: These hormones are not produced by the hypothalamus. Rather, the placenta produces and releases them during pregnancy.

27. Which statement describes the structure of the pituitary gland and the hormones it releases?
    A. The pituitary has an anterior portion (which is glandular tissue) and a posterior portion (which is neural tissue). The posterior releases GH, PRL and four "tropic hormones", the anterior releases oxytocin and ADH.
    B. The pituitary has an anterior portion (which is glandular tissue) and a posterior portion (which is neural tissue). The posterior releases oxytocin and ADH, the anterior releases GH, PRL and four "tropic hormones".
    C. The pituitary has a cortex (which is neural tissue) and a medulla (which is glandular tissue). The posterior releases oxytocin and ADH, the anterior releases GH, PRL and four "tropic hormones".
    D. The pituitary has an anterior portion (which is neural tissue) and a posterior portion (which is glandular). The posterior produces and releases GH, PRL and four "tropic hormones", the anterior releases oxytocin and ADH.

Answer is B: There are three things to get right: anterior/posterior or cortex/medulla; anterior = glandular or neural; anterior releases ADH or GH. The anterior pituitary is endocrine (glandular) tissue so makes hormones. The posterior pituitary is neural tissue (i.e. is part of the brain). The posterior stores and releases oxytocin and ADH, the anterior produces and releases GH, PRL and four "tropic hormones".

28. Peptide hormones are produced (and/or released) by which structure?
    A. The adrenal cortex
    B. The gonads
    C. The hypothalamus
    D. The kidneys

Answer is C: The hypothalamus produces peptide hormones.

29. Which of the following lists of two organs and one hormone that are involved in a control process that releases that hormone is **WRONG**?
    A. Hypothalamus, anterior pituitary, thyroid stimulating hormone
    B. Hypothalamus, posterior pituitary, antidiuretic hormone
    C. Hypothalamus, adrenal medulla, epinephrine
    D. Thalamus, hypothalamus, oxytocin

Answer is D: The thalamus is not involved in the chain of events that causes release of oxytocin from the posterior pituitary.

10.3 Organs and Their Hormones

10.3.1 Adrenals

1. What hormone(s) does the adrenal medulla produce?
    A. Aldosterone
    B. Epinephrine and norepinephrine
    C. Corticosteroids
    D. Glucocorticoids

    Answer is B: Epinephrine and norepinephrine are produced in the adrenal medulla (the deep or inside part). Aldosterone is a mineralocorticoid, which like corticosteroids and glucocorticoids, as the "cortico" in the name suggests, are all produced in the adrenal cortex.

2. What hormones are produced by the adrenal medulla?
    A. Epinephrine and norepinephrine
    B. Insulin and glucagon
    C. Aldosterone and erythropoietin
    D. Testosterone and oestrogen

Answer is A: The medulla is the deeper part of the adrenal gland. The cortex is more superficial.

3. Which hormones are produced by the adrenal medulla?
   A. Gonadocorticoids
   B. Steroid hormones
   C. Mineralocorticoids
   D. Catecholamines

Answer is D: Epinephrine and norepinephrine (=catecholamines) are produced in the adrenal medulla.

4. The adrenal medulla produces which of the following?
   A. Weak androgens
   B. Mineralocorticoids
   C. Testosterone and oestrogen
   D. Epinephrine and norepinephrine

Answer is D: The adrenal cortex produces weak androgens and mineralocorticoids.

5. Which hormones are produced by the adrenal medulla?
   A. Glucocorticoids
   B. Mineralocorticoids
   C. Adrenalin and noradrenalin
   D. Gonadocorticoids

Answer is C: Adrenalin and noradrenalin (also known as epinephrine and nor epinephrine). The other three choices all have "cortico" in their name indicating their adrenal cortex origin.

6. What is produced in the adrenal cortex?
   A. Cholesterol
   B. Catecholamines
   C. Adrenalin and noradrenalin
   D. Corticosteroids

Answer is D: "Cortico" steroids are made in the adrenal cortex. Choices B and C are the same—these hormones are made in the adrenal medulla.

7. Which structure produces epinephrine and norepinephrine?
   A. Adrenal pelvis
   B. The anterior pituitary
   C. Adrenal medulla
   D. Adrenal cortex

Answer is C: The medulla (interior) of the adrenal gland.

8. By what other name is adrenaline also known?
   A. Noradrenaline
   B. Epinephrine
   C. Androgen
   D. ANP

Answer is B: Perhaps to diminish the perception that the adrenal gland only produces adrenaline.

9. Which hormones are produced by the adrenal cortex and contribute to carbohydrate metabolism?
   A. Androgens
   B. Mineralocorticoids
   C. Glucocorticoids
   D. Gonadocorticoids

Answer is C: The "-cort-" part of the word refers to the adrenal cortex, while the "gluco-" part refers to the role carbohydrate (glucose) metabolism.

10. From which organ is aldosterone released?
    A. The hypothalamus
    B. The adrenal medulla
    C. The kidney
    D. The adrenal cortex

Answer is D: Aldosterone is a "mineralocorticoid" hormone which gives a clue that it comes from the cortex of the adrenal glands.

10.3.2 Kidneys

1. Which gland or organ releases erythropoietin?
   A. The kidneys
   B. The adrenal glands
   C. The anterior pituitary
   D. The pancreas

Answer is A: Kidneys produce erythropoietin (EPO)—which signals red bone marrow to increase production of RBC.

2. What is the function of erythropoietin (EPO)?
   A. Stimulate bone marrow to produce red blood cells
   B. Decrease the plasma concentration of Ca^{2+}
   C. Increase the plasma concentration of Ca^{2+}
   D. To raise blood sugar level

Answer is A: EPO is produced in the kidneys and stimulates active (red) marrow to produce blood cells.

3. What stimulates the release of erythropoietin?
   A. Low blood pressure or abnormally low blood oxygen concentration
   B. High blood pressure or abnormally low blood oxygen concentration
   C. Low blood pressure or abnormally high blood carbon dioxide concentration
   D. High blood pressure or abnormally high blood glucose concentration

Answer is A: EPO acts to constrict blood vessels, thereby increasing blood pressure. It also stimulates RBC production and maturation in bone marrow, thereby increasing blood's oxygen-carrying capacity.

10.3.3 Pancreas

1. What is produced by the beta cells of the pancreas?
   A. Angiotensin-converting enzyme
   B. Glucocorticoids
   C. Glucagon
   D. Insulin

Answer is D: Beta cells of the pancreatic islets produce insulin. The alpha cells produce glucagon.

2. What is the role of glucagon and insulin?
   A. Glucagon raises blood glucose level and inhibits gluconeogenesis.
   B. Glucagon lowers blood glucose level and stimulates glycogenolysis.
   C. Insulin raises blood glucose level and stimulates gluconeogenesis.
   D. Insulin lowers blood glucose level inhibits glycogenolysis.

Answer is D: Insulin promotes the uptake of glucose into cells from the blood. Hence it lowers blood glucose. It also inhibits glycogenolysis so less glucose is released from the glycogen store.

3. What effect does insulin have?
   A. It increases metabolic rate.
   B. It causes the breakdown of glycogen to glucose.
   C. It lowers blood sugar level.
   D. It stimulates gluconeogenesis.

Answer is C: Insulin stimulates enhanced membrane transport of glucose into body cells and inhibits glycogenolysis and gluconeogenesis. This lowers blood glucose level.

4. Which cells produce insulin?
   A. The acini cells of the pancreas
   B. Parafollicular cells of the thymus
   C. Alpha cells of the islets of Langerhans
   D. Beta cells of the islets of Langerhans

Answer is D: Islets of Langerhans = pancreatic islets. Alpha cells produce glucagon.

5. What is the function of insulin?
   A. Enhance the transport of glucose through the plasma membrane into the cell
   B. Promote glycogenolysis
   C. Promote gluconeogenesis
   D. To raise blood sugar level

Answer is A: Insulin lowers the level of glucose in the blood. The other three choices are about increasing the blood glucose level.

6. Which cells produce insulin?
   A. The acini
   B. The alpha cells
   C. The beta cells
   D. The islets of Langerhans

Answer is C: The beta cells of the islets of Langerhans.

7. What does insulin do?
   A. It lowers blood sugar level.
   B. It causes the breakdown of glycogen to glucose.
   C. It increases metabolic rate.
   D. It hydrolyses glucose into ATP.

Answer is A: Insulin stimulates the uptake of glucose into cells from the blood.

8. Which hormones does the pancreas produce?
   A. Epinephrine and norepinephrine.
   B. Oxytocin and antidiuretic hormone.
   C. Glucagon and insulin.
   D. Glucocorticoids and aldosterone

Answer is C: The pancreatic islets produce these hormones which manage the concentration of glucose in the blood.

9. What controls the blood glucose level?
   A. The action of insulin
   B. The action of glucagon
   C. The action of insulin and glucagon

D. The action of insulin, glucagon and glycogen

Answer is C: Insulin increases blood glucose, while glucagon decreases blood glucose. Glycogen is not a hormone, so it does not affect the level of glucose in the blood.

10. Which of the following glands have both exocrine and endocrine functions?
    A. The pancreas
    B. The thymus
    C. The parathyroids
    D. The pituitary

Answer is A: The pancreas produces digestive enzymes and secretes them via a duct into the duodenum—the exocrine function. It also produces insulin and glucagon for release into the circulation—its endocrine role. The other three choices are solely endocrine organs.

11. Which organ produces insulin?
    A. The pancreas
    B. The posterior pituitary
    C. The parathyroid glands
    D. The adrenal cortex

Answer is A: The beta cells of the islets of Langerhans (pancreatic islets) produce insulin.

10.3.4 Parathyroids

1. What effect does parathyroid hormone have?
   A. It increases plasma Ca^{2+} concentration.
   B. It decreases plasma Ca^{2+} concentration.
   C. It increases the rate of ATP formation.
   D. It stimulates the thyroid gland to produce thyroxine.

Answer is A: PTH increases plasma Ca^{2+}. (Calcitonin aids in lowering blood calcium.)

2. Which hormone(s) increases the reabsorption of Ca^{2+} from the filtrate in the kidney tubule?
   A. Calcitonin
   B. Mineralocorticoids
   C. Parathyroid hormone
   D. Aldosterone

Answer is C: PTH causes increased calcium absorption and hence increases blood calcium level.

3. Complete the sentence correctly. Parathyroid hormone:
   - A. Is produced by the parafollicular cells of the thyroid gland
   - B. Decreases the concentration of Ca^{2+} in the blood
   - C. Releases Ca^{2+} from the sarcoplasmic reticulum
   - D. Increases the concentration of Ca^{2+} in the blood

Answer is D: When the level of calcium in the blood is lower than required, PTH is released and causes the calcium concentration to increase.

4. What is a function of calcitonin?
   - A. Accelerating Ca^{2+} release from bone
   - B. Stimulating Ca^{2+} excretion by the kidneys
   - C. Reducing Ca^{2+} deposition in bone
   - D. Stimulating the formation of calcitriol in the kidneys

Answer is B: Calcitonin decreases the blood calcium concentration. B is the only choice compatible with this.

5. How do calcitonin or parathyroid hormone control blood calcium levels?
   - A. Calcitonin acts to increase blood calcium levels.
   - B. Parathyroid hormone release is inhibited by increased calcium levels.
   - C. Parathyroid hormone stimulates bone resorbing cells to take up calcium.
   - D. Calcitonin inhibits parathyroid hormone.

Answer is B: PTH acts to increase blood calcium levels. Hence less is required when blood calcium level is high.

6. Which endocrine gland produces parathyroid hormone?
   - A. The parafollicular gland
   - B. The thyroid gland
   - C. The parathyroid glands
   - D. The para cells of the pancreas

Answer is C: The four parathyroid glands, attached to the thyroid, produce parathyroid hormone.

7. Blood calcium levels are controlled by hormones from which structures?
   - A. The parathyroid glands and the thyroid gland
   - B. The anterior lobe of the pituitary gland and the thyroid gland
   - C. The parathyroid glands and the anterior lobe of the pituitary gland
   - D. The adrenal cortex and the hypothalamus

Answer is A: Parathyroid glands secrete parathyroid hormone (to increase blood Ca level) while the parafollicular cells of the thyroid produce calcitonin which aids in lowering blood Ca.

8. Where is parathyroid hormone produced?
   A. In the thyroid gland
   B. In the parathyroid glands
   C. In the parafollicular cells of the thyroid gland
   D. In the adrenal cortex

Answer is B: The four parathyroid glands, embedded in the dorsal surface of the thyroid gland, produce parathyroid hormone. (Parafollicular cells produce calcitonin.)

9. How many parathyroid glands are there and where are they located?
   A. One, placed superior to the thyroid
   B. Two, attached to the superior surface of the kidneys
   C. Three, in the mediastinum, superior to the heart
   D. Four, located on the dorsal part of the thyroid

Answer is D: They are embedded in the dorsal part of the thyroid, deep to the trachea.

10.3.5 Pineal

1. What hormone is produced by the pineal gland?
   A. Secretin
   B. Serotonin
   C. Somatostatin
   D. Melatonin

Answer is D: Melatonin "may" be a hormone, but its effects on humans are unclear. It may be involved in circadian rhythms and in sleep, and this has been enough to see it marketed as a drug to treat insomnia.

2. Which organ(s) is the pineal gland associated with?
   A. The brain
   B. The kidney
   C. The thyroid
   D. The reproductive organs

Answer is A: The pineal gland is adjacent to the diencephalon of the brain, but outside of the blood–brain barrier.

3. The pineal gland is part of the diencephalon and produces a hormone. This information should allow you to determine which of the following is true:
   A. The pineal gland produces a steroid hormone which is water soluble.
   B. The hormone produced is lipid soluble.

C. The pineal gland is adjacent to the kidney and its hormone affects urine production.
D. The pineal gland is in the brain but outside of the blood–brain barrier.

Answer is D: The pineal gland produces melatonin which is derived from an amino acid (tryptophan) which is released into the blood. Such a molecule cannot cross the BBB, hence the pineal gland, despite being located in the brain, is not behind the BBB. Fat-soluble molecules are able to pass through the BBB.

10.3.6 Thymus

1. What hormone does the thymus produce?
   A. Haemopoietin
   B. Tyrosine
   C. Thymopoietin
   D. Erythropoietin

Answer is C: Thymopoietin refers to a group of hormones that exert a regulatory effect on lymphocytes produced by the thymus. The suffix "-poietin" is used with words to indicate an agent with a stimulatory effect on growth or multiplication of cells.

2. In which part of the body is the thymus located?
   A. Superior to the kidneys
   B. Attached to the thyroid
   C. Adjacent to the spleen
   D. Superior to the heart

Answer is D: The thymus is located in the mediastinum, superior to the heart, deep to the sternum and medial to each lung.

3. To which system does the thymus belong in addition to the endocrine system?
   A. The nervous system
   B. The lymphatic system
   C. The renal system
   D. The cardiovascular system

Answer is B: The thymus is a lymphoid organ and contains actively dividing T-cell lymphocytes which contribute to our immunity.

4. What is the function of the hormones produced by the thymus?
   A. They stimulate the production of red blood cells.
   B. They promote the differentiation of white blood cells.
   C. They produce antibodies to foreign proteins.
   D. They promote the maturation of T cells.

Answer is D: The thymus is also part of the lymphatic system and is involved in immunity. Choice A refers to erythropoietin. T cells are a type of white blood cell, but choice B is wrong.

10.3.7 Thyroid

1. What hormone does the thyroid produce?
   A. Thyroid-stimulating hormone
   B. Calcitriol
   C. Thyroxine
   D. Parathyroid hormone

Answer is C: Thyroxine (or tetra-iodothyronine) is converted to tri-iodothyronine in target tissues.

2. Which hormone has the element iodine as part of its molecule?
   A. Calcitonin
   B. Haemoglobin
   C. Thyroxine
   D. Parathyroid hormone

Answer is C: Thyroxine or thyroid hormone contains iodine.

3. Iodine is an essential component of which hormone?
   A. Thyroid hormones
   B. Aldosterone
   C. Thyroid-stimulating hormones
   D. Parathyroid hormone

Answer is A: Thyroid hormones T3 and T4 contain three and four iodine atoms, respectively.

4. What hormone is produced by the parafollicular cells of the thyroid gland?
   A. Parathyroid hormone
   B. Calcitonin
   C. Thyroid hormone
   D. Thyroxine

Answer is B: Parafollicular (or "C") cells of the thyroid produce calcitonin which aids in lowering of blood calcium.

5. Which hormone is the one made in greatest quantity by the thyroid gland?
   A. Calcitonin
   B. Thyroid-stimulating hormone
   C. Tri-iodothyronine
   D. Thyroxine

Answer is D: Thyroxine is about 90% tetra-iodothyronine and 10% tri-iodothyronine.

6. The hormones thyroxine and tri-iodothyronine contain which element?
   A. Cobalt
   B. Iron
   C. Iodine
   D. Manganese

Answer is C: These thyroid hormones require iodine in their molecule.

7. Which cells of the thyroid gland produce thyroglobulin (thyroid hormone)?
   A. Parathyroid (chief) cells
   B. Follicular cells
   C. Oxyphil cells
   D. Parafollicular (C) cells

Answer is B: Follicular cells produce thyroglobulin which is stored within the follicles of thyroid tissue.

8. Which hormone is produced by the thyroid gland?
   A. Kalium
   B. Calcitriol
   C. Calcitonin
   D. Calmodulin

Answer is C: Calcitonin is produced by the C (parafollicular) cells of the thyroid gland.

9. Which endocrine organ has parafollicular (or C) cells?
   A. Parathyroid gland
   B. Thyroid gland
   C. The ovaries
   D. The thymus

Answer is B: The parafollicular cells of the thyroid produce calcitonin which acts to lower blood Ca. "Follicle" refers to a spherical group of cells that contain a cavity.

10. What type of hormone is thyroid-stimulating hormone?
   A. A peptide hormone released from the hypothalamus
   B. An amino acid-derived hormone released from the hypothalamus
   C. An amino acid-derived hormone released from the pituitary
   D. A peptide hormone released from the pituitary

Answer is D: TSH is more than 200 amino acids long, hence is a peptide hormone and is released from the anterior pituitary. Thyroid-releasing hormone comes from the hypothalamus.

11. Hormones are often released sequentially when directed to by the preceding hormone in the sequence. Which of the following is the initiating hormone?
    A. Thyroid-stimulating hormone (TSH)
    B. Thyrotropin-releasing hormone (TRH)
    C. Thyroid hormone (TH)
    D. Tetra-iodothyronine (T_4)

Answer is B: TRH from the hypothalamus signals the anterior pituitary to release TSH. TSH causes the release of TH from the thyroid. T_4 is one of the thyroid hormones, T_3 is the other. Hence choice C is a better answer than D.

Chapter 11
Renal System

Kidneys are the major excretory organs of the body. They excrete the organic wastes urea, uric acid and creatinine. They regulate the volume of blood (and hence blood pressure) by increasing or decreasing the volume of urine produced to match the body's requirement for water. They help regulate blood pH by excreting hydronium ions and by reabsorbing and producing bicarbonate ions. They maintain blood osmolarity at ~290 mosmol/L and individual ions (electrolytes) at their healthy concentrations. The kidneys produce the enzyme renin, the hormone erythropoietin and the "vitamin" calcitriol. In turn the physiological functions of the kidney are influenced by ADH, aldosterone, parathyroid hormone, ANP, BNP and angiotensin II.

Nephron is the blood processing functional unit of kidney. It consists of a glomerulus and a renal tubule. The glomerulus is a spherical capillary bed that is supplied with blood via the afferent arteriole and is drained by the efferent arteriole. The renal tubule consists of the "Bowman's" capsule (which surrounds the glomerulus and receives the fluid filtered from the glomerulus) and the proximal convoluted tubule, the descending and ascending limbs of the loop of Henle and the distal convoluted tubule. A collecting duct receives filtrate from several nephrons and delivers it to a calyx.

Macroscopically the kidney consists of a superficial cortex (where all of the glomeruli are located) and the deeper medulla. The medulla consists of renal "pyramids" separated by renal "columns". The renal tubules extend into the pyramids and the collecting ducts pass through them to deliver urine from their apex (the papilla) into a tube called a minor calyx. Minor calyces join to from calyces which then expand into the renal pelvis. The pelvis is drained by a ureter that delivers urine to the storage bladder. The efferent arteriole feeds into the peritubular capillary bed which surrounds the PCT and the DCT of the nephron and is located in the cortex. A small percentage of the blood from the efferent arteriole also flows into another capillary bed known as the vasa recta which surrounds the loop of Henle and extends into the pyramids of the medulla. The interlobar arteries are located in the renal columns and deliver blood from the renal artery to the arcuate arteries, which lie

M. Caon, *Examination Questions and Answers in Basic Anatomy and Physiology*, https://doi.org/10.1007/978-3-030-47314-3_11

superficial to the bases of the pyramids, and then to the cortical radiate arteries which in turn supply the afferent arterioles which enter the glomeruli.

To understand the physiology of the kidney, you must also understand how the juxtaglomerular apparatus (made up of the macula densa and the juxtaglomerular cells) works and how the vasa recta and the loop of Henle operate within the osmotic gradient of the medulla. In addition it is useful to know how the cells of the renal tubules absorb or secrete Na^+, K^+, H_3O^+, Cl^-, Ca^{++}, NH_4^+ and HCO_3^- ions. However, these processes will not be described here.

11.1 Renal Anatomy

1. What is the name for the entry point to the kidney for nerves, blood vessels, ureters and lymphatics?
   A. Calyx
   B. Hilus
   C. Pelvis
   D. Pyramid

Answer is B: Hilum is the name of the point of attachment between an organ and its supply services.

2. Where are all of the glomeruli of the kidney located?
   E. In the medulla
   F. In the columns
   G. In the pyramids
   H. In the cortex

Answer is D: The tubules and collecting ducts extend into the medulla, but all glomeruli are in the cortex.

3. What structure does the blood from the afferent arteriole enter?
   A. The peritubular capillaries
   B. The vasa recta
   C. The glomerulus
   D. Bowman's capsule

Answer is C: Afferent (incoming) arteriole enters the glomerulus, the efferent arteriole leaves the glomerulus.

4. Which part of the nephron is impermeable to water?
   A. Proximal convoluted tubule
   B. Distal convoluted tubule in the presence of ADH
   C. Ascending limb of the loop of Henle
   D. Descending limb of the loop of Henle

Answer is C: The ascending limb. Hence, volume of filtrate does not change as it passes through the ascending limb.

5. How are cortical nephrons different from juxtamedullary nephrons?
   A. Cortical nephrons lie almost entirely outside the renal medulla.
   B. Cortical nephrons have an associated vasa recta.
   C. Cortical nephrons have a longer tubule.
   D. There are fewer cortical nephrons.

Answer is A: Cortical nephrons are situated almost entirely within the cortex, while the far less numerous juxtamedullary nephrons have their glomeruli adjacent to the medulla and extend their loop of Henle into the medulla.

6. What is the entry point to the kidney for the renal artery, renal vein, lymphatics and nerves called?
   A. Renal pyramid
   B. Renal hilus
   C. Renal capsule
   D. Renal column

Answer is B: Hilus is a general term referring to the entry point (e.g. hilus of the lung).

7. In what part of the kidney are the glomeruli located?
   A. In the cortex
   B. In the medulla
   C. In the hilus
   D. In the minor calyces

Answer is A: Only the cortex contains glomeruli.

8. Four sections of the vasculature of the kidney tubule are listed below. Which one lists them in correct order of blood flow from left to right?
   A. Efferent arteriole, glomerulus, afferent arteriole, peritubular capillaries
   B. Afferent arteriole, glomerulus, efferent arteriole, peritubular capillaries
   C. Peritubular capillaries, afferent arteriole, glomerulus, efferent arteriole
   D. Glomerulus, afferent arteriole, peritubular capillaries, efferent venule

Answer is B: The afferent arteriole carries blood into the glomerulus so must be first.

9. What is the collective term applied to the proximal and distal convoluted tubules, the loop of Henle (i.e. the nephron loop) and the glomerular capsule?
   A. The renal corpuscle
   B. The renal tubule
   C. The nephron
   D. The renal capsule

Answer is B: The nephron includes the glomerulus, whereas the tubule does not.

10. Which list of structures is presented in the correct order in which urine passes through them on the way to the bladder?
    A. Ureter, minor calyx, major calyx, renal pelvis, papilla
    B. Renal pelvis, major calyx, minor calyx, papilla, ureter
    C. Papilla, minor calyx, major calyx, renal pelvis, ureter
    D. Minor calyx, major calyx, papilla, renal pelvis, ureter

Answer is C: Collecting ducts empty into the papilla, so the papilla must be first.

11. What part of the nephron performs the majority of the reabsorption of materials from the filtrate?
    A. The Bowman's capsule and glomerulus
    B. The loop of Henle (the nephron loop)
    C. The distal convoluted tubule and collecting duct
    D. The proximal convoluted tubule

Answer is D: The proximal convoluted tubule (the first part or the renal tubule) absorbs most of the required materials.

12. What is the name of the tube that connects the bladder to the kidney?
    A. Renal tubule
    B. Ureter
    C. Urethra
    D. Collecting duct

Answer is B: Not to be confused with the urethra through which urine exits the body.

13. Which of the following may be said of the renal medulla?
    A. It is the more superficial part of the kidney.
    B. It contains all of the glomeruli.
    C. It produces adrenaline and noradrenaline.
    D. It contains the pyramids and columns.

Answer is D: The medulla is deep to the cortex. In this region are many blood vessels (in the columns) and the collecting ducts (in the pyramids).

14. Which statement about kidney anatomy is correct?
    A. The cortex is superficial to the medulla and contains all of the glomeruli.
    B. The cortex is deep to the medulla and contains the collecting tubules.
    C. The pyramids are in the cortex and contain the collecting tubules.
    D. The pyramids are in the medulla and contain all of the glomeruli.

Answer is A: The cortex is superficial to the medulla. All glomeruli are in the cortex. The medulla has the collecting ducts.

15. The renal tubule of the nephron includes which of the following structures?
    A. Proximal convoluted tubule, vasa recta, Bowman's capsule, collecting duct
    B. Distal convoluted tubule, ascending limb of loop of Henle, Bowman's capsule, proximal convoluted tubule
    C. Descending limb of loop of Henle, collecting duct, distal convoluted tubule, ascending limb of loop of Henle
    D. Glomerulus, proximal convoluted tubule, distal convoluted tubule, Bowman's capsule

Answer is B: Glomerulus, collecting duct and vasa recta are not part of the tubule. The loop of Henle has both an ascending arm and a descending arm.

16. How does the descending limb of the loop of Henle differ from the ascending limb?
    A. The descending limb is impermeable to water but permeable to sodium chloride.
    B. The ascending limb is permeable to water but impermeable to sodium chloride.
    C. The descending limb is permeable to water but impermeable to sodium chloride.
    D. The ascending limb is permeable to both water and to sodium chloride.

Answer is C: The descending limb is permeable to water (the ascending limb is not). The descending limb is impermeable to solutes except urea.

17. Which one of the following is part of the renal tubule?
    A. Glomerulus
    B. Vasa recta
    C. Collecting duct
    D. Macula densa

Answer is D: The macula densa consists of cells in the wall of the distal convoluted tubule at the juxtaglomerular apparatus.

18. What does the renal system consists of?
    A. Two kidneys, two urethra, bladder, one ureter
    B. Two adrenal glands, two kidneys, one ureter, two urethra, bladder
    C. Two adrenal glands, two kidneys, two ureters, two urethra, bladder
    D. Two kidneys, one urethra, bladder, two ureters

Answer is D: There are two ureters (from kidney to bladder) and one urethra (from bladder to outside).

19. Which of the statements about the capillaries of the glomerulus is **NOT** true?
    A. Glomerular capillaries are fenestrated (i.e. porous).
    B. Blood enters and leaves the glomerulus via arterioles.

C. The blood pressure in glomerular capillaries is higher (55 mmHg) than in the capillaries in the rest of the body.
D. Glomerular capillaries have smooth muscle in their walls.

Answer is D: Capillary walls are made of endothelial cells, there is no smooth muscle.

20. From which arteriole does blood enter the peritubular capillaries of the nephron?
 A. Arcuate
 B. Efferent
 C. Afferent
 D. Renal

Answer is B: Efferent arteriole leaves the glomerulus and enters the peritubular capillaries.

21. What does the "juxtaglomerular apparatus" refer to?
 A. To those nephrons whose loop of Henle penetrate deep into the medulla
 B. To the lamina densa and podocytes that form filtration slits around the capillaries of the glomerulus
 C. To the capillaries that surround the loop of Henle of juxtamedullary nephrons
 D. To certain cells of the distal convoluted tubule where it touches the afferent arteriole

Answer is D: The juxtaglomerular = macula densa cells (of the DCT wall) and granular cells (of the afferent arteriole wall).

22. Which statement about kidney anatomy is correct? Renal pyramids are in the:
 A. Medulla and end in a papilla that empties into a minor calyx
 B. Medulla and end in a column that empties into a major calyx
 C. Cortex and end in a papilla that empties into a minor calyx
 D. Cortex and end in a column that empties into a minor calyx

Answer is A: Pyramids for the medulla. They contain collecting ducts that deliver urine in a minor calyx via a papilla.

23. In which structure does blood filtration in the kidney occur?
 A. Macula densa
 B. Renal corpuscle
 C. Major calyx
 D. Vasa recta

Answer is B: Solutes are filtered out of the blood in the glomerular capillaries into the Bowman's capsule. Together the two structures form the renal corpuscle.

24. From which part of the **nephron** is the greatest proportion of Na^+ absorbed from the filtrate?
    A. The proximal convoluted tubule
    B. The ascending limb of the loop of Henle
    C. The distal convoluted tubule
    D. The collecting duct in the presence of aldosterone

Answer is A: The PCT absorbs the greatest proportion of all things that are reabsorbed from the filtrate. It is the first part of the renal tube that filtrate passes through.

25. What is the place where the arteries, veins, lymphatics and nerves enter or leave the kidney called?
    A. The carina
    B. The reno-atrio notch
    C. The renal pelvis
    D. The hilus

Answer is D: Carina is in the trachea; renal pelvis refers to the ureter; reno-atrio notch does not exist.

26. The nephrons of the kidney consist of:
    A. Bowman's capsule, a loop of Henle, a collecting duct and a renal tubule
    B. A juxtaglomerular apparatus and collecting duct
    C. A glomerulus and a juxtaglomerular apparatus
    D. A glomerulus, a proximal convoluted tubule, loop of Henle and a distal convoluted tubule

Answer is D: The nephron includes the glomerulus and renal tubule. The later includes the Bowman's capsule. Choice D is best even though Bowman's capsule is not listed.

27. Which list has the blood vessels of the nephron in the correct order of blood flow?
    A. Afferent arteriole, glomerulus, efferent arteriole, peritubular capillaries, vasa recta
    B. Efferent arteriole, glomerulus, afferent arteriole, peritubular capillaries, vasa recta
    C. Afferent arteriole, vasa recta, efferent arteriole, peritubular capillaries, glomerulus
    D. Afferent arteriole, peritubular capillaries, efferent arteriole, glomerulus, vasa recta

Answer is A: Afferent arteriole brings blood into the glomerulus, then glomerular capillaries are the next.

28. Filtrate passes through each of the structures of the renal tubule listed below. Which list has the structures in the correct order?
    A. Proximal convoluted tubule, descending limb, ascending limb, distal convoluted tubule, collecting duct
    B. Bowman's capsule, proximal convoluted tubule, ascending limb, distal convoluted tubule, descending limb
    C. Collecting duct, proximal convoluted tubule, descending limb, ascending limb, distal convoluted tubule
    D. Proximal convoluted tubule, distal convoluted tubule, descending limb, ascending limb, collecting duct

Answer is A: The descending and ascending limbs should be between the PCT and DCT, while the collecting duct is last.

29. The nephron of the kidney consists of which of the following structures?
    A. Glomerulus, renal tubule and collecting duct
    B. Bowman's capsule, proximal convoluted tubule, loop of Henle and distal convoluted tubule
    C. Glomerulus and renal tubule
    D. Renal tubule and collecting duct

Answer is C: The glomerulus and renal tubule together constitute the nephron.

30. What name is given to the blood vessel that drains blood from the glomerulus after filtration?
    A. Vasa recta
    B. Afferent arteriole
    C. Efferent arteriole
    D. Efferent vein

Answer is C: Efferent means outgoing.

31. Through which structure must the filtrate move to enter the Bowman's capsule from the glomerulus?
    A. The capillary endothelial cell walls
    B. The capillary endothelial wall and basement membrane
    C. The capillary endothelial wall, basement membrane and podocytes
    D. The capillary endothelial wall, basement membrane, podocytes and the vasa recta

Answer is C: Together these three structures form the filtration membrane.

32. Which section of the renal tubule is permeable to urea?
    A. Descending limb of the loop of Henle
    B. Ascending limb of the loop of Henle
    C. Collecting duct in the presence of aldosterone
    D. Proximal convoluted tubule

Answer is D: PCT is permeable to urea, so as volume of filtrate in PCT decreases due to absorption of water, urea concentration in filtrate increases. This promotes passive reabsorption of urea along its concentration gradient.

33. What name is given to the blood vessel that connects the capillaries of the glomerulus to the vasa recta?
    A. Macula densa
    B. Afferent arteriole
    C. Efferent arteriole
    D. Afferent vein

Answer is C: The efferent arteriole transports blood from the glomerulus to the peritubular capillaries and vasa recta.

34. The functional unit of the kidney that filters blood and produces urine is called the:
    A. Medulla
    B. Glomerulus
    C. Neurone
    D. Nephron

Answer is D: The nephron performs all of the kidney functions.

35. In the kidney, the filtrate passes through several structures on its way to becoming urine. Which of the following lists presents these structures in the correct order?
    A. Collecting duct, glomerulus, proximal convoluted tubule, distal convoluted tubule, loop of Henle
    B. Proximal convoluted tubule, collecting duct, glomerulus, loop of Henle, distal convoluted tubule
    C. Glomerulus, proximal convoluted tubule, loop of Henle, distal convoluted tubule, collecting duct
    D. Glomerulus, collecting duct, proximal convoluted tubule, distal convoluted tubule, loop of Henle

Answer is C: Glomerulus must precede the collecting duct. Collecting duct must be after the DCT.

36. Which of the following statements about the structures in the loop of Henle is correct?
    A. Its ascending limb is permeable to water.
    B. Its descending limb is impermeable to urea.
    C. its descending limb is impermeable to water.
    D. Its ascending limb is impermeable to solutes.

Answer is B: The thin descending limb is impermeable to urea.

37. Which one of the following correctly lists the organs of the renal system?
   A. Two kidneys, two adrenals, one bladder, two urethras, one ureter
   B. Two kidneys, one urethra, one bladder, two ureters
   C. Two kidneys, one prostate, one bladder, two ureters, one urethra
   D. Two kidneys, two urethras, one bladder, one ureter

Answer is B: There are two ureters (one from each kidney), and one urethra from the bladder to the outside. The prostate is not part of the renal system and is present in males only.

38. The kidney turns blood into filtrate, then filtrate into urine. Which list has the structures in the correct order of liquid flow from proximal to distal?
   A. Glomerulus, papilla, renal cortex, major calyx
   B. Descending limb of the loop of Henle, distal convoluted tubule, renal pelvis, urethra
   C. Efferent arteriole, collecting duct, proximal convoluted tubule, renal medulla
   D. Afferent arteriole, ascending limb of the loop of Henle, ureter, Bowman's capsule

Answer is B: The renal cortex is before the papilla; the efferent arteriole is not part of flow of the filtrate; Bowman's capsule comes before the loop of Henle.

39. The following blood vessels are found in the kidney. Which list has the vessels in the correct order of blood flow?
   A. Glomerulus, efferent arteriole, vasa recta, arcuate arteries
   B. Glomerulus, afferent arteriole, peritubular capillaries, interlobular vein
   C. Peritubular capillaries, vasa recta, afferent arteriole, arcuate arteries
   D. Renal artery, arcuate arteries, glomerulus, interlobular vein

Answer is D: The efferent and afferent arterioles flow out from and into (respectively) the glomerulus, so the glomerulus is misplaced. In choice C, the afferent arteriole should precede the peritubular capillaries.

40. Where in the kidney are the collecting ducts found?
   A. In the renal cortex
   B. In the renal medulla
   C. In the renal pyramids
   D. In the renal columns

Answer is C: The collecting ducts are in the medulla, but so are the renal columns, so renal pyramids is a more specific and better choice.

41. What type(s) of cell may be found in the walls of the glomerular capillaries?
   A. Endothelial cells
   B. Podocyte cells
   C. Both podocytes and endothelial cells

D. Macula densa and juxtaglomerular cells

Answer is C: Endothelial cells make up the capillary walls, while podocytes surround the endothelial cells to make up the filtration membrane.

11.2 Renal Physiology

1. Solutes move from the blood in the glomerular capillaries into the Bowman's capsule due to which of the following influences?

   A. Osmotic pressure difference
   B. Diffusion down the concentration gradient
   C. Active transport
   D. Hydrostatic pressure difference

Answer is D: The efferent (outgoing) arteriole has a smaller diameter than the afferent arteriole; hence, the glomerulus is a high-pressure area. This aids the movement of dissolved substances through the filtration membrane.

2. Which material is actively reabsorbed from the filtrate in the kidney tubule?

   A. Na^+
   B. HCO_3^-
   C. Cl^-
   D. H_2O

Answer is A: All other substances are passively reabsorbed.

3. Which material is secreted into the filtrate in the kidney tubule?

   A. H_2O
   B. Urea
   C. Na^+
   D. Albumin

Answer is B: Urea (a waste product) is lipid soluble so can passively move out of the filtrate into the tubule cells. It is secreted back into the filtrate to effect its excretion.

4. Which of the following happens as we descend deeper into the kidney medulla?

   A. The concentration of the interstitial fluid does not change.
   B. The concentration of the interstitial fluid increases.
   C. The concentration of the filtrate within the tubule increases.
   D. The concentration of the interstitial fluid decreases.

Answer is B: Concentration of the interstitial fluid increases. Choice C is not correct as the nephron tubule does not necessarily descend into the medulla.

5. Which part of the renal tubule is impermeable to water?
   A. The ascending limb of the loop of Henle
   B. The collecting duct
   C. The proximal convoluted tubule
   D. The thin portion of the loop of Henle

Answer is A: Water does not pass through the ascending limb. However, solutes (Na and Cl) do pass through.

6. In the glomerulus, what is the method by which solutes are transferred from the blood to the Bowman's capsule?
   A. Diffusion
   B. Active transport
   C. Secretion
   D. Filtration

Answer is D: Filtration due to the pressure difference between the capillary blood and fluid in the Bowman's capsule.

7. How does the juxtaglomerular apparatus respond when systemic blood pressure is too high?
   A. The juxtaglomerular cells send a message to the afferent arteriole to dilate.
   B. The macula densa sends a message to the efferent arteriole to constrict.
   C. The macula densa sends a message to the afferent arteriole to constrict.
   D. The granular cells release renin which causes systemic arterioles to constrict.

Answer is C: Constricting the afferent arteriole will decrease the volume of blood entering the glomerulus and hence the blood pressure in it. All other choices will result in an increase in pressure within the glomerulus.

8. What is the kidney tubule's response to a rise in blood pH?
   A. Bicarbonate ions are created from carbonic acid and absorbed into the blood.
   B. Hydronium ions are secreted into the filtrate, where they are buffered by bicarbonate ions in the filtrate.
   C. Bicarbonate ions are secreted into the filtrate, while hydronium ions are absorbed from the filtrate into the blood.
   D. Hydronium ions are secreted into the filtrate, while bicarbonate ions are absorbed from the filtrate into the blood.

Answer is C: A rise in blood pH means it is getting more alkaline. That is there is too much bicarbonate in the blood. To counter this, some bicarbonate ions are secreted into the filtrate, while some hydronium ions are absorbed from the filtrate into the tubule cells.

9. How does the composition of the filtrate change as it travels through the loop of Henle?
   A. In the ascending limb, the volume decreases and in the descending limb, the concentration increases.
   B. In the descending limb, the volume decreases and in the ascending limb, the concentration decreases.
   C. In the descending limb, the volume decreases and in the ascending limb, the concentration increases.
   D. In the ascending limb, the volume decreases and in the descending limb, the concentration decreases.

Answer is B: The descending limb is permeable to water so water flows out of the tubule decreasing filtrate volume. The ascending limb is impermeable to water, so its volume does not change; however, sodium, potassium and chloride ions are reabsorbed which decreases the concentration of the filtrate.

10. What method does the glomerulus of the kidney nephron use to remove the dissolved substances from the blood to the filtrate?
    A. Active transport
    B. Diffusion along the concentration gradient
    C. High hydrostatic pressure
    D. Osmosis

Answer is C: The larger diameter of the afferent arteriole along with the smaller diameter of the efferent arteriole creates a high pressure in the glomerulus which facilitates filtration of dissolved material into the Bowman's capsule.

11. If the glomerular filtration rate is too high, the macula densa sends a message to the afferent arteriole. What is the effect of this message?
    A. Granular cells of arteriole walls release renin.
    B. Afferent arteriole dilates.
    C. It inhibits the action of ATP and adenosine on the afferent arteriole.
    D. Afferent arteriole constricts.

Answer is D: Constricting the afferent arteriole will decrease the rate of flow of blood into the glomerulus. This will decrease the GFR.

12. The descending limb of the loop of Henle is permeable to water so water diffuses out of the descending limb into the interstitial fluid. What happens to this water?
    A. It flows through the renal papillae into the minor calyces to become urine.
    B. It diffuses into the ascending limb of the Loop of Henle.
    C. It diffuses into the peritubular capillaries and ascending vasa recta for return to the blood.
    D. It diffuses into the filtrate for elimination from the body.

Answer is C: Water (and solutes) that are reabsorbed from the filtrate return to the blood stream via the peritubular capillaries and the vasa recta.

13. The concentration of blood is 280–300 mosmol/L, but may rise to 1200 mosmol/L in which situation?
    A. In the vasa recta of the kidney
    B. In severe dehydration
    C. In the peritubular capillaries of the kidney
    D. In severe over-hydration

Answer is A: Juxtamedullary nephrons have a capillary bed known as the vasa recta which descends into the kidney medulla. As it does so the concentration of the contained blood increases to 1200 mosmol/L.

14. When systemic blood pressure increases, how does the kidney respond to maintain glomerular filtration rate?
    A. The afferent arteriole dilates.
    B. The efferent arteriole constricts.
    C. The efferent arteriole dilates.
    D. The afferent arteriole constricts.

Answer is D: An increase in BP stretches the walls of the efferent arteriole which respond by constricting. The resulting decrease in diameter decreases glomerular blood flow which keeps GFR within normal limits.

15. What may correctly be said of the juxtaglomerular apparatus (or complex)?
    A. The juxtaglomerular cells are chemoreceptors.
    B. The granular cells are chemoreceptors.
    C. The macula densa cells are chemoreceptors.
    D. The macula densa cells are mechanoreceptors.

Answer is C: The macula densa cells are chemoreceptors that respond to changes in Na^+ and Cl^- ions. The granular cells are mechanoreceptors.

16. Which of the following mechanisms operates in the nephron to maintain pH balance in the body?
    A. Bicarbonate ions from the tubule cells are secreted into the filtrate, then $H^+ + HCO_3^- \rightarrow H_2CO_3 \rightarrow CO_2 + H_2O$.
    B. In tubule cells $CO_2 + H_2O \rightarrow H_2CO_3 \rightarrow H^+ + HCO_3^-$ then hydrogen ions are secreted into the filtrate.
    C. In tubule cells $CO_2 + H_2O \rightarrow H_2CO_3 \rightarrow H^+ + HCO_3^-$ then hydrogen ions are transported into the peritubular capillaries.
    D. In tubule cells $CO_2 + H_2O \rightarrow H_2CO_3 \rightarrow H^+ + HCO_3^-$ then bicarbonate ions are secreted into the filtrate.

Answer is B: Carbon dioxide and water form hydronium and bicarbonate ions, the hydronium ions being secreted into the filtrate for excretion in urine.

17. What influences and structures facilitate blood filtration in the renal corpuscle?
    A. High osmotic pressure in the capillaries and sinusoidal capillaries
    B. High hydrostatic pressure in the capillaries and fenestrated capillaries
    C. High osmotic pressure in the capillaries and fenestrated capillaries
    D. High hydrostatic pressure in the capillaries and sinusoidal capillaries

Answer is B: High hydrostatic pressure facilitates filtration and pores in the capillary walls (fenestre) allow easy passage.

18. What effect is achieved by having an arteriole that supplies blood and another that drains blood from the glomerulus?
    A. Oxygen-rich blood can be supplied to the nephron after blood leaves the glomerulus.
    B. The blood pressure within the glomerulus can be manipulated.
    C. Reabsorption of water and nutrients from the filtrate is facilitated.
    D. The concentration gradient within the kidney's medulla can be maintained.

Answer is B: Arterioles have smooth muscle in their wall which allows them to constrict and dilate, hence adjusting the flow through them and consequently the blood pressure within them.

19. In which part of the nephron does most of the reabsorption of water and solutes occur?
    A. Collecting duct
    B. Nephron loop (loop of Henle)
    C. Vasa recta
    D. Proximal convoluted tubule

Answer is D: The PCT absorbs the bulk of the water and solutes.

20. From which part of the nephron is the greatest proportion of Na^+ absorbed from the filtrate?
    A. Bowman's capsule (i.e. renal capsule)
    B. Proximal convoluted tubule
    C. Ascending limb of the loop of Henle
    D. Distal convoluted tubule

Answer is B: The PCT absorbs about 65% of the sodium from the filtrate.

21. What part of the renal tubule is **NOT** able to reabsorb water?
    A. The descending limb of the loop of Henle
    B. The proximal convoluted tubule
    C. The ascending limb of the loop of Henle
    D. The distal convoluted tubule

Answer is C: The ascending limb is impermeable to water.

22. The influence (or influences) that drives blood filtration in the kidney is:
    A. Difference in osmolarity between blood in the glomerulus and filtrate in the Bowman's capsule
    B. Fluid pressure difference between blood in the glomerulus and filtrate in the Bowman's capsule
    C. Osmotic pressure difference between blood in the glomerulus and filtrate in the Bowman's capsule
    D. Diffusion along the concentration gradient between blood and filtrate, and active transport

Answer is B: Filtration occurs because of a pressure difference. Osmotic pressure does not drive the process.

23. Which type of anti-hypertensive drug aims to prevent vasoconstriction?
    A. Beta-blockers
    B. Diuretics
    C. ACE inhibitors
    D. Calcium channel blockers

Answer is D: The contraction of smooth muscle causes vasoconstriction. Calcium is necessary for muscle contraction. A Ca channel blocker prevents Ca entering the cytoplasm and so prevents vasoconstriction.

24. The filtrate that is formed in the kidney contains all of the following except one. Which one?
    A. Metabolic wastes
    B. Electrolytes
    C. Plasma proteins
    D. Nutrients

Answer is C: Plasma proteins are too large to pass through the filtration slits of the glomerular capillaries.

25. What is the term applied to the first process in urine formation, where some components of blood pass into the Bowman's capsule?
    A. Filtration
    B. Active transport
    C. Dialysis
    D. Osmosis

Answer is A: Filtration is the first process driven by a fluid pressure gradient.

26. In what part (or parts) of the renal tubule reabsorb the least material from the filtrate?
    A. Distal convoluted tubule
    B. Loop of Henle
    C. Proximal convoluted tubule

D. Distal convoluted tubule and the collecting duct together

Answer is D: By the time the filtrate has passed through the PCT and loop of Henle, the majority of the reabsorption that is going to happen has happened. Nevertheless, the DCT and collecting duct do reabsorb some electrolytes and water.

27. By what name is the condition where nitrogenous wastes accumulate in the blood known?
    A. Anuria
    B. Uremia
    C. Polyuria
    D. Oliguria

Answer is B: Uremia. The other terms describe the volume of urine produced. Anuria = no urine; polyuria = far more than normal; oliguria = less than normal.

28. In the nephron, if the afferent arteriole dilates and the efferent arteriole constricts, which of the following would be true?
    A. The glomerular filtration rate would decrease.
    B. The pressure in the glomerulus will decrease.
    C. The absorption of sodium and chloride ions forms the filtrate would increase.
    D. Before these events, the granular cells would have released renin.

Answer is D: The granular cells of the afferent arteriole respond to low blood pressure by releasing renin. As a result, angiotensin II would form which causes systemic vasoconstriction, which increases blood pressure, dilates the afferent arteriole and increases glomerular filtration rate.

29. When is the majority of material reabsorbed from the filtrate reabsorbed from the renal tubule?
    A. After the loop of Henle
    B. In the descending limb of the loop of Henle
    C. Before the loop of Henle
    D. In the ascending limb of the loop of Henle

Answer is C: The majority of reabsorption occurs in the proximal convoluted tubule which is before the loop of Henle.

30. What is the term used to describe the production of an insufficient volume of urine?
    A. Polyuria
    B. Uremia
    C. Anuria
    D. Oliguria

Answer is D: Oliguria is the production of about 50–500 ml urine per day. Anuria is producing less than 50 ml per day.

31. By what process(es) do water and solutes move from blood in the glomerulus into the Bowman's capsule?
    A. Diffusion
    B. Osmosis and diffusion
    C. Filtration
    D. Dialysis

Answer is C: Filtration which is driven by the hydrostatic pressure difference between blood (at high pressure) in the glomerular capillaries and the filtrate (at lower pressure) in the Bowman's capsule.

32. What part of the renal tubule does **NOT** reabsorb water?
    A. The juxtaglomerular apparatus
    B. The ureter
    C. The ascending limb of the loop of Henle
    D. The collecting duct

Answer is C: The ascending limb is the only listed structure that is part of the renal tubule.

33. What is the influence(s) that drives blood filtration in the kidney?
    A. Dialysis through a semi-permeable membrane due to the different osmolarity of blood and filtrate
    B. Fluid pressure difference between blood and filtrate
    C. Osmotic pressure difference between blood and filtrate
    D. Diffusion along a concentration gradient between blood and filtrate, and active transport

Answer is B: The term "filtration" refers to a process that is driven by a pressure difference.

34. What does the term "oliguria" refer to?
    A. A daily urine production of much more than 2 L
    B. A daily urine production of less than 500 ml
    C. Production of less than 50 ml of urine in a day
    D. The condition of excessive concentration of urea in the blood

Answer is B: Oliguria refers to a less than normal daily volume of urine being produced. Normal is about 2 L per day.

35. Glomerular filtration rate can be altered by all of the following **EXCEPT** one. Which one?
    A. Constriction of renal tubule by macula densa
    B. Vasoconstriction of afferent arteriole
    C. Decrease in concentration of plasma proteins
    D. Vasoconstriction of efferent arteriole

Answer is A: Macula densa does not constrict the tubule. A decrease in plasma protein would increase the osmotic pressure opposing glomerular filtration.

36. One way to increase the glomerular filtration rate is to dilate:
    A. The afferent arteriole and to constrict the efferent arteriole.
    B. The efferent arteriole and to constrict the afferent arteriole.
    C. Both the afferent arteriole and the efferent arteriole.
    D. The efferent arteriole and to increase the permeability of the capillary endothelium.

Answer is A: Dilating the afferent arteriole increases the volume of blood flowing into the glomerulus, while constricting the efferent arteriole decreases the volume of blood leaving the glomerulus. This means water and solutes must leave the glomerulus and enter the Bowman's capsule, that is GFR increases.

37. What is the term meaning the production of urine?
    A. Oliguria
    B. Diuresis
    C. Hypouria
    D. Anuria

Answer is B: A drug that promotes urine production is a diuretic, while one that does the reverse is an antidiuretic.

38. The renal tubule reabsorbs all of the following ions except one. Which one?
    A. Na^+
    B. HCO_3^-
    C. Cl^-
    D. NH_4^+

Answer is D: Ammonium ions are a waste product so are not reabsorbed.

39. The gradient in osmotic concentration of the interstitial fluid in the medulla of the kidney is caused by:
    A. Blood flowing through the vasa recta
    B. Na^+ and Cl^- pumped out of the ascending limb of the loop of Henle and urea
    C. Na^+ and Cl^- pumped out of the descending limb of the loop of Henle and urea
    D. Water diffusing out of the descending limb of the loop of Henle and aldosterone

Answer is B: Sodium and chloride leave from the ascending limb not the descending limb.

40. Which of the following chemicals is produced by the kidney?
    A. Angiotensinogen
    B. Bicarbonate ions
    C. Sodium ions
    D. Vitamin C

Answer is B: Carbon dioxide and water react to form carbonic acid which dissociates into hydronium ions and bicarbonate ions. The bicarbonate ions are transported to the blood while the hydronium is excreted in acidic urine.

41. What could be concluded of a person who (during the previous 3 h) has produced a total of 100 ml of urine that is bright yellow and has a strong (but not unpleasant) odour?
    A. They are well hydrated.
    B. Their urine will have a high specific gravity.
    C. They are an uncontrolled diabetic.
    D. They have more than the usual concentration of bilirubin in their urine.

Answer is B: 100 ml/3 h is a low rate of urine production, indicating dehydration. Hence the urine will be concentrated as the body is conserving water and its SG will be high.

42. In which section of the renal tubule is most water reabsorbed?
    A. Descending limb of the loop of Henle
    B. Ascending limb of the loop of Henle
    C. Collecting duct
    D. Proximal convoluted tubule

Answer is D: About 2/3 of water in the filtrate is absorbed from the PCT.

43. The density of water is 1.00 g/ml, and it has a specific gravity of 1.000. Which of the following is most likely to be the specific gravity of a urine sample?
    A. 1.000
    B. 1.015
    C. 0.980
    D. 1.020 g/ml

Answer is B: SG has no unit so choice D is wrong. Choice C cannot be correct as the density is less than that of water.

44. Which of the following are organic wastes produced by the body?
    A. Uric acid and ammonium ions
    B. Amino acids and potassium ions
    C. Albumin and globulin
    D. Urea and sodium ions

Answer is A: Ammonium ions are produced when amino acids are deaminated. Sodium is not organic.

45. What is indicated if the specific gravity of a patient's urine is high?
    A. The patient has kidney disease.
    B. The urine's concentration is high.
    C. The patient is well hydrated.

D. The urine density is low.

Answer is B: High urine concentration correspond to a high SG.

46. If a urine specific gravity was measured to be 1.03, an interpretation would be that:

    A. The person was dehydrated.
    B. The person was well hydrated.
    C. The urine sample had a density less than water.
    D. The urine was dilute.

Answer is A: 1.03 is at the upper limit of urine SG, indicating that the kidney is conserving water so the person is dehydrated.

47. If a urine specific gravity was measured to be 1.003, an interpretation would be that:

    A. The person was dehydrated.
    B. The person was well hydrated.
    C. The urine sample had a density less than water.
    D. The urine was concentrated.

Answer is B: 1.003 is close to the lower limit of urine SG, indicating that the kidney is not conserving water so the person is well hydrated.

48. Which of the following statements about urine specific gravity is **WRONG**?

    A. If urine specific gravity is 1.003, the urine is dilute.
    B. A urine specific gravity value of 1.015 = 1015 mmol/L.
    C. If urine specific gravity is 1.030, the person is dehydrated.
    D. A specific gravity value of 1.010 is equal to a urine density of 1.010 g/ml.

Answer is B: Urine SG is the same value as urine density in g/ml, but without the units.

49. Given that the specific gravity of a urine sample is 1.009, which of the following statements is correct?

    A. The patient is dehydrated.
    B. The sample contains 1.009 mmol/L of dissolved particles.
    C. The sample contains 1.009 mmol of dissolved solutes.
    D. The urine has a density of 1.009 g/ml.

Answer is D: SG is a measure of urine density (or concentration). It is the ratio of urine density to water density, so has no units.

50. Glucose and other organic solutes are reabsorbed from the filtrate by which cells?

    A. Macula densa cells in the juxtaglomerular apparatus
    B. Cuboidal epithelial cells in the proximal convoluted tubules

C. Juxtaglomerular cells in the loop of Henle
D. Podocyte cells around the capillaries

Answer is B: The cells of the wall of the PCT absorb most wanted nutrients from the glomerular filtrate.

51. Only one of the following statements about the juxtaglomerular apparatus is **WRONG**. Which one?
    A. Macula densa cells are chemoreceptors that can secrete a paracrine message that causes the afferent arteriole to constrict.
    B. Macula densa cells are chemoreceptors that can secrete a paracrine message that results in the dilation of the afferent arteriole.
    C. Granular cells are mechanoreceptors that release renin that causes systemic arteries to constrict.
    D. Granular cells are chemoreceptors that release renin that causes systemic arteries to constrict.

Answer is D: Granular cells are not chemoreceptors. Systemic arteries constrict due to angiotensin II which is formed after the release of renin.

52. Which of the following would be indicative of urine with a high concentration of solutes?
    A. Specific gravity of 1.002
    B. Pale straw colour
    C. Measured volume of 500 ml on voiding
    D. Specific gravity of 1.028

Answer is D: A specific gravity of 1.028 is very high and indicates that the kidney is reabsorbing much water from the filtrate. This would leave the voided urine with a high concentration of dissolved solutes. Choice C is not correct, while 500 ml is a small volume of urine to be produced in 1 day, it is not necessarily a small volume for a single void.

53. What does the glomerulus do?
    A. Filters a solute-rich, protein-free liquid from blood to produce "filtrate".
    B. Absorbs required ions from the filtrate or secretes unwanted substances into the filtrate.
    C. Monitors blood pressure and sends paracrine message to afferent arteriole to constrict or dilate as required.
    D. Produces the enzyme renin.

Answer is A: The glomerulus is the vascular part of nephron. The endothelium of glomerular capillaries is fenestrated i.e. porous, to facilitate filtration. Choice B refers to cells of the renal tubule; C refers to the juxtaglomerular apparatus.

54. What happens in the vasa recta of the kidney?
    A. The volume of filtrate in this tube decreases as water is reabsorbed.

B. The osmolarity of blood in this vessel increases from 300 to 1200 mosmol/L (and vice versa).
C. Solutes and water reabsorbed from the proximal convoluted tubule return to circulation by entering this vessel.
D. The vasa recta surrounds cortical nephrons to carry away water reabsorbed from the filtrate.

Answer is B: The vasa recta is a bed of permeable capillaries that surrounds the long loop of Henle of juxtaglomerular nephrons. Together they penetrate deep into the renal medulla. Choice C refers to the peritubular capillaries. Choice D is wrong as the vasa recta is associated with juxtamedullary nephrons, not cortical ones.

55. What happens in the Bowman's capsule (aka renal capsule)?
    A. Bowman's capsule filters solutes from the blood.
    B. Bowman's capsule collects formed urine before it is directed into the ureter.
    C. It is composed of macula densa cells and granular cells which produce paracrine hormones and renin, respectively.
    D. Bowman's capsule receives the filtrate from the glomerulus.

Answer is D: Bowman's capsule surrounds the glomerulus (together they form the "renal corpuscle") and receives the raw filtrate solution before it enters the proximal convoluted tubule. Choice C refers to the juxtaglomerular apparatus.

56. With reference to the kidney tubule, which statement correctly distinguishes between filtration and osmosis? Filtration happens in the:
    A. Afferent arteriole, while osmosis occurs in the efferent arteriole.
    B. Peritubular capillaries, while osmosis occurs in the vasa recta.
    C. Glomerulus, osmosis happens in the descending limb of the loop of Henle.
    D. Ascending limb of the loop of Henle, osmosis happens in the glomerulus.

Answer is C: The glomerular capillaries contain blood at higher pressure than in other capillaries, so water and solutes are filtered out of blood (under pressure). Water leaves the descending limb osmotically because of the concentration gradient between filtrate in the tubule and the fluid in the interstitial space.

57. Which of the following is NOT one of the three basic processes in urine formation?
    A. Glomerular filtration.
    B. Juxtaglomerular renin production.
    C. Tubular reabsorption.
    D. Tubular secretion.

Answer is B: The granular cells of the juxtaglomerular apparatus do produce renin, but this enzyme does not directly affect urine production.

58. What is the function of the juxtaglomerular apparatus of the nephron?
    A. To regulate the secretion of urea and sodium ion reabsorption.
    B. To regulate water absorption in the collecting duct and the secretion of urea.
    C. To regulate the filtration rate of the glomerulus and water absorption in the collecting duct.
    D. To regulate blood pressure and the filtration rate of the glomerulus.

Answer is D: The JG apparatus contains granular cells which produce renin which catalyses the formation of angiotensin II and hence raises blood pressure. It also has macula densa cells which are chemoreceptors and function in stimulating renin release. Some macula densa cells produce paracrine messengers that signal the afferent arteriole to constrict to reduce the GFR.

59. How does the composition of the filtrate in the nephron change as it passes through the descending limb (DL) and then the ascending limb (AL) of the loop of Henle?
    A. In DL, filtrate concentration increases while volume of filtrate decreases; in AL, filtrate concentration decreases.
    B. In DL, volume of filtrate increases while concentration decreases; in AL, volume of filtrate decreases.
    C. In DL, filtrate concentration increases, while volume of filtrate increases; in AL, volume of filtrate decreases.
    D. In DL, volume of filtrate decreases, while concentration decreases; in AL, concentration of filtrate increases.

Answer is A: In the DL, filtrate concentration increases as water flows osmotically into the interstitial space. This causes the volume of filtrate to decrease. In the AL, filtrate concentration decreases as Na, Cl and K ions move out of the tubule and into the interstitial fluid. The AL is impermeable to water, so the volume of filtrate does not change in the AL.

60. What are the four main physiological processes that contribute to urine formation?
    A. Filtration of blood, reabsorption, secretion of substances, concentration of urine.
    B. Filtration of blood, reabsorption of water, secretion of substances, reabsorption of urea.
    C. Filtration of blood, reabsorption of water, secretion of ammonium, concentration of urine.
    D. Dialysis of blood, reabsorption of nutrients, secretion of substances, reabsorption of water.

Answer is A: Blood is filtered in the renal capsule; then a lot of water and nutrients and ions are reabsorbed in the PCF; then some ions, toxins, metabolites and molecules that were not filtered are secreted into the filtrate; finally the concentration of urine is adjusted to suit the body's needs. Either even more water is reabsorbed (in the DCT and collecting duct) to produce concentrated urine, or if the body is well hydrated, less water is reabsorbed and dilute urine is produced.

61. What does the specific gravity of urine measure?
    A. The density of urine.
    B. The density of urine compared to the density of water.
    C. Urine concentration.
    D. The concentration of dissolved particles in urine.

Answer is B: All answers have some truth to them. However, choices A, C and D require a unit after the measurement. SG measures the density of urine divided by the density of water (which is 1.000) and, being a ratio, has no units. Density of a solution indicates solution concentration. Adding solutes to water increases the mass of the solution faster than it increases volume; hence, density increases as solution concentration increases. Clinically, we are interested in urine concentration as an indicator of the level of hydration.

62. What is a typical value for the pH of urine, and WHY?
    A. 8.2 (the body usually needs to excrete base).
    B. 7.0 (urine is typically neutral).
    C. 6.2 (the body usually needs to excrete acid).
    D. 2.2 (concentrated urine is able to excrete a lot of acid).

Answer is C: Urine is almost always slightly acidic. Blood must maintain its pH between 7.35 and 7.45. Much of the food we eat is acidic, and many metabolic products are acidic, so we have a need to excrete acid. The kidneys do this.

63. The normal plasma concentration of Na^+ is about 140 mmol/L while that of K^+ is about 4 mmol/L. Which of the following is a reasonable concentration for these two ions in urine?
    A. 140 mmol/L for Na^+ and 4 mmol/L for K^+
    B. 1400 mmol/L for Na^+ and 40 mmol/L for K^+
    C. 72 mmol/L for Na^+ and 221 mmol/L for K^+
    D. 221 mmol/L for Na^+ and 72 mmol/L for K^+

Answer is C: Sodium is in much greater concentration than potassium in the blood. However, the kidney tubules reabsorb sodium from the filtrate in exchange for potassium. Consequently the concentration of potassium in urine is usually higher than is the concentration of sodium.

11.3 Hormones Affecting the Renal System (Angiotensin, Renin, Aldosterone, ADH, ANP)

1. What molecule catalyses the formation of angiotensin I?
    A. Carbonic anhydrase
    B. Calcitriol
    C. Erythropoietin

D. Renin

Answer is D: Renin is an enzyme that catalyses the reaction that forms angiotensin I from angiotensinogen.

2. What is the body's response to a rise in blood plasma osmotic pressure?
   A. The anterior pituitary releases ADH which makes the renal tubule permeable to water.
   B. The posterior pituitary releases ADH which makes the renal tubule permeable to water.
   C. The juxtaglomerular apparatus releases renin which promotes diuresis.
   D. The glomerular filtration rate increases so more urine is produced.

Answer is B: ADH, from the posterior pituitary, would allow more water to be reclaimed from the filtrate, so that plasma osmotic pressure does not increase further. Renin and its effects does not promote diuresis.

3. Which one of the following is **NOT** produced by the kidneys?
   A. Aldosterone
   B. Renin
   C. Erythropoietin
   D. Calcitriol

Answer is A: Aldosterone is produced in the adrenal cortex (not the kidneys). The kidneys along with the skin and liver contribute to producing calcitriol.

4. What is the effect of antidiuretic hormone on the kidney tubules?
   A. It causes Na^+ to be absorbed from the filtrate into the tubular cells.
   B. It causes the concentration of urine to decrease.
   C. It causes the filtrate volume to increase.
   D. It causes the walls of the collecting duct to become permeable to water.

Answer is D: ADH causes more water channels (aquaporins) to be inserted in the wall of the collecting duct. Hence the duct is more permeable to water.

5. What is the resulting effect of renin being released by the kidney?
   A. Angiotensin II is formed.
   B. Aldosterone is released.
   C. Macula densa sends paracrine message to afferent arterioles.
   D. Efferent arterioles are constricted.

Answer is A: Renin catalyses the formation of angiotensin I from angiotensinogen. ACE then converts angiotensin I to angiotensin II. Aldosterone is released from the adrenal cortex due to angiotensin II. Angiotensin II also promotes systemic vasoconstriction (not just in efferent arteriole).

6. What is the effect on the kidney caused by increasing the release of ADH?
   A. The collecting duct becomes more permeable to water.
   B. The ascending limb of the loop of Henle becomes impermeable to water.
   C. The descending limb of the loop of Henle becomes permeable to water.
   D. The collecting duct becomes impermeable to water.

Answer is A: Antidiuretic hormone prevents diuresis (decreases urine formation) by causing more water to be reabsorbed from the filtrate as it passes through the collecting duct.

7. The kidneys produce all of the following **EXCEPT** one. Which one?
   A. Erythropoietin
   B. Angiotensinogen
   C. Hydronium ions
   D. Bicarbonate ions

Answer is B: Angiotensinogen is produced in the liver. Renin (produced in the kidneys) converts it to angiotensin I.

8. Complete the following sentence correctly. Angiotensin II:
   A. Stimulates the adrenal glands to release aldosterone
   B. Increases potassium reabsorption from the filtrate
   C. Increases sodium (Na^+) excretion at the kidneys
   D. Reduces our thirst

Answer is A: Aldosterone is produced by the adrenal cortex in response to angiotensin II. Angiotensin II also stimulates thirst.

9. Which hormone causes an increase in permeability to water in the collecting ducts of the kidney?
   A. Antidiuretic hormone
   B. Aldosterone
   C. Angiotensin II
   D. Atrial natriuretic hormone

Answer is A: ADH. If the collecting ducts are permeable to water, more water will be reclaimed from the filtrate. This will decrease its volume, and there will be less diuresis. Hence antidiuretic hormone.

10. Which hormone causes increased sodium reabsorption in the kidney?
    A. Angiotensin I
    B. Antidiuretic hormone
    C. Vasopressin
    D. Aldosterone

Answer is D: Aldosterone stimulates the active absorption of sodium (and the secretion of potassium) from the filtrate.

11. The kidneys produce all of the following except one. Which one?
    A. Enzyme renin
    B. Hormone erythropoietin
    C. Antidiuretic hormone
    D. Vitamin calcitriol

Answer is C: ADH is produced in the hypothalamus and released from the posterior pituitary.

12. What effect does aldosterone have?
    A. Increases the absorption of Na^+ from the kidney tubules
    B. Makes the kidney tubules more permeable to water
    C. Catalyses the formation of angiotensin I
    D. Blocks the release of ADH

Answer is A: Aldosterone stimulates the active absorption of sodium from the filtrate (and the secretion of potassium).

13. The kidneys produce all of the following **EXCEPT** one. Which one?
    A. Erythropoietin
    B. Aldosterone
    C. Renin
    D. Active vitamin D

Answer is B: Aldosterone is produced in the adrenal cortex.

14. Which statement about the descending limb of the loop of Henle is true?
    A. It is freely permeable to water.
    B. It is impermeable to water.
    C. It is impermeable to water when ANP is present.
    D. It is permeable to water when ADH is present.

Answer is A: The ascending limb is impermeable to water. ADH and ANP have no effect on the loop of Henle.

15. Complete the following sentence correctly. Antidiuretic hormone:
    A. Stimulates our thirst
    B. Causes the wall of the collecting duct of the nephron to increase in permeability to water
    C. Increases sodium (Na^+) excretion at the kidneys
    D. Stimulates the adrenal glands to release aldosterone

Answer is B: Increased permeability of the collecting duct means that more water will be reclaimed from the filtrate. Angiotensin II stimulates thirst. ANP diminishes thirst and increases sodium secretion at the kidneys.

16. Complete the following sentence correctly. Angiotensin II:
    A. Stimulates the adrenal glands to release aldosterone
    B. Causes the wall of the collecting duct of the nephron to increase in permeability to water
    C. Increases sodium (Na^+) excretion at the kidneys
    D. Reduces our thirst

Answer is A: ADH increases the permeability of the collecting duct to water. ANP diminishes thirst and increases sodium secretion at the kidneys.

17. Complete the following sentence correctly. Atrial natriuretic peptide:
    A. Causes the wall of the collecting duct of the nephron to increase in permeability to water
    B. Increases sodium (Na^+) excretion at the kidneys
    C. Causes peripheral vasoconstriction
    D. Stimulates the adrenal glands to release aldosterone

Answer is B: ADH increases the permeability of the collecting duct to water. Angiotensin II stimulates the adrenal glands to release aldosterone and promotes peripheral vasoconstriction.

18. What is the role of aldosterone?
    A. To convert angiotensinogen into angiotensin I
    B. To inhibit the absorption of Na^+
    C. To promote the absorption of Na^+
    D. To promote the absorption of Ca^{++}

Answer is C: Sodium ions are always reabsorbed from the filtrate, Aldosterone promotes greater reclamation of sodium from the DCT.

19. Which one of the statements about the collecting ducts is true? The collecting duct absorbs:
    A. Calcium if parathyroid hormone is present
    B. Water if antidiuretic hormone is present
    C. Sodium if atrial natriuretic hormone is NOT present
    D. Urea

Answer is B: ADH causes aquaporins (water channels) to be inserted into the collecting duct walls. The collecting duct is impermeable to the other substances.

20. What is the function of angiotensin II?
    A. Causes constriction of systemic arteries
    B. Causes the collecting ducts to become permeable to water
    C. Causes the formation of atrial natriuretic hormone
    D. Causes constriction of the efferent arterioles

Answer is A: Angiotensin II promotes general vasoconstriction (which raises blood pressure) and stimulates the adrenal cortex to release aldosterone.

21. Under what conditions will the kidney produce concentrated urine? If:
    A. Glomerular filtration rate is low.
    B. Glomerular filtration rate is high.
    C. Atrial natriuretic peptide and aldosterone are present in blood.
    D. Antidiuretic hormone and aldosterone are present in the blood.

Answer is D: Aldosterone causes sodium to be absorbed from the filtrate at the DCT and collecting duct. This sets up an osmotic gradient. ADH makes the tubule more permeable to water. Hence water can move from the filtrate into the tubule cells along the osmotic gradient.

22. Which one of the following is a function of the renal system?
    A. Produce bile
    B. Produce the enzyme renin
    C. Produce the hormone aldosterone
    D. Produce vitamin K

Answer is B: Renin is produced in the kidney.

23. If aldosterone is present in the blood, what happens in the distal convoluted tubule?
    A. Sodium ions are reabsorbed from the filtrate.
    B. Calcium ions are reabsorbed from the filtrate.
    C. Sodium ions are secreted into the filtrate.
    D. Bicarbonate ions are reabsorbed from the filtrate.

Answer is A: Aldosterone promotes the absorption of sodium from the filtrate.

24. What is renin?
    A. An enzyme released by the juxtaglomerular cells of the kidney when arterial pressure falls.
    B. It catalyses the formation of angiotensin II in the lungs.
    C. It is a rapid acting, intense vasoconstrictor of arterioles.
    D. A protein that stimulates the adrenal glands to release aldosterone.

Answer is A: The JG cells produce renin which leads to angiotensin II being formed which will cause blood pressure to rise.

25. Which ion does aldosterone stimulate the kidneys to reabsorb?
    A. Calcium
    B. Sodium
    C. Potassium
    D. Bicarbonate

Answer is B: Sodium absorption is what aldosterone promotes.

26. If the blood concentration is rising towards 300 mosmol/L, more ADH would be released. What effect would this have on the concentration of blood?
    A. The concentration of blood would stop rising.
    B. The concentration of blood would fall.
    C. The rate of the rise in blood concentration would slow.
    D. ADH does not affect the concentration of blood.

Answer is C: Blood concentration rises as water is lost from the body by evaporation and urine is formed. ADH causes water to be reabsorbed from the kidney filtrate, so the rise in blood concentration would slow as more water is returned to the blood rather than being filtered out into urine. Blood concentration would not fall until some water has been drunk.

27. Which of the following molecules is NOT produced by the kidneys?
    A. Angiotensinogen
    B. Renin
    C. Erythropoietin
    D. Calcitriol

Answer is A: Angiotensinogen is produced in the liver. Renin (from the juxtaglomerular cells of the kidneys) converts angiotensinogen into angiotensin I.

28. Angiotensin II causes the following things to occur EXCEPT for one. Which one?
    A. It causes constriction of the systemic arterioles.
    B. It causes the release of aldosterone.
    C. It stimulates thirst.
    D. It encourages the absorption of water from the kidney filtrate.

Answer is D: Angiotensinogen II causes choices A, B and C to happen. While aldosterone does cause Na^+ to be reabsorbed from the filtrate and if ADH is present to promote the insertion of aquaporins into the tubule cell plasma membranes, water will be absorbed from the filtrate. But this is not a direct result, so “D” is the best choice of answer.

29. Caffeine is a diuretic. What is its mode of action?
    A. It inhibits Na^+ reabsorption from the filtrate, hence the water reabsorption that follows.
    B. It promotes the secretion of antidiuretic hormone (ADH), hence the insertion of aquaporins in the tubule cell membrane.
    C. It promotes the secretion of aldosterone, hence the absorption of Na^+.
    D. It inhibits the secretion of atrial natriuretic peptide (ANP), hence the absorption of Na^+.

Answer is A: A diuretic causes a greater volume of urine to be produced. This will happen if water is NOT reabsorbed from the filtrate. The other three choices will result in a decrease in the volume of urine produced.

30. Alcohol is a diuretic. How does it produce this effect?
    A. It promotes the production of aldosterone.
    B. It inhibits the production of antidiuretic hormone.
    C. It inhibits the production of atrial natriuretic peptide.
    D. It promotes the release of angiotensin II.

Answer is B: A diuretic causes a greater volume of urine to be produced. ADH promotes the reabsorption of water, so a lack of it will mean less water is reabsorbed from the filtrate and a greater volume of urine will be produced. The other three choices will cause a decrease in urine volume.

31. The kidney produces all but one of the following. Which one?
    A. Calcitriol
    B. Atrial natriuretic peptide
    C. Renin
    D. Bicarbonate ions

Answer is B: ANP is produced by the walls of the heart's atria. It inhibits the absorption of sodium from the filtrate.

32. What is the purpose of ANP in urine production?
    A. Stimulate the reabsorption of Na^+
    B. Stimulate the reabsorption of Ca^{++}
    C. Inhibit the reabsorption of Na^+
    D. Stimulate the reabsorption of water

Answer is C: ANP triggers the dilation of the afferent arterioles and constriction of efferent arterioles which increases GFR. It also decreases sodium reabsorption from the filtrate. Hence, urine volume increases (and blood volume and pressure decreases).

33. What does the presence of aldosterone in the blood cause?
    A. Calcium to be absorbed from the DCT
    B. The collecting duct to become permeable to water
    C. More bicarbonate to be formed in the tubule cells
    D. More sodium to be reabsorbed from the DCT

Answer is D: Sodium ions are reabsorbed from the filtrate. Aldosterone promotes greater reclamation of sodium from the DCT.

34. What is the role of antidiuretic hormone (ADH) in the body?
    A. ADH makes the collecting duct and proximal convoluted tubule of the nephron more permeable to water.
    B. ADH makes the collecting duct and distal convoluted tubule of the nephron more permeable to water.
    C. ADH makes the descending limb of the nephron more permeable to water and the ascending limb less permeable to water.

D. ADH makes the glomerulus and proximal convoluted tubule of the nephron more permeable to water.

Answer is B: Antidiuretic hormone causes aquaporins (pores for water) to be inserted into the plasma membranes of the cells that make up the collecting duct. This makes the collecting duct (and distal convoluted tubule) of the nephron more permeable to water. In this way, water can be reabsorbed from the filtrate (by osmosis), and the kidney will produce a more concentrated urine.

Chapter 12
Cardiovascular System

12.1 Blood

Humans have between 4 and 7 L of blood. Fifty-five percent of this volume is plasma which is an aqueous solution of proteins, blood clotting factors, dissolved ions, wastes, dissolved gases, glucose and cholesterol. Blood also contains red blood cells (erythrocytes), platelets and five types of white blood cells (leucocytes). They are neutrophils, lymphocytes, monocytes, eosinophils and basophils. In addition blood has three buffer systems: the phosphate, the carbonic acid/bicarbonate system and the protein buffer which maintains blood pH between 7.35 and 7.45.

Blood transports: O_2 from the lungs to the body's cells by binding O_2 to haemoglobin; CO_2 from cells to the lungs; nitrogenous waste from cells to kidneys and sweat glands; absorbed nutrients from the small intestine to the liver and to cells and hormones from endocrine system to target cells. Blood also transfers heat around the body. Blood is able to coagulate to minimise blood loss when a vessel is damaged and contains antibodies and complement proteins to defend against bacteria.

Blood cells are made in the active (red) bone marrow from stem cells called haemocytoblasts by the process known as haemopoiesis. The sequence of events that lead to blood clotting is known as haemostasis, while bleeding is known as a haemorrhage. Oxygen is transported attached to haemoglobin within the RBC—do you see a pattern yet in the words that refer to blood?

Blood coagulation occurs when a clot of insoluble fibrin is formed from fibrinogen. The transformation of fibrinogen is produced by the enzyme thrombin. However, thrombin is present in blood as the inactive prothrombin (factor II). Prothrombin is converted to thrombin by prothrombinase. However, this enzyme is not present in blood either until activated by factor X. Factor X is unable to activate prothrombin until it unites with molecules that are produced by either the "intrinsic pathway" or the "extrinsic pathway" following damage to a blood vessel. This complicated arrangement ensures that all of the factors required to coagulate blood are

M. Caon, *Examination Questions and Answers in Basic Anatomy and Physiology*, https://doi.org/10.1007/978-3-030-47314-3_12

present in plasma as inactive forms but are ready to go—they do not start the coagulation process until a blood vessel has been damaged.

Blood is such a complex liquid that it is easier to transfuse type-matched donated blood into a patient rather than to attempt to synthesise it. Eight different blood types are recognised within the ABO and Rh(D) system based on the antigens (proteins) that occur on the plasma membrane of the RBC: A antigen present or B antigen present or both or neither. Then, whether the D antigen of the Rh system is present (positive) or not (negative) doubles the four ABO types. If the donated blood is not matched to the recipient's type, the antibodies in the recipient's blood plasma will attack the antigens on the donated RBC causing them to agglutinate (clump together).

1. To which of the following would the term "white cell" **NOT** be applied?
   A. Erythrocyte
   B. Leucocyte
   C. Lymphocyte
   D. Monocyte

Answer is A: An erythrocyte is a red blood cell.

2. In the haemostasis process, what forms as a result of the extrinsic and intrinsic pathways?
   A. Fibrin
   B. Thrombin
   C. A platelet plug
   D. Prothrombinase

Answer is D: The extrinsic and intrinsic pathways form "prothrombinase" (also called prothrombin activator) from factor X.

3. The blood group known as the ABO system is based on the presence of what proteins on blood cells?
   A. Antibodies
   B. Antigens
   C. Agglutinins
   D. Immunoglobulins

Answer is B: Antigens are on the membrane of the RBC. The other three terms all describe the same thing, antibodies that are circulating in the plasma.

4. What is found in blood serum that is also in blood plasma?
   A. Blood cells
   B. Platelets
   C. Plasma proteins
   D. Clotting factors

Answer is C: Plasma proteins (except fibrinogen) are in plasma and in serum. Serum = plasma minus the clotting factors. Blood cells and platelets are not in plasma.

5. What is the term "formed elements" used to mean in a description of blood?
   A. White blood cells, red blood cells and platelets
   B. Blood plasma
   C. Blood serum
   D. The clotting factors in blood

Answer is A: Formed elements are the non-liquid or solute parts of the blood.

6. What is the SECOND step in the three phases of haemostasis listed below?
   A. The vascular phase
   B. The intrinsic pathway
   C. The extrinsic pathway
   D. The platelet phase

Answer is D: The three phases are: vascular phase, platelet phase, coagulation phase (which in turn has three steps).

7. What type of blood may a patient with blood type "B+" be infused with? Any blood that is:
   A. Positive for rhesus antigen D
   B. Negative for rhesus antigen D
   C. Negative for antigen B
   D. Negative for antigen A

Answer is D: A patient that is B+ may receive B+, B−, O+ or O− blood because those types do not have antigen A on the donated RBC.

8. What is the first process that occurs after a blood vessel is damaged?
   A. Coagulation
   B. Platelet plug formation
   C. Vasoconstriction
   D. Haemolysis

Answer is C: Almost immediately (within 2 s) after a blood vessel is cut, the vessel walls contract in a spasm to slow the flow of blood (vessel diameter decreases).

9. Which blood cells are involved in protecting the body from pathogens and foreign cells?
   A. Erythrocytes
   B. Leucocytes
   C. Platelets
   D. Haemoglobin

Answer is B: Leucocytes (white blood cells) include NK (natural killer), T and B lymphocytes and macrophages and microphages.

10. Which individuals can receive any type of blood and are considered universal recipients?
    A. A+
    B. O−
    C. AB+
    D. B−

Answer is C: People with AB+ blood do not have agglutinins (antibodies) against A, B or Rh D in their plasma. Hence can receive any blood without causing the RBC in the donated blood to clump.

11. Which is the most abundant plasma protein?
    A. Alpha- and beta-globulins
    B. Albumin
    C. Mitochondria
    D. Haemoglobin

Answer is B: About 58% of plasma proteins are albumins. Haemoglobin is a protein but it is contained within the RBC.

12. Which characteristic of blood refers to the concentration of solutes?
    A. Salinity
    B. pH
    C. Osmolality
    D. Viscosity

Answer is C: Osmolality is the number of osmoles (osmol) of solute per kilogram of solvent. (osmolarity (with an "r") is defined as the number of osmoles of solute per *litre* of solution).

13. Which type of white blood cell is responsible for engulfing pathogens during phagocytosis?
    A. Thrombocyte
    B. Neutrophil
    C. Erythrocyte
    D. Basophil

Answer is B: Neutrophils are microphages—phagocytes of bacteria. Thrombocytes and erythrocytes are not WBC.

14. What does "Rhesus positive" refer to?
    A. The presence of antigen D on the surface of red blood cells
    B. The final factor involved in blood clotting
    C. The presence of the rhesus antibody/agglutinin in the blood
    D. A deficiency of factor VIII that results in haemophilia

Answer is A: Rh factor, Rh positive and Rh negative refer to the D antigen only. If the antigen is present on your RBC, you are called Rh-positive (you have the Rh factor).

15. What are red blood cells primarily composed of?
    A. Alpha- and beta-globulins
    B. Albumin
    C. Mitochondria
    D. Haemoglobin

Answer is D: About one third of the mass of RBC is haemoglobin. Choices A and B are plasma proteins and are not in RBC.

16. Which is the **LEAST** common type of white blood cell?
    A. Lymphocyte
    B. Basophil
    C. Thrombocyte
    D. Neutrophil

Answer is B: Less than 1% of WBC are basophils. Neutrophils are the most common. Thrombocytes are not WBC, or even cells.

17. In the process of haemostasis, which phase involves the intrinsic and extrinsic pathways?
    A. The platelet phase
    B. The clot lysis phase
    C. The vascular phase
    D. The coagulation phase

Answer is D: The clotting (coagulation phase) has these two pathways.

18. In haemostasis, which molecule polymerises to become the insoluble blood clot?
    A. Factor X
    B. Thrombin
    C. Fibrin
    D. Plasmin

Answer is C: Fibrin is a monomer that polymerises to form a "soft clot", then cross-linking between fibrin produces a stable, web-like "hard clot".

19. Which enzyme converts fibrinogen to fibrin?
    A. Serotonin
    B. Thrombin
    C. Renin
    D. Secretin

Answer is B: Thrombin is the enzyme. It is not present until prothrombinase converts prothrombin to thrombin.

20. Which of the following is **NOT** a macrophage?
    A. Kupffer cell
    B. Monocyte
    C. Dendrocyte
    D. Megakaryocyte

Answer is D: A megakaryocyte is the cell that produces the membrane-covered cell fragments known as platelets. It is not a macrophage. A dendrocyte is also known as a Langerhans cell or a Granstein cell.

21. What can be said about a person who has the "A" antigen on their red blood cells?
    A. Their blood contains anti-B agglutinins.
    B. Their blood contains anti-A agglutinins.
    C. Their blood contains anti-A and anti-B agglutinins.
    D. Their blood contains neither anti-A nor anti-B agglutinins.

Answer is A: Their blood contains anti-B agglutinins. If you have the A antigen on your RBC, you cannot have the anti-A agglutinin in your plasma or you would agglutinate your own blood.

22. Which one of the following is **NOT** a plasma protein?
    A. Keratin
    B. Albumin
    C. Ferritin
    D. Globulin

Answer is A: Keratin is the protein that the stratum corneum of the skin and that hair and nails are made of.

23. What substance is produced by the first step in the blood clotting (coagulation) process?
    A. Thrombin
    B. Prothrombin
    C. Factor X
    D. Prothrombinase

Answer is D: Prothrombinase is made from factor X. It then acts on prothrombin to form thrombin.

24. Which statement about neutrophils is correct?
    A. They have no nucleus.
    B. They contain haemoglobin.
    C. They function as a body defence mechanism.
    D. Eosinophils are one type of neutrophil.

Answer is C: Neutrophils do have a nucleus but do not have haemoglobin. Eosinophils are a type of WBC as are neutrophils, but they are not the same.

25. What are red blood cells also known as?
    A. Erythrocytes
    B. Thrombocytes
    C. Monocytes
    D. Eosinophils

Answer is A: RBC = erythrocytes (erythroid means red).

26. In blood clotting, what activates "factor X"?
    A. Prothrombinase
    B. Thrombin
    C. The extrinsic pathway
    D. Tissue plasminogen activator

Answer is C: Factor X is a plasma protein produced by the liver. In the extrinsic pathway, factor III combines with factor VII to form an "enzyme complex" that activates factor X.

27. A person's blood group is determined by:
    A. The agglutinogens circulating in their plasma
    B. The antigens on the surface of their red blood cells
    C. The antibodies on the surface of their red blood cells
    D. The agglutinins circulating in their plasma

Answer is B: Antigens are on the surface of RBC. Antibodies (=agglutinins) circulate in the blood stream and react with (agglutinate) their appropriate antigen.

28. If a blood sample is taken for DNA testing, which of the following would be examined?
    A. Leucocytes
    B. Erythrocytes
    C. Thrombocytes
    D. Plasma proteins

Answer is A: Only leucocytes (WBC) possess a nucleus from which DNA may be sampled. Erythrocytes are anucleate.

29. What is the major task of red blood cells?
    A. To transport carbon dioxide
    B. To ensure haemostasis
    C. To provide immunity
    D. To transport oxygen

Answer is D: Red blood cells contain much haemoglobin which binds to oxygen to transport it to the tissues.

30. Careful blood matching is performed prior to transfusing blood in order to avoid which scenario?
    A. Newborn haemolytic disease
    B. The recipient's antigens attacking the red blood cells in the transfusion
    C. The recipient's antibodies attacking the red blood cells in the transfusion
    D. The antigens on the recipient's red blood cells reacting with the antibodies in the transfused blood

Answer is C: If the recipient's antibodies attack the donated RBC, agglutination will occur. The reverse (choice D) is not a problem.

31. Which cell in the list below is the MOST common white blood cell?
    A. Basophils
    B. Lymphocytes
    C. Monocytes
    D. Neutrophils

Answer is D: Neutrophils (also called polymorphonuclear leucocytes) constitute between 50–70% of WBC.

32. What substance is the product of the second step in the blood clotting process?
    A. Thrombin
    B. Prothrombin
    C. Prothrombin activator
    D. Fibrin

Answer is A: The second step is the formation of thrombin from prothrombin (first step is formation of prothrombinase (=prothrombin activator)).

33. A person whose blood group is "B positive" has which of the following?
    A. The rhesus D antigen and the B antigen on their RBC, and the anti-A agglutinin
    B. The rhesus D antigen and the B antigen on their RBC, and the anti-B agglutinin
    C. The rhesus D antigen and the A antigen on their RBC, and the anti-B agglutinin
    D. No rhesus D antigen and the B antigen on their RBC, and the anti-A agglutinin

Answer is A: "B+" means having the B antigen on RBC and being positive (i.e. having) the D antigen on the RBC as well.

34. The role of platelets in blood clotting includes all of the following EXCEPT one. Which one?
    A. To form a plug in the hole of the damaged blood vessel
    B. To convert prothrombin to thrombin

C. To release chemicals to attract other platelets
D. To adhere to exposed collagen fibres in damaged blood vessels

Answer is B: Prothrombinase is the enzyme that converts prothrombin to thrombin (it does not come from platelets).

35. If someone's ABO blood group is "type A", this means that:
    A. They have the type A antigen on their red blood cells.
    B. Their blood contains anti-A agglutinins.
    C. They can receive blood from a type B donor.
    D. They may donate blood to a type B recipient.

Answer is A: Type A means having the A antigen on the RBC (and having anti-B agglutinins in the plasma).

36. Which statement below about vitamin K is true?
    A. It is water soluble.
    B. It is essential for prothrombin production by the liver.
    C. It is part of the "extrinsic pathway" of formation of prothrombin activator.
    D. It destroys fibrin so allowing a clot to gradually dissolve.

Answer is B: Vitamin K is needed for the liver to produce prothrombin (inactive thrombin). Plasmin destroys fibrin.

37. What is the function of the plasma proteins in blood?
    A. To transport oxygen
    B. To regulate electrolyte balance
    C. To exert osmotic pressure and so help maintain blood volume
    D. To function as a non-specific body defence mechanism

Answer is C: Plasma proteins exist in the plasma but not in the interstitial fluid. Hence their osmotic influence draws water into the blood. Haemoglobin which is not a "plasma protein" transports oxygen.

38. The term "formed elements" used in relation to the blood include which of the following?
    A. Fibrinogen
    B. White blood cells
    C. Electrolytes
    D. Plasma proteins

Answer is B: White blood cells are one of the formed elements (i.e., substances that are not dissolved in plasma).

39. Which blood cell fits the following description: multi-lobed nucleus, inconspicuous cytoplasmic granules, most common type of blood cell except for red blood cells?
    A. Neutrophil
    B. Eosinophil

C. Basophil
D. Lymphocyte

Answer is A: Neutrophils are the most common WBC and have a large multi-lobed nucleus.

40. What constitutes blood plasma?

    A. Whole blood without the formed elements.
    B. Blood without the red blood cells.
    C. Whole blood without blood cells and clotting factors.
    D. Blood minus blood cells and proteins.

Answer is A: Plasma is blood without the "formed elements", that is, cells. Choice C refers to serum.

41. Which of the following statements about a person with blood group "A" is true? They have the:

    A. A antigen on their red blood cells
    B. Anti-A antibodies in their plasma
    C. Anti-A agglutinogen on their red blood cells
    D. A antibody on their red blood cells

Answer is A: Blood group A is named for having the A antigen on the RBC.

42. Which of the following statements concerning intracellular and extracellular fluids is **FALSE**?

    A. The concentration of sodium is higher in extracellular fluid than in intracellular fluid.
    B. The concentration of potassium is lower in extracellular fluid than in intracellular fluid.
    C. Blood plasma is an example of intracellular fluid.
    D. The volume of intracellular fluid is greater than that of extracellular fluid.

Answer is C: Blood plasma is external to cells hence is an extracellular fluid.

43. Which of the following is not a type of white blood cell?

    A. Leucocyte
    B. Eosinophil
    C. Erythrocyte
    D. Neutrophil

Answer is C: Erythrocytes are red blood cells.

44. Which of the following formed elements of the blood is important in the formation of clots?

    A. Erythrocytes
    B. Lymphocytes

C. Monocytes
D. Thrombocytes

Answer is D: Thrombocytes (platelets) are attracted by chemicals released by damaged endothelial cells and in turn release substances which cause other platelets to aggregate to form a platelet plug to cover the hole.

45. With which blood types can a person with blood type B be safely transfused?
    A. A or AB
    B. B or O
    C. A or O
    D. B or AB

Answer is B: A person with blood type B has anti-A agglutinins in their plasma so cannot receive cells with the A antigen on the RBC. Hence the blood types in choices A, C and D are not suitable.

46. Leucocytes may be correctly described as what?
    A. Cells with nuclei that do not contain haemoglobin
    B. Cells without nuclei that contain haemoglobin
    C. White blood cells with granules in their cytoplasm
    D. Neutrophilic

Answer is A: All leucocytes have a nucleus and do not contain haemoglobin.

47. What are lymphocytes? Blood cells that:
    A. Mature and proliferate in the bone marrow
    B. Contain haemoglobin
    C. Are involved in the body's immune response
    D. Mature into macrophages

Answer is C: Lymphocytes (NK, T and B types) are part of the body's specific defence mechanism.

48. Which of the following statements about platelets is **INCORRECT**? They:
    A. Adhere to collagen fibres of damaged tissue
    B. Release phospholipids which combine with "clotting factors" to produce prothrombin activator
    C. Are cell fragments derived from megakayoblasts
    D. Are part of the "extrinsic pathway" for the formation of prothrombin activator

Answer is B: Platelets do not release phospholipids.

49. Finish the sentence correctly. Plasma proteins:
    A. Help maintain blood volume due to colloid osmotic pressure
    B. Are regarded as formed elements of the blood

C. Are low molecular weight proteins
D. Are part of the blood serum

Answer is A: Plasma proteins are responsible for colloid osmotic pressure. Serum does not include clotting factors—since fibrinogen is a clotting factor and a plasma protein, choice D is incorrect.

50. The colloid osmotic pressure of blood is due to which of the following?

A. Proteins in the blood
B. Proteins in the interstitial fluid
C. Sodium and chloride ions dissolved in blood
D. The water component of the blood

Answer is A: Plasma proteins (also called "colloids") produce colloid osmotic pressure. There should be no proteins in the interstitial fluid.

51. Which one of the following terms refers to an abnormally low number of white blood cells?

A. Thrombocytosis
B. Haemostasis
C. Leukopenia
D. Cytokinesis

Answer is C: "Leuko-" refers to WBC; "-penia" refers to abnormally low, or lack of.

52. Which of the following three proteins are known as "plasma proteins"?

A. Albumin, globulin, haemoglobin
B. Insulin, glucagon, haemoglobin
C. Fibrin, globulin, albumin
D. Albumin, fibrinogen, globulin

Answer is D: Haemoglobin is not a plasma protein. Fibrinogen, not fibrin exists in blood.

53. Which are the two most common types of white blood cells?

A. Neutrophils and lymphocytes
B. Erythrocytes and neutrophils
C. Neutrophils and eosinophils
D. Monocytes and lymphocytes

Answer is A: 50–70% of WBC are neutrophils and another 25% are lymphocytes.

54. Blood plasma contains "plasma proteins". Which of the following lists the plasma proteins?

A. Insulin, kaolin, bilirubin
B. Cholesterol, urea, glucagon
C. Na^{+}, K^{+}, Ca^{2+}, Mg^{2+}

D. Albumins, fibrinogen, globulins

Answer is D: None of the other choices list any plasma proteins.

55. What causes the blood's osmotic pressure to be greater than the osmotic pressure of the surrounding interstitial fluid that is outside of the capillaries?
    A. There is a higher concentration of sodium and chloride ions in the blood than the interstitial fluid.
    B. There is a higher concentration of water in the blood than in the interstitial fluid.
    C. The plasma proteins in blood.
    D. The hydrostatic pressure produced by the heart's contractions.

Answer is C: Plasma proteins exist in blood but not in the interstitial fluid.

56. What does the term "neutrophil" refer to?
    A. An affinity for neutrons
    B. An abnormally low number of cells
    C. A type of white blood cell
    D. An immature cell that will become a neutrocyte

Answer is C: Neutrophils are a white blood cell (whose nucleus stains with dyes of neutral pH).

57. Which one of the following cells does **NOT** occur in blood?
    A. Erythrocytes
    B. Basophils
    C. Leucocytes
    D. Osteocytes

Answer is D: Osteocytes are bone cells.

58. What would a person with type A blood also have?
    A. Antibody A
    B. Antigen A
    C. Agglutinin A
    D. Agglutinogen B

Answer is B: Type A blood means that the A antigen is on the red blood cells. They also have antibody B (=agglutinin B) in their plasma. An agglutinogen is an antigen that stimulates production of agglutinin, hence antigen A is also known as agglutinogen A.

59. Which liquid below has the greatest volume in the body?
    A. Blood
    B. Plasma
    C. Serum

D. Extracellular fluid

Answer is D: Extracellular fluid includes blood and all other liquids not contained in cells. Plasma is serum plus the clotting factors. Blood is plasma plus RBC and WBC.

60. What is meant by the term "neutropenia" when applied to blood?
    A. Too few neutrophils
    B. Too many neutrophils
    C. Too few white blood cells
    D. Too many white blood cells

Answer is A: The suffix "-penia" means too few of that particular cell. Having too few white blood cells is described as leukopenia.

61. To what is the term "thrombocytosis" applied?
    A. An excessive number of red blood cells
    B. An excessive number of platelets
    C. An unusually high risk of internal bleeding
    D. An unusually low number of white blood cells

Answer is B: The suffix "-cytosis" denotes an excess of. The term thrombocyte is applied to platelets.

62. Blood clotting is a complicated process with many steps. Which of the following lists comprises substances that are already present in blood, ready to perform their role in clotting when needed?
    A. Thrombin, prothrombinase, tissue plasminogen activator
    B. Fibrinogen, prothrombin, plasminogen, factor X
    C. Platelets, fibrinogen, fibrin, fibrinase
    D. Fibrin, thrombin, plasmin, prothrombinase

Answer is B: These substances are all inactive (−ogen and pro-) until activated by an enzyme. The enzymes that can activate them are not produced until they are needed. (An enzyme complex converts factor X to prothrombinase. Prothrombinase converts prothrombin to thrombin. Thrombin converts fibrinogen to fibrin—which is insoluble. Tissue plasminogen activator converts plasminogen to plasmin. Plasmin dissolves fibrin.

63. Which are the most common circulating leucocytes?
    A. Haemocytoblasts
    B. Erythrocytes
    C. Basophils
    D. Neutrophils

Answer is D: Leucocytes are white blood cells, about 70% of which are neutrophils. Haemocytoblasts are stem cells from which white and red blood cells arise, erythrocytes are RBC.

64. Which of the following processes is **NOT** part of the vascular phase of haemostasis?
    A. The formation of prothrombinase
    B. The vessel walls contract in a spasm
    C. Endothelial cells release chemicals
    D. Endothelial cells become sticky allowing platelets to adhere

Answer is A: Prothrombinase formation is part of the coagulation phase.

65. How may the transfusion reaction that results from infusion with incompatible blood be described?
    A. Antibodies in the recipient's blood attack the donated RBC
    B. Antibodies in the donated blood attack the recipient's RBC
    C. Antigens on the recipient's RBC attack the donated blood
    D. Antigens on the recipient's blood attack the donated blood

Answer is A: Antigens are on the plasma membrane of RBC, antibodies are in the plasma. Antibodies attack antigens. While choice B can happen, it does not cause problems.

66. What substances produce "colloid osmotic pressure"?
    A. The clotting factors in blood
    B. The protein component of blood
    C. The cellular components of blood
    D. The serum component of blood

Answer is B: Plasma proteins (also called colloids) are present in blood but not in the interstitial fluid, so contribute "colloid osmotic pressure" to blood.

67. Which antigens would a person whose ABO blood group is "AB+" have on their red blood cells?
    A. Antigens A and B and D
    B. Antigens A and B
    C. Antigen D but neither of A or B
    D. Antigens A, B and plus

Answer is A: The AB blood group has both antigen A and B on the membrane of their RBC as well as rhesus antigen D (i.e. they are positive for D).

68. Why may plasma from a donor of any blood group be transfused into a patient with any blood group?
    A. Plasma does not contain white blood cells.
    B. Plasma does not contain plasma proteins.
    C. Plasma does not contain prothrombin or prothrombinase.
    D. Plasma does not contain red blood cells.

Answer is D: The antigens that react with the recipient's antibodies which result in agglutination reactions occur on the plasma membrane of the donor's red blood cells. As plasma does not contain any RBC, an agglutination reaction cannot occur.

69. Blood may be removed from a donor and centrifuged to separate plasma from cells (plasmapheresis). The cells are then returned to the donor mixed with a liquid. What liquid is used to return the cells?
    A. Sterile water
    B. 0.9% sodium chloride
    C. 9.0% sodium chloride
    D. Plasma

Answer is B: Red blood cells are mixed with a solution that is isotonic: that is, will not cause any net movement of water into or out of the cell. Choice D is wrong as using plasma would defeat the purpose of donating in the first place.

70. A mother with blood type O negative has a 2-year-old child of blood type A positive and is pregnant with a second child who is B positive. What should have been done to ensure the health of the people involved?
    A. The mother was administered anti-A antibodies after delivery of the first child.
    B. The second child should be administered anti-D antibodies while in utero.
    C. The mother was administered anti-D antibodies after delivery of the first child.
    D. The first child was administered anti-D antibodies after the birth.

Answer is C: During delivery of the **first** child, foetal blood may enter the maternal system from the placenta. The mother will then produce agglutinins against rhesus antigen D. These can cross the placenta in a subsequent pregnancy, and if the **second** baby is Rh+, attack the foetal erythrocytes causing newborn haemolytic disease. The disorder can be easily prevented by administration of anti-D antibodies to the mother after the first delivery.

71. Blood osmolarity is about 290 mosmol/L. What would happen to red blood cells if they were placed in a solution of 180 mosmol/L?
    A. The RBC would coagulate.
    B. An agglutination reaction would occur.
    C. They would crenate.
    D. Haemolysis would occur.

Answer is D: A solution of 180 mosmol/L is hypotonic, so osmosis would occur, moving water into the RBC. Consequently they would swell and many would lyse (split open). Haemolysis would occur.

72. Which of the following is NOT a function of blood plasma?
    A. Haemostasis (blood coagulation)
    B. Transport of plasma proteins

C. Transport of oxygen
D. Regulation of pH

Answer is C: The function of erythrocytes is to transport oxygen. (OK, some oxygen is transported and dissolved in plasma, but of the four choices, choice C is the more minor function.)

73. Which of the following is a function of erythrocytes?
A. Transport of hormones around the body
B. Phagocytosis of bacteria
C. Haemostasis (blood coagulation)
D. Transport of carbon dioxide

Answer is D: About 25% of carbon dioxide is transported bound to haemoglobin ($HbCO_2$). The other functions (choices A, B, C) are performed by plasma, leucocytes and platelets, respectively.

74. Which of the following is a function of platelets?
A. Phagocytosis of bacteria
B. Haemostasis (blood coagulation)
C. Regulation of blood pH
D. Electrolyte balance

Answer is B: The role of platelets is to commence blood clotting process. Platelets adhere to exposed collagen fibres from the damaged vessel wall to endothelial cells and to basal lamina and release substances which cause other platelets to aggregate forming a platelet plug.

75. What is the function of leucocytes?
A. Many function as microphages or macrophages.
B. Haemostasis (blood coagulation).
C. Regulation of blood pH.
D. Transport of oxygen.

Answer is A: Neutrophils are microphages (phagocytes of bacteria); eosinophils are microphages; many monocytes become macrophages.

76. Which of the following describes red blood cells?
A. Biconcave shape and 7–8 μm in diameter
B. Biconvex shape and 7–10 μm in diameter
C. Cytoplasm surrounded by membrane, 1–4 μm in diameter
D. Amoeboid in shape and 7–8 μm in diameter

Answer is A: Biconcave means shaped somewhat like a donut (but without the hole) with the centre portion of the RBC thinner than the rim.

77. Which of the following is NOT one of the processes of coagulation?
    A. Formation of prothrombinase
    B. Formation of thrombin
    C. Formation of plasmin
    D. Formation of fibrin

Answer is C: Plasmin is formed from plasminogen (by tissue plasminogen activator, t-PA) as the vessel repair proceeds. Plasmin destroys fibrin so that the clot dissolves.

78. What is the role of vitamin K?
    A. It is a fat-soluble vitamin.
    B. Needed by the liver to synthesise clotting factors.
    C. It is one of the antigens on the surface of red blood cells.
    D. It is a plasma protein that contributes to colloid osmotic pressure.

Answer is B: Vitamin K is needed for the liver to produce clotting factors II, VII, IX and X (prothrombin, proconvertin, PTC and Stuart–Prower factors). Choice A is true but does not answer the question.

79. What type of whole blood transfusion may be received by a person whose blood type is B− (B negative)?
    A. O−
    B. B−
    C. B+, B−, O+ or O−
    D. B− or O−

Answer is D: A person with B− blood can receive B− (of course) as well as O− as O blood has neither A nor B antigens on their RBC. An Rh− person does not have the rhesus D antigen on their RBC. They also do not usually contain antibodies against D. When an Rh− person receives Rh+ blood, their body will begin to produce antibodies against the foreign Rh antigen D. These antibodies (=agglutinins) will remain in the blood. If a second transfusion of Rh+ blood is given later, the agglutinins against D will now attack the donor erythrocytes, and a severe reaction may occur. Hence blood that is +ve for rhesus antigen D is incompatible.

80. What type of whole blood transfusion may be received by a person whose blood type is B+ (B positive)?
    A. O+ or O−
    B. B+
    C. B+, B−, O+ or O−
    D. B+ or O+

Answer is C: A person with B+ blood can receive B+ (of course) as well as B−, O+ and O− as O blood has neither A nor B antigens on their RBC. An Rh+ person has the rhesus D antigen and so (of course) does not contain antibodies against D. Hence can receive blood with RBC that are +ve or −ve for antigen D.

81. What type of whole blood transfusion is compatible for someone whose blood type is AB+ (AB positive)?
    A. AB+
    B. A+, A−, B+, B−, AB+, AB−
    C. A+, B+, AB+
    D. All blood types

Answer is D: A person with AB+ blood has both the A antigen and B antigen on their RBC hence does not possess antibodies to either A or B. This means that whatever blood type is infused, there will be no agglutination reaction. All blood types are compatible.

82. One microlitre of blood contains about five million (5×10^6) RBC; each RBC contains about 280 million (280×10^6) Hb molecules. Estimate how many Hb molecules there are in an adult human.
    A. 7000×10^6 Hb molecules/person
    B. 1.4×10^{18} Hb molecules/person
    C. 7000×10^{18} Hb molecules/person
    D. 7×10^{21} Hb molecules/person

Answer is D: An adult female contains 4–5 L whole blood. An adult male contains 5–6 L whole blood, so let us choose 5 L for the blood volume. Each litre of blood contains one million microlitres:

Hb molecules per RBC × # of RBC per microlitre × # of microlitres in 5 L of blood.

(280×10^6) Hb/RBC × (5×10^6) RBC/μL × (5×10^6) μL/person = 7×10^{21} Hb molecules/person.

Choice C is not wrong, but is not written in "scientific notation".

83. Blood plasma from which ABO blood types are compatible for donation to someone who is A+?
    A. A+ blood is compatible.
    B. All blood types are compatible.
    C. A+ and A−.
    D. A+, A−, AB+, AB−.

Answer is B: Blood plasma does not contain red blood cells. Hence the antibodies of the recipient have no antigens to attack (these are carried on the plasma membrane of the RBC). This means there are no compatibility issues, and this is why donating plasma is encouraged by blood banks.

84. Which cells account for almost half the volume of blood?
    A. Erythrocytes
    B. Leukocytes
    C. Platelets

D. Monocytes

Answer is A: Erythrocytes or red blood cells occupy about 45% of the blood's volume. That is, we say that the haematocrit is 45.

85. If red blood cells placed in a solution of unknown concentration suddenly swell and rupture, which of the following solutions is the likely one?
    A. 1% saline
    B. 9% saline
    C. 0.9% saline
    D. 0.1% saline

Answer is D: The solutions A and B are hypertonic to RBC so would cause the RBC to lose water and crenate. 0.9% is isotonic. Saline of 0.1% is hypertonic so would cause a net movement of water into the RBC which would cause them to swell and burst (haemolyse).

12.2 Heart

The left ventricle of the heart contracts about 60–80 times a minute to pump blood through the aortic valve into the aorta from where it flows throughout the systemic circulation. Simultaneously, the right ventricle pumps the same volume of blood through the pulmonary valve into the pulmonary trunk from where it flows into the lungs. As the combined length of the blood vessels in the rest of the body is much greater than the combined length of the blood vessels in the lungs, the left ventricle has to contract with greater force than does the right ventricle. Consequently the muscle of the LV is much stronger (thicker) than that of the RV. As the ventricles begin to contract, the mitral (bicuspid) valve closes to prevent blood "regurgitating" from the LV into the left atrium. Simultaneously the tricuspid valve on the right closes to prevent retrograde blood flow from the RV into the right atrium. The student should be able to name the eight blood vessels that enter or leave the heart, the four chambers of the heart, the four valves the blood passes through and the order in which it passes through them.

The myocardium is a tireless muscle in that it performs its cycle of contraction (systole) and relaxation (diastole) ceaselessly for the duration of a human lifespan. To perform this feature, it must be well supplied with energy and oxygen. Twenty-five percent of the volume of a cardiac muscle cell may be taken up by mitochondria which produce the required ATP. In turn the coronary arteries supply the cells with the oxygen that the mitochondria need to produce ATP. The left and right coronary arteries arise from the aortic sinus, just on the aorta side of the aortic valve. A blockage in some part of these arteries will produce the infamous "heart attack". Precisely where the blockage exists may be found by imaging the heart with a magnetic resonance scanner, by coronary angiography, or by investigating the difference between the affected heart's electrocardiogram and the expected normal.

12.2.1 Heart Anatomy and Physiology

1. Blood flow through the heart follows which of the sequences listed below?
   A. From left atrium, then mitral valve, right ventricle, aorta, left ventricle
   B. From right atrium, then mitral valve, right ventricle, pulmonary trunk, left ventricle
   C. From pulmonary trunk, then tricuspid valve, left atrium, aortic valve, aorta
   D. From vena cava, then right ventricle, pulmonary trunk, left ventricle, aorta

Answer is D: The mitral valve (bicuspid valve) comes after the right ventricle; the tricuspid valve comes before the pulmonary trunk.

2. What feature does cardiac muscle possess that is missing in skeletal muscle?
   A. Striations
   B. Multiple nuclei
   C. Voluntary control
   D. Intercalated discs

Answer is D: Intercalated discs join the membrane of one cardiac cell with its neighbour and promote rapid conduction of depolarisation between cells.

3. What is the name of the valve between the left atrium and the left ventricle?
   A. Mitral valve
   B. Tricuspid valve
   C. Semilunar valve
   D. Aortic valve

Answer is A: Also known as the bicuspid valve, the aortic valve is also a semilunar valve.

4. What is meant by a diastolic blood pressure of 100 mmHg?
   A. The maximum pressure at the start of the aorta during ventricular contraction
   B. The minimum pressure at the start of the aorta before the start of a ventricular contraction
   C. The maximum pressure at the start of the aorta and pulmonary trunk during ventricular contraction
   D. The minimum blood pressure measured when resting

Answer is B: Diastolic pressure refers to the lowest blood pressure measurement, and this occurs just before the start of ventricular contraction.

5. What is the main function of mitral valve?
   A. To increase the pressure inside the left atrium during systole
   B. To prevent a drop in pressure in the aorta during diastole
   C. To prevent backflow from left ventricle to left atrium during systole

D. To add additional blood from left atrium to left ventricle during atrial systole

Answer is C: During ventricular systole, blood should flow through the aortic valve into the aorta and not back into the left atrium. The mitral valve prevents blood going back into the left atrium.

6. The Frank–Starling law of the heart describes the proportional relationship between which of the following pairs?
   A. Stroke volume and cardiac output
   B. Stroke volume and end-diastolic volume
   C. The blood volume in the ventricles and stroke volume
   D. Systemic vascular resistance and stroke volume

Answer is C: The stroke volume of the heart increases in response to an increase in the volume of blood filling the heart. That is, stroke volume increases so that all the blood that enters the heart is pumped out.

7. If Sarah has a stroke volume of 70 mL and a cardiac output of 5950 mL/min, which of the following is her heart rate (in beats/min)?
   A. 70
   B. 75
   C. 80
   D. 85

Answer is D: 5950 mL/min divided by 70 mL/beat = 85 bpm.

8. What will cause the sinoatrial (SA) node to depolarise more frequently?
   A. Acetylcholine
   B. Norepinephrine
   C. Parasympathetic stimulation
   D. Vagus nerve

Answer is B: Norepinephrine (released by the sympathetic nervous system) acting on the SA node increases heart rate and blood pressure. Parasympathetic stimulation along the vagus nerve decreases heart activity.

9. How are cardiac cells mechanically attached to each other? By their:
   A. Mitochondria
   B. Intercalated discs
   C. Gap junctions
   D. Sarcolemma

Answer is B: Intercalated discs attach adjacent cardiac cells to each other.

10. Starting at the APEX of the heart and moving superiorly, what is the correct order in which you would encounter the four anatomical structures below?
   A. Valves, chordae tendineae, papillary muscle, ventricle
   B. Ventricle, papillary muscle, chordae tendineae, valves

C. Papillary muscle, chordae tendineae, ventricle, valves
D. Chordae tendineae, valves, ventricle, papillary muscle

Answer is B: The apex is the pointy end of the heart (the inferior end). So ventricles are first, then papillary muscle to which are attached the chordae tendineae, superior to which are the atrioventricular valves.

11. Which period of the heart cycle is completely occupied by the ventricles relaxing?
    A. Atrial systole
    B. Atrial diastole
    C. Ventricular systole
    D. Ventricular diastole

Answer is D: Diastole is the period of cardiac muscle relaxation.

12. Through which valve does blood flow when it moves from the right atrium into the right ventricle?
    A. The tricuspid valve
    B. The mitral valve
    C. The pulmonary valve
    D. The bicuspid valve

Answer is A: The tricuspid valve is on the right side of the heart between the atrium and ventricle. Bicuspid is on the left.

13. How is the fibrous pericardium attached to the surrounding structures?
    A. Laterally to the pleural surfaces of the lungs
    B. Posteriorly to the sternum
    C. Anteriorly to trachea, main-stem bronchi and oesophagus
    D. Inferiorly to the clavicles

Answer is A: The only sensible answer when anatomical position terminology is used. The lungs are lateral to the heart. The sternum is not posterior to the heart, the trachea and oesophagus are. The heart is not attached to the clavicles.

14. A drug, such as cocaine, which stimulates the heart but does directly inhibit the heart's ability to relax, would be considered a:
    A. Sympatholytic
    B. Sympathomimetic
    C. Parasympatholytic
    D. Parasympathomimetic

Answer is B: Sympathomimetic drugs are stimulant compounds which mimic the effects of agonists of the sympathetic nervous system such as epinephrine (adrenaline) and norepinephrine (noradrenaline), that is, they cause the heat to best faster. A sympatholytic (or sympathoplegic) drug is a medication which inhibits the functioning of the sympathetic nervous system. The choices C and D refer to slowing down the heart.

15. Why is the myocardium of the right ventricle (RV) thinner than that of the left ventricle (LV)?
    A. The RV pumps into the pulmonary circuit which has less resistance than the systemic circuit.
    B. The RV pumps a smaller volume of blood than the LV.
    C. The RV pumps blood out with a slower exit speed than the RV.
    D. The RV chamber has a smaller volume than the LV.

Answer is A: The RV pumps against a lesser resistance, it does not need to be as muscular as the LV.

16. Through which valve does blood flow when it moves from the left atrium into the left ventricle?
    A. The semilunar valve
    B. The mitral valve
    C. The tricuspid valve
    D. The aortic valve

Answer is B: The left atrioventricular valve is also known as the mitral (and bicuspid) valve.

17. Which period of the heart cycle is completely occupied by the ventricles contracting?
    A. Atrial systole
    B. Atrial diastole
    C. Ventricular systole
    D. Ventricular diastole

Answer is C: Systole is the term applied to contraction of the cardiac muscle.

18. Which statement below describes blood flow through the mitral valve?
    A. Blood flows from the right atrium into the right ventricle.
    B. Blood flows from the right ventricle into the pulmonary artery.
    C. Blood flows from the left ventricle into the aorta.
    D. Blood flows from the left atrium into the left ventricle.

Answer is D: The mitral valve is between the left atrium and left ventricle.

19. Which structure has the thickest wall?
    A. The aorta
    B. The inter-atrial septum
    C. The left ventricle
    D. The right ventricle

Answer is C: The LV has the thickest muscle wall.

20. Which tissue is supplied with blood via the coronary arteries?
    A. The lungs
    B. The myocardium
    C. The corona
    D. The aorta

Answer is B: The heart (myocardium) is supplied by the coronary circulation.

21. What is the innermost layer of the heart wall known as?
    A. Epicardium
    B. Pericardium
    C. Visceral pericardium
    D. Endocardium

Answer is D: "Endo-" means on the inside. Epicardium is also known as the visceral pericardium, is on the outside of the heart.

22. Which of the following is a difference between cardiac muscle and skeletal muscle?
    A. Cardiac muscle is not striated (and skeletal muscle is).
    B. Cardiac muscle fibres are branched (and skeletal muscle fibres are not).
    C. Skeletal muscle is involuntary and is uni-nucleate (and cardiac muscle is neither).
    D. Skeletal muscle has intercalated discs (and cardiac muscle does not).

Answer is B: Cardiac muscle cells have branches, but skeletal muscle cells are single fibres.

23. Where is the mitral valve of the heart located? Between the:
    A. Left atrium and left ventricle
    B. Left ventricle and the aorta
    C. Right ventricle and the pulmonary trunk
    D. Right atrium and right ventricle

Answer is A: The mitral valve is the left atrioventricular valve.

24. Choose the structure known as the pacemaker of the heart from the following.
    A. Atrioventricular node
    B. Sinoatrial node
    C. Atrioventricular bundle
    D. The bundle of His

Answer is B: The SA node, in the right atrial wall sets the pace of the heart rate.

25. Where is the aortic valve located?
    A. Between the right atrium and right ventricle
    B. Between the right ventricle and the pulmonary trunk

C. Between the left ventricle and the aorta
D. Between the left atrium and left ventricle

Answer is C: The aortic valve is at the start of the aorta. It prevents blood in the aorta flowing back into the left ventricle.

26. By what name is the heart muscle known?
   A. Epicardium
   B. Myocardium
   C. Pericardium
   D. Endocardium

Answer is B: "Myo-" is the prefix that refers to muscle.

27. The heart receives its own oxygenated blood supply via the:
   A. Coronary arteries
   B. The pulmonary veins
   C. The coronary sinus
   D. The foramen ovale

Answer is A: The coronary arteries supply oxygenated blood to the myocardium.

28. Which name is **NOT** applied to the valve between the left ventricle and the left atrium?
   A. Atrioventricular valve
   B. Semilunar valve
   C. The bicuspid valve
   D. The mitral valve

Answer is B: The left atrioventricular valve is not a semilunar valve.

29. Where does the pulmonary trunk deliver its blood to?
   A. The left atrium
   B. The right ventricle
   C. The lungs
   D. The left ventricle

Answer is C: Pulmonary refers to the lungs.

30. The heart can be made to beat faster by which of the following?
   A. Sympathetic stimulation of the SA node
   B. Sympathetic stimulation of the AV node
   C. Parasympathetic stimulation of the SA node
   D. Parasympathetic stimulation of the AV node

Answer is A: The sympathetic nervous system stimulates the heart, while the parasympathetic NS inhibits the heart.

31. What is the outermost layer of the heart wall known as?
    A. Epicardium
    B. Pericardium
    C. Parietal membrane
    D. Endocardium

Answer is A: "Epi-" refers to "on top of". It is the visceral part of the pericardium.

32. The valve between the atrium and the ventricle that pumps oxygenated blood is called:
    A. The right atrioventricular valve
    B. The semilunar valve
    C. The mitral valve
    D. The tricuspid valve

Answer is C: The left side of the heart pumps oxygenated blood, the mitral valve is the one.

33. What is the name given to the remnant of the opening in the foetal heart that allowed the foetal lungs to be bypassed?
    A. Coronary sinus
    B. Foramen ovale
    C. Interatrial septum
    D. Fossa ovalis

Answer is D: The fossa ovalis is the slight depression that remains when the foramen ovale closes.

34. The mitral valve of the heart is located between the:
    A. Right atrium and right ventricle
    B. Left ventricle and the aorta
    C. Right ventricle and the pulmonary trunk
    D. Left atrium and left ventricle

Answer is D: The mitral valve is the usual term used for the left atrioventricular valve.

35. Complete the sentence correctly. The left ventricle pumps:
    A. More blood than the right ventricle
    B. Blood at a lower pressure than the right ventricle
    C. Less blood than the right ventricle
    D. Blood at a higher pressure than the right ventricle

Answer is D: The LV produces a higher pressure than the right in order to overcome the greater resistance to flow that exists in the systemic circuit.

36. What is ventricular systole?
    A. It refers to contraction of the ventricles.
    B. It occurs at the same time as contraction of the atria.
    C. It occurs while the bicuspid valve is open.
    D. It refers to relaxation of the ventricles.

Answer is A: Ventricular systole refers to contraction of the ventricles. The other statements are wrong.

37. Which is correct? In its passage through the heart, blood is pumped into the pulmonary trunk:
    A. After leaving the left ventricle
    B. After leaving the left atrium
    C. After passing through the right AV valve
    D. After passing through the left AV valve

Answer is C: Blood passes through the left AV valve in order to enter the right ventricle from where it is pumped into the pulmonary trunk.

38. Cardiac muscle cells differ from skeletal muscle cells in that:
    A. Skeletal muscle cells are voluntary but cardiac muscle cells are not.
    B. Skeletal muscle cells are branched but cardiac muscle cells are not.
    C. Cardiac muscle cells are multinucleate but skeletal muscle cells are not.
    D. Cardiac muscle cells act as a syncytium, while skeletal muscle does not.

Answer is A: Skeletal muscle is voluntary, while cardiac muscle is not under our conscious control. A skeletal muscle cell is a syncytium (composed from the amalgamation of many small cells (myoblasts) while the myocardium behaves as a functional syncytium—the cells contract in unison.

39. Which chamber of the heart has the thickest myocardium?
    A. Left ventricle
    B. Right ventricle
    C. Left atrium
    D. Right atrium

Answer is A: The left ventricle has the thickest myocardium (muscle) as it needs to contract with the greatest force.

40. Why is the myocardium of the left ventricle thicker than that of the right ventricle?
    A. The left ventricle has to pump a greater volume of blood than the right ventricle.
    B. The resistance of the systemic circulation is greater than that of the pulmonary circulation.
    C. The left ventricle has to pump blood to the brain against gravity.
    D. The right ventricle is assisted by the “respiratory pump”.

Answer is B: The length of the blood vessels that make up the systemic circuit is much greater than the length of vessels in the pulmonary circulation. This greater length presents a greater resistance to the flow of blood. So the LV must be stronger—made of thicker muscle—to exert sufficient pressure on the blood to overcome the resistance.

41. What supplies blood to the myocardium?
    A. The coronary circulation
    B. The vena cavae
    C. The vasa recta
    D. The pulmonary circulation

Answer is A: The left and right coronary arteries supply the myocardium.

42. Which of the following heart structures are listed in the correct sequence of blood flow through them?
    A. Right atrium, bicuspid valve, pulmonary valve, left ventricle
    B. Tricuspid valve, right ventricle, left atrium, mitral valve
    C. Pulmonary valve, left atrium, tricuspid valve, left ventricle
    D. Right ventricle, left atrium, aortic valve, left ventricle

Answer is B: Blood enters the RV via the tricuspid valve, it leaves the LA via the mitral valve.

43. The tricuspid valve separates which two structures?
    A. Right ventricle and pulmonary trunk
    B. Right ventricle and right atrium
    C. Left ventricle and aorta
    D. Left ventricle and left atrium

Answer is B: The tricuspid valve is also known as the right atrioventricular valve, so it lies between the right atrium and right ventricle.

44. Why is the myocardium of the right ventricle thinner than that of the left ventricle?
    A. The left ventricle has to pump a greater volume of blood than the right ventricle.
    B. It results from left ventricular hypertrophy due to increased peripheral resistance.
    C. It pumps blood into the low resistance pulmonary circulation.
    D. It pumps blood into the high resistance systemic circulation.

Answer is C: The RV delivers blood to the lungs which has a relatively low resistance to blood flow. Hence a small pressure will suffice.

45. Which of the following events occur during late ventricular diastole?
    A. The atria are relaxed, the ventricles are filling passively, the atrioventricular valves are open.

B. The ventricles are starting to contract, the atrioventricular valves are closed, the semilunar valves are open.
C. The atria contract, the ventricles are relaxed, the atrioventricular valves are open.
D. The atria are relaxed, the ventricles are starting to relax, the atrioventricular valves are closed, the semilunar valves are closed.

Answer is C: Choice A describes early diastole. Towards the end of diastole, the atria contract to push any blood they contain into the ventricles.

46. What feature distinguishes pacemaker cardiac cells from other cardiac cells? Pacemaker cardiac cells:

 A. Require a stimulus from the vagus nerve in order to reach threshold; other cardiac cells do not
 B. Reach threshold with much weaker stimuli than other cardiac cells
 C. Have gap junctions, while other cardiac cells do not
 D. Do not require an external stimulus to reach threshold, while other cardiac cells do

Answer is D: Pacemaker cardiac cells are self-rhythmic. They will contract in the absence of stimulation from the nervous system. External stimulation will make them contract at a faster (or slower) rate.

47. What would be a possible consequence of the SA node failing to depolarise?

 A. The entire heart would not contract.
 B. The heart rate will decrease.
 C. The ventricles would not contract.
 D. The heart rate will increase.

Answer is B: The SA node contains cells that will depolarise at a faster rate than others in the heart. If it fails, these other places will depolarise, but their rate will be slower than that of the SA node.

48. By what means does an electrical signal travel from the atria to the ventricles?

 A. Gap junctions
 B. Purkinje fibres
 C. Intercalated discs
 D. Atrioventricular bundle

Answer is D: The conduction system of the heart passes the electrical signal via the AV node to the AV bundle (also known as the bundle of His) in the interventricular septum.

49. Which of the following does limb lead II of a typical electrocardiogram represent?

 A. A graph of the variation of voltage produced by the heart against time
 B. The voltage at right arm (RA) plus the voltage at left leg (LL)

C. The electrical events that precede the contraction of the ventricles
D. The projection of the electric dipole vector of the heart on the line from left arm (LA) to right arm (RA)

Answer is A: This is the best answer. “Lead II” is the voltage measured at the RA position *minus* the voltage measured at the LL position. “Lead II” is also the projection of the electric dipole vector of the heart on the line R_{II} (in Einthoven’s triangle) from right arm to the left *leg*.

50. Which of the following events occur during early ventricular systole?
   A. The atria are relaxed, the ventricles are filling passively, the atrioventricular valves are open.
   B. The ventricles are starting to contract, the atrioventricular valves are closed, the semilunar valves are closed.
   C. The atria contract, the ventricles are relaxed, the atrioventricular valves are open.
   D. The atria are relaxed, the ventricles are starting to relax, the atrioventricular valves are opening, the semilunar valves are closing.

Answer is B: In early systole, the pressure in the ventricles may not have been raised sufficiently to push open the semilunar valves yet. Choice D describes late systole.

51. When listening to the “lub-dup” sound of the heart with a stethoscope, what is the cause of the “dup” sound?
   A. The blood flowing through the open semilunar valves
   B. The blood flowing through the open atrioventricular valves
   C. The turbulent blood flow through closing atrioventricular valves
   D. The turbulent blood flow through closing semilunar valves

Answer is D: Turbulent blood flow generates audible sound that can be heard with a stethoscope placed on the skin above the heart. Such turbulent flow is produced when blood flow out of the heart is impeded by the cusps of the semilunar valves as they close.

52. What feature distinguishes pacemaker cardiac cells from other myocardial cells? Pacemaker cells:
   A. Require a stimulus from the vagus nerve in order to reach threshold, myocardial cells do not
   B. Reach threshold with much weaker stimuli than myocardial cells
   C. Have gap junctions, while myocardial cells do not
   D. Spontaneously generate action potentials, while myocardial cells do not

Answer is D: Pacemaker cells (generally in the SA node) generate action potentials 80–100 times per minute even without neural input. Neural input will slow or increase this rate. Myocardial cells are not autorhythmic.

53. What structure in the heart prevents backflow of blood into the right atrium?
    A. The tricuspid valve
    B. The bicuspid valve
    C. The mitral valve
    D. The foramen ovale

Answer is A: The tricuspid valve lies between the right ventricle and right atrium and closes when the ventricle contracts.

54. Which heart valve does blood pass through as it leaves the right ventricle?
    A. The tricuspid valve
    B. The aortic valve
    C. The pulmonary valve
    D. The right atrioventricular valve

Answer is C: Blood passes through the pulmonary valve as it leaves the right ventricle. Choices A and D are the same valve.

55. What two-layered membranous sac surrounds the heart?
    A. The pericardium
    B. The syncytium
    C. The visceral myocardium
    D. The peritoneum

Answer is A: "Peri-" refers to surrounding, while "-cardium" refers to the heart.

56. What is the function of the intercalated discs that exist in the heart?
    A. To initiate the contraction sequence of the heart
    B. To transmit the electrical impulse from the AV node to the Purkinje fibres
    C. To supply cardiac cells with energy for contraction
    D. To rapidly pass depolarisation from one cell to the adjacent cell

Answer is D: Intercalated discs attach a cardiac muscle cell to its neighbour. This connection allows rapid passage of the depolarisation signal from cell to cell, so that ventricular contraction is co-ordinated. Hence the ventricles behave as a "functional syncytium".

57. What is the name of the artery that leaves the left ventricle?
    A. Aorta
    B. Coronary artery
    C. Pulmonary trunk
    D. Pulmonary vein

Answer is A: It is the aorta.

58. Which ventricle of the heart (if any) pumps the greater volume of blood?
    A. The left ventricle
    B. The right ventricle

C. They pump the same volume of blood
D. During exercise, the left ventricle pumps the greater volume, during recovery, the right ventricle does.

Answer is C: Both ventricles pump the same volume of blood at all times. If for example the left ventricle pumped less blood than the right, blood would accumulate in the lungs. This would happen with congestive heart failure.

59. What term is used for the muscular part of the heart?
   A. Mediastinum
   B. Pericardium
   C. Myocardium
   D. Coronary endothelium

Answer is C: "Myo-" refers to muscle, while "cardium" refers to the heart. Pericardium is the serous membrane that surrounds the heart.

60. From which structure does blood that enters the right ventricle come?
   A. The left ventricle
   B. The left atrium
   C. The right atrium
   D. The vena cava

Answer is C: Blood passes from the right atrium into the right ventricle, by going through the tricuspid valve.

61. During which part of the cardiac cycle does blood enter the coronary arteries?
   A. Atrial diastole
   B. Ventricular diastole
   C. Atrial systole
   D. Ventricular systole

Answer is B: When the myocardium is relaxed (in diastole), and the aortic valve is shut, blood from the aorta flows into the coronary arteries.

62. Through which structures does the electrical signal that produces contraction travel once it is initiated by the heart's pacemaker?
   A. Sinoatrial node, sinoatrial bundle, atrioventricular node, atrioventricular bundle, left and right bundle branches
   B. Atrioventricular node, atrioventricular bundle, sinoatrial node, Purkinje fibres, left and right bundle branches
   C. Sinoatrial node, atrioventricular node, atrioventricular bundle, left and right bundle branches, Purkinje fibres
   D. Atrioventricular node, Purkinje fibres, sinoatrial node, left and right bundle branches, intercalated fibres

Answer is C: The signal that initiates contraction emanates from the SA node, but there is no sinoatrial bundle.

63. Where is the "apex" of the heart?
    A. It is in the right atrial wall.
    B. It is the most inferior part of the left ventricle.
    C. It is the start of the aorta.
    D. It is below the atria and between the two ventricles.

Answer is B: The apex is the "pointy" end of the heart at the most inferior part of the left ventricle. It rests on the diaphragm.

64. What is another name for the tricuspid valve of the heart?
    A. Left atrioventricular valve
    B. Right atrioventricular valve
    C. Aortic valve
    D. Pulmonary valve

Answer is B: Blood passes through the tricuspid valve as it leaves the right atrium and enters the right ventricle so is also known as the right atrioventricular valve.

65. What is another name for the mitral valve of the heart?
    A. Bicuspid valve
    B. Right atrioventricular valve
    C. Aortic valve
    D. Pulmonary valve

Answer is A: The mitral valve has two parts so is also called the bicuspid valve. Furthermore, blood passes through the mitral valve as it leaves the left atrium and enters the left ventricle so is also known as the left atrioventricular valve.

66. Which term refers to the contraction of heart muscle?
    A. Ventricular diastole
    B. Ventricular tachycardia
    C. Systole
    D. Diastole

Answer is C: Systole refers to contraction of the myocardium, while diastole refers to its relaxation. Tachycardia refers to faster than normal heartbeat, as occurs for example during particularly vigorous exercise.

67. Apart from the heart, what other structures lie within the mediastinum?
    A. Thymus, oesophagus, trachea, sciatic nerve, descending aorta
    B. Spinal cord, oesophagus, thoracic duct, descending aorta, inferior vena cava
    C. Thyroid, trachea, thymus, thoracic duct, inferior vena cava
    D. Oesophagus, trachea, thymus, inferior vena cava, descending aorta

Answer is D: These five structures all lie within the mediastinum (i.e. between the lungs). The sciatic nerve, spinal cord and thyroid are not within the mediastinum.

68. Of the following, which are all chambers of the heart?
    A. Right ventricle, right atrium, left ventricle
    B. Left ventricle, left auricle, right ventricle
    C. Right carotid, left atrium, left ventricle
    D. Right auricle, right atrium, left atrium

Answer is A: The right ventricle, right atrium, left ventricle are three of the four chambers of the heart—the fourth is the left atrium. Auricles and the carotid are not heart chambers.

69. Which of the listed blood vessels either return blood to the heart or allow blood to leave the heart for the general circulation?
    A. Aorta, coronary arteries, pulmonary arteries, the venae cavae
    B. Pulmonary veins, aorta, the venae cavae, carotid arteries
    C. Pulmonary arteries, aorta, the venae cavae, pulmonary veins
    D. The venae cavae, aorta, pulmonary veins, coronary sinus

Answer is C: Blood leaves the heart via the pulmonary arteries and aorta, and blood enters the heart via the inferior and superior vena cavae and pulmonary veins. The coronary arteries and coronary sinus serve the myocardium's own circulation, rather than the general circulation. The carotid arteries are in the neck.

70. What are the names of the four heart valves?
    A. The aortic semilunar valve, the pulmonary semilunar valve, the mitral and the bicuspid
    B. The mitral, aortic, pulmonary and tricuspid
    C. Right atrioventricular valve, left atrioventricular valve, mitral and bicuspid
    D. The tricuspid, the bicuspid, the right atrioventricular and the left atrioventricular valve.

Answer is B: The left atrioventricular valve, mitral and bicuspid are three names for the same valve. The tricuspid and the right atrioventricular are two names for the same valve. The aortic and the pulmonary valves are also known as semilunar valves.

71. What is the name the membrane that adheres to the outside of the myocardium?
    A. Parietal pericardium
    B. Endothelium
    C. Endocardium
    D. Epicardium

Answer is D: The epicardium is "on top of" the myocardium. Also known as the visceral pericardium. Together with the more superficial parietal pericardium, it forms the two-layered pericardium.

72. What features do cardiac and skeletal muscle cells share?
    A. Striations
    B. Multinucleated cells
    C. Branching cells
    D. Voluntary contraction

Answer is A: Cardiac cells and skeletal muscle cells both have striations when viewed under a microscope. SM cells are multinucleated and "voluntary" CM cells are not. CM cells are branched, but SM cells are not.

73. Where are the coronary arteries found?
    A. In the neck
    B. In the cortex of the kidneys
    C. In the heart
    D. In the coronal section of the body

Answer is C: The left and right coronary arteries both arise from the base of the aorta and supply the myocardium (the heart muscle).

74. What are the heart valves doing as the ventricles begin systole?
    A. Both atrioventricular valves are shut, and both semilunar valves are shut.
    B. Both atrioventricular valves are shut, and both semilunar valves are open.
    C. Both atrioventricular valves are open, and both semilunar valves are shut.
    D. Both atrioventricular valves are open, and both semilunar valves are open.

Answer is A: Ventricular systole refers to the contraction of the ventricular myocardium. The atrioventricular valves shut as soon as ventricular contraction starts. Both semilunar valves are also shut due to the "back-pressure" of blood in the pulmonary artery and aorta. They do not open until the ventricles have raised the pressure in the ventricular blood to greater than this back-pressure.

75. What are the heart valves doing at the time during which ventricular systole occurs is about two-thirds of the way through?
    A. Both atrioventricular valves are shut, and both semilunar valves are shut.
    B. Both atrioventricular valves are shut, and both semilunar valves are open.
    C. Both atrioventricular valves are open, and both semilunar valves are shut.
    D. Both atrioventricular valves are open, and both semilunar valves are open.

Answer is B: Ventricular systole refers to the contraction of the ventricular myocardium. The atrioventricular valves shut as soon as ventricular contraction starts. If systole has progressed to two-thirds, then both semilunar valves are open because the ventricles have exerted sufficient pressure to force blood to push the semilunar valves open the blood in the pulmonary artery and aorta away from the heart.

76. Why is the myocardium of the left ventricle so thick?
    A. It needs to produce sufficient force to push blood back to the heart.
    B. It pushes blood against gravity.

C. It pushes blood through the pulmonary circulation.
D. It pushes blood through the systemic circulation.

Answer is D: The LV pushes blood through the systemic circulation which has a higher resistance to blood flow than that of the pulmonary circulation. Hence the LV has a thicker (stronger) myocardium than the RV. Choice A is wrong as venous return is not influenced by the pumping action of the heart.

77. What is the general function of the heart valves?
    A. They prevent too much blood entering the chambers.
    B. They allow blood to flow in one direction only.
    C. They allow the chambers to fill in sequence.
    D. They prevent blood entering the ventricles during diastole.

Answer is B: A valve is a one-way device that prevents blood "regurgitating" from ventricles back into the atria. The contraction of the heart pushes blood in the correct direction. It is not possible for "too much" blood to enter the chambers.

78. How does the structure of the left ventricle wall of the heart differ from that of the right?
    A. The volume of the left ventricle is greater than that of the right.
    B. The muscle fibres of left ventricle are arranged in a transverse circular pattern, while those of the right are arranged longitudinally (in cephalic to caudal direction).
    C. The left ventricle has predominantly skeletal muscle, while the left ventricle is predominantly smooth muscle tissue.
    D. The left ventricle has a much thicker muscle wall than the right ventricle.

Answer is D: The thicker muscle allows a greater pressure to be generated by the LV. The other choices are nonsense.

79. During which part of the heart cycle does blood enter the coronary arteries?
    A. Atrial systole
    B. Ventricular systole
    C. Atrial diastole
    D. Ventricular diastole

Answer is D: When the ventricles are contracting, the coronary arteries are squashed shut preventing blood flow through them. During ventricular relaxation (diastole), the aortic valve shuts and blood can flow into the openings of the coronary arteries that are situated just superior to the valve.

80. What effect would a blockage in a coronary artery have on the myocardium?
    A. A blockage would prevent blood getting to part of the myocardium so some cells would die.
    B. It would cause a myocardial infarction and the person would die.

C. The myocardium would not receive sufficient oxygen and the pain of angina would be felt.
D. The person would faint as the heart cannot pump sufficient blood to the brain.

Answer is A: The person may not die if the MI is small. Choice C refers to a significant narrowing in the coronary artery rather than a blockage.

81. How many cusps does the RIGHT atrioventricular valve have?

A. One
B. Two
C. Three
D. Four

Answer is C: The right AV valve between the right atrium and right ventricle is also known as the tricuspid valve because it has three cusps or flaps.

82. What is a definition of fibrillation?

A. A resting heart rate that is too high (>90 bpm)
B. An arrhythmia where the myocardium is beating in an un-coordinated way and too fast
C. A heart rhythm that is not normal sinus rhythm
D. A resting heart rate that is low (<60 bpm)

Answer is B: The un-coordinated contraction of ventricular fibrillation means no blood is pumped and death is imminent. Atrial fibrillation maybe a chronic condition. Choices A and C are tachycardia and bradycardia, respectively.

83. What is considered to be a healthy resting adult heart rate?

A. Less than 60 beats/min
B. 70–90 beats/min
C. 65–85 beats/min
D. 50–70 beats/min

Answer is C: This is the best answer. The average for men is 70–75 bpm and for women 73–77 bpm, but some people may have a resting HR as high as 100 bpm. Below 60 bpm is considered to be bradycardia, and all well-trained athletes will have a resting HR below 60 bpm.

12.2.2 ECG

The myocardial cells of the heart can separate electrically charged ions by transporting some types (e.g. K^+) into the cells and others (e.g. Na^+) out of the cells using transporter proteins in the cell membrane. This separation of charge between the inside and outside of the cell is called the "resting potential". When this resting

potential is stimulated to change, the change is termed an "action potential" which then stimulates the myocardial cells to contract. These changes in potential may be measured. An electrocardiograph is the machine that produces the graphs known as electrocardiograms (ECG). These graphs display the variation of the electrical potential produced by the heart, against time. This electrical potential (voltage) is measured non-invasively at the body surface, using electrodes attached to the skin of the arms and legs (or the body trunk). The voltage changes occur when the membrane potential of the myocardial cells switch from their polarised state (in which the inside of the membrane is electrically negative while the outside of the membrane is positive) to the depolarised state, and then switch back again. After depolarisation, the myocardial cells contract. Six graphs are produced by combining the voltages measured at the four limb electrodes called Limb Lead I, II and II and augmented leads aVR, aVL and aVF. In addition six graphs are derived from the voltages measured by the six chest (precordial) electrodes. The three peaks and two "valleys" of the ECG are labelled with the letters P, Q, R, S, T. By knowing what the graphs look like in healthy individuals, changes from the expected patterns are diagnostic of damage to the cardiac muscle. The 12 ECG graphs are used to examine the normal functioning of different parts of the heart. The measured values for the time intervals between the five main features of these graphs have proved to be diagnostically valuable to cardiologists.

1. In a normal ECG trace, what does a QRS wave indicate?
   A. Depolarisation of the atria
   B. Repolarisation of the atria
   C. Depolarisation of the ventricles
   D. Repolarisation of the ventricles

Answer is C: The QRS "wave" is produced by the change in potential when the cells of the ventricles depolarise.

2. Which one of the following descriptions or statements about an electrocardiogram (ECG) is **NOT** correct?
   A. It is a record of the voltage changes (as measured at the body surface) due to the depolarisation of the muscle cells of the heart as it beats.
   B. The potentials measured by the ECG electrodes are combined in various ways to give 12 different ECGs.
   C. The ECG consists of the electrical events that follow the depolarisation of ventricles (PQR section), the depolarisation of atria (the S section) and the repolarisation of the ventricles (the T section).
   D. The value of the potential difference called "limb lead II" varies with time to produce the familiar ECG trace—a graph of voltage vs time.

Answer is C: The ECG also displays the depolarisation of the atria (the P wave), the depolarisation of the ventricles results in the QRS wave—not the PQR wave.

3. Why is it important to record the date, time and subject's name if taking an ECG recording in a clinical situation?
   A. To ensure that the right patient gets the right treatment.
   B. To allow action to be taken as soon as practicable to minimise the myocardial ischemia.
   C. This ensures that any differences between ECG recordings taken at different times are noticed.
   D. All pieces of data are required for the legal patient records.

Answer is C: It is expected that treatment will decrease the area of myocardial ischemia (oxygen deprivation), and so the ECG will show this as treatment progresses. Noting the date and time allows the changes to be tracked and treatment modified.

4. What is the duration (in seconds) of a 5 mm ECG trace? (the paper speed was 25 mm/s).
   A. 0.04
   B. 0.08
   C. 0.10
   D. 0.20

Answer is D: The paper moves 25 mm in 1 s, so moves 1 mm in 0.04 s ($1 \div 25 = 0.04$). Hence 5 mm is equivalent to $5 \times 0.04 = 0.2$ s.

5. Which of the following is NOT part of the normal conduction system of the heart?
   A. The right atrioventricular valve
   B. The atrioventricular node
   C. The sinoatrial node
   D. The right bundle branch

Answer is A: This is a valve and not part of the normal electrical conduction pathway of the heart.

6. Following which deflection seen on an ECG does ventricular systole occur?
   A. P wave
   B. R-T interval
   C. QRS wave
   D. T wave

Answer is B: The ventricles contract (ventricular systole) after the myocardial cells have depolarised. Depolarisation occurs during the QRS wave. So between this and the T wave, the muscle contracts.

7. Given that each millimetre square of the ECG graph paper corresponds to an interval on 0.04 s, if 32 mm separates consecutive QRS peaks, what is the heart rate?
   A. 128 bpm
   B. 77 bpm
   C. 64 bpm
   D. 47 bpm

Answer is D: If 1 mm = 0.04 s, then 32 mm = 32 × 0.04 = 1.28 s between beats. How many heart beats of this length will fit into 1 min (60 s)? bpm = 60 ÷ 1.28 = 47.

8. What are the letters ECG short for?
   A. Electroencephalogram
   B. Electrocolonogram
   C. Electrocardiogram
   D. Electrocryogram

Answer is C: An electrocardiogram is the diagram (the graph) of voltage against time that is produced by a machine called an "electrocardiograph".

9. The potential differences generated by the heart muscle as it contracts and relaxes can be measured by placing electrodes on the surface of the body. What is the record of these electrical events correctly called?
   A. An electrocardiogram
   B. Limb lead II
   C. An electrocardiograph
   D. An EEG

Answer is A: An electrocardiogram (ECG) is produced by an electrocardiograph machine.

10. What are the deflections seen on an electrocardiogram trace due to?
    A. Pressure differences created by ventricular contraction
    B. The closing and opening of heart valves
    C. The de- and repolarisation of the cells of the myocardium
    D. Variation in the electrical properties of oxygenated blood and deoxygenated blood as it moves through the heart

Answer is C: The P wave is due to the depolarisation of the atria, the QRS due to the depolarisation of the ventricles, the T due to the repolarisation of the ventricles.

11. The electrocardiogram chart is a graph. What is plotted against the vertical axis?
    A. The voltage produced by the myocardial cells
    B. The elapsed time
    C. The rate of contraction of the myocardium

D. The current flowing in the conduction system of the heart

Answer is A: The ECG records the voltage produced by the heart muscle measured at the body surface.

12. Consider a cardiac monitor whose time-base control is set so that the ECG trace moves horizontally at 25 mm/s. If the patient's heart rate is 75 beats/min, how far apart would the peaks of the ECG trace be?

    A. 15 mm
    B. 18 mm
    C. 20 mm
    D. 25 mm

Answer is C: 75 bpm means 75 beats take 60 s. Hence each beat must take less than one second. Divide by 75 to get length of one beat: 60/75 s = 0.8 s.

The ECG trace travels at 25 mm/s, so in 0.8 s would travel $\frac{4}{5}\text{s} \times 25\ \text{mm/s} = 20\ \text{mm}$.

13. The ECG trace on the screen of a cardiac monitor is measured to be 3 cm high. If the sensitivity control was set at 2 cm/mV, what is the value of the voltage being measured?

    A. 1.0 mV
    B. 1.5 mV
    C. 2.0 mV
    D. 2.5 mV

Answer is B: 3 cm ÷ 2 cm/mV = 1.5 mV.

14. During which deflection seen on an ECG does atrial DEpolarisation occur?

    A. P wave
    B. QRS wave
    C. T wave
    D. PQ interval

Answer is A: As the atria depolarise, the P deflection is recorded.

15. During which deflection seen on an ECG does ventricular DEpolarisation occur?

    A. P wave
    B. T wave
    C. ST interval
    D. QRS wave

Answer is D: As the abundant muscle cells of the ventricles depolarise, the large QRS deflection is produced.

16. During which deflection seen on an ECG does atrial REpolarisation occur?

    A. T wave
    B. QRS wave

C. P wave
D. ST segment

Answer is B: Atrial repolarisation is a small electrical event that happens to occur at the same time as the depolarisation of the ventricles (i.e. during the QRS wave). The small deflection that would be produced by atrial depolarisation is rendered undetectable by the comparatively large QRS wave.

17. During which deflection seen on an ECG does ventricular REpolarisation occur?

A. PQ interval
B. ST interval
C. P wave
D. T wave

Answer is D: As the ventricles repolarise they produce the T wave, a smaller electrical event than ventricular depolarisation. Hence the T peak is much smaller than the QRS peak.

18. Which ECG feature is recorded at the same time as the mechanical event of atrial systole is occurring?

A. QT segment
B. QRS complex
C. PR segment
D. ST interval

Answer is C: First the atrial muscle cells must depolarise (the P wave), after which the cells can contract (systole). This contraction occurs after the P wave, and while the PR segment is being recorded.

19. Which ECG feature is recorded at the same time as the mechanical event of ventricular systole is occurring?

A. QT segment
B. QRS complex
C. QT segment
D. ST segment

Answer is D: First the ventricle muscle cells must depolarise (the QRS complex), after which the cells can contract (systole). This contraction occurs after the QRS wave, and while the ST segment is being recorded, and before the T wave which signifies the repolarisation of the ventricles.

20. Given that the ECG trace of lead aV_R is the potential difference given by: [potential at right arm – ½ (potentials at left arm + left leg)], which ECG trace is most similar to an inverted version of lead aV_R?

A. Limb lead I = potential at left arm – potential at right arm
B. Limb lead II = potential at left leg – potential at right arm

C. Limb lead III = potential at left leg – potential at left arm
D. V_1 = potential at lead I – ⅓ (potentials at left arm + left leg + right arm)

Answer is B: An inverted aV_R may be produced by taking the negative of the trace. The negative of aV_R is: [½ (potential at left arm + potential at left leg) – potential at right arm], which, if the "potential at left arm" is ignored, is very similar to the equation for limb lead II.

21. Why can the electrical activity of the heart as measured by the ECG be determined from electrodes placed on the skin instead of needing to be placed directly on the heart muscle?
    A. The skin is not a barrier to electrical current produced within the body.
    B. Placing electrodes on the cardiac muscle is too invasive, hence they are placed on the skin.
    C. The heart generates a large voltage which is not dissipated by the distance between the heart and the skin.
    D. Human tissue is an electrical conductor so current easily flows from the heart to the skin surface.

Answer is D: Human tissue both intracellular and extracellular is largely a solution of dissolved ions which allow electrical current to flow with little resistance. Provided that the skin electrodes are coated with a conducting gel which allows current to cross the barrier presented by the skin, placing electrodes to the skin is equivalent to connecting them directly to the heart.

22. Why can an ECG be recorded by using skin electrodes attached to the wrists and ankles rather than being attached to the chest where the heart is located?
    A. These four locations are approximately equidistant from the heart and thus contribute equally to the electrical signal.
    B. The wrists and ankles are narrow and so concentrate the electrical potential.
    C. If the electrodes are too close to the heart the signal is too large to be faithfully recorded.
    D. The arms and legs behave like electrically conducting wires and wrists and ankles are easily accessible.

Answer is D: Extracellular fluid is an aqueous solution of ions and therefore conducts electricity. The tissues between the wrist and the heart and between the ankles and the heart behave like electrically conducting wires and conduct the changes in potential produced by the heart to the electrodes, wherever they may be placed. Before conducting plastic that could adhere to the skin was available, it was easier to clamp metal electrodes around the wrists and ankles than it was to clamp the chest.

12.3 Blood Vessels

Blood vessels that carry blood away from the heart are called arteries. The largest is the aorta with a diameter of 2–3 cm. Additional arteries arise as the branches of the arterial tree divide and divide again to reach every part of the body. Eventually their diameter is small enough for the vessels to be called arterioles, and they direct blood into a capillary bed. Artery walls are elastic and expand in diameter when the heart pumps blood into them and recoil to a smaller diameter during diastole as blood continues its flow towards the capillaries. This expansion and contraction may be felt as the "pulse" by palpating a superficial artery.

Blood vessels that carry blood towards the heart are called veins and have thinner walls but larger diameters than arteries. Venules drain blood away from capillary beds and join with each other to form larger veins and eventually become the inferior vena cava, if from that part of the body inferior to the diaphragm, or the superior vena cava if coming from the head, neck, arms and chest. Veins have valves (every few centimetres in the legs) which allow blood to flow towards the heart but not away from the heart. Furthermore, veins are compressed when the skeletal muscles that they pass by contract, and this compression propels the contained blood through the valves. By this means, blood is lifted towards the heart against gravity when a human is not lying down. Both veins and arteries have smooth muscle within their vessel walls whose contraction produces vasoconstriction and whose relaxation produces vasodilation. In this way blood may be directed away from or towards a tissue as its requirements change.

Capillaries are the place at which nutrients and oxygen enter the blood (from the gut and lungs). They are also the destination for the materials carried in the blood. Capillaries in the muscles and other organs are the place to which oxygen, water, nutrients and electrolytes are delivered. From here they pass out of the capillaries and make their way by diffusion to the neighbouring cells. Wastes from the cells move into the capillaries to be transported and in turn delivered to the capillaries of the lungs, liver and kidneys. Consequently the structure of capillaries, be they continuous, fenestrated or sinusoidal capillaries, facilitates exchanges between their lumen and the surrounding interstitial fluid.

1. Why does the lumen of a large vein have a larger diameter than the lumen of a large artery?
   A. They need to withstand higher pressure than arteries.
   B. Veins contain the majority of the blood volume.
   C. This allows blood to return to the heart rapidly.
   D. So that vasoconstriction can produce a greater change in diameter than for arteries.

Answer is B: Veins contain about 60% of the blood volume (and because they have thinner walls, they are also more distensible than arteries).

2. Different types of capillaries are distinguished by their structure. Which of the following is **NOT** a structural difference between capillaries?
   A. The endothelial cells of some capillaries are joined by tight junctions while for others there are gaps between the cells.
   B. Some capillaries have pores that allow movement between the plasma and interstitial fluid.
   C. Some capillaries have an arterial end and a venous end, while others begin in the tissues and drain into a venule.
   D. Some capillaries are surrounded by a basement membrane while others are not.

Answer is C: No capillaries "start" in the tissues. Continuous capillaries are joined by tight junctions; fenestrated capillaries have pores; sinusoid capillaries may have no basement membrane.

3. The "systemic circulation" refers to which of the following?
   A. The movement of blood from the pulmonary trunk, through capillaries into the pulmonary veins.
   B. The movement of blood into the coronary arteries, through capillaries into the coronary sinus.
   C. The flow of blood into the right atrium and eventually out of the left ventricle.
   D. The movement of blood from the aorta, through capillaries, then eventually into the vena cavae.

Answer is D: The "system" is supplied with blood from the aorta. Choice A refers to the pulmonary circulation, B to the coronary circulation, C to the flow of blood through the heart.

4. What causes venous blood to return to the heart?
   A. The pumping action of the heart
   B. The squashing action of muscles and valves in the veins
   C. Rhythmic vasoconstriction and valves in the veins
   D. Gravity, valves and the negative pressure generated by the atria emptying

Answer is B: Valves prevent blood in veins from flowing away from the heart, while contracting skeletal muscle compresses the veins and this pushes blood through them towards the heart (as the valves prevent blood form being pushed in the other direction).

5. When cardiac ejection ceases during diastole, what is the most important factor maintaining blood flow in arteries of the body?
   A. Contraction of skeletal muscle
   B. Closing the venous valves
   C. Elastic recoil of the arteries close to heart
   D. Contraction of the atria

Answer is C: Arteries store elastic potential energy in their distended elastic walls, so that when ventricular contraction ceases, the elastic artery walls recoil, forcing blood to continue moving away from the heart.

6. In which organs would be found continuous, fenestrated, and sinusoid capillaries, respectively?
   A. Brain, small intestine, liver
   B. Bone marrow, brain, spleen
   C. Liver, bone marrow, brain
   D. Small intestine, liver, brain

Answer is A: The blood supply to the brain has continuous capillaries (in the blood–brain barrier), the SI has fenestrated capillaries (to facilitate absorption of digestion products), the liver has sinusoids.

7. What is the advantage of having a wide lumen in veins?
   A. It provides less resistance to the blood on its way to heart.
   B. It controls the opening and closing of the valves.
   C. It produces high pressure on the blood on its way to heart.
   D. It helps the pre-capillary sphincters to stay open for a longer time.

Answer is A: The larger is the vessel diameter, the less will be the resistance to blood flow. The other choices are nonsense.

8. Given that we have the following blood pressures at the arterial end of a capillary: blood pressure: 28 mmHg, and osmotic pressure: 20 mmHg; what is the net filtration pressure at this end?
   A. 0.71 mmHg
   B. 1.4 mmHg
   C. 8 mmHg
   D. 48 mmHg

Answer is C: BP promotes the outflow of liquid and solutes from the arterial end, while the blood's osmotic pressure promotes the inflow of water. That is, these two influences are opposing. Hence $28 - 20 = 8$ mmHg.

9. What is the pulse that we feel at the anterior medial part of the elbow (when in anatomical position) called?
   A. Radial pulse
   B. Ulnar pulse
   C. Carotid pulse
   D. Brachial pulse

Answer is D: Because at this position, we are palpating the brachial artery.

10. Which capillaries allow cells and plasma proteins to enter or leave their lumen?
    A. Continuous
    B. Fenestrated

C. Sinusoidal
D. Anastomatic

Answer is C: Sinusoidal capillaries have some gaps between the adjacent endothelial cells (and a thin or absent basement membrane). The gaps are large enough for plasma proteins and some cells to pass through.

11. Which of the following does **NOT** assist in returning the blood to the heart through the veins?

    A. Valves in the veins
    B. The "respiratory pump"
    C. The effect of gravity
    D. The pumping action of the heart

Answer is D: The heart's contractions do not affect blood movement after blood passes through the capillaries. The contraction of the diaphragm increases pressure in the abdomen, and this creates a pressure gradient that pushes blood towards the heart against gravity. Gravity assists blood to return from the head and neck.

12. Which of the following materials is found in the walls of capillaries?

    A. Endothelium
    B. Elastic fibres
    C. Collagen fibres
    D. Smooth muscle

Answer is A: Only endothelium (endothelial cells and the basement membrane) compose the walls of capillaries.

13. Which of the following arteries do **NOT** arise from the ascending aorta?

    A. Brachiocephalic trunk
    B. Left brachiocephalic
    C. Left common carotid
    D. Left subclavian

Answer is B: There is no left brachiocephalic. The left arm is supplied with blood via the left subclavian.

14. What is the pulse we feel in the anterior lateral wrist (in anatomical position—the normal wrist pulse) called?

    A. Radial pulse
    B. Ulnar pulse
    C. Dorsalis pedis pulse
    D. Brachial pulse

Answer is A: Radial pulse because we are palpating the radial artery at that position.

15. Which type of capillary is required to allow the liver to perform its function of producing plasma proteins?
    A. Continuous
    B. Fenestrated
    C. Sinusoidal
    D. Anastomatic

Answer is C: Plasma proteins are very large molecules, which are too big to pass out of (or into) capillaries. Sinusoidal capillaries have gaps between the endothelial cells of their walls which are large enough to allow proteins to enter the blood.

16. What is it called when plaque dislodges from a lesion in a blood vessel wall and then moves "downstream" to lodge in the capillary bed feeding the digestive tract?
    A. Myocardial infarction
    B. Stroke
    C. Pulmonary embolism
    D. Mesenteric embolism

Answer is D: Mesenteric because the membrane surrounding the gut is the mesentery. An embolism is a dislodged blood clot that eventually gets trapped in a capillary.

17. Which of the following is found in the walls of capillaries?
    A. Endothelial cells and basement membrane
    B. Tunica externa
    C. Tunica media
    D. Smooth muscle

Answer is A: Capillary walls are very thin to facilitate passage of water and solutes. They have only a tunica intima, that is, endothelial cells and a basement membrane.

18. In which of the four lists below are the types of capillaries listed in order of permeability with the first being the least permeable and the last being the most permeable type of capillary?
    A. Blood–brain barrier capillaries, continuous capillaries, fenestrated capillaries, sinusoids
    B. Continuous capillaries, blood–brain barrier capillaries, fenestrated capillaries, glomerular capillaries
    C. Fenestrated capillaries, blood–brain barrier capillaries, continuous capillaries, sinusoids
    D. Liver sinusoids, blood–brain barrier capillaries, continuous capillaries, fenestrated capillaries

Answer is A: The capillaries of the BBB are least permeable (they form a "barrier") while sinusoids are the most permeable.

19. Which of the following assists in returning the blood to the heart through the veins?
    A. Valves in the veins, the effect of breathing, gravity
    B. Valves in the veins, the effect of breathing, squashing action of muscles, gravity
    C. The effect of gravity, the pumping action of the heart
    D. The effect of breathing, squashing action of muscles, the right ventricle

Answer is B: Each of the four assist in some way. Choice A omits the action of muscular movement. The heart has no role in vascular return.

20. Why do arteries have more elastic and muscular tissue than veins?
    A. Arteries need to expand and contract as blood flows through them.
    B. Arteries need carry a greater volume of blood than do veins.
    C. To ensure that blood flows only in the direction away from the heart.
    D. In order to support the larger diameter of arteries compared to veins.

Answer is A: Arteries distend to accept the volume of blood pumped into them, then elastically recoil to assist blood flow during diastole. The ability to contract enables arteries to direct blood away from tissues that require less blood. Arteries contain less volume than veins; the heart ensures forward flow; veins have a larger diameter than arteries.

21. Which capillaries have walls that allow the easiest passage of materials through them?
    A. Fenestrated capillaries
    B. Capillaries of the blood–brain barrier
    C. Sinusoids
    D. Continuous capillaries

Answer is C: Sinusoids have spaces between endothelial cells and may have an absent basement membrane. This allows for the easiest movement of material.

22. Which factor below does **NOT** assist venous return of blood?
    A. Breathing
    B. Gravity
    C. Smooth muscle contraction
    D. Skeletal muscle contraction

Answer is C: While contraction of smooth muscle can decrease the volume of veins and so shift more blood to the arterial side (and increase blood pressure). This is not the same as assisting the continuous return of blood to the right atrium. Breathing (the respiratory pump) assists the return of blood from the abdomen; skeletal muscle contraction (and valves) assist return from the limbs; and gravity assists return from the head.

23. The usefulness of having elastic arteries is in:
    A. Their ability to regulate blood pressure
    B. Their ability to expand as the heart pumps blood into them
    C. The assistance they give to venous return
    D. Their ability to produce vasoconstriction and vasodilation

Answer is B: The expansion of arteries as blood is pumped into them limits resistance to flow and blood pressure. Smooth muscle in the wall produces vasoconstriction and vasodilation.

24. What are capillaries that have endothelial cells joined by "tight junctions" and have "intercellular clefts" between cells called?
    A. The blood–brain barrier
    B. Continuous capillaries
    C. Fenestrated capillaries
    D. Sinusoids

Answer is B: An intercellular cleft is the space between adjacent cells. A tight junction is where the lipid portions of the membranes of adjacent cells are bound together preventing water and solutes from passing between the cells.

25. Vasoconstriction and vasodilation of blood vessels is facilitated by the:
    A. Elastic fibres in vessel walls
    B. Parasympathetic division of the nervous system
    C. Smooth muscle in vessel walls
    D. Tunica intima of the blood vessel

Answer is C: Contraction and relaxation of smooth muscle produces constriction and dilation of blood vessels. This is under the control of the sympathetic nervous system.

26. Which one of the following does **NOT** help blood to move through arteries?
    A. The influence of gravity
    B. The action of breathing and the movement of the diaphragm
    C. The elastic recoil of artery walls
    D. The pumping action of the heart

Answer is B: Movement of the diaphragm during breathing assist venous return but not movement of blood through arteries.

27. Veins in the limbs have which one of the following characteristics?
    A. Thin walls composed of epithelium
    B. Vessel walls with a thick layer of smooth muscle
    C. Pulsatile flow
    D. Valves

Answer is D: Veins have valves but not pulsatile flow or a thick layer of smooth muscle.

28. Exchange between the blood and the interstitial fluid occurs most readily through which type of capillary?
    A. Venules
    B. Fenestrated capillaries
    C. Sinusoids
    D. Arterioles

Answer is C: Sinusoids have the largest gaps (intercellular clefts) between cells.

29. What is meant by the "pulmonary circulation"? The flow of blood:
    A. Out the aorta and back through the vena cavae
    B. From the heart through the lungs and back to the heart
    C. Into the coronary arteries and back through the coronary sinus
    D. Into the vena cavae and out to the pulmonary trunk via the right ventricle

Answer is B: The pulmonary circulation refers to the flow of blood from the heart (via the pulmonary trunk) through the lungs and back to the heart.

30. What is a small artery called?
    A. An anastomosis
    B. An arteriole
    C. An efferent artery
    D. A distributing artery

Answer is B: A small artery or arteriole may be 30 μm (0.03 mm) in diameter.

31. How do arteries differ from veins?
    A. Arteries have a larger diameter than veins.
    B. Arteries have more elastic tissue than veins.
    C. There is a greater volume of blood in the arteries than in the veins.
    D. Arteries have valves but veins do not.

Answer is B: The walls of arteries have more elastic tissue than veins allowing them to expand as blood is pumped into them and then to passively recoil.

32. The wall of a capillary consists of one layer (or coat). What is it called?
    A. Tunica intima
    B. Tunica externa
    C. Lamina propria
    D. Tunica media

Answer is A: Tunica means "coat" while "intima" means inner most. In the case of capillaries this is the only tunica.

33. Which part(s) of the cardiovascular system has/have a single layer of endothelial cells as the innermost layer of the structure(s)?
    A. Arteries, capillaries and veins
    B. Arteries and veins

C. Heart, arteries, veins and capillaries
D. Capillaries

Answer is C: Heart, arteries, capillaries and veins all have an endothelium—a single layer of endothelial cells as their most internal layer, in contact with the blood they contain.

34. Which of the following parts of the cardiovascular system contain the greatest volume of blood?

    A. Heart
    B. Systemic arteries
    C. Pulmonary circulation
    D. Systemic veins

Answer is D: Systemic veins contain about 64% of the blood volume, while systemic arteries contain about 13%.

35. Which of the following describes an artery? An artery is a blood vessel that:

    A. Has a thinner wall than a vein of comparable size
    B. Carries blood away from the heart
    C. When located in the limbs, has valves
    D. Carries oxygenated blood

Answer is B: Arteries carry blood away from the heart. In the case of the pulmonary arteries, they carry de-oxygenated blood.

36. Some capillaries allow plasma proteins and phagocytes to enter and leave the blood. Which type?

    A. Sinusoids
    B. Fenestrated capillaries
    C. Continuous capillaries with intercellular clefts
    D. Lacteals

Answer is A: Sinusoids are capillaries that have gaps between their endothelial cells that are large enough to allow the passage of cells.

37. Arteries may be characterised as:

    A. Elastic tubes that carry oxygenated blood
    B. Elastic tubes that carry blood away from the heart
    C. Muscular tubes that have valves
    D. Muscular tubes with a larger diameter than veins

Answer is B: Arteries are both elastic, and carry blood away from the heart. They do not have valves, and they are smaller in diameter than veins.

38. Some capillaries are called "sinusoids". These are capillaries that:

    A. Are fenestrated
    B. Have endothelium without gaps between cells

C. Have endothelium with gaps between adjacent cells
D. Have pores in the endothelial cell walls

Answer is C: Sinusoids have incomplete walls around them, so large molecules and phagocytes can enter and leave the lumen.

39. Which of the following definitions best describes veins?
    A. A vessel that carries blood towards the heart
    B. A vessel that carries oxygenated blood
    C. Vessels that carry blood away from the heart
    D. Vessels whose walls are composed of three tunics

Answer is A: Veins carry blood towards the heart.

40. What can be said about the endothelium of fenestrated capillaries?
    A. The endothelial cells have pores (windows) to allow rapid movement of solutes and water.
    B. Gaps between adjacent endothelial cells allow free exchange between blood and interstitial fluid.
    C. Their tunica intima is composed of a complete lining of endothelial cells.
    D. They do not have an endothelium.

Answer is A: Fenestrated capillaries have pores (windows/fenestre) which penetrate the endothelial lining which facilitates the passage of material through them.

41. Which of the following definitions best describes veins?
    A. A vessel that carries de-oxygenated blood
    B. A vessel that carries blood towards the heart
    C. Vessels that contain valves
    D. Vessels through which blood flows under the influence of pressure produced by the heart

Answer is B: Veins carry blood towards the heart. In the case of the pulmonary veins only, the blood is oxygenated.

42. Fenestrated capillaries permit the exchange of nutrients and wastes between cells and the blood because of which feature of their structure?
    A. Gaps between adjacent endothelial cells allow free exchange between blood and interstitial fluid.
    B. They are in close proximity to most cells.
    C. Their tunica intima is composed of a single layer of endothelial cells.
    D. Their endothelium has pores (windows) to allow the rapid movement of water and solutes.

Answer is D: The pores (holes) in the endothelium allows passage of material through the capillary wall.

43. Which of the following is a difference between arteries and veins?
    A. Artery walls have more elastic tissue and smooth muscle than veins.
    B. Veins have three distinct "tunics" in their walls, whereas arteries have only two.
    C. Blood flow in veins is pulsatile while that is arteries is continuous.
    D. The walls of veins have more elastic tissue and smooth muscle than in arteries.

Answer is A: The walls of arteries are thicker than the walls of veins due to arteries having more smooth muscle and elastic tissue.

44. Capillaries permit the exchange of nutrients and wastes between cells and the blood because of which feature of their structure?
    A. Adjacent endothelial cells are separated by gaps which allow free exchange between blood and interstitial fluid.
    B. They are in close proximity to most cells.
    C. They have a single layer, the tunica intima, consisting of a single layer of endothelial cells in their walls.
    D. Their endothelial cells have pores (windows) to allow the rapid movement of water and solutes.

Answer is C: The walls of capillaries are very thin and have only one tunic, not three in their wall.

45. People with occupations that involve standing all day are at risk of varicose veins. Why should this be true?
    A. Veins have a larger lumen than arteries so hold more blood which distends them.
    B. Veins have only two tunics so their walls are more easily stretched into varicosity.
    C. Gravity acting on blood in the legs puts stress on the vessel walls increasing the likelihood that they will stretch and distort.
    D. When standing still, the valves stay open and blood accumulates in the leg veins.

Answer is C: Gravity acting on the blood in the legs of a standing person increases the strain on the walls of a vein. A weakness will lead to the walls becoming distended and the valve not operating correctly.

46. Which tunic(s) are ABSENT from the walls of blood capillaries?
    A. Tunica intima
    B. Tunics externa and intima
    C. Tunics media and externa
    D. Tunics interna and media

Answer is C: Capillaries are thin-walled vessels that only have the tunica intima (=interna) in their walls. Hence the tunica media and tunica externa are absent.

47. Which blood vessels allow blood to leave the heart and which allow blood to enter the heart?
    A. The aorta and pulmonary trunk leave, the venae cavae and pulmonary veins enter.
    B. The aorta and venae cavae leave, the pulmonary trunk and pulmonary veins enter.
    C. The aorta and pulmonary veins leave, the venae cavae and pulmonary trunk enter.
    D. The venae cavae and pulmonary trunk leave, the aorta and pulmonary veins enter.

Answer is A: The aorta and pulmonary trunk carry blood away from the LV and RV (respectively), while the two vena cavae and four pulmonary veins bring blood into the RA and LA, respectively.

48. Which statement about the aorta is true?
    A. The aorta carries blood into the left ventricle.
    B. The aorta carries blood that is leaving the left ventricle.
    C. The aorta carries blood into the right ventricle.
    D. The aorta carries blood that is leaving the right ventricle.

Answer is B: The aorta is an artery so carries blood away from the LV. The pulmonary trunk carries blood from the right ventricle.

49. Which of the following descriptions refers to the "microcirculation"?
    A. From aorta arteries to arterioles to systemic capillaries to venules to veins
    B. From coronary arteries to arterioles to capillaries to anterior cardiac veins and coronary sinus
    C. From pulmonary trunk to arteries to arterioles to alveolar capillaries to venules to veins
    D. From capillaries to interstitial fluid to cells and lymph and back again

Answer is D: Microcirculation is the exchange of solution between the capillary blood and the cells via the interstitial fluid. It occurs by diffusion.

50. Which of the following descriptions refers to the "pulmonary circulation"?
    A. From aorta arteries to arterioles to systemic capillaries to venules to veins
    B. From coronary arteries to arterioles to capillaries to anterior cardiac veins and coronary sinus
    C. From the right ventricle arteries to arterioles to alveolar capillaries to venules to veins
    D. From capillaries to interstitial fluid to cells and lymph and back again

Answer is C: The pulmonary circulation is the circuit from the heart through the lungs and back again. It occurs because of the pumping action of the right ventricle sending blood through the pulmonary artery.

51. What is the term applied to the balloon-like sac that develops when the weakened wall of a blood vessel bulges out?
    A. Arteriosclerosis
    B. Aneurysm
    C. Varicose veins
    D. Deep vein thrombosis

Answer is B: An aneurysm describes a blood vessel distended beyond normal and in danger of rupturing.

52. Which of the following list of characteristics distinguishes a vein from an artery? A vein:
    A. Has a larger diameter, has walls that are more elastic than those of arteries and carries blood at a higher pressure.
    B. Lacks a tunica media, carries oxygenated blood and carries blood at a lower temperature.
    C. Has pulsatile blood flow, carries deoxygenated blood and has a tunica muscularis.
    D. Has valves, carries blood towards the heart and carries blood at low pressure.

Answer is D: A vein is not more elastic than an artery; carries blood at the same temperature as arteries and flow in a vein is not pulsatile.

53. What are the blood vessels called that allow exchange of materials though their walls to occur?
    A. Capillaries
    B. Sinusoids
    C. Arterioles
    D. Anastomoses

Answer is A: Sinusoids are one type of capillary. Anastomoses are a network of capillaries.

54. If the hydrostatic blood pressure at one point within a capillary is 35 mmHg while the colloid osmotic pressure of blood is 26 mmHg, what is the value of the pressure moving fluid into or out of the vessel? (a positive value indicates fluid movement out of the vessel, while a negative value indicates movement into the vessel).
    A. $35 + 26 = 61$ mmHg, and fluid moves from blood to the interstitial fluid
    B. $35 - 26 = 9$ mmHg, and fluid moves from blood to the interstitial fluid
    C. $26 - 35 = -9$ mmHg, and fluid moves from the interstitial fluid into blood
    D. $-26 - 35 = -61$ mmHg, and fluid moves from the interstitial fluid into blood

Answer is B: Hydrostatic blood pressure is an influence that moves fluid out of a capillary (so is +ve), while colloid osmotic pressure is an influence that moves water into the capillary (so is −ve). Hence $+35 - 26 = 9$.

55. Which layer in the walls of blood vessels is responsible for their ability to vasoconstrict and vasodilate?
    A. Tunica intima
    B. Tunica media
    C. Tunica externa
    D. Tunica muscularis

Answer is B: The tunica media of a blood vessel has most of the smooth muscle. It is tempting to choose "tunica muscularis", but a layer with that name does not exist.

56. Which blood vessel supplies oxygenated blood to the liver?
    A. Hepatic vein
    B. Renal artery
    C. Hepatic artery
    D. Pulmonary vein

Answer is C: The term hepatic refers to the liver. Arteries supply oxygenated blood (except for the pulmonary artery).

57. Which artery is palpated when a heart rate measurement is made from the wrist?
    A. The brachial artery
    B. The radial artery
    C. The ulnar artery
    D. The dorsalis pedis artery

Answer is B: The radial artery lies over the radius bone which is on the same side of the hand as the thumb (the lateral side). The dorsalis pedis is in the foot.

58. From which type of capillary is it easiest for a white blood cell to leave the blood (to extravasate) in order to move into the tissues?
    A. Sinusoids
    B. Fenestrated capillaries
    C. Continuous capillaries
    D. Those in the blood–brain barrier

Answer is A: Sinusoidal capillaries have gaps between their endothelial cells large enough for cells to pass through. In addition, their basement membrane, being incomplete, presents a lower barrier to extravasation.

59. What are the three tunics that comprise the walls of veins and arteries?
    A. Tunica intima, tunica externa, tunica adventitia
    B. Tunica intima, tunica externa, tunica interna
    C. Tunica media, tunica externa, tunica adventitia
    D. Tunica media, tunica externa, tunica intima

Answer is D: The three tunics (coats) are: (from external to internal) tunica externa, tunica media, tunica intima. Tunica externa and tunica adventitia are two terms for the same tunic.

60. What is the difference between veins and arteries?

    A. Arteries have a tunica media, veins do not.
    B. Arteries carry blood away from the heart, veins carry blood towards the heart.
    C. Veins have a smaller diameter than the corresponding and adjacent arteries.
    D. Veins carry deoxygenated blood, arteries carry oxygenated blood.

Answer is B: Arteries carry blood away from the heart. In the aorta the blood is oxygenated, but in the pulmonary trunk the blood is deoxygenated. Veins carry blood towards the heart. In the vena cavae, the blood is de-oxygenated but in the pulmonary veins the blood is oxygenated.

61. What is the difference between muscular and elastic arteries?

    A. The largest arteries are "elastic" arteries.
    B. Muscular arteries are called conducting arteries, while elastic arteries are called distributing arteries.
    C. Muscular arteries are able to bulge more than elastic arteries as blood is pumped into them.
    D. Medium-sized arteries (approximate diameter 0.5 cm) are called elastic arteries.

Answer is A: The largest arteries (aorta, brachiocephalic, left common carotid, left subclavian, pulmonary trunk) are called elastic or conducting arteries. Choice B is the reverse of the truth. Elastic arteries bulge to receive blood as it is pumped into them. The diameter of muscular arteries is more controllable (due to the smooth muscle in their wall) hence are able to distribute blood as required.

62. What is the function of the smooth muscle in blood vessel walls?

    A. To assist with blood flow by performing peristaltic movements
    B. To allow the vessel wall to recoil after a surge of blood flow through it
    C. To control the diameter, hence blood capacity of the vessel
    D. To provide rigidity to the vessel wall

Answer is C: Smooth muscle is under autonomic control and can contract or relax to constrict or dilate the blood vessel. The diameter of the vessel determines the volume that flows through it. Arteries do not perform peristalsis.

63. What are the names of the three types of blood capillaries?

    A. Continuous, choroid, sinusoids
    B. Fenestrated, lymphatic, continuous
    C. Fenestrated, continuous, sinusoids
    D. Sinusoids, glomeruli, fenestrated

Answer is C: In order of increasing permeability: continuous, fenestrated and sinusoids. Lymphatic capillaries exist, but carry lymph. Glomeruli are the capillary beds at the start of a nephron.

64. The rate at which fluid is filtered across the capillary endothelium is determined by the sum of two outward pressures (capillary pressure, P_c and interstitial protein osmotic pressure, π_i), and two absorptive pressures (plasma protein osmotic pressure, π_p and interstitial fluid pressure, P_i). Which of these four pressures changes from the value at the arterial end to the value at the venous end of a capillary?
    A. Capillary pressure
    B. Interstitial protein osmotic pressure
    C. Plasma protein osmotic pressure
    D. Interstitial fluid pressure

Answer is A: The pressure of blood in the capillary changes from about 35 mmHg at the arterial end to about 15 mmHg at the venous end. The other three variables do not change.

65. What causes venous return (blood flow towards the heart) in the venous side of the systemic circulation?
    A. The elastic recoil of artery walls and valves in veins
    B. Smooth muscle contraction and diaphragm movement during breathing
    C. Gravity and the pumping action of the heart
    D. Skeletal muscle contraction and valves in veins

Answer is D: Skeletal muscle contraction exerts a squashing action on blood in veins, which moves blood away from the squashing muscle. Valves in veins prevent the movement of blood away from the heart so blood is forced towards the heart. The heart does not have any influence on blood in the veins.

66. What causes blood to flow in the arterial side of the systemic circulation?
    A. The breathing action of the lungs
    B. The pumping action of the left ventricle
    C. The pumping action of the right ventricle
    D. Gravity and the elastic recoil of the artery walls

Answer is B: The left ventricle pumps blood out of the heart through the aorta and into the systemic circulation.

12.4 Pressure: The Physics of Pressure

The concept of pressure may be applied to solids resting on a surface, liquids held in a container, or to gases. The pressure exerted by a solid is determined by dividing the weight of the solid (weight = mass × 9.8) by the area of the solid that is in contact with the surface upon which it rests. Thus pressure has the SI units "newton per square metre" and the "pascal" (1 N/m^2 = 1 Pa). Pressure in liquids may be determined by the depth below the liquid's surface at which the pressure measurement is made. This depth, being in units of length, gives rise to some of the units used to measure pressure within liquids: "mmHg" and "cmH_2O". Gases, like liquids, are fluids, and so the pressure exerted by the atmosphere may also be expressed as mmHg. Thus at sea level, "normal" atmospheric pressure is 760 mmHg (or 1 atm, or 14.7 psi, or 1016 mbar, or 1016 hPa). If the pressure within an enclosed gas is less than the surrounding atmospheric pressure, it is called a "negative pressure" (or suction). If it is greater than the atmospheric pressure, it is called a positive pressure—both are medically and physiologically useful. Fluids flow because of a pressure difference. That is, fluids flow away from regions of higher pressure towards regions of lower pressure. The body, by creating a region of high (or low) pressure can induce the blood to flow through the blood vessels and air to flow into and out of the lungs.

When fluids are flowing, their behaviour is complex but reasonably predictable. The rules or "laws" that allow us to predict such behaviour are Boyle's law, Henry's law, Poiseuille's law, Pascal's principle, Bernoulli's theorem, Fick's law of diffusion, the law of partial pressures and the ideal gas equation. Fluid flow in the body is complicated by the dilation and constriction of the airways and blood vessels, by the ability of constituents in the gas or blood to move out of and into the tubes and vessels and by the body altering the pressure differences generated by the muscular efforts of the heart and respiratory muscles or by changing its posture.

1. How much pressure in units of centimetres of water pressure is equivalent to 10 mmHg pressure?
   A. 1.36 cm of water
   B. 10 cm of water
   C. 13.6 cm of water
   D. 136 cm of water

Answer is C: 1 mmHg ~= 1.36 cmH_2O, so 10 mmHg ~= 13.6 cmH_2O. Even without knowing the conversion factor between cmH_2O and mmHg, you should know that mercury is much denser than water. Hence choices A and B are not correct.

2. The pressure inside a working suction bottle is most likely to be:
   A. 740 mmHg
   B. 760 mmHg
   C. 780 mmHg
   D. 800 mmHg

Answer is A: Given that atmospheric pressure is about 760 mmHg, for suction to occur, the pressure in the suction bottle must be a "negative" pressure. That is, less than 760 mmHg.

3. The volume of gas trapped within a space will increase as pressure decreases, the physics law that refers to this is:
   A. Henry's law
   B. Poiseuille's law

C. Hugh's law
D. Boyle's law

Answer is D: Boyle's law relates gas pressure and gas volume.

4. Increasing the number of red cells per millilitre of blood will:
   A. Increase the viscosity and increase the rate of flow of the blood
   B. Decrease the viscosity and increase the rate of flow of the blood
   C. Increase the viscosity and decrease the rate of flow of the blood
   D. Decrease the viscosity and decrease the rate of flow of the blood

Answer is C: Increasing the number of RBC would increase blood viscosity, and if viscosity increases, then rate of flow would decrease (Poiseuille's law).

5. A scuba diver swimming at a depth of 10 m will experience a pressure of how many atmospheres?
   A. 1
   B. 1.5
   C. 2
   D. 3

Answer is C: A head of water of 10 m will produce an atmosphere of pressure, which when added to the 1 atm of pressure due to the air of the atmosphere gives 2 atm pressure.

6. What law is being "obeyed" by a gas bubble as it expands inside a person's colon as the aeroplane ascends to cruising height?
   A. Dolphin's law
   B. Henry's Law
   C. Boyle's Law
   D. Aames's law

Answer is C: Boyle's law relates *P* and *V*: the volume of an enclosed gas expands as pressure decreases. The air pressure in an aeroplane cabin is allowed to decrease to the equivalent of the air pressure at about 2700 m above sea level.

7. 100 mmHg pressure equates to how many cm of water?
   A. 134
   B. 13.4
   C. 760
   D. 76

Answer is A: 1 mmHg ~= 1.34 cmH_2O, so 100 mmHg ~ 134 cmH_2O.

8. According to Boyle's law, what will happen in a fixed amount of gas if its volume decreases?
   A. The pressure of the gas will decrease.
   B. The pressure of the gas will increase.

C. The temperature of the gas will decrease.
D. The temperature of the gas will increase.

Answer is B: Pressure is inversely proportional to the volume, so if volume DEcreases, the pressure will INcrease. Boyle's law applies to gas at a constant temperature.

9. Consider Boyle's law. If the pressure being exerted on an enclosed volume of gas (at a constant temperature) is increased by 20%, by what percentage would its volume change?

   A. Volume would increase by 20%.
   B. Volume would increase by 44%.
   C. Volume would decrease by 44%.
   D. Volume would decrease by 20%.

Answer is D: Pressure and volume are inversely proportional. So if P increases by 20%, V will decrease by 20%.

10. A suction pump when used to clear an obstructed airway expands the volume in its chamber to:

    A. Produce a negative pressure and obstructing liquid is pushed out by air pressure in the lungs
    B. Produce a positive pressure and obstructing liquid is pushed out by air pressure in the lungs
    C. Produce a negative pressure and obstructing liquid is sucked up into the chamber
    D. Produce a positive pressure and obstructing liquid is sucked by atmospheric pressure into the chamber

Answer is A: A suction pump produces a "negative" pressure (one that is less than atmospheric pressure). Hence air in the lungs will flow out into the suction pump, hopefully pushing the obstructing liquid out too.

11. A walker carried a balloon filled with 10 L of air, from sea level (where the atmospheric pressure was 101 kPa), up a mountain to a height of 3000 m (where the atmospheric pressure was 70 kPa). What would be the approximate volume of the balloon at 3000 m?

    A. 7 L
    B. 10 L
    C. 14 L
    D. 30 L

Answer is C: Boyle's law states that volume must increase if pressure decreases, so choices A and B are wrong. If pressure decreases by 70/101, then volume must increase by 101/70 = 1.4. so 10 L × 1.4 = 14 L.

12. What formula or law or principle refers to the partial pressure of a gas?
    A. Boyle's law
    B. Henry's law
    C. Pascal's principle
    D. Poiseuille's law

Answer is B: Only Henry's law: the amount of dissolved gas is proportional to its partial pressure; refers to gas partial pressure.

13. Approximately how much pressure does a science lecturer who has a mass of 70 kg and is standing on 1 foot (which has an area of 0.0165 m^2), exert on the ground? (acceleration due to gravity is 10 m/s^2)
    A. 4.2 kPa
    B. 42 kPa
    C. 420 kPa
    D. 2.4×10^{-5} Pa

Answer is B: Pressure = force divided by area (and force is mass times acceleration due to gravity). $P = (70 \times 10) \div 0.0165 = 42{,}424$ Pa ~ = 42 kPa.

14. Which of the following is a statement of Boyle's law?
    A. If the pressure applied to an enclosed fluid is doubled, then that same pressure is transmitted to all parts of the fluid.
    B. If the pressure gradient between two places in an artery is doubled, then the rate of blood flow will double.
    C. If the pressure of a fixed amount of gas is doubled, then its volume will be halved.
    D. If the pressure of the gas above a liquid is doubled, then twice as much gas will dissolve.

Answer is C: Boyle's law relates pressure in a gas to its volume.

15. In which of the following situations would the greatest pressure be produced?
    A. A force of 500 N acts on an area of 0.1 m^2.
    B. A force of 800 N acts on an area of 0.1 m^2.
    C. A force of 300 N acts on an area of 0.2 m^2.
    D. A force of 500 N acts on an area of 0.2 m^2.

Answer is B: Pressure = force ÷ area. The higher the force acting on an area (or the smaller the area that a force acts upon), the greater will be the pressure. The pressure in choice A is greater than in choice D.

16. Using "suction" to drain fluid is described correctly by which choice below? Creating a:
    A. Positive pressure so that it can suck the fluid in the direction of the positive pressure

B. Negative pressure so that it can push the fluid towards the positive pressure
C. Negative pressure so that atmospheric pressure will push the fluid towards the negative pressure
D. Positive pressure so that atmospheric pressure will push the fluid towards the positive pressure

Answer is C: Fluids will flow towards the lower pressure, so creating a pressure below atmospheric pressure will cause fluid to be pushed by atmospheric pressure towards the negative pressure.

17. Pressure is a quantity derived from force and the area in contact with the force. What is it that pressure measures?

A. The force being exerted on an area.
B. The force divided by the area.
C. The force multiplied by the area.
D. The force divided by the area squared.

Answer is B: Pressure is force divided by area and is measured in N/m^2 (=pascal). Choice A is wrong as pressure is not a force.

18. Which of the following is **NOT** used as a unit of pressure?

A. Pascal
B. Newton per square metre
C. Millimetre of mercury
D. Millilitre of water

Answer is D: Millilitre of water is a volume and is not a form of pressure unit.

19. When performing cardio-pulmonary resuscitation (CPR) why do you use the "heel" of the hand rather than the whole palm and fingers?

A. A greater force can be exerted by the "heel" of the hand which results in a greater pressure being exerted.
B. Using the "heel" places a smaller area of the hand in contact with the sternum, hence allows a greater force to be exerted.
C. The same force can be applied to a smaller area of the sternum which is less able to resist the greater downward pressure.
D. The resuscitator's wrist is less likely to be damaged by applying CPR in this way.

Answer is C: Maximum force during CPR is determined by the strength of the resuscitator. Maximum pressure in addition depends on minimising the area upon which the force is being applied. The heel of the hand allows a small area to be used.

20. Which choice correctly completes the sentence? Pressure is a:

A. Force and is measured in newton (N).
B. Force multiplied by an area and is measured in newton-square metres ($N\ m^2$).

C. Force per unit area and is measured in newton per square metre (N/m^2).
D. Length and is measured in millimetres of mercury (mmHg).

Answer is C: Pressure is a unit derived from dividing force by area.

21. If a blood pressure is stated as 120 over 80 (in units of mmHg) this means:
    A. Pulse pressure is 120 mmHg.
    B. Diastolic pressure is 80 mmHg below atmospheric pressure.
    C. Systolic pressure is 80 mmHg above atmospheric pressure.
    D. Systolic pressure is 120 mmHg above atmospheric pressure.

Answer is D: The larger value (120) is the systolic pressure (i.e. at end of ventricular contraction) and states by how much the blood pressure is above atmospheric pressure.

22. The pressure, *P*, exerted by an intravenous infusion of saline at the level of the cannula must be great enough to overcome the venous blood pressure in the vein. Its magnitude is given by: $P = 10 \times 1000\ kg/m^3 \times$ head of liquid (in m), where $1000\ kg/m^3$ is the density of saline. If the venous pressure is 5 kPa, what is the smallest head of liquid required for the infusion to flow into the vein?
    A. 25 cm
    B. 35 cm
    C. 45 cm
    D. 55 cm

Answer is D: Convert cm to m and kPa to Pa. The pressure at the cannula must be greater than 5 kPa (5000 Pa). Use the value in choice A: $P = 10 \times 1000\ kg/m^3 \times 0.2$ $5\ m = 2500\ Pa = 2.5\ kPa$. The answer is only half of the pressure required (5 kPa), hence a head of at least twice 25 cm is required. Choice D is the only one that will do.

23. Which of these statements about pressure in static liquids is **NOT** true? Pressure at a point in a liquid:
    A. Depends on the height of liquid above it
    B. Acts equally in all directions
    C. Depends on the depth it is below the surface
    D. Depends on the volume of liquid above it

Answer is D: Volume (and even mass) of liquid above a point does not affect the pressure felt at that point. It is the length of the column of liquid above the point, and the effect of gravity that determines the pressure in the liquid.

24. Which of the following statements about stationary liquids is **FALSE**?
    A. At the liquid surface, pressure equals atmospheric pressure.
    B. The volume of liquid in an enclosed space can be compressed.
    C. Pressure increases with depth.
    D. Pressure due to weight of liquid is called "head of pressure".

Answer is B: Liquids are incompressible.

25. Which of the following is closest to a statement of Poiseuille's law?
    A. Volume flow rate is proportional to the fourth power of the radius of the tube.
    B. The pressure in flowing fluids is lowest where the speed of flow is greatest.
    C. The amount of gas that will dissolve in a liquid is proportional to its partial pressure.
    D. When the flow of fluid is producing sound, then turbulent flow is present.

Answer is A: This is one of the factors in Poiseuille's law upon which volume flow rate depends: $V \propto R^4$.

26. The symbols in Poiseuille's law have the following meanings: V represents the volume flow rate; ΔP the pressure drop; R the radius of the tube; η the viscosity of the liquid; l the length of the tube. Which of the following relationships from Poiseuille's law is **NOT** correct?
    A. $V \propto R^4$
    B. $V \propto l$
    C. $V \propto 1/\eta$
    D. $V \propto \Delta P$

Answer is B: Volume flow rate decreases with length of tube as greater length means greater resistance to flow. It should be $V \propto 1/l$.

27. The quantity called "pressure" is defined as the:
    A. Mass per unit area
    B. Force per unit area
    C. Height of mercury supported by the atmosphere
    D. Newton per square metre

Answer is B: Pressure = force ÷ area. Force and mass are different. Newton per square metre is the unit but not a definition of pressure.

28. The unit of pressure called the "pascal" (symbol Pa) is the name given to which of the following combination of units?
    A. Newton per metre (N/m).
    B. Newton per square metre (N/m^2).
    C. Millimetres of mercury (mmHg).
    D. Force per area (F/A).

Answer is B: Pressure units are derived from force and area units. Choices C and D are not combinations of units.

29. The millimetre of mercury (mmHg) is one of the non-SI units used in the measurement of pressure. It is used because:
    A. Mercury is much denser than water so a much shorter tube can be used than if cm H_2O was the unit.

B. One of the factors that determine pressure is the length of the object supplying the force.
C. A manometer that utilises mercury does not require an electrical power supply.
D. Historically the first measurement of air pressure was made using a glass tube containing mercury.

Answer is D: The best answer is the historical use of liquid mercury (hydrogyrum) to measure pressure. The reason that Hg was used is as stated in choice A.

30. Which of the following statements about pressure is true?
    A. As boxes are stacked on top of each other, the pressure that they exert on the floor decreases.
    B. Very small forces will exert small pressures.
    C. By standing on wide flat boards, the pressure exerted on the floor by your body is decreased.
    D. Objects of large mass will produce large pressures.

Answer is C: Increasing the area of contact between a mass and the floor will decrease the pressure being exerted.

31. When an 80 kg person is seated on a stool so that the area of contact is 400 cm^2, the pressure on the stool due to the person's weight is very nearly
    A. 2 kPa
    B. 20 kPa
    C. 200 kPa
    D. 2000 kPa

Answer is B: Remember to multiply the mass by 10 (the acceleration due to gravity).

$$P = (80 \times 10)\ N \div 0.04\ m^2 = 20{,}000\ Pa = 20\ kPa.$$

32. A bed-ridden patient (whose mass is 60 kg) is lying on a firm mattress, so that the area of contact between the mattress and the patient's bony prominences (shoulders, buttocks and heels) is 600 cm^2. Given that pressure is force divided by area and that the acceleration due to gravity is 10 m/s^2, what is the pressure (in kPa) that would be experienced by the bony prominences?
    A. 1 kPa
    B. 10 kPa
    C. 100 kPa
    D. 600 kPa

Answer is B: $P = (60 \times 10)\ N \div 0.06\ m^2 = 10{,}000\ Pa = 10\ kPa$.

33. The measurement of cerebrospinal fluid (CSF) pressure is made while the patient is lying down rather than sitting up. The horizontal posture prevents a false high reading due to:
    A. The weight of the "head" of CSF in the spinal cord

B. Possible movement of the patient while sitting up
C. Pressure on the lumbar vertebrae when the back is vertical
D. The greater muscle tone needed to maintain a sitting position

Answer is A: In the sitting position, the CSF pressure as measured by tapping into the lumbar region would be increased by the head of pressure of about 40 cm of water.

34. A patient lying on a mattress of enclosed air is less likely to develop decubitus ulcers because:

    A. The mass of the patient is decreased.
    B. The force exerted by the patient is decreased.
    C. The patient's surface area in contact with the mattress is increased.
    D. The pressure exerted by the patient on the mattress is increased.

Answer is C: An air mattress flexes around a patient's body, so that it conforms more closely to their curved shape. Consequently, the area of contact between the patient and the supporting surface is increased. This increased area means that the pressure will be decreased. This lower pressure is less likely to compress the blood vessels that supply the skin, and so circulation will not be cut off. Hence ulcers will be avoided.

35. A bedridden patient is less likely to develop bed sores while lying on a water bed. This is because:

    A. The force applied to their bony projections is acting over a tiny area.
    B. The flexibility of the bed assists the blood circulation.
    C. The weight of the patient is being supported by parts of the body that are adapted for weight-bearing.
    D. The patient experiences a uniform pressure over most of the lower surface of their body.

Answer is D: A water bag mattress flexes around a patient's body, so that it conforms more closely to their curved shape. Consequently, the area of contact between the patient and the supporting surface is increased. This increased area means that the pressure will be decreased. This lower pressure is less likely to compress the blood vessels that supply the skin, and so circulation will not be cut off. Hence bed sores (ulcers) will be avoided.

36. Pascal's principle ("pressure applied to an enclosed fluid at rest is transmitted to every portion of the fluid and to the walls of the containing vessel") may be used to understand which of the following phenomena?

    A. An air mattress minimises the pressure applied to a bed-ridden patient's body.
    B. A worker can walk on wet concrete without sinking by standing on wide boards.
    C. A pulse can be felt as blood flows through a superficial artery.

D. The collapsing of a plastic intravenous bag as the liquid runs out.

Answer is C: The blood may be considered an enclosed fluid. As the left ventricle exerts pressure on the blood, the pressure can be felt throughout the arterial system. So this pressure can be detected by palpating a pulse point.

37. The consideration of pressure in gases differs from pressure in liquids because:
    A. Pressure at any point in a gas acts differently in different directions.
    B. In a gas mixture, the pressure exerted by each different gas must be considered.
    C. The pressure in a liquid increases with depth but does not for a gas.
    D. Liquids are virtually incompressible, whereas gases are compressible.

Answer is D: There is a lot of space between the particles in a gas, so the volume of an enclosed can be decreased. Boyle's law.

38. Boyle's law states: "when the volume (*V*) of a fixed amount of gas decreases, its pressure (*P*) will increase, and vice versa (provided that the temperature (*T*) does not change)". In symbols, this law is:
    A. $P \propto V$
    B. $P \propto T$.
    C. $P \propto 1 \div V$
    D. $P \propto V \times T$

Answer is C: Pressure is inversely proportional to volume. So as volume decreases, pressure does the opposite—it increases.

39. Which of the following statements from kinetic molecular theory can be used to explain why a region of low pressure will result in gas particles rushing into that region?
    A. Particles of a gas are widely separated, consequently can be easily compressed.
    B. Gas particles are moving at very high speed in random directions.
    C. There are a great many sub-microscopic particles per unit volume.
    D. Gas particles will exert a force on colliding with the walls of their container.

Answer is B: As particles of a gas are moving at high speed (hundreds of metres per second) and in random directions, every direction will have some particles moving in that direction. Hence a region with fewer particles (i.e. at a lower pressure) will have more particles moving into it than are moving away from it. This will tend to eliminate volumes with different pressures.

40. The kinetic molecular theory of gases provides us with some valuable insights into the behaviour of gases. One correct prediction of the theory is that gas pressure is:
    A. Due to the force exerted by gas molecules as they collide with the walls of their container

B. Due to the forces of attraction between gas particles
C. Inversely proportional to the gas temperature
D. Proportional to the partial pressures of the different gases that make up the gas

Answer is A: Gas molecules exert a force on walls as they bounce off. Gas pressure may be understood as the cumulative effect of the force exerted by millions of particles, divided by the area of the walls.

41. Given that 20% of the air at the normal pressure of 100 kPa is oxygen, then what is the partial pressure of oxygen?

A. 0.2 kPa
B. 2.0 kPa
C. 20 kPa
D. 80 kPa

Answer is C: The partial pressure of oxygen is proportional to its percent concentration in the mixture of gases. 20% of 100 kPa = 20 kPa.

42. What will be the partial pressure of oxygen in a hyperbaric chamber where the atmosphere contains 30% oxygen and is at a pressure of three atmospheres (300 kPa)?

A. 20 kPa
B. 30 kPa
C. 60 kPa
D. 90 kPa

Answer is D: The partial pressure of oxygen is proportional to its percent concentration in the mixture of gases. 30% of 300 kPa = 90 kPa.

43. Which of the following is a statement of Henry's law?

A. The partial pressure of a gas, in a mixture of gases, is the contribution it makes to the total pressure of the mixture.
B. In a mixture of gases, the total pressure is the sum of the pressures exerted by each of the gases alone.
C. Pressure applied to any point in a gas is transmitted equally and undiminished to all parts of the gas and to the walls of the container.
D. The quantity of gas that will dissolve in a liquid at a given temperature is proportional to the partial pressure of the gas and to its solubility coefficient.

Answer is D: Henry's law relates the amount of gas that will dissolve in a liquid to its partial pressure (or pressure).

44. Which of the following sentences best represents a statement of the Bernoulli effect? Pressure in a"

A. Flowing fluid is greatest where its speed is greatest.
B. Fluid acts equally in all directions.

C. Flowing fluid is lowest where its speed is lowest.
D. Flowing fluid is lowest where its speed is greatest.

Answer is D: Faster speeds of flow indicate that pressure in the liquid at that point is low.

45. An ophthalmologist using a tonometer measures the pressure in an eyeball to be 10 mmHg. What does this mean?
    A. Gauge pressure is atmospheric pressure plus 10 mmHg.
    B. Actual pressure is atmospheric pressure plus 10 mmHg.
    C. Actual pressure is atmospheric pressure minus 10 mmHg.
    D. 10 mmHg is a negative pressure.

Answer is B: At sea-level on earth, atmospheric pressure is about 760 mmHg. This value is taken to be "zero" and is often ignored by pressure measuring devices. When such devices show the pressure to be 10 mmHg on their gauge, this means that the actual pressure is 10 mmHg above atmospheric pressure. Hence total pressure is 760 + 10 mmHg.

46. A statement of Boyle's law could be:
    A. As pressure of a gas increases, its solubility increases.
    B. As volume of a gas increases, its pressure decreases.
    C. As volume of a gas decreases, its pressure decreases.
    D. As pressure of a gas increases, its solubility decreases.

Answer is B: Boyle's law relates gas volume to its pressure. The relationship is an inverse one, so choice B is correct.

47. If the composition of air is 20% oxygen with an atmospheric pressure of 700 mmHg, what would the partial pressure of oxygen be?
    A. 14 mmHg
    B. 20 mmHg
    C. 140 mmHg
    D. 700 mmHg

Answer is C: 10% of 700 mmHg is 70, so 20% is 140 mmHg.

48. Which of the following statements uses Pascal's principle to describe a physical function of synovial fluid in the knee?
    A. Synovial fluid is enclosed by the joint capsule. Hence the pressure exerted by the femur on the articular surfaces of the tibia during weight-bearing is distributed over the entire joint.
    B. The total pressure exerted by the femur on the tibia is the sum of the partial pressures exerted on each of the tibia's superior articular surfaces.
    C. Synovial fluid provides the lubrication between the articulating femur and tibia that allows free movement of the knee.
    D. Synovial fluid supplies nutrients and oxygen to the chondrocytes in the knee cartilages and removes carbon dioxide and wastes.

Answer is A: Pascal's principle applies to stationary liquids that are enclosed. Pressure applied to any point in such liquids is transmitted undiminished and equally to all parts of the liquid. This is what allows the body's weight to be distributed across all the surfaces within the joint capsule, rather than just to the points of contact between the femur and tibia. The other statements are true but do not rely on Pascal's principle.

49. Which of the following is the statement of the behaviour of enclosed liquids when a pressure is applied to them, that is known as Pascal's principle? "When pressure is applied to an enclosed liquid that is at rest, the pressure:
   A. In the liquid falls uniformly throughout the volume"
   B. Throughout the liquid is unchanged"
   C. In the liquid rises uniformly throughout the volume"
   D. In the liquid oscillates between a higher value and a lower value until the source of pressure is removed"

Answer is C: Pascal's principle states that the pressure is transmitted undiminished to all parts of the liquid and its container. That is, the pressure rises uniformly.

50. What is the standard international unit of pressure?
   A. The standard atmosphere (atm)
   B. Millimetre of mercury (mmHg)
   C. Pascal (Pa)
   D. Kilopascal (kPa)

Answer is C: Pascal (after Blaise Pascal) is the SI unit. It is a small unit so often the multiple kPa is used.

51. How is "pressure" different from a "force"?
   A. A force causes motion to start, stop or change. Pressure is due to gravity.
   B. Pressure and force are the same quantity.
   C. A single force of a particular magnitude may exert many different pressures.
   D. Force has units of N/m^2 while pressure has units of Pa (pascals).

Answer is C: Depending on the area on which the force acts, many different pressures can be exerted. Choice A is wrong as you can exert pressure on something by squeezing it with your fingers—that pressure is not due to gravity. Choice D is wrong: the unit of force is N (newton). The pascal is the same unit as N/m^2.

52. Given that 1 mmHg = 0.133 kPa, convert a blood pressure measurement of 130 mmHg/80 mmHg to units of kPa:
   A. 17.3/10.6 kPa
   B. 977/601 kPa
   C. 0.216 kPa
   D. 10.6/17.3

Answer is A: 130 mmHg × 0.133 kPa/mmHg = 17.3 kPa, and 80 × 0.133 = 10.6.

53. Use $P = F/A$ to determine the pressure exerted during a CPR for the case where the resuscitator is using the whole hand held flat on the sternum (area = 140 cm^2 = 0.014 m^2) and a force of 200 N can be applied.
    A. 2.8 kPa
    B. 14.3 kPa
    C. 1.43 kPa
    D. 28 kPa

Answer is B: $P = F/A = 200$ N ÷ 0.014 m^2 = 14285.7 N/m^2 = 14285.7 Pa = 14.3 kPa.

54. Use $P = F/A$ to determine the pressure exerted during a CPR for the case where the resuscitator is using only the "heel" of the hand the sternum (area = 40 cm^2 = 0.004 m^2) and a force of 200 N can be applied.
    A. 8 kPa
    B. 0.8 kPa
    C. 5 kPa
    D. 50 kPa

Answer is D: $P = F/A = 200$ N ÷ 0.004 m^2 = 50,000 N/m^2 = 50,000 Pa = 50 kPa. Note this pressure is greater than if the whole hand (which has a greater area) is used.

55. What is meant by the term: "positive pressure"?
    A. An air pressure greater than atmospheric pressure
    B. An air pressure that is less than atmospheric pressure
    C. An air pressure that is equal to the surrounding air pressure
    D. The pressure exerted by an object, rather than the pressure exerted by the surroundings on that object

Answer is A: Positive pressure is the amount by which a pressure exceeds the ambient air pressure. Hence if air pressure is 101 kPa, and the absolute pressure in a tyre is 190 kPa, the positive pressure is 89 kPa. This is the pressure that would be shown on the tyre gauge.

56. Consider what happens to the foetus if the abdomen of a pregnant woman is struck by a football (say). Which description of events applies Pascal's principle correctly? The football applies a pressure to the abdomen that is:
    A. Transmitted by the amniotic fluid to that part of the foetus immediately deep to the area struck (and to the mother's internal organs). Hence the pressure felt by the foetus is the same as felt by the mother.
    B. Transmitted by the amniotic fluid to that part of the foetus immediately deep to the area struck. The foetus will be moved by the blow and bump into the other side of the abdomen. Hence the pressure felt by the foetus is on the opposite side from where the blow occurred.
    C. Spread by the amniotic fluid to the entire surface area of the foetus and the mother's internal organs. Hence the pressure felt by the foetus is reduced.

D. Spread by the amniotic fluid to the entire surface area of the foetus and the mother's internal organs. Hence the pressure felt by the foetus is increased.

Answer is C: Pascal's principle refers to an enclosed fluid (such as amniotic fluid within its sac). The pressure applied by the football is transmitted to all parts of the fluid and surfaces in contact with the fluid (i.e. to the foetus and to the mother's organs that surround the sac). In this way the force of the football's blow is spread over a larger area and hence the pressure felt by the foetus is reduced.

57. Which of the following uses Boyle's law to describe inhalation?
    A. Air in the lungs is an enclosed fluid. As the chest moves outward, it exerts a pressure on the surrounding air. This extra pressure causes air to move into the lungs.
    B. The motion of the chest wall and diaphragm causes the volume within the lungs to expand which reduces the pressure within. Hence atmospheric air is pushed into the lungs.
    C. As the volume of the chest expands, the pressure also increases; this pushes air out of the lungs.
    D. The motion of the diaphragm causes the volume within the lungs to expand which reduces the pressure within. Hence atmospheric air is sucked into the lungs.

Answer is B: Boyle's law: pressure and volume are inversely proportional. Air movement occurs because of the greater density of molecules in one place moving (and being pushed) to the place with the lower density (they are not "sucked").

58. Which definition below refers to gas pressure?
    A. The pressure within the gas container divided by the container's height
    B. The pressure exerted by the atmosphere at sea level
    C. The pressure exerted by an enclosed gas on the walls of its container
    D. The influence causing the gas to move one place to another

Answer is C: A gas exerts pressure on its containing walls due to the collisions of its moving particles with the walls. Choice B refers to atmospheric pressure and choice D to a pressure gradient.

59. Which statement refers to a pressure gradient? Pressure gradient is:
    A. The difference in pressure between two points
    B. The difference in pressure between the start point of fluid motion and its destination point
    C. The pressure difference between two points, divided by their distance apart
    D. The difference in pressure that is driving fluid motion

Answer is C: Pressure gradient = pressure drop ÷ distance. Hence pressure difference is measured in Pa, while pressure gradient is in Pa/m. It is the pressure difference per metre.

60. Which statement refers to "partial pressure" of a gas in a mixture of gases?
    A. It is the pressure of a constituent gas as shown on a gauge, that is, the actual pressure minus atmospheric pressure.
    B. The partial pressure of a constituent gas is always less than the total gas pressure.
    C. The partial pressure of a constituent gas is always less than the ambient atmospheric pressure.
    D. Partial pressure is the notional pressure of that constituent gas if its molecules alone occupied the entire volume of the original mixture.

Answer is D: In a mixture of gases, each constituent gas has a partial pressure which is the notional pressure of that constituent gas if its molecules alone occupied the entire volume of the original mixture at the same temperature. The total pressure of an ideal gas mixture is the sum of the partial pressures of the gases in the mixture. Choice B is true, but is not as good an answer as D.

61. Which statement refers correctly to "negative pressure"?
    A. Any value of pressure that is greater than atmospheric pressure
    B. Any value of pressure that is less than atmospheric pressure
    C. Negative pressure exists at a point if gas is flowing towards that point
    D. Negative pressure "sucks" gas towards it

Answer is B: Average atmospheric pressure (101 kPa) is always present so is taken to be a "relative zero". Pressure gauges display the value of "ambient pressure minus atmospheric pressure". Hence pressures greater than 101 kPa are displayed as positive pressures, while those less than 101 kPa are negative. Places with negative pressure do not "suck", rather the higher pressure nearby "pushes" air towards the lower pressure region.

62. Which statement refers correctly to "gauge pressure"?
    A. Gauge pressure is usually greater than atmospheric pressure.
    B. Gauge pressure is usually less than atmospheric pressure.
    C. Gauge pressure is total (actual) pressure minus atmospheric pressure.
    D. Gauge pressure and total (actual) pressure are the same thing.

Answer is C: The measurement that is displayed on the dial or readout of pressure measuring device (the pressure gauge) is called the "gauge pressure". It displays the value of pressure that is in excess of atmospheric pressure. Hence if a systolic blood pressure is displayed as 120 mmHg, it is actually 120 + 760 mmHg in absolute (actual) pressure terms.

63. If atmospheric pressure is 1020 hPa and 80% of the atmospheric gas is nitrogen, what is the partial pressure of nitrogen?
    A. 816 hPa
    B. 1100 hPa
    C. 940 hPa

D. 1275 hPa

Answer is A: 80% of 1020 hPa must be a number less than 1020, so choices B and D are wrong. ppN = 80% of 1020 hPa = 0.8 × 1020 = 816 hPa.

64. Why does a bottle of carbonated soft drink generate bubbles in the liquid when the lid is removed? (hint: use Henry's law)
   A. With the lid off, the pressurised liquid can push the dissolved carbon dioxide out of the bottle.
   B. With the lid off, the liquid is at atmospheric pressure which has a low partial pressure of carbon dioxide, so less dissolves.
   C. With the lid off, the plastic bottle which was previously under tension, can recoil and squeeze some carbon dioxide out of solution.
   D. When removed from the fridge, the bottle begins to warm. As temperature rises, less carbon dioxide can dissolve, so bubbles appear.

Answer is B: Henry's law states that the amount of gas that dissolves in a liquid (at a given temperature) is proportional to the partial pressure of that gas in the "atmosphere" above the liquid. Carbonated drinks have a lot of CO_2 dissolved in them because the bottle was sealed with a high partial pressure of CO_2 (much higher than its PP in normal atmospheric air). That is, there is CO_2 under high pressure trapped in the bottle by the air-tight lid. When the lid is removed, the air space above the liquid is suddenly equalised to atmospheric pressure. As amount of dissolved gas is directly proportional to its PP and the PP is suddenly decreased, the amount of gas that can dissolve is much less. So excess CO_2 bubbles out of solution.

65. Given that *pressure* = *force* ÷ *area*, what is an atmospheric pressure of 1016 hPa when converted to units of newton per square metre?
   A. 1.016×10^5 N/m^2
   B. 101.6 kPa
   C. 101.6 N/m^2
   D. 1016 N/m^2

Answer is A: 1 Pa is equal to 1 N/m^2. 1016 hPa = 101,600 Pa = 1.016×10^5 N/m^2.

66. Use the equation $P = F \div A$ to determine the pressure on the knee joint (area = 100 cm^2) when a 70 kg person is standing on one leg (assume that 70 kg is the body mass above the knee). Hint: convert cm^2 to m^2.
   A. 7 kPa
   B. 70 kPa
   C. 700 kPa
   D. 7000 kPa

Answer is C: Using $P = F \div A$ and remembering that the weight, F of a person with mass 70 kg is 70 × 10 and converting 100 cm^2 to 0.001m^2, putting in the numbers gives:

$P = 70 \times 10 \div 0.001 = 700{,}000$ Pa = 700 kPa.

12.5 Pressure Applied to the Cardiovascular System

Blood moves from the heart to the capillaries due to the pressure exerted by the heart muscle (and by gravity). The ventricles contract to force blood into the aorta and the pulmonary trunk and relax to allow blood from the vena cavae and pulmonary veins to enter them via the atria. The number of cycles per minute is called the heart rate. Aortic blood pressure oscillates between the systolic pressure (the maximum value) of about 120 mmHg—for a male at rest—and the diastolic value (the minimum value) of about 80 mmHg. These values increase as the level of physical activity increases and return to their resting values after exercise. This constantly oscillating blood pressure causes the arteries to bulge and recoil and can be felt as "the pulse" in superficial arteries.

Hypertension is the diagnosis if resting systolic and diastolic blood pressures are greater than 140 and 90 mmHg (respectively). This indicates that the arteries are presenting a greater than expected resistance to the flow of blood and that the heart is working harder to move blood to the capillaries. The extra resistance may be due to the arteries being unable to expand as blood is pumped into them (they have lost elasticity) or because the lumen of the artery has been narrowed due to a plaque growing in the wall. This hypertension is a cause of medical concern.

Blood moves from the capillaries to the heart due to gravity (for blood returning from anatomy superior to the heart) and when contracting muscles squash against veins which forces the contained blood through the venous valves in the arms and leg veins. The valves ensure that blood can only flow towards the heart. The act of inhalation increases pressure in the abdomen, and this also helps to push blood in the inferior vena cava towards the heart.

1. When a person is standing upright, what can be said about the arterial blood pressure in their feet?
   A. It will be greater than arterial pressure in the aorta.
   B. It will be less than the arterial pressure in the aorta.
   C. It will be the same as the arterial pressure in the aorta.
   D. It will be the same as the venous blood pressure in the feet.

Answer is A: Due to the head of liquid above the feet contributing hydrostatic to the arterial pressure produced by the heart, arterial blood pressure in the feet will be greater than in the aorta.

2. A plaque in a coronary vessel wall will result in a decrease in the lumen diameter of a coronary artery and in less oxygen being delivered to the heart muscle. Why is this?
   A. A protruding plaque increases the distance blood needs to travel, and this increases resistance to flow.
   B. A narrow artery restricts blood flow by increasing blood's viscosity.
   C. The constricted artery causes the pressure gradient to decrease which results in lower blood flow.

D. The decrease in artery radius will cause blood volume flow rate to decrease.

Answer is D: Poiseuille's law states that volume flow rate is proportional to radius the fourth power so a decrease in radius results in a decrease in flow rate and hence less oxygen being delivered to the myocardium.

3. When the AV valve between the left atrium and left ventricle is open, the blood pressure in both chambers is zero. When the left ventricle contracts the AV valve shuts and blood pressure in the left ventricle rises while blood pressure in the left atrium reaches and stays at ~5 mmHg. Blood pressure in the aorta is ~80 mmHg at the end of diastole. Which of the following will happen next?
   A. When blood pressure in the left ventricle reaches 6 mmHg, blood will flow into the left atrium.
   B. When blood pressure in the left ventricle reaches 80 mmHg, blood will flow into the aorta.
   C. When blood pressure in the left ventricle reaches 81 mmHg, blood will flow into the aorta.
   D. Blood volume in the left ventricle will decrease as pressure increases until there is sufficient pressure to push blood into the aorta.

Answer is C: Pressure in the LV will rise as the myocardium contracts until pressure in the LV is greater than the pressure in the aorta. When this happens, the aortic valve will be forced open and blood will move from the LV into the aorta through the aortic valve.

4. What units is a blood pressure of 120/80 is expressed in?
   A. Millimetres of mercury
   B. Centimetres of blood
   C. Centimetres of mercury
   D. Pascal

Answer is A: Despite the mmHg not being an SI unit, blood pressure is still commonly reported in this unit.

5. What effect will increasing the concentration of red cells in blood have? It will:
   A. Decrease blood viscosity and increase aortic pressure
   B. Increase blood viscosity and decrease aortic pressure
   C. Increase blood viscosity and increase aortic pressure
   D. Decrease blood viscosity and decrease aortic pressure

Answer is C: Viscosity will increase as the number of RBC increases. This will also increase the resistance to flow so BP in the aorta will increase a bit.

6. By which means will the sympathetic response raise blood pressure?
   A. Vasoconstriction due to stimulation of alpha receptors
   B. Vasoconstriction due to stimulation of beta receptors
   C. Increase cardiac output due to stimulation of alpha receptors

D. Increased stroke volume due to stimulation of alpha receptors

Answer is A: The neurotransmitter epinephrine will attach to the alpha receptor sites of smooth muscle in blood vessels and cause vasoconstriction (except for blood vessel in skeletal muscle). Choices C and D are wrong as beta, not alpha, receptors increase CO. Choice B is wrong as alpha (not beta) produce vasoconstriction.

7. What is a normal pressure in the venous system returning blood to the heart?
   A. 35 mmHg
   B. 35 cm water
   C. 80 mmHg
   D. 10 cm water

Answer is D: 35 mmHg is BP at *arterial* end of capillary, so C is also wrong. 17 mmHg is BP at venous end of capillary, while 35 cmH_2O is ~26 mmHg so choice B is too high.

8. The intravenous cannula that will give the largest flow rate is:
   A. 0.7 mm diameter 32 mm length
   B. 0.7 mm diameter 19 mm length
   C. 1.8 mm diameter 32 mm length
   D. 1.8 mm diameter 19 mm length

Answer is D: The largest diameter gives the largest flow rate and the shortest length provides the lesser resistance to flow.

9. Consider Pascal's principle. If the heart exerts a pressure of 120 mmHg on the blood in the aorta, where else in the body will the blood pressure be 120 mmHg?
   A. In the brachial artery of a seated person
   B. In the superior vena cava of a standing person
   C. In the capillaries of the feet of a supine person
   D. In the anterior tibial artery of a person who is standing

Answer is A: Resting BP is usually measured while the subject is seated. In this posture, the brachial artery is at the level of the heart. Hence gravity will not affect the BP in the brachial artery. In a standing person, gravity would increase the BP in the legs. BP in the capillaries and vena cava is not comparable to BP in the aorta.

10. If an atheroma reduces the diameter of an artery from 2 to 1 mm, what effect would this have on the blood flow through the artery? (According to Poiseuille's law, volume flow rate through a blood vessel is proportional to its radius to the fourth power, i.e. $V \propto R^4$).
    A. Blood flow would decrease to half (1/2) the value for an artery with diameter 2 mm.
    B. Blood flow would decrease to a quarter (1/4) of the value for an artery with diameter 2 mm.

C. Blood flow would decrease to an eighth (1/8) of the value for an artery with diameter 2 mm.
D. Blood flow would decrease to one sixteenth (1/16) of the value for an artery with diameter 2 mm.

Answer is D: As volume flow rate is proportional to radius to the fourth power, if radius decreases by half, volume flow rate decreases by ½ × ½ × ½ × ½ = 1/16.

11. If the radius of a blood vessel is halved, the blood flow through it drops to one sixteenth its previous value. This statement could be describing which of the following?
    A. A decrease in pressure gradient
    B. Atherosclerosis
    C. Vasoconstriction
    D. Poiseuille's law

Answer is D: The statement also describes vasoconstriction, but the reference to a quantity means the Poiseuille's law is the best answer.

12. A man who is standing has a resting systolic blood pressure of 120 mmHg at the start of his aorta. What will be the pressure in the arteries of his feet?
    A. About 40 mmHg, as arterial blood pressure decreases with distance from the heart.
    B. About 120 mmHg as arterial blood pressure does not fall appreciably until just before blood enters the capillaries.
    C. About 200 mmHg as the "head" of liquid increases the blood pressure in the feet.
    D. About 80 mmHg as blood pressure will drop in the absence of venous return when the "skeletal muscle pump" is not operating.

Answer is C: When standing, the BP in the feet results from the pumping of the heart (120 mmHg) as well as the hydrostatic effect of the head of liquid—another 80 mmHg or more.

13. To what does the term "cardiac output" refer?
    A. The speed of blood flow through the aorta
    B. The volume of blood flowing through the aorta per minute
    C. The volume of blood pumped by the heart with each beat
    D. The number of heart beats per minute

Answer is B: CO is the volume flow rate. The volume of blood pumped per minute. Choice C is the stroke volume. Choice D is the heart rate.

14. A resting blood pressure stated as 120/80 (in units of mmHg) refers to:
    A. Maximum pressure in the aorta/minimum pressure in the aorta
    B. Maximum pressure in the aorta/maximum pressure in the vena cavae
    C. Diastolic pressure/systolic pressure

D. Left ventricular systolic/right ventricular systolic

Answer is A: Arterial blood pressure cycles between a maximum value and a minimum value in time with the heartbeat. It is recorded as the maximum value adjacent to the minimum value.

15. Which of the following does **NOT** influence the resistance to blood flow?
    A. Diameter of the arterioles
    B. Temperature of the blood
    C. Haematocrit
    D. Radius of the veins

Answer is D: Resistance to blood flow is of interest on the arterial side of the circulation. Radius of veins is much greater than radius of arteries. Temperature does affect blood flow for example in the situation of extremities with "frostbite".

16. Colloid osmotic pressure:
    A. Is due to all of the dissolved particles in blood
    B. Is the difference in pressure between the arterial end and the venous end of a capillary
    C. Is the difference in pressure between the inside of a capillary and the interstitial fluid
    D. Is due to the plasma proteins in the blood

Answer is D: Plasma proteins (also known as colloids) are responsible for the colloid osmotic pressure of blood.

17. If a person has an arterial blood pressure measurement of 16 kPa (120 mmHg), this equates to which of the following pressures in newton per square centimetre?
    A. 16,000
    B. 160
    C. 1.6
    D. 1600

Answer is A: The "newton per square metre" is renamed as pascal. 1 N/m^2 = 1 Pa. 16,000 N/m^2 = 16,000 Pa = 16 kPa.

18. An intravenous (IV) infusion flows into a vein because of hydrostatic pressure. This depends **LEAST** on which of the following?
    A. Gravity
    B. The density of the IV solution
    C. The volume of liquid in the IV bag
    D. The "head" of liquid

Answer is C: Gravity acting on the head of liquid provides the pressure that forces an IV solution into a vein. As long as the head of liquid is maintained, the volume contained in the IV bag has little influence.

19. Due to an atheromatous plaque, a man's coronary artery has narrowed to one third of its healthy diameter, all other things being equal, his blood flow rate (mL/min) in that artery would:
    A. Be one third of the unobstructed value
    B. Be one eighty-first of its healthy value
    C. Be one ninth of its healthy value
    D. Be one twenty-seventh of its healthy value

Answer is B: Poiseuille's law states that volume flow rate is proportional to radius to the fourth power, so blood flow rate would be $\frac{1}{3} \times \frac{1}{3} \times \frac{1}{3} \times \frac{1}{3} = 1/81$.

20. According to Poiseuille's law, volume flow rate through a blood vessel is proportional to its radius to the fourth power ($V \propto R^4$). If an atheroma reduces the radius of an artery from 4 to 2 mm, what effect would this have on the blood flow through the artery?
    A. Blood flow would decrease to half (1/2) the value for an artery with radius 4 mm.
    B. Blood flow would decrease to a quarter (1/4) of the value for an artery with radius 4 mm.
    C. Blood flow would decrease to an eighth (1/8) of the value for an artery with radius 4 mm.
    D. Blood flow would decrease to one sixteenth (1/16) of the value for an artery with radius 4 mm.

Answer is D: 2 mm is half of 4 mm and according to Poiseuille's law: $V \propto (1/2)^4 = 1/16$.

21. Say a diastolic blood pressure reading was 80 mmHg. This is consistent with which one of the following statements?
    A. The diastolic reading is 80 mmHg greater than atmospheric pressure.
    B. Pressure is measured in length units.
    C. The diastolic reading is 80 mmHg less than atmospheric pressure.
    D. This is the maximum pressure produced by contraction of the myocardium.

Answer is A: Because atmospheric pressure (about 760 mmHg) is always present, it is taken to be a "relative zero" and other pressures are stated as an amount greater than atmospheric pressure. We say 80 implying 80 more than 760, rather than stating 840 mmHg (760 + 80).

22. If mean arterial pressure is kept constant while a small artery changes its radius from 1 to 2 mm, what will happen?
    A. Blood volume flow rate through the artery will double.
    B. Blood volume flow rate through the artery will increase to four times its previous value.
    C. Blood volume flow rate through the artery will be 16 times the original value.

D. Blood speed will halve so there will be no change in volume flow rate.

Answer is C: If an artery dilates to twice its previous value, 16 times (2^4) as much blood will flow per minute.

23. Complete the following sentence. When standing, the arterial blood pressure in the feet is:

    A. Less than the arterial blood pressure of the hands
    B. Reduced due to the action of valves in supporting the blood in the veins of the legs
    C. Increased by the hydrostatic pressure due to the "head of blood" in the vessels
    D. Less than the venous blood pressure in the feet

Answer is C: Because of gravity, the position of the body in space affects blood pressure. When standing, hydrostatic pressure increases the BP in the blood vessels of the feet.

24. When "colloid osmotic pressure" is used in relation to the blood. What is being referred to?

    A. The movement of water molecules across the membrane of a red blood cell
    B. The filtration pressure in the glomeruli of the kidneys
    C. The osmotic pressure forcing water and solute out of capillaries
    D. The osmotic pressure due to the plasma proteins

Answer is D: Plasma proteins are known as colloids. They are present in blood but not in the interstitial fluid. That part of the osmotic pressure of blood that they produce is known as colloid osmotic pressure.

25. According to Poiseuille's law of fluid flow, the volume flow rate, V, is proportional to radius to the fourth power. What would happen to the volume flow rate of blood if the diameter of an artery decreased to 20% (1/5) of its original diameter? V would:

    A. Be one fifth (0.2) of its original value.
    B. Be one twenty-fifth (0.04) its original value.
    C. Be 1/625 (0.0016) of its original value.
    D. Five times its original value.

Answer is C: $(1/5)^4 = (0.2)^4 = 0.0016$. Narrowing of arteries has dire consequences for blood flow.

26. If the blood pressure at the arterial end of a blood capillary is 4000 Pa (30 mmHg) and at the venule end of the capillary is 2000 Pa (15 mmHg), and the capillary has a length of 1 mm, what will the pressure gradient in the capillary be?

    A. 4000 Pa/mm
    B. 2000 Pa/mm

C. 15 mmHg
D. 45 mmHg

Answer is B: Pressure gradient = (pressure difference)/length = (4000 – 2000)/1 = 2000 Pa/mm. Choice C states the pressure difference but not the gradient!

27. If a blood pressure is stated as 16 over 10.6 (in units of kilopascals), this means:
    A. Pulse pressure is 16 kPa.
    B. Diastolic pressure is 10.6 kPa below atmospheric pressure.
    C. Systolic pressure is 5.4 kPa above atmospheric pressure.
    D. Systolic pressure is 16 kPa above atmospheric pressure.

Answer is D: The larger value (16) is the systolic pressure (i.e. maximum blood pressure at end of ventricular contraction) and states by how much the blood pressure is above atmospheric pressure.

28. As a result of the pumping action of the heart, we are able to feel pulsations in our superficial arteries (e.g. the radial pulse). This effect is an example of which of the following?
    A. Pascal's principle
    B. Torricelli's law
    C. The Bernoulli effect
    D. Starling's law of the heart

Answer is A: Pascal's principle states that pressure applied to an enclosed fluid is transmitted undiminished throughout the fluid. The LV pushes blood into the aorta at pressure. This pressure distends the artery walls of the brachial artery—even though it is 70 cm away from the heart—and we can feel the pulsations as pressure rises and falls.

29. Suppose that a person who is standing still has a mean arterial pressure in the aorta of 13 kPa (98 mmHg). The blood pressure at the start of the arterioles of the feet will be:
    A. About the same since healthy arteries present very little resistance to blood flow
    B. Less because blood pressure decreases along arteries as distance from the heart increases
    C. More because of the hydrostatic pressure exerted by the column of blood in the arteries
    D. Less because of the decreased venous return which results from the person's inactivity

Answer is C: When standing, the vertical column of blood in the arteries exerts pressure on the blood in the arterioles in the feet due to the head of liquid. Hence mean arterial BP in the feet will be considerably greater than 13 kPa.

30. A systolic blood pressure which is stated as 16 kPa (120 mmHg) means that the pressure in the arterial blood is:
    A. Negative 16 kPa
    B. 16 kPa above 0 kPa
    C. 16 kPa greater than atmospheric pressure
    D. 16 kPa less than atmospheric pressure

Answer is C: The pressure exerted by the atmosphere is disregarded when reading the pressure on the gauge. 16 kPa actually means that blood is actually at a pressure of 16 + 101 kPa.

31. Given that a healthy resting blood pressure may be stated as 16/10.6 (in units of kPa). What would the reading on the pressure gauge of a sphygmomanometer that was measuring blood pressure in the posterior tibial artery of a standing patient be closest to?
    A. 6 kPa, the average blood pressure at the arterial end of the capillaries
    B. 12 kPa, the hydrostatic pressure due to head of blood
    C. 28 kPa, the hydrostatic pressure plus average aortic pressure
    D. 107 kPa, atmospheric pressure plus average pressure at the start of the capillaries

Answer is C: Both pressure due to the head of liquid between the heart and the feet (about 11 kPa) and pressure due to the pumping action of the heart (16 kPa) must be added to arrive at BP in the feet when standing. Alternatively, the value in choice D is too high, and choices A and B are values that are too low and do not take into account the two components that are operating in a standing person.

32. Consider an arteriole that is 50 cm distant from the aorta. If blood pressure in the aorta is 130 mmHg and in the arteriole is 30 mmHg, what is the pressure GRADIENT between the two?
    A. 2 mmHg/cm
    B. 30 mmHg/cm
    C. 100 mmHg/cm
    D. 130 mmHg/cm

Answer is A: Pressure gradient = pressure drop ÷ distance = (130 − 30) ÷ 50 = 2 mmHg/cm.

33. In a resting heart pumping 5 L of blood per minute, the average aortic blood pressure is 13.3 kPa. Suppose arterial blood pressure falls to 3.3 kPa at the start of a capillary that is 50 cm from the heart. What is the pressure gradient along the path from heart to capillary?
    A. 5 L/min
    B. 200 Pa/cm
    C. 2000 Pa/m
    D. 10,000 Pa

Answer is B: Pressure gradient = pressure drop ÷ distance = (13,300 − 3300) ÷ 50 = 200 Pa/cm. Choices A and D are not gradients.

34. If a stenosis reduces the size of a blood vessel to half of the original diameter, the volume flow rate through the vessel will be reduced. Which relationship below determines the extent of the decrease in flow?

    A. Poiseuille's law
    B. Bernoulli's theorem
    C. Dalton's law
    D. Pascal's principle

Answer is A: Poiseuille's law relates diameter (radius) of blood vessel to volume flow rate.

35. During an auscultatory blood pressure determination, the Korotkoff sounds that are listened for are produced because:

    A. The partial pressure of the blood has been increased.
    B. The blood flow is turbulent.
    C. Of the viscosity of the blood.
    D. The volume flow rate has decreased.

Answer is B: Turbulent blood flow produces sound that is audible with a stethoscope. Auscultatory systolic BP measurement requires the artery to be squashed flat. At the point just before blood flow is stopped, the blood squirts through the squashed artery in a turbulent fashion and this noisy squirting blood produces a sound audible with a stethoscope.

36. Consider a capillary where the blood hydrostatic pressure is 3300 Pa at the arterial end and 2000 Pa at the venous end. If the difference between the osmotic pressures inside and outside the capillary is 2900 Pa, what would be the net pressure difference between the surrounding tissue and blood in the venous end of the capillary?

    A. 400 Pa
    B. 900 Pa
    C. 1300 Pa
    D. 4900 Pa

Answer is B: Blood hydrostatic pressure tends to move water and solutes out of the capillary, while osmotic pressure tends to move water into the capillary. Taking the outward direction as positive: net pressure at the venous end is = 2000 − 2900 = −900 Pa and so is directed into the capillary.

37. When taking a subject's blood pressure, you are actually measuring the difference between total pressure and which other pressure?

    A. Gauge pressure
    B. Blood pressure

C. Atmospheric pressure
D. Standard atmospheric pressure

Answer is C: The pressure on the gauge used to measure blood pressure displays the difference between atmospheric pressure and blood pressure. That is, total pressure is atmospheric pressure plus the reading on the gauge.

38. The resting systolic blood pressure of a person has been measured in the brachial artery (when it is at the level of the heart) to be 115 mmHg. What is the likely systolic pressure of this person when standing at rest, measured in the posterior tibial artery at the level of the ankle?
    A. 40 mmHg
    B. 100 mmHg
    C. 115 mmHg
    D. 215 mmHg

Answer is D: When standing, there is a column of blood filling the posterior tibial artery (PTA) that rises up to the aorta. The column is about 130 cm tall for a man of about 1.75 m in height. This "head" of liquid exerts a hydrostatic pressure on the wall of the PTA that is in addition to the systolic pressure produced by the pumping action of the left ventricle. A column of 130 cm of blood exerts a pressure of about 100 mmHg (1 mmHg = 1.36 cmH_2O) which is in addition to the 115 mmHg original systolic value.

39. Why is a resting arterial blood pressure (BP) measurement stated as a maximum value and a minimum value?
    A. BP has two values depending on whether the myocardium is in systole or in diastole. Hence these two values are required.
    B. BP varies continuously during the heart's contracting or relaxing. It is not necessary to continuously read all values, so only the maximum and minimum are recorded.
    C. The maximum is produced by the left ventricle in systole, while the minimum is produced by the right ventricle in systole. These two values provide the necessary BP information.
    D. BP must be within this range for optimal health. BP outside this range is deemed to be hypertension.

Answer is B: BP oscillates between maximum systolic pressure when the myocardium contracts and minimum diastolic pressure as the myocardium relaxes. These two values measured while at rest, provide information about the health of the arterial tree which is why BP is measured. BP pressure rises above the resting systolic value during exercise as it should. The other choices are wrong as they contain a nonsense statement.

40. Why does the heart rate increases during exercise?
    A. To supply more lactic acid to the muscles for anaerobic respiration
    B. To supply more oxygen and glucose to the respiring cells

C. To keep up with the pulse rate
D. To supply carbon dioxide to respiring cells

Answer is B: Heart rate and hence cardiac output must rise to increase blood pressure. A higher BP increases the rate of blood flow, so blood can supply the active muscles with more oxygen and glucose, so these materials can be used to produce ATP which supply energy for the exercise.

41. Consider that at a point in time, an arterial blood pressure—stated in mmHg—is measured to be 115. Given that the atmospheric or air pressure was 760 mmHg at the time of blood pressure measurement, what is the actual blood pressure?
   A. 760 − 115 = 645 mmHg.
   B. The blood pressure is actually 115 mmHg.
   C. 115/760.
   D. 115 + 760 = 875 mmHg.

Answer is D: Atmospheric pressure is a "relative" zero, with values above atmospheric being termed "positive" while values below are termed "negative". A blood pressure of 115 mmHg means 115 above atmospheric pressure, so the actual pressure is 115 + 760 = 875 mmHg.

42. What is meant by the statement that the amount of dissolved O_2 in blood is 100 mmHg?
   A. The oxygen dissolved in blood contributes 100 mmHg to the blood pressure.
   B. It is a statement of the concentration of dissolved oxygen in the solution.
   C. Blood has been in close contact with air that had contained oxygen at a partial pressure of 100 mmHg.
   D. The part of blood's osmotic pressure that is due to dissolved oxygen is 100 mmHg.

Answer is C: When a liquid is in contact with a mixture of gases, the amount of a gas that will dissolve in the liquid (at equilibrium) is stated as the partial pressure of that gas in the mixture. Hence if there is 100 mmHg of dissolved O_2 in a liquid, then the gas that was in contact with the liquid must have contained oxygen at a partial pressure of 100 mmHg. Choice B is also correct but not as good an answer as C. Choices A and D are nonsense.

43. Which of the following statements correctly states the relationship between mean arterial pressure (MAP) and the other variables?
   A. MAP = stroke volume × heart rate
   B. MAP = cardiac output + stroke volume
   C. MAP = heart rate × cardiac output
   D. MAP = cardiac output × total peripheral resistance

Answer is D: Cardiac output is stroke volume × heart rate. MAP is the product of how much blood is pumped into the arteries (CO) and how easy it is for the blood to flow through the arteries (TPR).

44. What is the mean arterial pressure determined from a resting blood pressure of 118/79 mmHg?
    A. 39
    B. 92
    C. 98.5
    D. 118

Answer is B: MAP = diastolic pressure + $^{1}/_{3}$ × pulse pressure
= 79 + $^{1}/_{3}$ (118 − 79)
= 79 + $^{1}/_{3}$ × 39
= 79 + 13 = 92

45. What is the pulse pressure determined from a resting blood pressure reading of 124/92 mmHg?
    A. 1.35
    B. 32
    C. 124
    D. 216

Answer is B: Pulse pressure = systolic pressure − diastolic pressure = 124 − 92 = 32 mmHg.

46. In the nephron glomerulus, the hydrostatic pressure of the blood in the glomerulus is 8.0 kPa, the hydrostatic pressure of the filtrate in the Bowman's capsule is 2.7 kPa and the blood colloidal osmotic pressure is 4.0 kPa. What is the net filtration pressure?
    A. 1.3 kPa
    B. 5.7 kPa
    C. 9.3 kPa
    D. 14.7 kPa

Answer is A: Glomerular HP is positive (fluid moves from glomerulus into Bowman's capsule), while filtrate HP and colloidal OP are negative (causing fluid movement back into the glomerulus). Hence performing the arithmetic gives: 8.0 − 2.7 − 4.0 = 1.3 kPa. It is this pressure difference that drives the filtration.

47. How may mean arterial blood pressure be calculated?
    A. (Diastolic pressure + systolic pressure) ÷ 2
    B. Systolic pressure + $^{1}/_{3}$ × (systolic pressure − diastolic pressure)
    C. (Systolic pressure − diastolic pressure) ÷ 2
    D. Diastolic pressure + $^{1}/_{3}$ × (systolic pressure − diastolic pressure)

Answer is D: Mean arterial pressure is greater than diastolic pressure but less than systolic pressure. It is approximated by adding one third of the pulse pressure to the diastolic pressure.

48. Why is the systolic blood pressure in the arteries of the feet greater when standing still than when supine?
    A. When standing, the head of blood in the arteries between the feet and heart adds to the BP produced by the pumping action of the heart.
    B. By standing still, the squashing action of muscles on veins in assisting blood to return to the heart is disabled, hence the greater pool of blood raises the BP at the feet.
    C. The arterioles of the feet are narrow in diameter, which increases their resistance to blood flow. It is the increase in resistance that causes the increase in BP.
    D. When standing, blood has to work against gravity in order to return to the heart. This increases the heart's workload and the pressure that it exerts.

Answer is A: The head of liquid (blood) in the arteries exerts a pressure on the arteries of the feet when standing of about 140 cm of water (=the vertical distance from the heart to the feet). This static pressure adds to the pressure that is created by the contraction of the heart's left ventricle. This additional pressure due to head of liquid is not present when lying supine as the feet are approximately at the same level as the heart.

12.6 Blood Pressure and Its Control

The heart produces the pressure difference that is required to lift blood (with its contained oxygen) to the top of the head against gravity and to move blood through the other arteries despite blood's viscosity and the friction provided by the vessel walls. The body's need for oxygen changes as the level of bodily activity changes, so the movement of blood must change to match these requirements. That is, cardiac output must be sufficient to ensure that the demand for oxygen by the tissues is met. Cardiac output will increase if the heart rate increases and if the volume pumped by each stroke increases. These rates are set by the sinoatrial node (the heart's pacemaker) in the wall of the right atrium and may be increased or decreased by stimulation from the "cardiovascular centre" of the medulla oblongata via the autonomic nervous system.

Blood experiences resistance to its flow which is overcome by the body's ability to adjust cardiac output, to dilate and constrict its blood vessels and, to some extent, to alter its blood volume. A further complication is that the body's posture as it changes from supine to sitting to standing will affect the movement of blood. Baroreceptors (pressure receptors) are located in the walls of the carotid sinus, the aortic arch and wall of most large arteries in neck and thorax. Over the short term, if BP falls, baroreceptor reflexes increase in heart rate and force of contraction and promote vasoconstriction. The reverse happens if BP rises. Chemoreceptors are located in the two carotid bodies and several aortic bodies, close to the baroreceptors. They respond to an increase of CO_2 and to the decreased pH of blood that occurs as arterial pressure falls. They transmit signals to the vasomotor centre which excites it to produce vasoconstriction. This increases blood pressure. They also signal the cardioaccelerator centre which increases cardiac output. Hence more blood moves through the lungs, and this allows the excretion of CO_2 and the intake of O_2 to increase.

Long-term control of blood pressure is achieved by the kidneys as they alter the amount of water that is filtered from blood and excreted in urine. Water loss in urine decreases the volume of blood, and this in turn decreases blood pressure. Four hormones (angiotensin II, ADH, aldosterone, ANP) are associated with or act on the kidney to regulate urine volume and hence blood volume and BP.

1. When arterial blood pressure is stated as 120/80, what do the numbers refer to?
   A. 120 ÷ 80 = 1.5 = mean arterial pressure
   B. Pulse pressure/mean arterial pressure
   C. Systolic pressure/diastolic pressure
   D. Arterial pressure/venous pressure

Answer is C: Writing BP as 120/80 is an abbreviated way of writing systolic pressure of 120 mmHg and diastolic BP of 80 mmHg.

2. An increase in which of the following would **NOT** produce an increase in cardiac output?
   A. Heart stroke volume
   B. Heart rate
   C. Peripheral resistance
   D. Venous return

Answer is C: Increasing peripheral resistance would not increase CO unless there was a compensating increase in BP.

3. Which of the following would increase arterial blood pressure?
   A. A decrease in sympathetic impulses along the cardioaccelerator nerves
   B. An increase in parasympathetic impulses along the vagus nerve
   C. A decrease in sympathetic impulses along vasomotor nerves
   D. An increase in sympathetic impulses along vasomotor nerves

Answer is D: Increase the rate of sympathetic impulses along vasomotor nerves would result in vasoconstriction which would increase BP. An increase in **parasympathetic** impulses would slow the heart rate, which would decrease BP.

4. Which of the following chemicals would cause blood pressure to **decrease** when they appear in blood?
   A. Antidiuretic hormone
   B. Angiotensin II
   C. Aldosterone
   D. Atrial natriuretic peptide

Answer is D: ANP acts to reduce blood volume by increasing Na+ excretion at kidneys (blocking release aldosterone), increasing volume of urine produced, reducing thirst, blocking release of ADH. ANP acts to reduce blood pressure by blocking release of the vasoconstrictors adrenaline and noradrenaline and stimulating peripheral vasodilation.

5. Which of the following statements about resting blood pressure is correct? According to the Australian Heart Foundation Classifications:
   A. High blood pressure is considered greater than 180/110 mmHg.
   B. A systolic reading alone of greater than 140 mmHg is classed as high blood pressure.
   C. If the pulse pressure remains around 1/3 of the systolic reading, then BP is considered normal regardless of the systolic and diastolic readings.
   D. Blood pressure is classed as high if it is greater than 140/90 mmHg.

Answer is D: If resting systolic BP is above 140 mmHg **and** diastolic BP is above 90 mmHg, then a diagnosis of hypertension may be made. Choice A is very high BP.

6. Which statement regarding the regulation of blood pressure by the endocrine system is correct?
   A. In comparison to other physiological regulatory processes, it is slow.
   B. Aldosterone is responsible for the conversion of angiotensin I into angiotensin II.
   C. In response to a drop in blood pressure, the renin–angiotensin–aldosterone system leads to parasympathetic stimulation.
   D. Angiotensin-converting enzyme (ACE) is responsible for converting angiotensinogen into angiotensin I.

Answer is A: Hormone control mechanisms are slower than those controlled by the nervous system. ACE converts angiotensin I to angiotensin II. Parasympathetic stimulation decreases CO and hence BP.

7. Blood flow is largely regulated at a tissue level. Which of the following could be said regarding this process?
   A. A rise in the blood level of O_2 will result in vasodilation.
   B. A raised CO_2 level results in vasodilation.
   C. Acidaemia directly increases vasopressin (ADH) release.
   D. A raised CO_2 blood level will result in an increased serum alkalinity.

Answer is B: A high level of CO_2 in the tissues will promote vasodilation to remove it and to provide more O_2. ADH produces vasoconstriction which is the opposite of what is required to correct acidaemia. Normally (in the absence of ventilation problems) the blood buffers prevent any change in blood pH so choice D is wrong.

8. Which of the following statements regarding antihypertensive medication is correct?
   A. Calcium channel blockers are a class of drug used to reverse a decrease in blood volume.
   B. Diuretic medication principally affects peripheral resistance.
   C. Beta-blockers target cardiac sympathetic innervation.
   D. ACE inhibitors promote the effects of the renin–angiotensin–aldosterone system.

Answer is C: The neurotransmitters epinephrine and norepinephrine attach to the beta-1 receptor sites of the heart and cause HR and contractility to increase. This sympathetic stimulation promotes an increase in CO (and hence BP). Blocking (preventing the binding of E and NE) their effect prevents sympathetic stimulation.

9. Which of the following is a class of antihypertensive medication which specifically target a reduction in blood volume?
   A. Beta-blockers
   B. Calcium Channel Blockers
   C. Diuretics
   D. Anticoagulants

Answer is C: Diuretics promote the production of urine—which is filtered from blood. Hence blood volume decreases as more urine is produced.

10. Which of the following events would you expect to observe in response to a drop in a patient's blood pressure?
    A. Renin is converted to angiotensinogen.
    B. Angiotensin II is converted into aldosterone
    C. ADH (antidiuretic hormone) will be released by the posterior pituitary gland.
    D. Baroreceptors signal the SA node to slow.

Answer is C: ADH causes more water to be reclaimed from the urine filtrate. This return of water to the blood volume prevents it from decreasing. Maintain blood volume is a way of preventing BP from falling further.

11. Which of the following responses best describes term "pulse pressure"?
    A. A mean measurement of the systolic and diastolic readings
    B. A measurement calculated from 1/3 of the diastolic added to the systolic value
    C. The lowest audible Korotkoff sound when recording blood pressure
    D. A measurement of the difference in pressure between systolic and diastolic readings

Answer is D: Pulse pressure is defined as the difference between the maximum and minimum BP values.

12. Which of the following readings (in mmHg) would be considered within a healthy resting blood pressure range?
    A. 115/70
    B. 120/30
    C. 100/90
    D. 145/90

Answer is A: A reading above 140/90 is considered hypertensive. A diastolic reading of 30 is too low; a systolic reading of 100 is low-ish; a systolic reading of 145 is too high.

13. What does measuring a patient's blood pressure using the auscultatory method do?
    A. It offers valuable information related to cardiac preload.
    B. It involves the reporting of audible venous turbulence created by an inflated arm cuff.
    C. It relates to a systolic relaxation of the ventricles.
    D. It reports the Korotkoff sounds heard from a partially compressed artery.

Answer is D: Korotkoff sounds are made as blood flows turbulently through a compressed artery. They cease when the artery is completely squashed and no blood flows (this is taken to be the systolic pressure) and also when the artery is not compressed at all. In this case, the air pressure in the cuff is equal to or less than the diastolic pressure.

14. What may correctly be said about baroreceptors?
    A. They are located in the walls of the aortic arch and the inferior vena cava.
    B. A drop in blood pressure triggers the baroreceptor reflex which causes vasodilation and an increased heart rate.
    C. They promote vasoconstriction and an increased force of myocardial contraction in the hypotensive patient.
    D. They respond directly to alterations in circulating oxygen levels.

Answer is C: Vasoconstriction and increased contractility promotes an increase in BP. Baroreceptors are located in the walls of the carotid sinus and aortic sinus, among other places. Choices B and D are wrong.

15. What does angiotensin II do?
    A. It is a weak vasoconstrictor and requires activation by angiotensinogen.
    B. It acts via several mechanisms that cause blood pressure to increase.
    C. It reduces blood pressure through decreasing vascular smooth muscle tone.
    D. It causes an increase in urine output by triggering ADH release.

Answer is B: Angiotensin II is a vasoconstrictor. It also stimulates thirst, the secretion of ADH and the adrenal glands to secrete aldosterone. All of which cause BP to increase (provided that we do drink).

16. What does administering beta-blocking medication do?
    A. It targets adrenergic neurotransmission to beta receptors.
    B. It targets cholinergic neurotransmission to decrease blood pressure.
    C. It acts principally upon beta-2 and alpha-1 receptors.
    D. It exposes the beta-1 receptors to enhance neurotransmission.

Answer is A: "Beta-blockers" block beta-1 receptors. These receptors are "adrenergic" as they bind adrenaline (and noradrenaline). Blocking them diminishes neurotransmission.

17. Cardiac output does **NOT** depend on one of the following, which one?
    A. The rate of venous return to the heart
    B. The blood viscosity
    C. The volume flow rate through the circulatory system
    D. The pressure drop (between start of aorta and start of capillaries)

Answer is C: This is the required answer as CO is the same thing as volume flow rate.

18. Which of the following is most **unlikely** to increase blood viscosity?
    A. Leucocytosis
    B. Dehydration
    C. Hypothermia
    D. An infusion of packed red blood cells

Answer is A: Leucocytosis is an increase in the number of WBC above the normal range. Such a rise would have a minimal effect on viscosity as viscosity is due largely to the number of the much more numerous RBC in normal blood.

19. Which of the following would cause blood pressure to DECREASE?
    A. An increase in heart rate
    B. An increase in total peripheral resistance
    C. An increase in heart stroke volume
    D. An increase in parasympathetic impulses along the vagus nerve

Answer is D: Parasympathetic activity decreases heart rate and decreases force of myocardial contraction, hence CO and BP would decrease.

20. Which of the following is the best definition of hypertension?
    A. A systolic pressure of more than 140 mmHg and a diastolic pressure of more than 90 mmHg
    B. A diastolic pressure of more than 140 mmHg and a systolic pressure of more than 90 mmHg
    C. A blood pressure of more than 140/90 mmHg measured after 5 min of inactivity
    D. A blood pressure of less than 110/70 mmHg measured after 5 min of inactivity

Answer is C: Hypertension refers to a resting BP of more than 140/90. Choice A is correct except that there is no reference to being “at rest”.

21. In which part of the brain is the cardiovascular control centre located?
    A. The neurohypophysis
    B. The cerebrum
    C. The hypothalamus
    D. The medulla oblongata

Answer is D: The CVC is in the brain stem, specifically in the medulla oblongata.

22. When blood pressure drops, which of the following responses would happen?
    A. Atrial natriuretic peptide is released from the heart.
    B. The kidneys release renin which catalyses the formation of angiotensin I.
    C. The rate of sodium excretion by the kidneys increases.
    D. The secretion of antidiuretic hormone is inhibited.

Answer is B: Renin acts on angiotensinogen and eventually angiotensin II would be formed. Angiotensin has four effects that will cause BP to rise.

23. Which of the following are two of the factors that influence arterial blood pressure?
    A. Peripheral resistance and gravity
    B. Cardiac output and the partial pressure of oxygen dissolved in blood
    C. Blood volume and blood osmotic pressure
    D. Cardiac output and peripheral resistance

Answer is D: Mean arterial blood pressure, MAP = CO × TPR (TPR = total peripheral resistance).

24. What would be the effect produced if the cardiovascular centre increased the rate of parasympathetic impulses it sends out?
    A. Heart rate would increase.
    B. Heart rate would decrease.
    C. Vasoconstriction of blood vessels would increase.
    D. Vasodilation of blood vessels would increase.

Answer is B: Increased parasympathetic causes a decreased heart rate. Vasoconstriction and vasodilation are controlled by the sympathetic division not the parasympathetic.

25. Too much salt (sodium chloride) in the diet increases blood osmolarity and can cause increased arterial blood pressure. Which of the following is **NOT** a way that salt affects blood pressure?
    A. Salt causes peripheral vasoconstriction.
    B. Salt stimulates thirst.
    C. Salt stimulates ADH secretion.

D. Salt causes fluid shift from the intracellular fluid to the extracellular fluid.

Answer is A: Sodium chloride does not cause peripheral vasoconstriction.

26. 120/80 is recorded as a blood pressure measurement. What does this mean?
    A. Systolic pressure is 120 and diastolic pressure is 80.
    B. Left ventricular systolic pressure is 120 and left ventricular diastolic pressure is 80.
    C. The average of the left and right ventricular systolic pressure is 120 and the average of the left and right ventricular diastolic pressure is 80.
    D. Left ventricular diastolic pressure is 120 and left ventricular systolic pressure is 80.

Answer is B: Although not always specifically stated, a blood pressure of 120/80 refers to the BP in the aorta, that is, caused by the left ventricle. Hence choice A is not as good an answer as choice B.

27. Which of the following does **NOT** influence the resistance to blood flow?
    A. Diameter of the arterioles
    B. Temperature of the blood
    C. Haematocrit
    D. Diameter of the veins

Answer is D: The arterial side of the circulation contributes to resistance to blood flow. Hence the veins do not contribute to resistance.

28. Which will produce a decrease in arterial blood pressure?
    A. Vasoconstriction
    B. Increased parasympathetic stimulation
    C. Increased blood osmolarity
    D. Increased cardiac output

Answer is B: Increased parasympathetic stimulation will decrease heart rate and myocardial contractility. This will cause CO and hence BP to decline.

29. What achieves short-term control of blood pressure?
    A. Hormonal mechanisms
    B. The kidneys
    C. Changes in concentration of chemicals such as O_2, CO_2, H^+, K^+
    D. Neural mechanisms

Answer is D: In the short term, it is the cardiovascular centre (of the brain). It receives input from higher brain centres (hypothalamus and cerebrum), baroreceptors and chemoreceptors.

30. What is the class of antihypertensive drug that prevents smooth muscle contraction and hence promotes vasodilation called?
    A. ACE inhibitors
    B. Beta blockers

C. Diuretics
D. Calcium channel blockers

Answer is D: In order for muscle to contract, calcium is released from the sarcoplasmic reticulum and binds to troponin, which changes its shape and pulls tropomyosin away which exposes actin's binding site. A blocker prevents the release of Ca^{++} into the muscle cell sarcoplasm.

31. What does the term "systolic pressure" refer to?
    A. The value, in mmHg, that appears in the denominator of a blood pressure measurement
    B. The peak pressure in the blood due to the contraction of the left ventricle
    C. The minimum pressure in the aorta prior to left ventricular contraction
    D. The difference between maximum and minimum arterial blood pressures

Answer is B: It is the maximum pressure produced by the LV during contraction (while the body is at rest).

32. What are the receptors that are sensitive to blood pressure called?
    A. Pacinian corpuscles
    B. Nociceptors
    C. Baroreceptors
    D. Chemoreceptors

Answer is C: The prefix "baro" refers to air pressure where the unit 1 bar is the pressure of one standard atmosphere.

33. Which situation below would make the heart beat faster?
    A. An increase in sympathetic impulses along the cardioaccelerator nerves
    B. An increase in parasympathetic impulses along the cardioaccelerator nerves
    C. An increase in sympathetic impulses along the vagus nerves
    D. An increase in parasympathetic impulses along the vagus nerves

Answer is A: Increased sympathetic stimulation cause an increase in heart rate. The vagus nerve carries only parasympathetic impulses.

34. Which of the following will **NOT** increase cardiac output?
    A. Increasing strength of contraction
    B. Increasing stroke volume
    C. Increasing heart rate
    D. Increasing total peripheral resistance

Answer is D: If peripheral resistance increases, all else being unchanged, CO will fall.

35. Which of the following will increase cardiac output?
    A. An increase in sympathetic impulses
    B. An increase in parasympathetic impulses

C. A faster stream of impulses from the baroreceptors
D. An increase in vasodilation

Answer is A: Sympathetic impulses will increase CO. Increased baroreceptor activity would decrease CO.

36. Angiotensin II does all of the following except one. Which one?
    A. Stimulates thirst
    B. Causes the release of aldosterone
    C. Causes the release of ADH
    D. Stimulates peripheral vasodilation

Answer is D: Angiotensin II promotes an increase in blood volume and vasoconstriction, but not vasodilation.

37. What is the consequence when ADH is released?
    A. Blood osmolarity increases
    B. The permeability of the collecting ducts to water is increased
    C. Peripheral vasodilation increases
    D. Blood pressure decreases

Answer is B: Antidiuretic hormone decreases the volume of urine formed by causing water to be reclaimed from the DCT and collecting duct. Hence blood osmolarity will not increase and BP will not decrease.

38. If dietary salt intake is excessive, which of the following will **NOT** occur?
    A. Less ADH will be secreted.
    B. The osmolarity of extracellular fluids will increase.
    C. The thirst centre will be stimulated.
    D. The extracellular fluid volume increases.

Answer is A: If a greater than recommended amount of salt is in the diet, **more** ADH will be secreted, so that additional water is reclaimed from the nephron filtrate. This will tend to decrease blood osmolarity. Hence it will not occur that "less ADH will be secreted".

39. The antihypertensive drugs known as "ACE inhibitors" function by doing which of the following?
    A. Preventing the release of ADH
    B. Blocking the formation of angiotensin II
    C. Blocking the release of renin
    D. Preventing the entry of Ca^{2+} to vascular smooth muscle

Answer is B: Angiotensin-converting enzyme is known as "ACE". An ACE inhibitor will prevent this enzyme from converting angiotensin I into angiotensin II. Hence the blood pressure raising effects of angiotensin II will be prevented.

40. Which hormone produces a decrease in arterial blood pressure?
    A. Vasopressin
    B. ANP
    C. ADH
    D. Angiotensin II

Answer is B: Atrial natriuretic enzyme (ANP) exerts effects that result in a decrease in BP.

41. What will be the result of an increase in sympathetic impulses along the vasomotor nerves?
    A. Increased heart rate and force of contraction
    B. Decreased heart rate and force of contraction
    C. Generalised vasoconstriction
    D. Generalised vasodilation

Answer is C: Vasomotor nerves should give the clue that vessels will be set in motion. An increase in impulses will result in vasoconstriction.

42. Given that a heart pumps out 70 mL of blood with each stroke and beats 70 times per minute, what is the cardiac output?
    A. 70 mL/min
    B. 490 mL/min
    C. 700 mL/min
    D. 4900 mL/min

Answer is D: Cardiac output, CO = SV × HR = 70 mL/beat × 70 beat/min = 4900 mL/min.

43. Which completed statement is **NOT** true? Peripheral resistance:
    A. Increases if diameter of blood vessels increases
    B. Is greater if the total length of blood vessels is greater
    C. Increases if viscosity of blood increases
    D. Is greater than pulmonary resistance

Answer is A: If the diameter of blood vessels increases, there will be less friction with the vessel wall and peripheral resistance will decrease.

44. Complete the following sentence correctly for a person at rest. Hypertension:
    A. Occurs when blood volume is too low
    B. Refers to the increased blood pressure in the legs while standing
    C. Indicates that resistance to blood flow is low
    D. Is when systolic blood pressure is more than 140 mmHg

Answer is D: Hypertension means a resting BP that is too high. The value beyond which hypertension is identified is 140 mmHg for the systolic reading.

45. An increase in parasympathetic impulses along the vagus nerve causes:
    A. Dilation of the arterioles
    B. A decrease in blood pressure
    C. An increase in vasoconstriction
    D. Increases the force of myocardial contraction

Answer is B: Parasympathetic stimulation causes a decrease in BP as heart rate and force of contraction are decreased. Dilation and constriction of blood vessels are controlled by the sympathetic nervous system.

46. Which of the following would cause a rise in mean arterial blood pressure?
    A. Sympathetic impulses along the cardioaccelerator nerves
    B. Changing from a standing position to a supine position
    C. Generalised vasodilation of blood vessels
    D. A severe haemorrhage

Answer is A: Sympathetic stimulation increases heart rate and contractility which increases cardiac output which increases BP. The term cardioaccelerator nerve should provide a clue.

47. What is the role of angiotensin II?
    A. Decrease blood pressure by promoting vasodilation of veins
    B. Increase blood pressure by promoting vasoconstriction of arterioles
    C. Decrease blood pressure by promoting excretion of water in urine
    D. Increases blood pressure by promoting absorption of Na^+

Answer is B: Angiotensin II promotes vasoconstriction which increases peripheral resistance which increases BP. It also stimulates the adrenal glands to produce aldosterone. It is aldosterone that causes an increase in Na^+ absorption, hence choice B is a better answer than D.

48. Which one of the following is a vasodilator?
    A. Atrial natriuretic peptide (ANP)
    B. Angiotensin II
    C. Epinephrine (adrenaline)
    D. An increase in sympathetic impulses

Answer is A: ANP causes a decrease in BP and vasodilation is one of the ways by which this is achieved.

49. To what does the term ventricular systole refer?
    A. Relaxation of the ventricles
    B. Relaxation of the atria
    C. Contraction of the myocardium
    D. Contraction of the ventricles

Answer is D: Ventricular systole is contraction of the ventricles, which does not coincide with relaxation of the atria.

50. Which of the following is **NOT** true about turbulent blood flow?
    A. Turbulent flows exist when blood is flowing in smooth streamlines.
    B. It may occur when blood passes through a constriction or stenosis.
    C. It occurs if blood speed is high.
    D. Turbulent flow is noisy.

Answer is A: When blood flows in smooth streamlines parallel to the vessel walls, it is known as streamline flow. In turbulent flow the liquid moves in eddies and will not always be parallel to the vessel walls.

51. What is the pressure gradient produced by the left ventricle equal to?
    A. The mean arterial pressure
    B. The difference between mean arterial pressure and pressure at the start of the capillaries
    C. The difference between mean arterial pressure and pressure at the start of the capillaries, divided by the distance between the start of the aorta and capillaries
    D. The mean arterial pressure divided by the distance between the start of the aorta and capillaries

Answer is C: Pressure gradient between two points is the pressure difference between two points, divided by their distance apart.

52. What is one of the determinants of the resistance to blood flow?
    A. Blood viscosity
    B. Cardiac output
    C. Heart rate
    D. The blood osmolarity

Answer is A: Blood viscosity, largely determined by the concentration of RBC (the haematocrit), affects blood's resistance to flow.

53. Which one of the following will make blood pressure fall?
    A. Increased cardiac output
    B. Increased heart rate
    C. Increased vasodilation
    D. Increased peripheral resistance

Answer is C: Vasodilation decreases resistance to blood flow, which means the same CO can be maintained by a lower BP.

54. Hypertension in adults may be defined as:
    A. Excessive decrease in blood pressure
    B. Mean arterial pressure greater than 110 mmHg
    C. Systolic blood pressure less than 100 mmHg when resting
    D. Systolic blood pressure persistently greater than 140 mmHg when resting

Answer is D: Choice B would also be correct if reference was made to "resting". The hypertensive value for children is less.

55. Which of the following statements about the cardiovascular control centre of the brain is **FALSE**?
    A. Increased impulses along parasympathetic fibres cause vasoconstriction.
    B. Increased output along the sympathetic fibres INcreases heart rate.
    C. Output along of the parasympathetic fibres DEcreases heart rate.
    D. Decreased output along the sympathetic fibres causes dilation of arterioles.

Answer is A: The parasympathetic system does not have a role in vasoconstriction or vasodilation.

56. Which of the following does angiotensin II cause to happen?
    A. Atrial natriuretic peptide to be released
    B. The collecting ducts in the kidney to become permeable to water
    C. The release of antidiuretic hormone to be suppressed
    D. Aldosterone to be released

Answer is D: Angiotensin II stimulates the adrenal glands to release aldosterone.

57. The definition of mean arterial pressure (MAP) may be written as:
    A. MAP = stroke volume × heart rate
    B. MAP = (diastolic pressure + systolic pressure) ÷ 2
    C. MAP = cardiac output × peripheral resistance
    D. MAP = diastolic pressure + pulse pressure

Answer is C: Mean arterial pressure, MAP = CO × TPR. MAP is also diastolic pressure + $^{1}/_{3}$ × pulse pressure.

58. Which one of the following does **NOT** contribute to peripheral resistance?
    A. Heart rate
    B. Blood viscosity
    C. Diameter of blood vessels
    D. Length of blood vessels

Answer is A: Heart rate contributes to cardiac output but not peripheral resistance.

59. Which of the following statements about the cardiovascular control centre of the brain is **TRUE**?
    A. Increased impulses along parasympathetic fibres cause vasoconstriction.
    B. Output along the sympathetic fibres INcreases heart rate.
    C. Output along the parasympathetic fibres INcreases heart rate.
    D. Output along the sympathetic fibres DEcreases heart rate.

Answer is B: Impulses travelling in the sympathetic fibres increases heart rate (in preparation for increased physical activity).

60. Which three hormones have a role in regulating blood pressure?
    A. Angiotensin II, ADH and ANP
    B. Renin, angiotensin II and ADH

C. Vasopressin, ADH and ANP
D. Angiotensin II, ACE and ADH

Answer is A: Renin is an enzyme that coverts angiotensinogen into angiotensin I, while ACE is an enzyme that converts angiotensin I into angiotensin II. Vasopressin and ADH are the same hormone.

61. Which of the following statements about the cardiovascular centre of the brain is **FALSE**?
    A. It consists of the cardiac centre and the vasomotor centre.
    B. Output along fibres of the sympathetic nervous system DEcreases heart rate.
    C. Output along fibres of the parasympathetic nervous system DEcreases heart rate.
    D. Output along fibres of the sympathetic nervous system INcreases heart rate.

Answer is B: Output along fibres of the sympathetic nervous system actually **increases** heart rate. So choice B is FALSE and answers the question that was asked.

62. Which statement about cardiac output is correct?
    A. Cardiac output is peripheral resistance multiplied by stroke volume.
    B. Mean arterial pressure multiplied by peripheral resistance is cardiac output.
    C. Cardiac output is heart rate multiplied by stroke volume.
    D. Cardiac output is blood volume multiplied by heart rate.

Answer is C: Cardiac output (in millilitre per minute) = stroke volume (in millilitre per beat) times heart rate (in beats per minute).

63. Which of the following would produce a **DECREASE** in heart rate?
    A. Sympathetic impulses along the cardioaccelerator nerves
    B. Increased sympathetic impulses along the vasomotor nerves
    C. Decreased sympathetic impulses along the vasomotor nerves
    D. Parasympathetic impulses along the vagus nerve

Answer is D: Parasympathetic stimulation (which almost always utilises the vagus nerve) causes a decrease in heart rate.

64. A rise in arterial blood pressure stretches the vessel walls which contain baroreceptors. Which of the following responses does this produce? The baroreceptors send a:
    A. Slower stream of impulses to the vasomotor centre which inhibits it
    B. Faster stream of impulses to the vasomotor centre which inhibits it
    C. Slower stream of impulses to the vasomotor centre which stimulates it
    D. Faster stream of impulses to the vasomotor centre which stimulates it

Answer is B: When a blood vessel wall is stretched, it stimulates baroreceptors to **increase** their stream of impulses to the vasomotor centre which **inhibits** it. This means the smooth muscle in blood vessel walls will relax, and the vessel diameter will increase and so BP is reduced.

65. Why is the blood pressure in the pulmonary arteries less than in the aorta?
    A. The expansion and contraction of the lungs pump blood through its blood vessels.
    B. Blood flowing to the lungs does not need to overcome gravity.
    C. Pulmonary resistance is greater than the peripheral resistance.
    D. Pulmonary resistance is less than peripheral resistance.

Answer is D: As pulmonary resistance is much lower than systemic resistance, the BP required to overcome it is much less too. Hence the right ventricle does not need to produce the high pressure of the left ventricle to move blood through the pulmonary circuit.

66. When blood pressure is measured in the brachial artery by the auscultatory method, why should the arm be at the same level as the heart?
    A. If the arm is lower than the heart, brachial artery pressure will be lower as blood flow is assisted by gravity.
    B. If the arm is higher than the heart, brachial artery pressure will be higher as more force is required to pump blood up hill.
    C. To avoid any hydrostatic pressure effects on the brachial artery pressure.
    D. There is no reason for it, the practice is part of "nursing ritual".

Answer is C: If the arm is higher, the pressure in the brachial artery will be lower than aortic pressure (and vice versa).

67. Why is blood pressure stated in units of "mmHg"? Because:
    A. The first barometers are operated with mercury.
    B. Standard International (SI) units are not required in human biology.
    C. The haemoglobin molecule contains an atom of mercury.
    D. Blood pressure is a length.

Answer is A: Mercury (also known as hydrogyrum) was the liquid used in manometers (pressure measuring devices) because of its very high density compared with water.

68. Blood may flow in the aorta with a speed of 30 cm/s and in the capillaries with a speed of only 0.1 m/s. Why is there such a large difference in speed?
    A. The capillaries are much further from the heart than the aorta.
    B. The very narrow capillaries present a large resistance to blood flow compared to the large diameter aorta.
    C. The total cross-sectional area of the lumens of the capillaries is much greater than the cross-sectional area of the aorta.

D. The length of the aorta is short compared to the length of a capillary.

Answer is C: The "equation of continuity" may be stated: volume flow rate = cross-sectional area × speed of flow. Hence if the total cross-sectional area of capillaries is very large (as it is), the speed of flow can be very slow while still allowing the required volume flow rate.

69. Which statement about cardiac output is correct? Cardiac output is:
    A. The sum of volume of blood pumped by left and right ventricles per minute
    B. The mean arterial pressure divided by total peripheral resistance
    C. The mean arterial pressure multiplied by stroke volume
    D. The blood volume multiplied by heart rate

Answer is B: MAP = CO × TPR, so CO = MAP ÷ TPR.

70. In which of the following situations would blood pressure be increased?
    A. Antidiuretic hormone (ADH) secretion is inhibited.
    B. The kidneys absorb less water before it is excreted as urine.
    C. The extracellular fluid volume decreases.
    D. The extracellular fluid volume increases.

Answer is D: Extracellular volume includes the blood. If its volume increases, that means blood volume increases and this will increase BP.

71. In the circulatory system, why does an increase in cardiac output cause an increase in volume flow rate (of blood)? Because:
    A. An increased cardiac output causes vasoconstriction.
    B. As cardiac output increases so blood viscosity increases.
    C. As cardiac output increases, the resistance of the systemic circulation to blood flow decreases.
    D. Cardiac output and volume flow rate are the same thing.

Answer is D: The number of millilitres of blood per minute is volume flow rate and cardiac output. They are the same thing.

72. The Korotkoff sounds that are listened for during a blood pressure measurement by the auscultatory method are caused by:
    A. Turbulent blood flow in the aorta
    B. The difference between systolic pressure and cuff pressure
    C. Turbulent flow in the collapsed brachial artery
    D. The opening and closing of the heart valves

Answer is C: Squashing the brachial artery will cause blood flow to be turbulent as it passes through the constriction. This turbulence creates sound that can be detected with a stethoscope.

73. What is “systolic pressure”?
    A. The peak pressure in the blood due to contraction of the left ventricle.
    B. The minimum pressure in the blood prior to contraction of the left ventricle.
    C. It may be obtained by multiplying cardiac output by total peripheral resistance.
    D. It is also known as mean arterial pressure.

Answer is A: Systole means “contraction” in Greek. So systolic pressure is the pressure produced by the contraction of the left ventricle. Choice C refers to mean arterial pressure, which is less than systolic pressure.

74. What is “autoregulation”? It is the adjustment of blood flow to each tissue due to:
    A. Hormonal control
    B. Neural control
    C. Systemic factors
    D. Local factors

Answer is D: Autoregulation refers to the tissue regulating its own blood flow depending on the concentration of chemicals (oxygen, carbon dioxide, metabolites) within the tissue.

75. Which three factors affect blood pressure?
    A. Heart rate, stroke volume and total peripheral resistance
    B. Cardiac output, heart rate and pulse pressure
    C. Total peripheral resistance, cardiac output and blood volume
    D. Diastolic pressure, systolic pressure and pulse pressure

Answer is C: Blood pressure rises and falls as blood volume rises and falls. If volume does not change, then just cardiac output and total peripheral resistance determine BP. Choice A is correct for a constant blood volume.

76. Which of the following occurrences would cause blood pressure to DECREASE?
    A. An increase in cardiac output
    B. An increase in vasodilation
    C. An increase in heart rate
    D. An increase in blood volume

Answer is B: Vasodilation increases the diameter of blood vessels, and this will cause BP to decrease. The other events will raise blood pressure.

77. What name is given to the pressure sensitive neurones that monitor stretching in blood vessel walls?
    A. Baroreceptors
    B. Muscle spindles
    C. Sympathetic receptors
    D. Integral proteins

Answer is A: The "bar" was used as unit of air pressure: 1 mbar = 1 hPa. It is now used as a prefix in a word to denote an association with pressure. A barometer is a device to measure air pressure; hence, baroreceptor refers to a pressure receptor.

78. Which of the following would result in an increase in blood pressure?
    A. An increase in glomerular filtration rate
    B. An increase in parasympathetic stimulation of the heart
    C. A decrease in antidiuretic hormone secretion
    D. An increase in sympathetic stimulation of the heart

Answer is D: Sympathetic stimulation prepares the body for vigorous activity and such stimulation would result in an increase in HR would increase BP. Parasympathetic stimulation would do the reverse. An increase in GFR would filter more water from the blood thereby decreasing its volume and BP.

79. What effect does vasoconstriction have on arterial blood pressure and why?
    A. It increases BP. As vessel diameter narrows, peripheral resistance increases; hence, a greater BP is required to overcome this resistance.
    B. It increases BP. As vessel diameter increases, peripheral resistance increases; hence, a greater BP is required to overcome this resistance.
    C. It decreases BP. As vessel diameter narrows, peripheral resistance decreases; hence, a lower BP is sufficient to overcome this resistance.
    D. It decreases BP. As vessel diameter increases, peripheral resistance decreases; hence, a lower BP is sufficient to overcome this resistance.

Answer is A: Vasoconstriction cause narrowing of vessel diameter. A narrower blood vessel presents a greater resistance to blood flow through it. In order to maintain blood flow at the same rate, blood pressure must increase to overcome the greater resistance.

80. What are the effects on blood volume and arterial pressure of an INcreased glomerular filtration rate?
    A. Blood volume would increase and arterial pressure would decrease.
    B. Both blood volume and arterial pressure would increase.
    C. Blood volume would decrease and arterial pressure would increase.
    D. Both blood volume and arterial pressure would decrease.

Answer is D: An increase in GFR would produce more filtrate. This would result in a lower blood volume and hence a decreased BP.

81. What are the effects on blood volume and arterial pressure of a glomerular filtration rate that is DEcreased below the average healthy level?
    A. Blood volume would increase and arterial pressure would decrease.
    B. Neither blood volume nor arterial pressure would decrease.
    C. Blood volume would decrease and arterial pressure would increase.
    D. Both blood volume and arterial pressure would decrease.

Answer is B: A decrease in GFR would produce less filtrate. This would result in less blood being removed from circulation; hence, blood volume would not decrease. As blood volume is not decreasing, blood pressure would not decrease (and it may increase!).

82. What are the effects on blood volume and arterial pressure of an increase in sympathetic stimulation to the heart and vascular smooth muscle?
    A. Blood volume would decrease, but arterial blood pressure would increase.
    B. Both blood volume and arterial pressure would decrease.
    C. Blood volume would not change, but arterial blood pressure would increase.
    D. Blood volume would increase and arterial pressure would decrease.

Answer is C: An increase in sympathetic impulses along cardioaccelerator nerves would increase heart rate via the SA node, would shorten conduction time at the AV node and would increase the force of myocardial contraction; hence, cardiac output would increase. This would increase BP. An increase in sympathetic impulses along vasomotor nerves to blood vessel walls causes generalised vasoconstriction (i.e. increased peripheral resistance). This would increase BP.

83. What would be the effects on blood volume and arterial pressure of an increase in parasympathetic stimulation to the heart?
    A. Blood volume would not change, but arterial blood pressure would decrease.
    B. Blood volume would not change, but arterial blood pressure would increase.
    C. Blood volume would increase and arterial pressure would decrease.
    D. Blood volume would decrease, but arterial blood pressure would increase.

Answer is A: An increase in parasympathetic impulses along vagus nerve would decrease heart rate and decrease force of contraction in the myocardium; hence, cardiac output would decrease. This would cause BP to decrease.

84. What would be the effects on blood volume and arterial pressure of antidiuretic hormone (ADH) secretion?
    A. Blood volume would increase and arterial pressure would decrease.
    B. Blood volume would increase and arterial pressure would increase.
    C. Blood volume would not decrease and arterial pressure would not decrease.
    D. Both blood volume and arterial pressure would decrease.

Answer is C: ADH causes the walls of the collecting ducts of the kidney to become permeable to water. Consequently water is reabsorbed from the filtrate and moves to the interstitial fluid. This return or water to the blood means that blood volume would not fall. The maintenance of blood volume would also mean that blood pressure would not be altered by change to blood volume.

85. What would be the effects on blood volume and arterial pressure of drinking water?
    A. Blood volume would increase and arterial pressure would decrease.
    B. Blood volume would increase and arterial pressure would increase.

C. Blood volume would decrease and arterial pressure would increase.
D. Both blood volume and arterial pressure would decrease.

Answer is B: As water is absorbed from the gastrointestinal tract, it moves into the blood stream and so blood volume would increase. This increased blood volume would cause blood pressure to also increase. Of course, homeostatic mechanisms would come into play, so that if too much water was drunk, then a greater volume of urine would be produced and appropriate levels of blood volume and BP would be restored.

86. What is the stimulus for renin release by the juxtaglomerular cells?
    A. An increase in parasympathetic stimulation
    B. An increase in renal blood flow
    C. A decrease in the level of circulating angiotensin II
    D. A decrease in renal blood pressure

Answer is D: Renin results in an increase in circulating angiotensin II which in turn has a lot of effects that increase blood pressure. An increase in sympathetic stimulation (not parasympathetic) also stimulates renin release.

87. What effects does angiotensin II have?
    A. It stimulates the right atrium to release ANP.
    B. It makes the juxtaglomerular cells release renin.
    C. It makes the adrenal glands release aldosterone.
    D. It stimulates peripheral vasodilation.

Answer is C: Aldosterone increases reabsorption of Na^+ from the filtrate by kidney tubules. Angiotensin II works to increase blood pressure (or prevent it falling further) by retaining Na^+ and water at the kidneys and promoting vasoconstriction.

88. How do the drugs known as angiotensin-converting enzyme (ACE) inhibitors work?
    A. They prevent the enzyme renin from being released.
    B. They prevent renin from acting on angiotensinogen.
    C. They prevent the liver from making angiotensinogen.
    D. They prevent the AC enzyme from acting on angiotensin I

Answer is D: ACE converts angiotensin I into angiotensin II. If the enzyme is inhibited, the production of angiotensin II is prevented, and so its blood pressure raising effects are blocked.

89. Why do "calcium channel blocker" drugs help to reduce blood pressure?
    A. Ca^{++} is required for smooth muscle contraction, so veins dilate.
    B. Ca^{++} is required for smooth muscle contraction, so arteries dilate.
    C. Ca^{++} is required for smooth muscle relaxation, so veins constrict.
    D. Ca^{++} is required for smooth muscle relaxation, so arteries constrict.

Answer is B: Ca^{++} is required for smooth muscle contraction. If it is prevented from entering the cell, the artery cannot constrict. Dilation of the arteries reduces peripheral resistance and thereby the effort the heart must exert to pump blood. This reduces blood pressure. Dilation of veins does not affect peripheral resistance but may reduce venous return and so the volume of blood pumped by the heart.

90. Which is a mechanism for the short-term regulation of blood pressure?
    A. The baroreceptor reflex
    B. The release of ADH
    C. The release of vasopressin
    D. The stretch reflex

Answer is A: Baroreceptors are pressure-sensitive neurones that monitor stretching of blood vessel walls and atria. Located in wall of carotid sinus, aortic arch and wall of most large arteries in neck and thorax. If BP falls, baroreceptor reflexes increase heart rate, force of contraction and promote vasoconstriction.

91. The ankle brachial index (ABI) is the systolic BP measured at the ankle (posterior tibial artery) divided by the systolic BP measured at the arm (brachial artery) while the subject is supine. What problem may be indicated if the ABI measured with the left arm and leg is not about the same as for the right limbs?
    A. The subject has hypertension.
    B. The subject has a dominant right hand (or left hand).
    C. One leg may have an obstruction in its blood vessels.
    D. The inflatable cuff has not completely occluded (squashed) the artery.

Answer is C: ABI should be the same for each leg if the two legs have the same anatomy. A difference means that one leg has a different level of blood circulation than the other. This may be due to a narrowing or hardening of the arteries in one leg. Hypertension is a systolic BP > 140 mmHg and is diagnosed by measuring BP at the level of the heart, not the ABI.

92. Why does arterial blood pressure and heart rate increase when we perform exercise?
    A. Blood pressure increases because heart rate increases.
    B. The exercise performed by skeletal muscle during exercise is supported by an increase in blood flow to the muscle. The increased blood flow results from increased blood pressure. BP is increased by increasing the heart rate.
    C. Exercise requires conscious effort, and this effort is stimulated by the release of epinephrine. Epinephrine causes heart rate and blood pressure to increase.
    D. Exercise causes metabolic waste products and CO_2 to be released into the blood. To clear these wastes, heart rate and blood pressure need to increase.

Answer is B: At rest leg muscles (for example) are inactive and receive little blood. During running activities (say), the leg muscles are being vigorously used. The vigorous use requires increased blood flow to deliver the oxygen required to produce the ATP needed for muscular activity. Increased blood flow will be achieved if blood pressure increases. Increasing the heart rate will produce the required increase in BP.

Chapter 13
Respiratory System

13.1 Anatomy and Physiology

Respiration is a term that is variously applied to the acts of inhalation and exhalation, to the movement of gas molecules in the lungs between alveolar air and blood in the alveolar capillaries, to the exchange of dissolved gases in the tissues between the systemic capillaries and the surrounding interstitial fluid and to the process conducted within the mitochondria of cells that results in the production of ATP (and CO_2) from small organic molecules by using O_2. The respiratory system is a set of tubes that branch to increase in number and decrease in size, within an elastic structure that is moved by muscles. The lungs and chest wall together act like a bellows to move air into and out of the alveoli. The epithelial cells of the walls of the alveoli are part of the respiratory membrane that separates the air in the alveoli from the blood in the alveolar capillaries. The endothelial cells of the capillary walls are also part of the respiratory membrane.

Air passes through the nostrils, the meatus of the nasal cavity, the pharynx, the larynx, the glottis, the trachea, the bronchi, then into the secondary and tertiary (and smaller) bronchi eventually into the bronchioles, then into smaller airways to finally reach the alveoli. Here oxygen diffuses through the respiratory membrane from the alveoli to the blood in capillaries and carbon dioxide diffuses from the blood to the alveoli. Bronchi are held open by the cartilage in their walls, while bronchioles are without cartilage but may dilate and constrict their diameter as the smooth muscle in their wall relaxes or contracts.

At the end of an inhalation, the volume of air that is within the respiratory tree but that has not reached the alveoli is not in contact with the respiratory membrane and so does not participate in gas exchange. This volume of "fresh" air is called the anatomical dead space and is then exhaled. At the end of an exhalation, there is a volume of "stale" air within the bronchial tree that has yet to pass out of the body. Instead this stale air re-enters the alveoli, pushed in by the fresh air of the next inhalation. Hence, alveolar air is not completely refreshed with each inhalation. Carbon

M. Caon, *Examination Questions and Answers in Basic Anatomy and Physiology*, https://doi.org/10.1007/978-3-030-47314-3_13

dioxide produced in the tissues dissolves in water to form carbonic acid ($CO_2 + H_2O \rightleftharpoons H_2CO_3$). This reaction is catalysed within the RBC. Then the carbonic acid molecule disassociates to form the hydronium and bicarbonate ions ($H_2CO_3 \rightleftharpoons HCO_3^- + H^+$). The hydronium ions are buffered by haemoglobin in the RBC while the bicarbonate ions move into the blood plasma. When blood reaches the lungs, carbon dioxide moves from the blood to the alveoli, and the two previous equations proceed from right to left. Hence, breathing out carbon dioxide causes acid (hydronium ions) to be converted to water. If the ability to breathe out is impaired, then acid is not excreted, so blood pH will fall and the body may be in respiratory acidosis.

1. Which of the following statements could be applied to "external respiration"?
   A. Exchange of gases between alveolar air and the blood in pulmonary capillaries
   B. Exchange of dissolved gases between blood in tissue capillaries and the body tissues
   C. The production of CO_2 from organic molecules in the cells by using O_2
   D. The inhalation of atmospheric air into the lungs followed by exhalation

Answer is A: External respiration refers to the movement of oxygen from the alveoli into the capillary blood and carbon dioxide from the capillary blood into the alveoli. Choice B is "internal respiration" and choice C is "cellular respiration".

2. Which anatomical structures does the "conducting zone" of the lower respiratory tract contain?
   A. Eustachian tube, larynx and trachea
   B. Primary, secondary and tertiary bronchi and bronchioles
   C. Nares, conchae, olfactory mucosa and sinuses
   D. Nasopharynx and larynx

Answer is B: The conducting zone is distal to the trachea and before the alveoli.

3. What is the function of the cilia on the cells that line the bronchial tree?
   A. They help mix the inhaled fresh air with the residual air contained in the bronchial tree.
   B. They slow the movement of air to allow for efficient exchange of gases.
   C. They move the mucus on the cell surface up out of the bronchial tree.
   D. They filter particles from inhaled air.

Answer is C: The beating of the cilia moves mucus lying on the surface of the epithelium of the conducting zone, and any contained dust, up out of the bronchial tree.

4. Which one of the following statements is correct?
   A. The visceral pleura is attached to the chest wall, and the parietal pleura is attached to the lung.

B. The two lungs and their associated structures are known as the pneumothorax.
C. The hilum is a serous membrane that surrounds each lung separately.
D. A negative pressure is maintained between the two lung pleura.

Answer is D: The parietal pleura (attached to the chest wall) and the visceral pleura (attached to the lung) are in very close contact but "separated" by pleural fluid within which there is a pressure that is less than atmospheric pressure (i.e. negative). This means that the lungs are stuck to the chest wall and expand when it does.

5. What term is applied to the volume of air that moves into the lungs while breathing at rest?
   A. Anatomical dead space
   B. Inspiratory reserve capacity
   C. Tidal volume
   D. Residual volume

Answer is C: Tidal volume moves into the lungs (and out) with each completed inhalation/exhalation at rest.

6. Severing the nerves that innervate the breathing muscles may lead rapidly to death. Will a spinal cord break between the level of cervical vertebrae 6 and 7 leave the victim able to breathe? Choose the answer with the correct reason.
   A. No. The breathing muscles are innervated by spinal nerves that leave the spinal cord at the level of each thoracic vertebra.
   B. Yes. The diaphragm will work as it is innervated by nerves arising from C3 to C5.
   C. No. The breathing muscles are innervated by autonomic impulses from the respiratory centre which is located in the brain stem.
   D. Yes. The muscles of breathing are innervated by the sympathetic nervous system which is unaffected by damage to the somatic nervous system.

Answer is B: The diaphragm, but not the intercostal muscle, will still receive innervation as the diaphragmatic nerves leave the spinal cord superior to the break at C6–C7.

7. Which molecule or ion dissolved in blood is able to stimulate the central chemoreceptors of the brain's respiratory centre?
   A. CO_2
   B. H_3O^+
   C. O_2
   D. Ca^{2+}

Answer is A: Blood-borne hydronium cannot cross the blood–brain barrier. However, carbon dioxide can and it produces hydronium ions when it is on the brain side of the BBB. As there is no buffer in the CSF, these hydronium ions stimulate the central chemoreceptors of the respiratory centre.

8. What term is applied to the exchange of dissolved gases between capillary blood and body tissues?
   A. Internal respiration
   B. External respiration
   C. Ventilation
   D. Cellular respiration

Answer is A: Internal respiration refers to the movement of oxygen from the capillary blood into the tissues and carbon dioxide from the tissues into capillary blood. Choice B occurs between the alveoli and pulmonary capillaries, and choice C occurs within the mitochondria.

9. Between which two anatomical structures does the Larynx lie?
   A. The nares and the choanae
   B. The epiglottis and the trachea
   C. The choanae and the glottis
   D. The glottis and the epiglottis

Answer is B: The larynx is the tube that surrounds the vocal apparatus. It consists of the epiglottis, thyroid cartilage and cricoid cartilage and is located at the top of the trachea.

10. What is the function of the ciliated cells of the respiratory epithelium?
    A. To trap inhaled particles not removed by the nasal cavity
    B. To secrete a mucus layer onto the epithelium
    C. To move mucus and trapped particles up the bronchial tree
    D. To secrete surfactant that decreases water surface tension

Answer is C: Ciliated cells have cilia that beat rhythmically to move inhaled dust particles that have been caught by the mucus out of the respiratory tract.

11. Why is it that bronchioles can constrict and so reduce their diameter while secondary bronchi and respiratory bronchioles cannot constrict?
    A. Bronchioles have smooth muscle but no cartilage in their walls while secondary bronchi are supported by cartilage.
    B. Bronchioles have smooth muscle but no cartilage in their walls while respiratory bronchioles are supported by cartilage.
    C. Bronchioles have cartilage but no smooth muscle in their walls while secondary bronchi are supported by cartilage.
    D. Bronchioles have cartilage but no smooth muscle in their walls while respiratory bronchioles only have smooth muscle in their walls.

Answer is A: Because bronchioles have no cartilage in their walls, they are able to dilate and constrict to alter the amount of air that enters alveoli. Respiratory bronchioles that feed into alveolar ducts have little smooth muscle and no cartilage.

12. A person with severe trauma to the cervical region has damage to the spinal cord. If the spinal cord is severed between C3 and C4, what is the likely outcome?
    A. The person will be able to breathe but will have paralysis of the lower limbs.
    B. The person will be able to breathe but will have paralysis of the upper and lower limbs.
    C. The person will be able to breathe with the intercostal muscles, but will lose the use of the diaphragm have paralysis of the upper and lower limbs.
    D. The person will be unable to breathe and will have paralysis of the upper and lower limbs.

Answer is D: The diaphragm is innervated by the phrenic nerves which arise from the C3–C5 vertebrae. Hence the person will be unable to use their diaphragm (much). The upper limbs are innervated by spinal nerves that leave the spinal cord between C5 and T2, so those nerves will have been interrupted as well.

13. The goal of respiration is to control the concentration of which substances dissolved in the blood?
    A. Oxygen
    B. Oxygen and carbon dioxide
    C. Oxygen, carbon dioxide and hydrogen ions
    D. Oxygen, carbon dioxide, hydrogen ions and ATP

Answer is C: Respiratory activity responds to changes in the dissolved concentration of these three substances in blood.

14. The walls of the following structures are all supported by cartilage except for one of them. Which one?
    A. Bronchioles
    B. Trachea
    C. Bronchi
    D. Larynx

Answer is A: Bronchioles have smooth muscle but no cartilage in their walls.

15. What constitutes the respiratory membrane?
    A. The parietal and visceral pleurae and enclosed pleural fluid
    B. Capillary and alveolar epithelial cells, their basement membranes and adjacent fluid
    C. The alveolar surface fluid and epithelial cells
    D. Alveolar epithelial and septal cells, ciliated cells, macrophages and surfactant

Answer is B: The cells lining the alveoli and the capillaries and their fused basement membrane, along with the fluid (surfactant) lining the alveoli together are the membrane. Pleurae, ciliated cells and macrophages are not.

16. What is the “cribriform plate”?
    A. That part of the nose with three folds of tissue called conchae
    B. The structure that separates the nose from the nasopharynx
    C. Part of the ethmoid bone through which olfactory nerves pass
    D. The nose structure through which air is warmed and humidified as it passes

Answer is C: The cribriform plate of the ethmoid bone has many foramina through which pass olfactory nerves that transmit sensory information from inhaled air to the olfactory bulbs.

17. What is a good definition of “internal respiration”?
    A. The exchange of gases between body tissues and capillary blood
    B. Ventilation of the lungs
    C. The production of ATP and CO_2 from small molecules using O_2
    D. The exchange of gases between the alveoli and pulmonary capillaries

Answer is A: Choice C refers to cellular respiration, choice D to external respiration.

18. Which statement may be used to define a bronchiole?
    A. They are the airways that branch from the left and right primary bronchi.
    B. They are kept open by “C” shaped rings of cartilage.
    C. Their walls have supporting cartilage between smooth muscle.
    D. Their walls contain smooth muscle but no cartilage.

Answer is D: Bronchioles differ from all the larger airways in having no supporting cartilage. They do have smooth muscle.

19. Which one statement below about the larynx is correct?
    A. It has walls lined with ciliated cells.
    B. It has walls made of cartilage.
    C. It has walls made of bone.
    D. It has walls made of muscle.

Answer is B: The larynx is a cartilaginous tube that surrounds the glottis and consists of the thyroid and cricoid cartilages.

20. How is the diaphragm innervated?
    A. By the parasympathetic division arising from the sacral region
    B. By the spinal nerves arising from T5 to T10
    C. By the phrenic nerves arising from C5 to C7
    D. By the phrenic nerves arising from C3 to C5

Answer is D: The two phrenic nerves innervate the diaphragm, and they arise from spinal nerves that leave the spinal cord between vertebrae C3 and C5.

21. The respiratory centre of the brain controls the respiratory muscles. Which of the following does **NOT** happen?
    A. An increase in carbon dioxide blood concentration stimulates peripheral chemoreceptors to signal the respiratory centre.
    B. A decrease in blood pH stimulates peripheral chemoreceptors to signal the respiratory centre.
    C. Hydronium ions cross the blood–brain barrier to directly stimulate the central chemoreceptors of the respiratory centre.
    D. A decrease in oxygen blood concentration stimulates peripheral chemoreceptors to signal the respiratory centre.

Answer is C: Hydronium ions cannot cross the BBB. Carbon dioxide does and forms hydronium ions which go on to stimulate the central chemoreceptors as the CSF has no buffer to remove them.

22. What does the term "external respiration" refer to?
    A. Ventilation of the lungs (breathing)
    B. Exchange of gases between alveolar air and lung capillaries
    C. The production by cells of ATP from small molecules and oxygen
    D. Exchange of dissolved gases between capillary blood and body tissues

Answer is B: When gas moves from the alveoli ("outside") to the blood in pulmonary capillaries ("inside"), this is known as external respiration.

23. What distinguishes bronchioles from the larger bronchi?
    A. Bronchioles have no cartilage in their walls.
    B. Bronchioles have smooth muscle in their walls.
    C. Bronchioles collapse between exhalation and inhalation.
    D. The alveoli open onto these air passages.

Answer is A: Bronchioles have no cartilage so may constrict and dilate. They do not collapse. Terminal bronchiole and respiratory bronchioles lie between the bronchioles and the alveoli.

24. Which structures are included in the respiratory membrane?
    A. Alveolar fluid and surfactant
    B. Alveolar fluid, surfactant and epithelial cells of alveoli
    C. Alveolar fluid, surfactant, epithelial cells of alveoli and basement membrane of epithelial cell
    D. Alveolar fluid, surfactant, epithelial cells of alveoli, basement membrane of epithelial cell and endothelial cell of capillary

Answer is D: All these structures form the thin (0.5 μm) respiratory membrane.

25. How is the diaphragm innervated?
    A. From the respiratory centre in concert with chemoreceptors that detect blood oxygen level

B. By the spinal nerves arising from thoracic vertebrae at the same level
C. By the phrenic nerve arising from vertebrae C3 to C5
D. By the vagus nerve arising from the medulla oblongata

Answer is C: The phrenic nerve (not thoracic spinal nerves) innervates the diaphragm.

26. What mechanism transports the largest portion of oxygen around the body?
   A. Oxygen is carried bound to plasma proteins.
   B. Oxygen is transported in solution dissolved in blood plasma.
   C. Oxygen is bound to haemoglobin within red blood cells.
   D. Oxygen is transported as bicarbonate after reacting with water to form carbonic acid.

Answer is C: Haemoglobin within RBC is the protein that binds to oxygen enabling its transport.

27. Which structures constitute the "upper respiratory tract"?
   A. Nose, pharynx and larynx
   B. Larynx, epiglottis and bronchi
   C. Trachea, bronchi and bronchioles
   D. Terminal bronchioles, alveoli and pleurae

Answer is A: The upper respiratory tract includes the larynx and superior structures.

28. The lists below include four respiratory tract structures. Which list has them in the order that inhaled air would pass through them on the way to the lungs?
   A. Glottis, pharynx, conchae, trachea
   B. Nares, pharynx, larynx, conchae
   C. Conchae, pharynx, larynx, trachea
   D. Pharynx, conchae, trachea, glottis

Answer is C: The conchae are adjacent to the external nares; the pharynx is superior to the larynx, and the larynx is at the entrance to the trachea.

29. What passes through the foramina of the cribriform plate of the ethmoid bone?
   A. Inhaled air on its way through the nose
   B. Tubes that drain the sinuses of the facial bones
   C. Nerve fibres associated with the sense of smell
   D. Blood vessels that supply the nasal mucosa

Answer is C: The olfactory bulbs rest upon the cribriform plate and nerves from the bulbs descend through it to the top of the nasal cavity.

30. What function is served by the goblet/mucus cells of the bronchial "tree"?
   A. They trap small inhaled particles.
   B. They secrete mucus onto the surface of the airways.

C. They increase the surface area available for gas exchange.
D. They move mucus up the bronchial tree.

Answer is B: Goblet cells produce mucus to trap any inhaled particles before they reach the alveoli.

31. The central chemoreceptors in the brain stem increase breathing rate in response to which stimulus?
    A. An increase in CO_2 concentration in the CSF
    B. An increase in CO_2 and H+ concentration in the CSF
    C. A decrease in O_2 concentration in the CSF
    D. A decrease in O_2 concentration in the blood

Answer is B: Carbon dioxide that enters the CSF will react with water to form hydronium ions and these ions stimulate the central chemoreceptors.

32. In what form is the majority of carbon dioxide that is generated by cellular respiration transported to the lungs?
    A. As dissolved carbon dioxide in the blood plasma
    B. Bound to haemoglobin in red blood cells
    C. As carbonic acid inside red blood cells
    D. As bicarbonate ions in the blood plasma

Answer is D: CO_2 leaves cell as dissolved gas. About 7% is transported in solution in plasma, 23% bound to haemoglobin ($HbCO_2$) in RBC, while 70% reacts with water to form carbonic acid, which forms bicarbonate ions.

33. A bronchiole differs from tertiary (and smaller) bronchi in that it:
    A. Has cartilage in its wall (and bronchi do not)
    B. Does not have cartilage in its wall (and bronchi do)
    C. Has smooth muscle in its wall (and bronchi do not)
    D. Does not have smooth muscle in its wall (and bronchi do)

Answer is B: While bronchi are held rigidly open by their cartilage, bronchioles do not have cartilage in their wall.

34. The term "cellular respiration" is applied to:
    A. Exchange of gases in the lungs
    B. Ventilation of the lungs (breathing)
    C. Exchange of gases in the body tissues
    D. The production of ATP in the cells

Answer is D: Cellular respiration is applied to the use of oxygen in mitochondria to metabolise glucose into ATP and carbon dioxide.

35. The respiratory centre in the brain is sensitive to:
    A. An increase in H+ concentration in the CSF
    B. A decrease in O_2 concentration in the blood

C. An increase in H+ concentration in the blood
D. A decrease in O_2 concentration in the CSF

Answer is A: The CSF is not buffered so is very sensitive to changes in hydronium ion concentration.

36. Trauma that severs the spinal cord between C6 and C7 will:
    A. Mean that the phrenic nerve has been severed
    B. Mean that artificial ventilation will be required to sustain life
    C. Cause the diaphragm to lose innervation but allow the intercostal muscles to operate
    D. Cause the intercostal muscles to lose innervation but allow the diaphragm to operate

Answer is D: The phrenic nerve that stimulates the diaphragm will not be severed as the spinal nerves that form it leave the spinal cord above C6. The intercostal muscles will be paralysed.

37. What happens when carbon dioxide levels in the blood decrease to below normal?
    A. pH of the blood decreases.
    B. The blood becomes more acidic.
    C. The concentration of hydrogen ions in the blood decreases.
    D. pH does not change.

Answer is C: Carbon dioxide reacts with water to form carbonic acid and thence hydronium ions. So a decrease in CO_2 will decrease the concentration of hydronium ions (and pH will increase).

38. In a healthy person, which of these lung volumes should be the largest?
    A. Tidal volume
    B. Vital capacity
    C. Expiratory reserve volume
    D. Residual volume

Answer is B: Vital capacity = expiratory reserve volume + tidal volume + inspiratory reserve capacity. So VC is the largest (RV is ~1.2 L).

39. What term refers to the exchange of gases between alveolar air and blood in the pulmonary capillaries?
    A. Inhalation
    B. Internal respiration
    C. Ventilation
    D. External respiration

Answer is D: This is external respiration as it is an exchange that occurs with the atmosphere outside the body.

40. What is a cavity in a skull bone that is lined with mucus membrane?
    A. Sinus
    B. Bronchiole
    C. Glottis
    D. Larynx

Answer is A: The frontal sinus, sphenoid and maxillary sinuses are in bones that surround the nasal cavity.

41. What is the function of ciliated cells in the lungs?
    A. They form part of the respiratory membrane
    B. To move mucus out of the bronchial tree
    C. To secrete surfactant onto the lining of the alveoli
    D. To phagocytose inhaled bacteria

Answer is B: Ciliated cells have cilia which beat in a coordinated fashion to move mucus out of the airways.

42. What ensures that the lungs expand as the chest wall expands?
    A. Secreted surfactant
    B. Negative pressure between the pleura
    C. Serous liquid secreted by the pleura
    D. The elastic recoil of the alveolar tissue

Answer is B: Negative pressure (suction) causes the visceral pleura to remain in contact with the parietal pleura, so that when the chest wall and parietal pleura moves, the parietal pleura takes the visceral pleura and lungs with it.

43. Which of the following is responsible for increasing respiratory activity under normal conditions?
    A. Decreased CO_2 level in blood
    B. Decreased O_2 level in blood
    C. Increased CO_2 level in blood
    D. Increased blood pH

Answer is C: If blood CO_2 is high, then more will cross the BBB into the CSF. This will decrease the pH of the CSF (because CO_2 will produce carbonic acid). This decreased pH will cause breathing rate to increase.

44. What does "internal respiration" refer to?
    A. Inhalation
    B. Exchange of gases between alveolar air and blood in the pulmonary capillaries
    C. Exchange of gases between capillary blood and interstitial fluid
    D. The production of ATP from organic molecules using oxygen

Answer is C: Internal respiration refers to the movement of oxygen from the blood into the interstitial fluid and CO_2 from interstitial fluid to blood. It also refers to the diffusion of these gases through the interstitial fluid to and from the cells.

45. Which structures are called bronchioles?
    A. Respiratory passageways that have cartilage in their walls
    B. The tubes that open from the left and right primary bronchi
    C. The tubes that enter a lobule
    D. The tubes that enter an alveolar sac

Answer is C: Terminal bronchioles supply air to a lobule. Respiratory bronchioles supply air to an alveolar sac. Secondary bronchi diverge from primary bronchi.

46. Which of the following is **NOT** part of the respiratory membrane?
    A. The basement membrane of alveolar epithelial cells
    B. The plasma membrane of red blood cells
    C. Capillary endothelial cells
    D. Alveolar fluid and surfactant

Answer is B: The plasma membrane of RBC can only be reached by dissolved gases after they have crossed the respiratory membrane into the capillary blood.

47. How is the diaphragm innervated?
    A. By the phrenic nerves arising from C3 to C5
    B. By the vagus nerves from the respiratory centre
    C. By cranial nerves arising from C5 to C7
    D. By nerves arising from the spinal cord from T5 to T9

Answer is A: Unusually the diaphragm is not innervated by spinal nerves from between vertebrae adjacent to the diaphragm, but from spinal nerves arising from C3 to C5.

48. Which molecule has the greatest effect in controlling lung ventilation?
    A. Oxygen in the blood
    B. Hydrogen ions in the blood
    C. Carbon dioxide in the blood
    D. Oxygen in the cerebrospinal fluid

Answer is C: It is carbon dioxide and not oxygen that stimulates lung ventilation. Respiratory centre of the brain responds to incoming signals from peripheral chemical receptors (when CO_2 ↑), by sending signals to respiratory muscles via nerves.

49. Which is the incorrect statement among the four below?
    A. More oxygen is carried bound to haemoglobin than dissolved in plasma.
    B. More carbon dioxide is carried bound to haemoglobin than dissolved in plasma.
    C. Haemoglobin buffers hydrogen ions derived from carbon dioxide.
    D. Carbonic anhydrase is the enzyme that binds oxygen to haemoglobin.

Answer is D: Oxygen does not need an enzyme to become bound to haemoglobin. About 23% of carbon dioxide is bound to haemoglobin while only about 7% is transported in solution in plasma.

50. Which structures comprise the lower respiratory tract?
    A. Pharynx, larynx, trachea
    B. Larynx, trachea, bronchi
    C. Nose, pharynx, larynx
    D. Trachea, bronchi, lungs

Answer is D: The lower respiratory tract begins below the pharynx.

51. A bronchiole differs from a bronchus in that it is:
    A. Unable to change its diameter (and a bronchus can)
    B. Able to change its diameter (and a bronchus cannot)
    C. A smaller diameter tube than is a bronchus
    D. A larger diameter tube than is a bronchus

Answer is B: Bronchiole can constrict and dilate while bronchi cannot (they are held open by cartilage). Bronchioles are more distal than bronchi in the respiratory "tree".

52. With regard to lung ventilation, what does "dead space" refer to?
    A. Air in the conducting zone of the bronchial tree
    B. Air remaining in the alveoli after an exhalation at maximal effort
    C. Air between the parietal and visceral pleura
    D. The difference between the volume of a maximum inhalation and the tidal volume

Answer is A: Dead space refers to air that is inhaled into the airways but does not reach the alveoli, hence is exhaled again without participating in gas exchange.

53. If the spinal cord is severed below C4, what will be the effect on breathing?
    A. Paraplegia will result, but breathing will not be affected.
    B. The intercostal muscles are able to effect ventilation, but the diaphragm will not work.
    C. The diaphragm is able to affect ventilation, but the intercostal muscles will not work.
    D. Independent breathing will not be possible.

Answer is C: The diaphragm is innervated by spinal nerves that leave the spinal cord between C3 and C5, so a break below C4 will leave some the nerves above C4 intact and some diaphragm movement is possible.

54. When is the respiratory centre of the brain stimulated to increase ventilation? When:
    A. Plasma CO_2 concentration increases
    B. Plasma pH increases
    C. Plasma O_2 concentration decreases
    D. Plasma HCO_3^- decreases

Answer is A: Concentration of carbon dioxide in plasma stimulates ventilation. The presence of more dissolved carbon dioxide would tend to decrease pH were it not for the buffers in blood.

55. In which form is the majority of CO_2 transported in the blood?
    A. As a dissolved solute
    B. Bound to plasma proteins
    C. As carbonic acid molecules
    D. As bicarbonate (HCO_3^-) ions

Answer is D: CO_2 leaves the cell as dissolved gas. About 7% is transported in solution in plasma, 23% bound to haemoglobin ($HbCO_2$) in RBC, while 70% reacts with water to form carbonic acid, which disassociates to form bicarbonate ions.

56. Which of the following lists the components of the respiratory membrane?
    A. Alveolar epithelial cells, capillary endothelial cells and their basement membranes
    B. Fluid and surfactant alveolar epithelial cells, capillary endothelial cells and their basement membranes
    C. Visceral pleura, parietal pleura and serous fluid
    D. Ciliated epithelial cells, mucus cells and secreted mucus

Answer is B: The epithelium of the alveoli and capillary (including the basement membrane) and the fluid covering the alveolar surface (which includes surfactant) together form the respiratory membrane through which gases must pass.

57. To which of the following does the term "respiration" **NOT** apply?
    A. The conversion of carbon dioxide to bicarbonate ions for transport to the lungs
    B. The exchange of gases between alveolar air and capillary blood
    C. The derivation of energy from organic molecules in the cells
    D. The exchange of gases between capillary blood and body tissues

Answer is A: Carbon dioxide reacts with water to form carbonic acid which disassociates to form bicarbonate ions. But this is not referred to a respiration.

58. The "lower respiratory tract" includes all those structures below which of the following?
    A. The internal nares
    B. Larynx
    C. Trachea
    D. Conducting zone

Answer is B: Below the larynx. That is, the LRT includes the trachea, the conducting airways and the alveoli.

59. What are the cells that produce surfactant called?
    A. Mucus cells
    B. Ciliated cells
    C. Alveolar macrophages
    D. Type II pneumocytes

Answer is D: Type II pneumocytes produce the surface active agent (surfactant) that reduces the surface tension of the water that lines the alveoli.

60. In red blood cells, carbonic anhydrase catalyses the formation of carbonic acid which then disassociates into bicarbonate ions and hydrogen ions. What happens next?
    A. Bicarbonate ions bind to haemoglobin.
    B. Hydrogen ions move into the plasma to be buffered by bicarbonate ions.
    C. Hydrogen ions are buffered by haemoglobin.
    D. Chloride ions enter the RBC to form hydrochloric acid (HCl).

Answer is C: Haemoglobin is a protein so can act as a buffer, thus removing the hydrogen ions from solution. The bicarbonate ions move out of the RBC into the plasma in exchange for chloride ions.

61. Which of the following statements is correct?
    A. Tidal volume is maximum volume that can be inhaled and exhaled.
    B. FEV1 is the maximum volume of air that can be forcefully exhaled in 1 s.
    C. Expiratory reserve volume is the maximum volume of air that can be exhaled after a deep inhalation.
    D. Vital capacity is the expiratory reserve volume added to the inspiratory reserve volume.

Answer is B: Forced expiratory volume in 1 s is a measure of the degree of pulmonary obstruction. Choice C refers to vital capacity. VC = ERV + TV + IRV.

62. What happens when carbon dioxide levels in the blood increase?
    A. pH of the blood increases.
    B. The blood becomes more alkaline.
    C. The number of hydrogen ions in the blood decreases.
    D. The blood becomes more acidic.

Answer is D: Choice D is the best answer (the other three are wrong). However, the buffer systems in the blood would prevent all but a slight change in pH.

63. Chemoreceptors in the medulla oblongata are sensitive to:
    A. Increases in blood oxygen content
    B. Increases in blood carbon dioxide
    C. Increases in blood pH
    D. Both choices A and B

Answer is D: Peripheral chemical receptors send signals to respiratory centres of the brain located in medulla oblongata when blood carbon dioxide rises or blood oxygen concentration falls below 60 mmHg.

64. What does sympathetic nervous system stimulation to the smooth muscle layers in the bronchioles cause?
    A. Bronchoconstriction
    B. Bronchodilation
    C. An increase in tidal volume
    D. Increase in activity of the cilia

Answer is B: Bronchodilation is stimulated by sympathetic nerve impulses, adrenalin and noradrenalin. Bronchoconstriction is triggered by parasympathetic nerve impulses.

65. Patients with diabetes mellitus who neglect insulin therapy rapidly metabolise lipids, and there may be an accumulation of the acidic by-products of lipid metabolism in the blood. What effect would this have on respiration?
    A. Increase in respiratory rate
    B. Decrease in respiratory rate
    C. Decrease in respiratory rate if oxygen is reduced
    D. No influence on respiratory rate

Answer is A: Respiratory rate increases if blood pH decreases in order to breathe out carbon dioxide (and hence reduce acid). In uncontrolled diabetics, this is known a Kussmaul breathing.

66. Which of these structures has no cartilage in it?
    A. Epiglottis
    B. Trachea
    C. Bronchi
    D. Alveoli

Answer is D: Alveoli are formed by a single layer of squamous epithelium. There is no cartilage.

67. Where does the actual gas exchange between inspired air and the blood in the capillaries occur?
    A. Bronchi
    B. Bronchioles
    C. Alveolar ducts and alveoli
    D. Respiratory bronchioles

Answer is C: Gas exchange occurs across the respiratory membrane of the alveoli and ducts.

68. Which of these statements concerning ventilation is **NOT** correct?
    A. During inspiration, the pressure in the alveoli is less than atmospheric pressure.
    B. Contraction of the neck muscles decreases the volume of the thoracic cavity.
    C. When the diaphragm contracts, thoracic cavity volume increases.
    D. During quiet breathing, passive recoil of the lung and thoracic wall cause expiration.

Answer is B: Deep inspiration is helped by contracting the neck muscles (scalene, sternocleidomastoid), but this increases, not decreases lung volume.

69. With regard to the respiratory centre, which of the following is TRUE?
    A. Blood oxygen concentration affects the respiratory centre.
    B. Anaesthetics do not affect respiration.
    C. Raised intracranial pressure increases ventilation.
    D. Narcotic drugs may depress ventilation.

Answer is D: Anaesthesia or narcotics (e.g. morphine) can depress respiratory centres (which decreases ventilation). Raised intracranial pressure can also depress respiration. It is CO_2 concentration rather than O_2 concentration that affects the respiratory centre of the brain.

70. With respect to adrenaline and noradrenaline, which of the following is true?
    A. They relax smooth muscle resulting in dilation of the airways.
    B. They contract smooth muscle resulting in constriction of the airways.
    C. They relax smooth muscle resulting in constriction of the airways.
    D. They have no effect on the bronchioles.

Answer is A: Adrenaline and noradrenaline cause bronchioles to dilate which allows for greater ventilation.

71. In the control of respiration, which of the following is **NOT** true?
    A. Peripheral chemoreceptors respond to changes in oxygen and carbon dioxide concentration in the blood.
    B. Central chemoreceptors respond to changes in carbon dioxide concentration in the blood.
    C. Respiration responds to smaller changes in the blood concentration of oxygen than carbon dioxide.
    D. Central chemoreceptors are sensitive to changes in the pH of the cerebrospinal fluid.

Answer is B: Chemoreceptors in the central nervous system respond to changes in pH as there is no buffer in CSF. The presence of carbon dioxide in CSF generates hydronium ions which affect the central chemoreceptors.

72. Gas exchange takes place in the:
    A. Larynx
    B. Bronchioles
    C. Alveoli
    D. Pleura

Answer is C: The alveoli are part of the respiratory membrane.

73. The walls of the trachea are held open by which of the following?
    A. Nerve impulses
    B. Rings of cartilage
    C. Fine bones
    D. Smooth muscle contractions

Answer is B: Bands of cartilage shaped like a "C" keep the trachea rigid and open.

74. Oxygen and carbon dioxide cross the respiratory membrane by the process of:
    A. Counter-current exchange
    B. Diffusion
    C. Active transport
    D. Oxygen–carbon dioxide pump

Answer is B: The gases move by diffusion along their concentration gradients.

75. Which of the following is **NOT TRUE**?
    A. Raised intracranial pressure may depress respiration.
    B. Anaesthetics never affect respiration.
    C. Receptors in the airways trigger sneezing.
    D. Narcotic drugs may suppress respiration.

Answer is B: In fact anaesthetics do affect respiration and that is why artificial ventilation is necessary during a surgical procedure under general anaesthetic.

76. The greatest amount of air which can be moved in and out of the respiratory system is the:
    A. Tidal volume
    B. Vital capacity
    C. Ventilatory volume
    D. Pulmonary capacity

Answer is B: Vital capacity includes inspiratory reserve volume, tidal volume and expiratory reserve volume. All can be measured separately by our conscious effort.

77. Normal expiration in a person at rest is due to:
    A. Elastic tissue in the lung
    B. Contraction of abdominal muscles
    C. Contraction of the expiratory muscles

D. Diffusion

Answer is A: Expiration is accomplished by passive elastic recoil of the expanded lung.

78. A molecule that is important in maintaining normal lung structure is:
    A. Immunoglobulin
    B. Haemoglobin
    C. Peroxidase
    D. Surfactant

Answer is D: By reducing the surface tension of water, surfactant allows the alveoli to be expanded more easily and prevents their collapse.

79. The volume of air which moves in and out of the lungs during a normal quiet respiratory cycle is called the:
    A. Tidal volume
    B. Vital capacity
    C. Ventilatory volume
    D. Pulmonary capacity

Answer is A: Tidal volume is the volume of air that is inhaled and exhaled while "at rest".

80. What are the main muscles involved in normal inspiration?
    A. Muscles of the neck
    B. Abdominal muscles
    C. Intercostal muscles
    D. Intercostals and the diaphragm

Answer is D: Normal inspiration is achieved by contraction of the external intercostal muscles and the diaphragm. The neck and abdominal muscles are used for deep inspiration and forced exhalation, respectively.

81. What is the number of breaths per minute called?
    A. Respiratory rate
    B. Respiratory speed
    C. Pulmonary index
    D. Respiratory volume

Answer is A: Respiratory rate is stated in breaths per minute.

82. A man runs up a flight of stairs. His respiratory rate rose from 18 breaths per minute (bpm) (resting) to 36 bpm (at the top of stairs). The increase was probably a result of:
    A. Increased blood pH
    B. Increased concentration of CO_2 in the blood

C. Decreased concentration of O_2 in the blood
D. Decreased concentration of CO_2 in the blood

Answer is B: Muscular activity would have increased the rate of oxygen consumption and carbon dioxide production by cells. A higher breathing rate is required to take in the needed oxygen and to excrete the produced carbon dioxide. It is increased CO_2 rather than decreased O_2 that stimulates respiratory rate.

83. Why is oxygen therapy—allowing a patient to breathe in an atmosphere in which the proportion of oxygen is greater than 20%—beneficial?
   A. The partial pressure of CO_2 in the inhaled air is decreased, making it easier to clear CO_2 from the lungs.
   B. The partial pressure of oxygen in the lungs is increased, allowing more oxygen to dissolve in the alveolar fluid.
   C. The oxygen molecule is smaller than the nitrogen molecule, so a greater number of moles of air can be drawn into the lungs with each breath.
   D. The patient is required to inhale less frequently, and this reduces the strain on the respiratory system.

Answer is B: Breathing an atmosphere where oxygen is at greater than 20% concentration (by volume) will mean more oxygen will cross the respiratory membrane to dissolve in plasma. In many cases this will be of benefit to the patient.

84. What are the membranes that surround each lung called?
   A. Parietal and visceral membranes
   B. Parietal and visceral meninges
   C. Pleura
   D. Peritoneum

Answer is C: The pleura surround each lung. There is an outer (parietal) layer and an inner layer, the visceral pleura.

85. Which membrane surrounds the lungs?
   A. The pericardium
   B. The pleura
   C. The mediastinum
   D. The diaphragm

Answer is B: The pleura surround each lung. There is an outer (parietal) layer and an inner layer, the visceral pleura.

86. Which of the following buffer systems of the body is affected by the action of the lungs?
   A. Protein
   B. Monohydrogen phosphate/dihydrogen phosphate
   C. Ammonia/ammonium

D. Carbonic acid/bicarbonate

Answer is D: Exhaled CO_2 comes from the disassociation of the carbonic acid molecule which in turn forms when a hydronium ion (acid) combines with a bicarbonate ion. Hence, exhaling CO_2 reduces the amount of acid in the blood.

87. When a peak flow meter is used to measure maximum expiratory flow rate, why is the measurement repeated many times?
    A. In order to determine the average value.
    B. Many values are required to draw a graph of expiratory flow behaviour.
    C. We want to see how the value drops as the expiratory muscles become fatigued.
    D. The highest number, the "peak flow", is desired.

Answer is D: Maximum speed of expired air flow measures the airflow through the bronchi and thus the degree of obstruction in the airways. Peak flow readings are higher when patients are well, and lower when the airways are constricted. The highest value contains the required information while values taken with less than maximum effort do not provide any reliable information.

88. Which of the following terms refers to normal, unlaboured breathing?
    A. Eupnoea
    B. Dyspnoea
    C. Bradypnoea
    D. Tachypnoea

Answer is A: Dyspnoea is laboured, difficult breathing, bradypnoea is abnormally slow breathing (<12 bpm for ages 12+), tachypnoea is abnormally fast breathing (>20 bpm at rest for an adult).

89. What are the names of the air passages in the nose?
    A. Conchae
    B. Meatus
    C. Sinuses
    D. Glottis

Answer is B: The three nasal meatus are air passages. They are surrounded by the fleshy conchae. Sinuses are air spaces in bone, while the glottis is at the top of the trachea rather than in the nose.

90. What is the function of the goblet cells that make up part of the lining of the respiratory tree?
    A. To engulf ("eat") inhaled infections organisms
    B. To move mucus and trapped particles out to the respiratory tree
    C. To secrete mucus
    D. To secrete surfactant

Answer is C: Goblet cells secrete mucus. This traps small inhaled particles which are then moved out of the respiratory tree by the action of the ciliated cells (which also line the respiratory tree). Type II pneumocytes secrete surfactant, macrophages eat bacteria, etc.

91. What is the "pleura"?
    A. The notch in the left lung that accommodates the heart
    B. The respiratory membrane
    C. The point where the primary bronchus and arteries and veins enter the lung
    D. The membrane that surrounds each lung

Answer is D: The two-layered pleura surround the lung. The visceral pleura is attached to the lung while the parietal pleura is attached to the inside of the body wall. Choice C is the hilum, choice A is the cardiac notch.

92. Given that the lung contains a residual air volume of ~1.2 L, has an expiratory reserve volume of ~1.2 L and the dead space is about 150 mL, while resting tidal volume is about 500 mL, approximately what percentage of the volume of air in the lung is turned over during one normal tidal inhalation at rest?
    A. 5%
    B. 15%
    C. 60%
    D. 90%

Answer is B: Some people may remember the value. Or by calculation, lung volume at end of a tidal exhalation = RV + ERV = 1.2 L + 1.2 L = 2.4 L. The next inhalation brings in (TV – dead space) mL of fresh air, that is: (500 – 150) mL = 350 mL. 350 mL is 14.6% of 2400 mL (or 12.7% of 2750 mL if the 350 mL inhalation is added to the 2.4 L already present). Hence choice B is the closest.

93. During inhalation, air passes through the four listed structures. Which list has them in the correct order?
    A. Choanae, meatus, pharynx, larynx
    B. Pharynx, larynx, meatus, choanae
    C. Meatus, choanae, pharynx, larynx
    D. Larynx, pharynx, meatus, choanae

Answer is C: The meatus are the air passages behind the nostrils (external nares), the choanae (internal nares) are at the start of the pharynx, while the larynx starts at the epiglottis.

94. The respiratory system helps the body regulate the pH of blood. How is this done?
    A. The alveolar epithelial cells secrete the monohydrogen/dihydrogen phosphate buffer which is able to maintain blood pH.

B. The CO_2 exhaled causes carbonic acid in the blood to disassociate into water and CO_2 which in turn allows bicarbonate to react with hydronium ions to form more carbonic acid.
C. The alveolar epithelial cells produce angiotensin-converting enzyme which promotes the formation of the bicarbonate buffer in the blood.
D. The inhaled O_2 causes dihydrogen phosphate to convert to monohydrogen phosphate which is able to destroy acid.

Answer is B: $H_3O^+ + HCO_3^- \leftrightarrow H_2CO_3 + H_2O \leftrightarrow 2H_2O + CO_2$ so when carbon dioxide is breathed out, the reaction "progresses to the right". Hydronium ions react to form carbonic acid which in turn dissociates into water and carbon dioxide, thus removing acid. The other choices are nonsense.

95. What mechanism allows the lungs to expand along with the chest wall when it is moved outwards during an inhalation?
A. The negative pressure between the visceral and parietal pleura
B. The surfactant produced by the septal cells
C. The movement of the cilia of the ciliated cells that line the bronchial "tree"
D. The contraction of the internal intercostal muscles

Answer is A: Negative pressure ensures that the visceral pleura adheres to the parietal pleura when it is moved outwards by the action of the intercostal muscles. As the visceral pleura is attached to the lungs, so the lungs are expanded which allows atmospheric pressure to push air into the alveoli.

96. Given that the partial pressures of oxygen and carbon dioxide in blood entering the alveolar capillaries are $pO_2 = 40$ mmHg and $pCO_2 = 46$ mmHg, respectively, and that the air in the alveoli contains $pO_2 = 104$ mmHg and $pCO_2 = 40$ mmHg: which of the following will happen?
A. O_2 will diffuse from alveoli into blood and CO_2 will diffuse into alveoli from blood.
B. Both O_2 and CO_2 will diffuse from alveoli into blood.
C. Both O_2 and CO_2 will diffuse from blood into alveoli.
D. CO_2 will diffuse from alveoli into blood and O_2 will diffuse into alveoli from blood.

Answer is A: Oxygen will diffuse from where it is in high concentration (104 mmHg in alveolar fluid) to where it is in lower concentration (40 mmHg in venous blood). CO_2 will diffuse from blood (where it is at 46 mmHg) into the alveolar fluid (where it is at 40 mmHg). From the alveolar fluid, CO_2 will vaporise into the alveoli for exhalation.

97. When discussing lung volumes, the term "anatomical dead space" is used. What does this term refer to?
A. The difference between the volume of an inhalation during tidal breathing and a maximal inhalation

B. The air that is in the bronchial tree at the end of an exhalation just before an inhalation begins
C. The air still remaining in the lungs after a maximal exhalation
D. The inhaled air in the bronchial tree that has not yet reached the alveoli before an exhalation begins

Answer is D: This inhaled air never reaches the alveoli, it is exhaled along with "stale" air behind it that is expelled from the alveoli. Choice C refers to residual volume. Choice A refers to inspiratory reserve volume.

98. How is oxygen transported around the body?
A. Attached to haemoglobin in red blood cells
B. Attached to haemoglobin in red blood cells and dissolved in plasma
C. Attached to haemoglobin in red blood cells, dissolved in plasma and as bicarbonate ions
D. Bound to haemoglobin as $HbCO_2$ in addition to the ways stated in choice "C" above

Answer is B: Most oxygen is carried bound to haemoglobin and some is dissolved in plasma, so B is a better choice than A. Bicarbonate ions and $HbCO_2$ are not involved in oxygen transport.

99. Gas exchange occurs across the respiratory membrane. Which structures constitute the membrane?
A. Capillary endothelial cells, alveolar epithelial cells and the hilum
B. Alveolar epithelial cells, surfactant and the hilum
C. Visceral and parietal pleura and pleural fluid
D. Alveolar epithelial cells, capillary endothelial cells and the basement membrane

Answer is D: Gas must pass through two cells and their basement membrane to move from alveoli into blood and vice versa. The hilum and pleura are not involved.

100. To which structures are the lung's alveoli attached?
A. Alveolar ducts
B. Terminal bronchioles
C. Respiratory bronchioles
D. Alveolar sacs

Answer is D: Terminal bronchioles end in respiratory bronchioles, which branch into alveolar ducts from which open alveolar sacs. These alveolar sacs are lined with alveoli.

101. When performing cardiopulmonary resuscitation ("mouth to mouth"), what air enters the recipient from the resuscitator?
A. Stale air with some residual oxygen from the resuscitators lungs
B. Fresh air from the resuscitators lungs

C. Stale air from the lungs and fresh air from the anatomical dead space
D. Fresh air and stale air from the anatomical dead space

Answer is C: After an inhalation, the anatomical dead space (~150 mL air in the trachea and bronchi that has not reached the alveoli) contains air from the atmosphere, albeit now saturated with water vapour that has not had its oxygen content depleted by gas exchange in the alveoli, hence is "fresh". This air, plus some "stale" air from the alveoli, but with some residual oxygen, is exhaled into the recipient's lungs.

102. What does the term "lung compliance" refer to?
A. The ease with which the lungs can be expanded
B. The ease with which the volume of air in the lungs can be exhaled
C. The resistance to airflow presented by the airways
D. The total air volume that can be forcefully exhaled after a maximum effort inhalation

Answer is A: Lung compliance is the ease with which lungs can be expanded. The greater the compliance, the easier it is to fill the lungs. An inadequate level of surfactant will reduce compliance. Choice D refers to vital capacity.

103. What does the term "respiratory distress syndrome" refer to?
A. The alveolar damage resulting from emphysema
B. The condition that makes it hard to exhale all the air in the lungs
C. A condition resulting from insufficient surfactant in the alveoli
D. The condition that makes it difficult to fully expand the lungs

Answer is C: Lung surfactant is a "detergent" which reduces surface tension of alveolar fluid. It allows the alveolus to expand more easily against the tendency of water's surface tension to contract alveoli. Its absence in premature babies leads to "respiratory distress syndrome". Choice D, while not incorrect, is not as good an answer.

104. How does the construction of the walls of bronchi and bronchioles differ?
A. Bronchioles have cartilage in their walls while bronchi do not.
B. Bronchi have cartilage, mucosa and smooth muscle while bronchioles have smooth muscle and ciliated cuboidal epithelium.
C. The walls of bronchioles are thinner than the walls of bronchi.
D. Bronchi have only cartilage in their wall while bronchioles have only smooth muscle.

Answer is B: The salient point is that as well as mucosa and smooth muscle, bronchi have cartilage which functions to keep their internal diameter constant. Bronchioles, with smooth muscle but without cartilage, may dilate and constrict their diameter to alter airflow.

105. What can cause bronchodilation and what can cause bronchoconstriction?
    A. Sympathetic nerve impulses and parasympathetic impulses, respectively
    B. Sympathetic nerve impulses and adrenalin, respectively
    C. Parasympathetic nerve impulses and sympathetic impulses, respectively
    D. Parasympathetic nerve impulses and histamine, respectively

Answer is A: Bronchodilation is stimulated by sympathetic nerve impulses (and adrenalin and noradrenalin) while bronchoconstriction is triggered by parasympathetic nerve impulses, histamine and airborne irritants such as smoke and dust.

106. Why does hyperventilation (= rapid shallow breaths) result in an increase in dissolved CO_2 in the blood?
    A. Rapid breathing does not allow enough time for alveolar gas to exchange with blood gases.
    B. Hyperventilation requires extra muscular effort which generates more CO_2 which increases the blood concentration.
    C. Stale air in the airway dead space re-enters the alveoli with each breath. Shallow breathing means very little fresh air can enter.
    D. A shallow breath inhales insufficient air volume to flush out the alveolar air, so CO_2 is not expelled.

Answer is C: Rapid shallow breathing will not ventilate the lungs adequately as a large proportion of each intake of breath will be the 500 mL of stale exhaled air that remains in the airways and re-enters the alveoli—shallow breaths do not have sufficient volume to refresh the alveolar volume with fresh air. Hence CO_2 in the blood will not be "blown off". CO_2 will accumulate in the blood, and decrease the blood pH.

107. What do the chemoreceptors in the respiratory centre of CNS respond to?
    A. To a decrease in oxygen and an increase in carbon dioxide molecules
    B. To a decrease in oxygen molecules
    C. To an increase in carbon dioxide molecules
    D. To an increase in Hydronium ions (H_3O^+)

Answer is D: H+ cannot cross blood–brain barrier, but CO_2 does. Dissolved CO_2 in cerebrospinal fluid (CSF) produces H_3O^+. The central chemoreceptors are very responsive to change in pH as there is no buffer in CSF. Thus CO_2 (via H+) is the main controller of ventilation.

108. How is the composition of alveolar air different from atmospheric air?
    A. Alveolar air has a higher concentration of nitrogen than atmospheric air.
    B. Alveolar air has a higher proportion of oxygen than atmospheric air.
    C. Alveolar air has a lower proportion of carbon dioxide that atmospheric air.
    D. Alveolar air is saturated with water vapour but atmospheric air is not.

Answer is D: Because alveolar air is saturated with water vapour its composition is different from that of the atmosphere. Furthermore, because of the gas exchange that occurs in the lungs, alveolar air is continually being depleted of oxygen and enriched with carbon dioxide.

109. What chemical change occurs in the RBC that facilitates carbon dioxide transport?
   A. H^+ is buffered by combining with haemoglobin.
   B. HCO_3^- moves out of RBC into plasma in exchange for Cl^-.
   C. Dissolved CO_2 diffuses into RBC where carbonic anhydrase converts CO_2 to H_2CO_3.
   D. CO_2 combines directly with haemoglobin to form a carbaminohaemoglobin compound.

Answer is C: The enzymatic change of carbon dioxide to carbonic acid is the major chemical change. Then carbonic acid disassociates. Then the bicarbonate moves out of the RBC, and the hydronium is buffered by haemoglobin.

110. What is the function of the mucous glands and ciliated epithelial cells that line the airway tubes of the respiratory system?
   A. To assist in keeping the airways from collapsing which would restrict airflow.
   B. To trap inhaled particles and move them out of the airways.
   C. To protect the airways against inhaled pathogens.
   D. To reduce the friction between inhaled air and the lung airways.

Answer is B: Mucous glands secrete mucous which traps inhaled solid particles. Epithelial cells are ciliated and covered by the produced mucous. The cilia beat in order to move the mucous along with trapped particles up the walls of airways towards the glottis to be swallowed.

111. What is the difference between a bronchus and a bronchiole?
   A. A bronchiole has a smaller diameter than a bronchus.
   B. There are more bronchioles than there are bronchi.
   C. A bronchiole has no cartilage in its wall while a bronchus does.
   D. Inhaled air passes through a bronchus prior to passing through a bronchiole.

Answer is C: While all of the four choices are true, the best physiological answer is that bronchioles have no cartilage in their walls. This means they are not held open by the stiff cartilage. They do have smooth muscle in their wall which means their diameter may dilate or decrease as the muscle relaxes or contracts.

112. What is meant by each of the terms: apnoea, dyspnoea, eupnoea and tachypnoea (respectively)?
   A. Rapid breathing rate, normal breathing, difficulty breathing, absence of breathing

B. Normal breathing, absence of breathing, rapid breathing rate, difficulty breathing
C. Difficulty breathing, rapid breathing rate, absence of breathing, normal breathing
D. Absence of breathing, difficulty breathing, normal breathing, rapid breathing rate

Answer is D: Only this choice lists the definitions in the correct order.

113. Why are people encouraged to take deep breaths rather than short shallow ones, to relieve anxiety or panic?
   A. The proportion of air in each inhalation that reaches the alveoli is greater for deep breaths than for shallow ones. Hence more oxygen is available to be dissolved in blood.
   B. Tidal volume is greater for shallow breaths than for deep breaths. Hence more fresh air is available during deep breaths.
   C. Fewer deeper breaths involve less muscular effort than rapid shallow breaths which relieves anxiety.
   D. Taking deep breathes shifts the mental focus to controlling the skeletal muscles that perform breathing, and this distracts the person from whatever is causing them to be anxious.

Answer is A: Short shallow breaths contain a small volume of air which may be only slightly more than the anatomical "dead space"—the volume of inhaled air that fills the airways that lead to the alveoli. Hence short shallow breaths do not ventilate the alveoli with much fresh air. Deep breaths have sufficient volume to not only fill the dead space but also to ensure abundant fresh air ventilates the alveoli. This delivery of oxygen to the alveoli allows gas exchange with blood in pulmonary capillaries and often promotes clear thinking and alleviates feelings of anxiety. Choice B is wrong. Tidal volume consists of dead space plus air entering the alveoli.

114. What would be the effect on breathing of a spinal cord injury at the level of the sixth cervical vertebra?
   A. Both the diaphragm and the intercostal muscles would not be innervated, so the person will NOT be able to breathe.
   B. The diaphragm and the intercostal muscles would still be innervated, so the person WILL be able to breathe.
   C. The diaphragm would still be innervated, but the intercostal muscles would not, so the person WILL be able to breathe.
   D. The most superior intercostal muscles would still be innervated, but the diaphragm would not, so the person WILL be able to breathe.

Answer is C: If the spinal cord is severed below the sixth cervical vertebra, the spinal nerves inferior to that will not work and the person will be paraplegic. Hence the intercostal muscles will not be able to be stimulated. However, the diaphragm is innervated by the phrenic nerve which arises from spinal nerves at C3 to C5. As the injury is below C5, the phrenic nerve will still work, the diaphragm will still be innervated and breathing will be possible.

115. What would be the effect of a pneumothorax (air in the intrapleural space) on the lungs and breathing?

   A. Breathing would be painful until the lung spontaneously reinflates.
   B. The affected lung would collapse or not be able to expand fully, so lung ventilation would decrease.
   C. The affected person would experience respiratory alkalosis.
   D. The affected person would experience hypercapnia (abnormally high level of CO_2 in the blood).

Answer is B: Choice A is wrong as spontaneous reinflation may not occur. Respiratory **acidosis** is the more likely condition. Choice D may occur in severe cases.

116. What is FEV_1 and what is it used for?

   A. The maximum volume of air that can be exhaled in 1 s. To compare with predicted value for healthy lung function.
   B. The maximum volume of air that can be exhaled at the first attempt. To compare with predicted value for healthy lung function.
   C. The amount of air that remains in your lungs after a maximum exhalation. To determine total lung capacity.
   D. The maximum velocity (speed) with which air in the lungs can be exhaled. To compare with healthy age and gender-based averages.

Answer is A: Choice B is wrong—it usually requires several attempts to achieve maximum volume. Choice C describes residual volume. Choice D is PEF.

117. Given the terms used in spirometry: TV = tidal volume; IRV = inspiratory reserve volume; ERV = expiratory reserve volume; VC = vital capacity; RV = residual volume; TLC = total lung capacity. Which of the following equations is correct?

   A. TLC = VC + IRV
   B. VC = ERV + IRV + RV
   C. TLC = RV + ERV + IRV
   D. VC = ERV + TV + IRV

Answer is D: Both choices for TLC are incorrect (TLC = VC + RV), while RV is not included in VC.

118. When a person is aware that their breathing is being monitored, why is their resting respiratory rate difficult to measure?

   A. Breathing is effected by skeletal muscles, which are under our conscious control. Hence we can alter breathing rate when we know we are being watched.
   B. At rest, the amount of oxygen we consume is minimal. Hence tidal volume during inspiration and expiration is low which makes chest movements difficult to discern.

C. Breathing is controlled by the somatic division of our nervous system. This is under our conscious control hence is controllable by our will.
D. The respiratory centres are automatically stimulated by dissolved CO_2 and blood pH. However, input from the cerebral cortex can affect respiration and occurs when we are conscious of being observed.

Answer is D: Choice C is wrong. Breathing is under the control of the autonomic division of the nervous system, specifically the respiratory centres in the medulla oblongata. Choice B is true but is not an explanation. While skeletal muscles do cause chest expansion during breathing, they are controlled by autonomic centres in the medulla oblongata, rather than consciously. In turn, these autonomic centres can also receive input from our conscious cerebral cortex which happens when we know someone is observing us. So our resting breathing rhythm will be disrupted by our thoughts which will modify our resting respiratory rate.

119. Which answer is the **WORST** at explaining why respiratory rate increases when we perform exercise?
    A. To deliver more oxygen to the active muscles
    B. To allow more oxygen to be inhaled to be dissolved in the greater flow of blood through the lungs
    C. Exercising muscle produce more CO_2 which dissolves in blood. The increased concentration of dissolved CO_2 stimulates an increase in breathing rate
    D. Exercise causes enhanced venous return, which in turn increases heart rate which causes more blood to be pumped to the lungs. Respiratory rate increases in order to oxygenate the greater volume of blood moving past the alveoli

Answer is B: Choices A, C and D are certainly reasons why respiratory rate increases. Choice B, while not totally wrong, is the worst at explaining the increase in respiratory rate.

120. What is considered to be a healthy resting adult respiratory rate?
    A. 8–20 breaths/min
    B. 12–18 breaths/min
    C. 10–20 breaths/min
    D. 15–25 breaths/min

Answer is B: The normal respiration rate for an adult at rest is 12–18 bpm (sometimes you see 12–20). A respiration rate under 12 or over 25 bpm while resting is considered abnormal.

13.2 Pressure Applied to the Respiratory System

A decreasing pressure is created within the lungs as the diaphragm contracts (and moves caudally) and the chest wall moves outwards (see Boyle's law). This pressure difference allows atmospheric pressure to push air into the lungs. Since (by using the ideal gas law) air molecules are travelling at more than 500 m/s (at 28 °C), it does not take long for the air molecules to enter the lungs. When the diaphragm and external intercostal muscles relax, the chest and lungs recoil to a smaller volume which increases the pressure of the air within the alveoli (see Boyle's law). This provides the pressure gradient to push the now CO_2-rich and O_2-poor air out of the lungs.

At the end of an inhalation, the air pressure in the alveoli is about 101 kPa (i.e. equal to atmospheric pressure). This is made up of about 75kPa N_2, 14 kPa O_2, 5.3 kPa CO_2, 6.2 kPa H_2O. These values add to 101 kPa, and each one is called the partial pressure of that gas. The concentration of dissolved gases in the alveolar fluid is characterised by these same numbers (and units). The solution concentration of these same gases in venous blood entering the alveolar capillaries is about 75 kPa N_2, 5.3 kPa O_2, 6.1 kPa CO_2, 6.2 kPa H_2O (adding to ~93 kPa). Hence O_2 will diffuse down its concentration gradient from alveolar fluid (where oxygen is at partial pressure 14 kPa) across the respiratory membrane into the capillary blood (6.1 kPa). Similarly, CO_2 will diffuse down its concentration gradient from the venous blood through the respiratory membrane into the alveoli. Note that 1 kPa = 7.52 mmHg.

Each lung is surrounded and enclosed by a two-layer membrane called the pleura. The parietal pleura is attached to the inside of the chest wall, while the visceral pleura is attached to the surface of the lungs. The "space" between the two layers of the pleura is occupied by a few millilitres of lubricating pleural fluid which is at a negative pressure (i.e. a pressure less than atmospheric pressure) which is always less than the pressure within the alveoli. This ensures that as the chest expands during inhalation, the lungs are pushed outwards by the air pressure within them and inflate. In effect the two layers of pleura are pressed together.

1. If the partial pressure of oxygen in alveolar air is 100 mmHg and the concentration of dissolved oxygen in capillary blood arriving at the alveoli is 40 mmHg, what will be the concentration of dissolved oxygen in blood leaving the alveoli?

   A. 140 mmHg
   B. 100 mmHg
   C. 70 mmHg
   D. 60 mmHg

Answer is B: Oxygen in alveolar air will diffuse down its concentration gradient into capillary blood until blood in the capillary is also at 100 mmHg. After this there is no concentration gradient to effect a net diffusion.

2. What does it mean if oxygenated blood leaving the alveolar capillaries has an oxygen partial pressure of 100 mmHg?
   A. Oxygen will have diffused from the blood into the alveoli
   B. The partial pressure of oxygen in the alveoli was 100 mmHg
   C. The partial pressure of oxygen in the alveoli was less than 100 mmHg
   D. The partial pressure of oxygen in the alveoli was more than 100 mmHg

Answer is B: A PP of 100 mmHg is a way of expressing the concentration of oxygen that is dissolved in blood. Oxygen will continue to move into blood from the alveoli as long as the oxygen concentration in blood is less than that in the alveoli. If ppO_2 in the alveoli is 100 mmHg, then the ppO_2 in the blood will continue to increase until it is equal to that in the alveoli.

3. Consider Henry's law. If a patient is given air with 30% oxygen to breathe (i.e. at one and a half times normal oxygen partial pressure), by how much would the amount of dissolved oxygen in the blood change?
   A. It would increase by 150%.
   B. It would increase by 50%.
   C. It would increase by 30%.
   D. It would increase by 10%.

Answer is B: Henry's law states that the amount of gas that will dissolve in water is proportional to its partial pressure. If its PP increases by 50% (i.e. by 1.5 times), then 50% more gas will dissolve.

4. When the diaphragm contracts which of the following will happen in the lungs?
   A. Air pressure will increase, volume will decrease and exhalation will occur.
   B. Air pressure will decrease, volume will increase and exhalation will occur.
   C. Air pressure will decrease, volume will increase and inhalation will occur.
   D. Air pressure will increase, volume will increase and inhalation will occur.

Answer is C: Contracting the diaphragm will increase the volume of the lungs, which will decrease the pressure of the contained air. Consequently air from the room will be pushed into the lungs (inhalation occurs).

5. What does it mean when the concentration of dissolved oxygen in blood is 100 mmHg?
   A. The oxygen dissolved in blood exerts a pressure of 100 mmHg over and above the blood pressure generated by the heart.
   B. 100 mmHg of the blood pressure is due to the dissolved oxygen within it.
   C. The blood had been exposed to air in the lungs that contained oxygen at a partial pressure of 100 mmHg.
   D. The dissolved oxygen exerts an osmotic pressure of 100 mmHg.

Answer is C: The concentration of a gas dissolved in water is expressed in terms of the partial pressure of that gas in the mixture of gases that were in contact with the water. Gas will continue to dissolve until the water is saturated. At this point, we say that the concentration of the dissolved gas is the same as the partial pressure of that gas in the atmosphere in contact with the water. The same unit mmHg is used.

6. A pump used to clear an airway of obstructing liquid expands the volume in its chamber to produce a:
   A. Negative pressure, and liquid is pushed by atmospheric pressure into the chamber.
   B. Positive pressure, and liquid is pushed by atmospheric pressure into the chamber.
   C. Negative pressure, and liquid is sucked up into the chamber.
   D. Positive pressure, and liquid is sucked by atmospheric pressure into the chamber.

Answer is A: Once a negative pressure is established by expanding the volume within the pump, atmospheric pressure acting on the patient's body will force air in their lungs out through the airway, along with the obstructing liquid.

7. According to Boyle's law, the pressure in a fixed amount of gas will increase as its volume decreases. Which one of the statements that follow is consistent with Boyle's law?
   A. When the diaphragm contracts, pressure in the lungs increases.
   B. When the chest recoils during exhalation, the air in the lungs is at a negative pressure.
   C. When the chest recoils during exhalation, the air in the lungs increases in volume.
   D. When the diaphragm contracts, pressure in the lungs decreases.

Answer is D: The diaphragm moves inferiorly as it contracts. This expands the volume of the lungs.

8. Deoxygenated blood entering the alveoli has dissolved oxygen with partial pressure of 40 mmHg, while alveolar air has oxygen present at a partial pressure of 90 mmHg. What will be the partial pressure of dissolved oxygen in blood that leaves the lungs?
   A. 40 mmHg
   B. 65 mmHg
   C. 90 mmHg
   D. 100 mmHg

Answer is C: Oxygen will continue to move from the alveoli into the blood while a pressure gradient (a concentration gradient) exists.

9. Boyle's law may be stated as: "when the volume of an enclosed gas expands, its pressure decreases". Thus the pressure in our lungs will:
   A. Increase as our diaphragm contracts
   B. Be negative as we breathe out
   C. Be negative as our rib cage moves up and out
   D. Be positive as our rib cage moves up and out

Answer is C: The volume of our chest increases as the ribcage moves outwards. Hence the air pressure within the lungs will decrease below atmospheric pressure (= negative pressure).

10. Boyle's law may be stated: Provided that the temperature does not change, the volume of a fixed amount of gas decreases as its pressure increases (and vice versa). Which statement concerning the pressure of the air in the lungs is consistent with Boyle's law?
    A. It will decrease as the chest expands.
    B. It decreases as we breathe out.
    C. It increases when we contract our diaphragm.
    D. It decreases as our intercostal muscles relax.

Answer is A: As the chest expands, its volume will increase. This means that the pressure of the contained air will decrease in line with Boyle's law.

11. A statement of Boyle's law is 'the volume of a fixed amount of gas is inversely proportional to the pressure of the gas, as long as temperature does not change'. This means that:
    A. A balloon would expand to a larger volume in a hyperbaric chamber.
    B. The air pressure in the lungs would decrease if the diaphragm is contracted.
    C. The pressure inside a gas cylinder remains constant while some gas is let out, because the volume of the cylinder has not changed.
    D. If the lungs could expand to twice their volume, they would contain air at twice the pressure.

Answer is B: Contracting the diaphragm will increase the volume of the chest and lungs above the diaphragm. As volume has increased, the pressure will decrease. Choice C is wrong as the amount of gas has changed (it is not fixed).

12. Boyle's law states that pressure multiplied by volume is a constant value. Hence which of the following statements is correct?
    A. The amount of air in the lungs will increase when the pressure inside them increases.
    B. A positive pressure is produced in the lungs when the chest expands.
    C. A negative pressure is produced in the lungs when the chest expands.
    D. The amount of air in the lungs will decrease when the pressure in them decreases.

Answer is C: Boyle's states: $P \times V$ = constant. This means that as one of P or V increases, the other decreases enough for their product to be unchanged. Hence as V increases with chest expansion, P decreases to below atmospheric pressure (a negative pressure).

13. If the partial pressure of oxygen in the atmosphere was halved, what effect would this have on the amount of oxygen that would now dissolve in the alveolar fluid? It would:
    A. Decrease to one quarter its former value
    B. Decrease to one half its former value
    C. Be about the same as before
    D. Increase to one and a half times its former value

Answer is B: The amount of gas that will dissolve in water is proportional to its partial pressure (Henry's law). Hence if the partial pressure in the atmosphere halves, the partial pressure in the alveoli will be approximately half, so the amount that will dissolve will also approximately halve.

14. At a height of 3000 m above sea level atmospheric pressure is 30 kPa less than at sea level. A consequence of this lower pressure is that less oxygen will dissolve in the alveolar fluid. The reason for this is that:
    A. The solubility coefficient of oxygen is lower.
    B. The partial pressure of oxygen in the atmosphere is lower.
    C. Water vapour evaporating from the lung will exert a greater partial pressure.
    D. There will be more carbon dioxide leaving the blood and entering the alveoli.

Answer is B: At 3000 m, atmospheric pressure is 70 kPa, the oxygen partial pressure is 20% of this (about 14 kPa). 14 kPa is less than 20 kPa (oxygen's partial pressure at sea level), so less oxygen will dissolve in the alveolar fluid at 3000 m than at sea level.

15. If the partial pressure of oxygen in the air contained in the alveoli of the lungs is 14 kPa, then the partial pressure of oxygen dissolved in the alveolar fluid will be:
    A. Very close to 14 kPa
    B. Significantly greater than 14 kPa
    C. Significantly less than 14 kPa
    D. Unable to be determined without the solubility coefficient of oxygen and the temperature of the alveolar fluid

Answer is A: The amount of a gas that will dissolve in a liquid that is in contact with the gas is determined by the partial pressure of that gas. If PP of O is 14 kPa, then the concentration of O that is dissolved in the alveolar fluid is also 14 kPa.

16. Suppose that the partial pressure of oxygen in the alveoli is doubled from its normal value of 14 kPa. What would then be the partial pressure of oxygen dissolved in the blood leaving the alveoli?
   A. 7 kPa
   B. 14 kPa
   C. 21 kPa
   D. 28 kPa

Answer is D: The amount of a gas that will dissolve in a liquid that is in contact with the gas is determined by the partial pressure of that gas. If PP of O is 14 kPa × 2, then the concentration of O that is dissolved in blood in the alveolar capillaries is also 28 kPa.

17. One of the reasons that the mixture of gases in the air contained in the alveoli of the lungs differs from atmospheric air is that alveolar air is:
   A. Saturated with water vapour whereas atmospheric air is not
   B. At a higher temperature than atmospheric air
   C. At a higher pressure than atmospheric air
   D. Enriched with nitrogen as it diffuses into the alveoli from the blood

Answer is A: Alveolar air is saturated with water vapour, and this has a partial pressure greater than the PP of water in atmospheric air. Alveolar air and atmospheric air are at the same pressure except when inhalation and exhalation is occurring, so if water is in greater proportion within the alveoli (than in the atmosphere), then the other gases must be in lesser proportion.

18. If the pressure between the visceral pleura of the lungs and the parietal pleura of the thoracic cage is −6 mmHg, then:
   A. The lung will collapse.
   B. Exhalation is occurring.
   C. The pressure is above atmospheric pressure.
   D. The lungs will fill with air.

Answer is D: The intra-pleural pressure must be negative (= below atmospheric pressure) if the lungs are to expand as the chest wall expands. It oscillates between about −2 mmHg (−270 Pa) when inhalation is about to begin and about −6 mmHg (−800 Pa) when inhalation has ended and exhalation is about to occur. Hence a value of ~6 mmHg indicates that inhalation is occurring.

19. Boyle's law states that the pressure in an enclosed gas is inversely related to its volume. Which of the following descriptions is compatible with Boyle's law and with respiratory physiology?
   A. As the sternum and ribs move outward, the volume of the chest increases. This causes an increase in the enclosed air pressure. Hence air is exhaled along the pressure gradient.

B. Bronchodilation expands the volume enclosed by the bronchioles. This decreases the air pressure within them and so atmospheric air moves into the lungs along the pressure gradient.
C. Diffusion of oxygen from alveolar air into the blood decreases the partial pressure of oxygen in the lungs. This decreased pressure allows atmospheric air to enter the lungs along the pressure gradient.
D. As the diaphragm moves down, the chest volume expands which decreases the pressure in that volume. Consequently atmospheric air is forced into the lungs along the pressure gradient.

Answer is D: As the chest volume expands, we inhale. The downward movement of the diaphragm expands the lung volume and by Boyle's law the air pressure within decreases. This creates a pressure gradient between the alveoli and the outside air, so air will move into the lungs.

20. Given that de-oxygenated blood entering the alveolar capillaries has an oxygen partial pressure of 42 mmHg and a carbon dioxide partial pressure of 47 mmHg, while the partial pressures of O_2 and CO_2 in the alveolar air are 103 and 40 mmHg, respectively; which of the following gas movements will occur?
    A. Both CO_2 and O_2 move from alveolar air to the capillaries.
    B. CO_2 moves from alveolar air to capillaries and O_2 moves from capillaries to the alveolar air.
    C. Both CO_2 and O_2 move from capillaries to alveolar air.
    D. CO_2 moves from capillaries to alveolar air and O_2 moves from alveolar air to the capillaries.

Answer is D: Dissolved substances will diffuse from where they are in high concentration to where they are at lower concentration. So O_2 moves from alveolar air (where it is at 103 mmHg) to the capillaries (at 42 mmHg), while CO_2 moves from capillaries (47 mmHg) to alveolar air (40 mmHg).

21. What are the conditions under which O_2 diffuses from the blood plasma into red blood cells and attaches to haemoglobin? When partial pressure of oxygen dissolved in plasma is:
    A. 40 mmHg in the systemic capillaries
    B. 100 mmHg in the alveolar capillaries
    C. 40 mmHg in the alveolar capillaries
    D. 100 mmHg in the systemic capillaries

Answer is B: Movement of oxygen from the plasma into RBC occurs in the alveolar capillaries. The higher number (100 mmHg) should be chosen to ensure that a concentration gradient exists.

22. After an inhalation, the partial pressure of O_2 in the alveoli is 104 mmHg and of CO_2 is 40 mmHg. What will be the concentration of these gases dissolved in blood in the capillaries leaving the alveoli?
    A. $O_2 = 40$ mm, $CO_2 = 104$ mmHg

B. O_2 = 40 mmHg, CO_2 = 46 mmHg
C. O_2 = 90 mmHg, CO_2 = 35 mmHg
D. O_2 = 100 mmHg, CO_2 = 40 mmHg

Answer is D: Oxygen will diffuse into the plasma from the alveoli, and carbon dioxide will diffuse the other way, while a concentration gradient exists. The gradient will exist until the plasma concentrations of both gases equal the concentrations in the alveoli.

23. Some partial pressures for **oxygen** (in mmHg) in different parts of the respiratory system are given below. Given that atmospheric air has O_2 at 160 mmHg, which ones are reasonable values?
    A. Capillary blood entering the alveoli has 40 mmHg, the alveolar air has 160 mmHg, capillary blood leaving the alveoli has 100 mmHg.
    B. Capillary blood entering the alveoli has 40 mmHg, the alveolar air has 104 mmHg, capillary blood leaving the alveoli has 100 mmHg.
    C. Capillary blood entering the alveoli has 40 mmHg, the alveolar air has 104 mmHg, capillary blood leaving the alveoli has 40 mmHg.
    D. Capillary blood entering the alveoli has 104 mmHg, the alveolar air has 100 mmHg, capillary blood leaving the alveoli has 40 mmHg.

Answer is B: Choice A is wrong as alveolar air will have a lower PP than atmospheric air. Choice C and D are wrong as capillary blood leaving the alveoli should have oxygen PP nearly the same as that in the alveoli.

24. Blood entering the capillaries of the alveoli has dissolved CO_2 at 46 mmHg and is separated from air in the alveoli, which contains CO_2 at a partial pressure of 40 mmHg, by the respiratory membrane. What will be the concentration of dissolved CO_2 of the blood capillaries leaving the alveoli?
    A. 40 mmHg
    B. 43 mmHg
    C. 46 mmHg
    D. 86 mmHg

Answer is A: Carbon dioxide will continue to diffuse from blood across the respiratory membrane into the alveoli while the concentration gradient exists. The gradient will exist until the dissolved CO_2 in the blood in the capillaries falls to 40 mmHg.

25. If arterial blood entering the lung has pCO_2 = 46 mmHg and alveolar air has pCO_2 = 40 mmHg then
    A. CO_2 will move out of blood into the alveoli.
    B. CO_2 will move out of alveoli into the blood.
    C. Blood leaving the alveoli will have pCO_2 = 43 mmHg.
    D. There will be an increase in the concentration of carbonic acid in the blood.

Answer is A: Carbon dioxide will diffuse in the direction of its concentration gradient. From high concentration in blood (46 mmHg) to alveoli (40 mmHg).

26. If the partial pressure of oxygen in the alveolar air is 104 mmHg, approximately what will be the concentration of dissolved oxygen in the blood leaving the pulmonary capillaries?
    A. 21 mmHg
    B. 40 mmHg
    C. 100 mmHg
    D. 160 mmHg

Answer is C: Oxygen will continue to diffuse from the alveoli across the respiratory membrane into the blood while the concentration gradient exists. The gradient will exist until the plasma concentration reaches 104 mmHg. In reality blood travelling away from the alveoli will be at slightly less than 104 mmHg as some oxygen will have been used by the lung tissue.

27. Recall Boyle's law applied to the lungs. Which of the following would occur as air pressure in the lungs increases?
    A. Volume increases
    B. Volume decreases
    C. The lungs expand
    D. The diaphragm contracts

Answer is B: As the diaphragm contracts, the lungs expand and their volume increases. All these things are inconsistent with increased air pressure in the lungs.

28. If the pO_2 of oxygen in the alveoli is 104 mmHg and in the blood in alveolar capillaries is 40 mmHg, while the pCO_2 in the alveoli 40 mmHg and in the blood is 46 mmHg, which of the following will happen?
    A. Carbon dioxide will diffuse from the alveoli to the blood.
    B. Oxygen will diffuse from the blood to the alveoli.
    C. Oxygen will diffuse from the red blood cells to the plasma.
    D. Carbon dioxide will diffuse from the blood into the alveoli.

Answer is D: Dissolved gases will diffuse from where they are in high concentration to where they are in low concentration. Hence, CO_2 can go from 46 mmHg to 40 mmHg, and pO_2 can move from where it is at 104 mmHg to where it is at 40 mmHg. Choice D is the only choice consistent with this.

29. Inhalation of air into the lungs is correctly described by which of the following?
    A. The action of the diaphragm and the ribs create a positive pressure in the thoracic cavity which causes air to move into the lungs.
    B. The volume of the thoracic cavity is increased as muscles relax, thus increasing the pressure and air is forced into the lungs.
    C. The thoracic cavity decreases in volume as muscles relax and pressure decreases so external air is forced into the lungs.

   D. The lungs expand as muscles contract, this creates a negative pressure so air is forced into the lungs.

Answer is D: Inhalation is an active process that requires muscles to contract. Their contraction increases lung volume which produces a pressure in the lungs that is lower than atmospheric pressure (a negative pressure).

30. The act of breathing in (inhalation) involves expanding the thoracic cage. What does this cause the lung volume to do?
   A. Increase, the pressure of the air in the lungs to decrease and atmospheric air to move into the lungs.
   B. Decrease, the pressure of the air in the lungs to decrease and atmospheric air to move into the lungs.
   C. Increase, the pressure of the air in the lungs to increase and atmospheric air to move into the lungs.
   D. Decrease, the pressure of the air in the lungs to decrease and atmospheric air to move out of the lungs.

Answer is A: Expanding the thoracic cage increases lung volume which (by Boyle's law) decreases the air pressure in the lungs.

31. Henry's law relates the amount of gas that will dissolve in a liquid to the partial pressure of the gas in contact with the liquid. In which situation would the greatest amount of oxygen dissolve in the liquid lining the alveoli?
   A. Breathing in an atmosphere at normal pressure but enriched to 30% oxygen
   B. Resting in a mountain range at a height of 3000 m above sea level
   C. Seated inside a hyperbaric chamber at a pressure of 3 atm
   D. Scuba diving at a depth of 10 m underwater

Answer is C: The PP of oxygen in the situations described by choices B, A, D and C are 0.7×, 1.5×, 2× and 3×, respectively, the PP of oxygen at 1 atm pressure. Hence choice C would result is the greatest amount of oxygen dissolving in liquid.

32. Air flows from a region of high pressure to a region of lower pressure. To explain the process of inhalation and exhalation, you need this fact and which other?
   A. Boyle's law
   B. Dalton's law
   C. Henry's law
   D. Charles' law

Answer is A: Boyle's law states that the pressure within an enclosed gas decreases as its volume increases. So by increasing the volume of your lungs, a low pressure is created which allows air to move into this region of lower pressure and vice versa.

33. Mt. Everest "base camp" is located 5300 m above sea level. Here the atmospheric pressure is 403 mmHg (rather than 760 mmHg at sea level) and the oxygen partial pressure in atmospheric air at this altitude is about 79 mmHg. What is the likely value for dissolved oxygen concentration in arterial blood for someone at base camp?
    A. 149 mmHg
    B. 104 mmHg
    C. 79 mmHg
    D. 50 mmHg

Answer is D: This value has been measured in climbers at base camp. If the partial pressure of oxygen gas in the atmosphere is 79 mmHg, the ppO_2 in the alveoli will be less than this, because there is substantial carbon dioxide and water vapour in alveolar air. The ppO_2 in the atmosphere at sea level is 149, while ppO_2 in alveolar air at sea level is 104.

34. Hyperbaric oxygen therapy is breathing 100% oxygen while under about three atmospheres of pressure (normal air is 20% oxygen). Normal arterial blood oxygen concentration when at 1 atm of pressure is about 100 mmHg. Use Henry's law to decide the likely value from the list below, of arterial blood oxygen concentration of a patient undergoing hyperbaric oxygen therapy.
    A. 1500 mmHg
    B. 500 mmHg
    C. 300 mmHg
    D. 100 mmHg

Answer is A: Henry's law states "the amount of gas that will dissolve in water is proportional to its partial pressure". The ppO_2 in an atmosphere of 100% oxygen is about five times normal. In addition, air at three times normal atmospheric pressure has ppO_2 about three times normal. Hence the blood oxygen concentration will be $5 \times 3 = 15$ times normal. 15×100 mmHg = 1500 mmHg.

35. Instead of the Heimlich manoeuvre, self-administered abdominal thrusts against the back of a chair may be used to dislodge food obstructing your airway. This thrusting increases the pressure in the lungs by decreasing their volume. The increase in pressure in the oesophagus has been measured to be about 10 kPa. Given that normal atmospheric pressure is 100 kPa, use Boyle's law to determine the change (from 100%) in lung volume required to produce the pressure increase.
    A. Lung volume decreases to 91%.
    B. Lung volume increases to 110%.
    C. Lung volume decreases to 10%.
    D. Lung volume decreases to 76%.

Answer is A: Boyle's law states $P_1V_1 = P_2V_2$ rearranging this gives: $V_2 = (P_1/P_2) \times V_1$. Putting in numbers: $V_2 = (100/110) \times V_1 = 0.91 \times V_1$ which is 91% of volume V_1.

36. If a patient is given air enriched to 40% oxygen to breathe (i.e. at two times normal atmospheric oxygen partial pressure), use Henry's law to determine by how much the amount of dissolved oxygen in the blood would change compared to breathing air with 20% oxygen:
    A. It would increase by 150%.
    B. It would increase by 100%.
    C. It would increase by 50%.
    D. It would increase by 40%.

Answer is B: Henry's law states that the amount of gas that will dissolve in water is proportional to its partial pressure. If its PP increases by 100% (i.e. to two times), then 100% more oxygen will dissolve.

Chapter 14
Nervous System

14.1 Cells and Action Potential

Nerve cells are called neurons and are able to conduct an electrical impulse. They may be interneurons, or anaxonic, unipolar, bipolar or multipolar neurons. All neurons (except for anaxonic ones) have an axon that carries an impulse away from the cell body and one, two or many dendrites that carry an impulse towards the cell body. Other cells, collectively called neuroglia are support cells for neurons, but do not carry impulses. Within the CNS, they are ependymal cells, astrocytes, microglia and oligodendrocytes, while in the PNS they are satellite cells and Schwann cells. Oligodendrocytes and Schwann cells form the myelin sheath around axons.

Within the neuron, the concentration of Na^+ ions is much less than their concentration in the interstitial fluid, while the concentration of K^+ ions is the reverse. The great difference in the distribution of these ions (and of other ions) results in the electrical potential (measured in millivolts) on the inside of the neuron's membrane being negative with respect to the electrical potential on the outside of the plasma membrane. When an appropriate stimulus is received by a neuron, sodium channels open and Na^+ ions rush into the neuron (down their concentration gradient). This changes the potential at that spot from negative to positive. Almost immediately, potassium channels open, so that K^+ ions can rush out. This again changes the potential at that spot back to negative from positive. This inflow of Na^+ ions followed by the outflow of K^+ ions is termed the "action potential" and propagates along the axon as the nerve impulse.

A nerve impulse will travel from the axon hillock of the cell body to the end of the axon where it reaches the axon terminals. These terminals will in turn stimulate the cell to which they are attached. This attachment is called a synapse. If the axon stimulates a muscle fibre, the meeting place is also called the neuromuscular junction. The gap between the axon terminal and the adjoining cell is called the synaptic cleft and is crossed by a chemical (a neurotransmitter) released from the axon terminal. The neurotransmitter molecule binds to a receptor protein on the plasma

M. Caon, *Examination Questions and Answers in Basic Anatomy and Physiology*, https://doi.org/10.1007/978-3-030-47314-3_14

membrane of the adjoining cell. In this way, the neurotransmitter passes on the action potential to the next cell. That is, the next cell is stimulated.

1. Which word correctly completes the statement: "All motor neurons are…"?
   A. Interneurons
   B. Multipolar
   C. Bipolar
   D. Unipolar

Answer is B: Motor neurones (that innervate muscles) are multipolar.

2. In the peripheral nervous system, which cells form the myelin sheath?
   A. Ependymal cells
   B. Schwann cells
   C. Astrocytes
   D. Oligodendrocytes

Answer is B: Schwann cells surround the axon to form the myelin sheath in peripheral neurones. Oligodendrocytes perform a similar function for neurones in the CNS.

3. A difference in the amount and type of ions between the two sides of a plasma membrane or a charge difference that occurs when ions move along a membrane is called an "electrical potential". What does the term "action potential" refer to?
   A. The distribution of ions that results in the inside of the cell being at about −70 mV compared to outside the cell
   B. A movement of sodium ions into the cell following a stimulus and the ions spreading out along the inside of the cell membrane
   C. The rapid movement of sodium ions into the cell followed by potassium ions moving out of the cell, with the movement being repeated along the length of the neuron
   D. Sodium ions being pumped out of the cell, while potassium ions are moved into the cell

Answer is C: The influx of sodium ions followed by the efflux of potassium ions describes the action potential. Choice A describes the "resting potential".

4. What is the gap between the plasma membranes of a neuron that conducts an incoming signal and the cell that is going to receive the signal called?
   A. Neuromuscular junction
   B. Intercellular cleft
   C. Synaptic cleft
   D. Intercalated disc

Answer is C: The gap between cells at a synapse is called a synaptic cleft. A neuromuscular junction is one type of synapse.

5. Which of the following substances **CANNOT** pass through the "blood–brain barrier"?
   A. Steroid hormones
   B. O_2 molecules
   C. Alcohol
   D. Potassium ions

Answer is D: Fat-soluble molecules can pass the BBB, but most charged particles cannot.

6. What name is given to the cells in the nervous system that produce nerve impulses?
   A. Neurotransmitters
   B. Nerves
   C. Neurons
   D. Neuroglia

Answer is C: Neurons produce nerve impulses, nerves are bundles of neurones. Neuroglia are cells that support and protect neurones.

7. Which structure carries incoming impulses towards the nerve cell body?
   A. Axon hillock
   B. Axon
   C. Dendrite
   D. Synaptic knobs

Answer is C: Dendrites receive stimuli for nerve cells and transmit them towards the cell body. An axon transmits a nerve impulse away from the neurone cell body.

8. Which neurons are unipolar?
   A. Neurons in the central nervous system
   B. Neurons in the retina
   C. Sensory neurons
   D. Motor neurons

Answer is C: Most sensory neurones are unipolar, that is the axon and dendrites are not separated by the cell body but are the one strand (are fused), with the cell body attached to it by a single process.

9. Which glial cells are responsible for forming the myelin sheath around peripheral nerve cells?
   A. Astrocytes
   B. Schwann cells
   C. Satellite cells
   D. Oligodendrocytes

Answer is B: The oligodendrocytes perform a similar function for neurones within the CNS.

10. Inactive muscle and nerve cells maintain a resting membrane potential. This potential results in:
    A. The outside of the cell being negative
    B. The inside of the cell being positive
    C. The inside and outside of the cell having the same charge
    D. The inside of the cell being negative

Answer is D: The resting potential has the inside of the cell at about −70 mV compared to the outside.

11. When an action potential arrives at a synapse, what happens first?
    A. A neurotransmitter is released into the synaptic cleft.
    B. Extracellular Na^+ crosses the postsynaptic membrane.
    C. Choline in the synaptic cleft enters the nerve cell and is converted to acetyl choline.
    D. Extracellular Ca^{++} enters the nerve cell.

Answer is D: As the axon terminal depolarises when the action potential arrives, calcium channels open and extracellular calcium enters the axon terminal. This stimulates the release of a neurotransmitter (e.g. ACh) into the synaptic cleft.

12. What is the last part of a nerve cell that is involved when a nerve impulse passes to another cell?
    A. Synaptic knob
    B. Axon hillock
    C. Dendrite
    D. Axon

Answer is A: Synaptic knobs are at the distal end of an axon. The axon hillock is at the proximal end.

13. Which of the following statements is true of neuroglia?
    A. They are the cells that link motor neurons to sensory neurons.
    B. It is the non-cellular material that lies between neurons.
    C. They have only one dendrite and one axon.
    D. They produce the myelin sheath.

Answer is D: Neuroglia refers to several types of cells of the nervous system that are not neurones. Some of them form the myelin sheath.

14. What can correctly be said about somatic motor neurons?
    A. They are unipolar neurons.
    B. Their cell bodies are in the dorsal root ganglia.
    C. Their cell bodies are located in the central nervous system.
    D. They are bipolar neurons.

Answer is C: Somatic motor neurons are multipolar. Their cell bodies are located in the CNS—within the spinal cord.

15. What is the effect of the movement of Na^+ into a nerve cell followed very soon by the movement of K^+ out of the nerve cell?
    A. This establishes the resting membrane potential.
    B. These movements are known as depolarisation and repolarisation.
    C. These movements repolarise the cell.
    D. It changes the membrane potential from about −70 mV to about −50 mV.

Answer is B: When Na^+ moves into a nerve cell, the cell depolarises (from −70 mV to about +30 mV); when soon afterwards K^+ moves out of the nerve cell, it repolarises (from +30 mV back to −70 mV).

16. There is a space between a neuron and the cell it stimulates, that is crossed by a neurotransmitter. What is it called?
    A. Synaptic cleft
    B. Voltage-gated channel
    C. Synapse
    D. Postsynaptic membrane

Answer is A: The synaptic cleft lies between the presynaptic membrane of the neuron and the postsynaptic membrane of the cell about to be stimulated.

17. Which of the following CAN cross the blood–brain barrier to enter the brain?
    A. K^+
    B. O_2
    C. Proteins
    D. Most pharmaceuticals

Answer is B: All cells require oxygen, so of course it can cross the BBB most pharmaceuticals cannot.

18. What name is used for a nerve cell?
    A. Neuron
    B. Neuroglia
    C. Ganglion
    D. Astrocyte

Answer is A: Neurons (or neurones) are nerve cells. Neuroglia are cells in the nervous system, but they do not produce nerve impulses.

19. What is the name of the nerve cell structure that carries incoming impulses towards the cell?
    A. Dendrite
    B. Axon
    C. Cell body
    D. Ganglion

Answer is A: Dendrites receive incoming (efferent or stimulating) impulses.

20. Which is the major type of nerve cell in the CNS?
    A. Anaxonic
    B. Unipolar
    C. Bipolar
    D. Multipolar

Answer is D: Multipolar are most common in the CNS. Motor neurons, whose cell bodies lie within the spinal cord, are multipolar.

21. What is the purpose of the myelin sheath around an axon?
    A. To control the chemical environment around the nerve cell
    B. To phagocytose microbes
    C. To prevent movement of ions through the nerve cell membrane
    D. To form the blood–brain barrier

Answer is C: The myelin sheath insulates the axon. That is, it prevents ions from crossing the plasma membrane except at the nodes of Ranvier, where the sheath is absent.

22. Which nerve cells carry impulses from the brain to the muscles?
    A. Sensory
    B. Motor
    C. Afferent
    D. Association

Answer is B: Motor (or efferent) nerves carry impulses away from the CNS to the muscles.

23. Which of the following describes an "action potential"?
    A. The high concentration of Na^+ and Cl^- outside the cell and of K^+ inside the cell
    B. The voltage change that moves along the cell membrane until it reaches the axon hillock
    C. The movement of a neurotransmitter from the presynaptic membrane to the postsynaptic membrane.
    D. The movement of Na^+ across the cell membrane into the cell, followed by the movement of K^+ out of the cell.

Answer is D: Choice B describes a graded potential. Choice C describes synaptic transmission. Choice A describes the distribution of those ions that contribute to the resting potential.

24. What part of the neurone carries the "action potential"?
    A. The cell body
    B. The dendrites
    C. The synaptic knobs
    D. The axon

Answer is D: The action potential travels along the axon, away from the cell body.

25. What type of neurones are motor neurones?
    A. Anaxonic
    B. Multipolar
    C. Bipolar
    D. Unipolar

Answer is B: All motor neurones are multipolar.

26. What event during the action potential causes the resting membrane potential to change from about −70 mV to about +30 mV?
    A. K^+ ions moving into the cell
    B. K^+ ions moving out of the cell
    C. Na^+ ions moving into the cell
    D. Na^+ ions moving out of the cell

Answer is C: It takes positive ions moving INTO the cell to change the resting potential from −70 to +30 mV. The ions are sodium.

27. Where are the cell bodies of somatic motor neurones found?
    A. In the peripheral nervous system
    B. In the central nervous system
    C. In the dorsal root ganglia
    D. In the spinal cord

Answer is D: They are found in the spinal cord (in the ventral grey horns). Choice B is also correct, but choice D is more specific. The dorsal root ganglia contain cell bodies of sensory neurons).

28. What is an "action potential"? It is:
    A. When the resting potential changes from −70 mV to +30 mV and then back again
    B. The name given to the difference in electrical charge between the inside and outside of the plasma membrane of a neurone
    C. The name given to the stimulus that changes the resting potential from −70 to −50 mV
    D. The voltage produced by a stimulus which causes a nerve impulse to be generated

Answer is A: The action potential is the name given to the change in voltage of the inside of the cell from −70 to +30 mV and back to −70 mV again. This voltage change propagates along the axon.

29. What does the term "synapse" refer to?
    A. The plasma membrane of the axon terminal of a nerve cell
    B. That part of the plasma membrane of the cell being stimulated, that is opposite the axon terminal
    C. The gap between the stimulating nerve cell and the receiving cell

D. The place where signal transmission between a nerve cell and the cell it is stimulating occurs

Answer is D: The impulse carried by a nerve cell is transferred by neurotransmitter to the stimulated cell by this structure at the point of contact. Choice C describes the synaptic cleft.

30. Which statement is true of a multi-polar neuron?

   A. Has many axons attached to the cell body.
   B. Is the major type of neuron in the peripheral nervous system.
   C. All sensory neurons are multi-polar.
   D. Has many dendrites attached to the cell body.

Answer is D: The many dendrites give the "multi-" part of the name of this type of neuron. Sensory neurones are unipolar.

31. Which of the following is a true statement about an "action potential"?

   A. It refers to the movement of a neuro-transmitter along an axon.
   B. It travels away from the cell body along the axon.
   C. It causes K^+ to rush into the cell.
   D. It travels between the dendrite and the axon hillock.

Answer is B: An action potential is an efferent impulse that travels along the axon away from the cell body.

32. Complete the sentence. Neuroglia:

   A. Are bundles of axons
   B. Contain cell bodies outside the central nervous system
   C. Are a type of neuron
   D. Include ependymal cells, astrocytes and satellite cells

Answer is D: Neuroglia are nervous system cells that are not neurons. Choice D gives three types of these cells.

33. Which of the three structures listed below constitute a nerve cell?

   A. Dendrites, ganglion, myelin sheath
   B. Dendrites, cell body, axon
   C. Neuron, neuroglia, synaptic process
   D. Cell body, synaptic knobs, efferent fibre

Answer is B: Neuroglia and ganglion are not part of a neuron. Efferent fibre is an ambiguous term that may refer to an axon or to a motor neuron.

34. What are the major type of nerve cells in the CNS?

   A. Multipolar
   B. Sensory
   C. Interneurons

D. Unipolar

Answer is A: Multipolar are most common in the CNS. Motor neurons, whose cell bodies lie within the spinal cord, are multipolar.

35. Which sequence of ion movements describes the action potential?
    A. Na^+ moves out of cell then K^+ moves in.
    B. K^+ moves in to cell then Na^+ moves out.
    C. K^+ moves out of cell then Na^+ moves in.
    D. Na^+ moves into cell then K^+ move out.

Answer is D: Sodium ions are in high concentration outside the cell (and in low concentration inside the cell). The correct sequence is Na in followed by K out.

36. What is the space between a neuron and the following neuron, muscle or gland that it stimulates called?
    A. Synaptic vesicle
    B. Ion channel
    C. Synaptic cleft
    D. Receptor

Answer is C: The synaptic cleft is the space over which an incoming nerve impulse is transmitted by a molecule called a neurotransmitter.

37. What feature do the dendrites of a nerve cell have?
    A. They transmit an action potential.
    B. They contain the cell nucleus and organelles.
    C. They carry incoming impulses to the cell body.
    D. They are connected to the cell body by the axon hillock.

Answer is C: Dendrites are the afferent pathway by which a stimulus is brought to a cell body.

38. Most sensory neurones may be described as which of the following?
    A. Multipolar
    B. Bipolar
    C. Having cell bodies within the CNS
    D. Unipolar

Answer is D: Sensory neurons are unipolar. And their cell bodies are outside the CNS in the dorsal root ganglia.

39. What is the type of neuroglia that forms the myelin sheath on neurons outside of the CNS?
    A. Oligodendrocytes
    B. Satellite cells
    C. Schwann cells

D. Microglia

Answer is C: Schwann cells surround the axon to form the myelin sheath in peripheral neurones. Oligodendrocytes perform a similar function for neurones in the CNS.

40. The action potential occurs when one of the following events occurs. Which one?
    A. Na^+ rushes into the cell followed by Cl^-.
    B. Na^+ rushes out of the cell followed by PO_4^{3-} rushing in.
    C. K^+ rushes into the cell followed by Na^+ rushing out.
    D. Na^+ rushes into the cell followed by K^+ rushing out.

Answer is D: Sodium rushing in followed by potassium rushing out produces the action potential.

41. A multipolar neuron has more than one what?
    A. Dendrite attached to the cell body
    B. Axon attached to the cell body
    C. Synaptic terminal attached to the axon
    D. Cell body

Answer is A: Many dendrites are attached to the cell body of multipolar neurones.

42. What is the depolarisation and repolarisation of a nerve cell membrane called?
    A. Graded potential
    B. Action potential
    C. Threshold potential
    D. Resting membrane potential

Answer is B: Depolarisation of the resting membrane potential leads to an action potential.

43. Which best describes a nerve?
    A. Dendrites, cell bodies, axons, Schwann cells
    B. Dendrites, cell bodies, axon hillock, axon terminals, vesicles
    C. Dendrites, cell bodies, axon hillock, axon terminals, Schwann cells, neurotransmitters
    D. Axons, blood vessels, connective tissue, Schwann cells

Answer is D: A nerve is a bundle of axons of individual neurons, hence does not contain nerve cell bodies. A nerve does include the Schwann cells around axons, BV and CT.

44. Which of the following would conduct an action potential with the greatest speed?
    A. Myelinated, large diameter fibres
    B. Myelinated, small diameter fibres

C. Unmyelinated, large diameter fibres
D. Unmyelinated, small diameter fibres

Answer is A: Myelinated fibres allow for faster conduction speed as does greater diameter of axon.

45. Which of the following is a characteristic of an action potential?
   A. The signal is graded.
   B. It results due to an influx of potassium ions.
   C. It is an all or none response.
   D. It results from an initial outflow of sodium ions.

Answer is C: When a graded potential exceeds the threshold for producing an action potential, the AP happens regardless of the strength of the graded potential. This is an all or none event. (Choice D is wrong as sodium ions should be moving IN).

46. What is meant by an absolute refractory period?
   A. At least 5 ms must elapse from the time of the first action potential before a second can be initiated.
   B. An action potential cannot be initiated during this period regardless of the strength of the stimulus.
   C. An action potential can be initiated if the strength of the stimulus is higher than normal (>70 mV).
   D. An action potential can be initiated if the strength of the stimulus is lower than normal (<70 mV).

Answer is B: During the ~0.5 ms period of the action potential in which sodium ions are rushing into a nerve cell, the nerve cannot respond to any stimulus no matter how strong the stimulus. Because of the impossibility of response, this period is called "absolutely" refractive (refractory = resisting ordinary methods).

47. What is meant by the "absolute refractory period" of a nerve cell membrane? The time during which:
   A. A larger than normal stimulus is required to cause an action potential.
   B. A smaller than normal stimulus will produce an action potential.
   C. No stimulus will produce an action potential.
   D. Two stimuli in quick succession are required to add to an above threshold stimulus.

Answer is C: During the ~0.5 ms period of the action potential in which sodium ions are rushing into a nerve cell, the nerve cannot respond to any stimulus no matter how strong the stimulus. Because of the impossibility of response, this period is called "absolutely" refractive (refractory = resisting ordinary methods).

48. What would happen if a neuron lost its myelin sheath?
   A. Na^+ would leak out of the axon leaving too few ions to stimulate the Na channels at the next node to open.

B. The neuron would die.
C. More Na^+ channels would be exposed allowing freer entry, so conduction speed would increase.
D. More K^+ channels would be exposed allowing freer exit, so the cell would hyperpolarise.

Answer is A: The myelin sheath prevents sodium ions from leaking back into the intracellular fluid; hence, they can travel with minimum diminution to the next node to trigger the opening of abundant voltage-gated sodium channels.

49. Some cells in the body can maintain an electric potential across their cell membrane. How do they do this?
    A. By using the sodium–potassium pump to continually eject positive sodium and potassium ions from the cell
    B. By allowing negative chloride ions to enter the cell along their concentration gradient
    C. By trapping large cations inside the cell membrane
    D. By keeping unequal concentrations of various ions on each side of the cell membrane

Answer is D: It is the distribution of ions of different charge on either side of the membrane that determines the membrane potential.

50. In nerve fibres with myelin sheaths, which of the following is true about the electrical conduction?
    A. It is "saltatory", so propagates at higher speed.
    B. It requires more energy to send an impulse.
    C. The conduction between adjacent axons is enhanced ("cross talk" is increased).
    D. It is slower due to the separation between the "nodes of Ranvier".

Answer is A: The myelin sheath increases the speed of conduction along an axon as the action potential is regenerated when migrating Na ions reach the next node. It "jumps" (saltates) from one node to the next.

51. When we say that the cell membrane is polarised we mean that:
    A. The outside of the cell is negative with respect to the inside.
    B. The inside of the cell is negative with respect to the outside.
    C. There are more Na^+ ions and less K^+ ions inside the cell than outside.
    D. Na^+ ions have moved out of the cell and K^+ ions have moved in.

Answer is B: Polarised means that the charge is different on the two sides of the membrane, with the inside being negative with respect to the outside.

52. Depolarisation of the cell membrane involves:
    A. Sodium channels opening to allow Na^+ to flow in
    B. Potassium channels opening to allow K^+ to flow in

C. Chloride pumps quickly pumping large amounts of Cl^- outside
D. Electrical attraction between K^+ inside and Cl^- outside

Answer is A: The influx of positive Na ions changes the polarisation of the cell from negative inside to positive inside at that location on the membrane. This reversal is called depolarisation.

53. Which of the following statements about the action potential is **FALSE**?
    A. The action potential lasts about 4 ms.
    B. It is triggered by anions crossing the cell membrane.
    C. The sequence: "Na ions moving in, K ions moving out" constitutes the action potential.
    D. Repolarisation follows depolarisation of the cell membrane.

Answer is B: Anions (+ve ions) crossing the membrane is the first part of the action potential, but this movement is triggered by a graded potential that is above threshold arriving at the start of the axon (the axon hillock). So choice D is false.

54. When is an action potential initiated? When:
    A. The resting membrane potential changes from −70 to +30 mV.
    B. A nerve impulse has caused some muscle action to be produced.
    C. The potassium "gates" in the cell membrane open and potassium ions flood into the cell.
    D. A stimulus, which is above the threshold level, is applied to a receptor.

Answer is D: A stimulus which exceeds the threshold level is required to generate an action potential.

55. The sequence of events that constitute an action potential is correctly described by which of the following? Resting potential of:
    A. +35 mV, stimulus above the threshold, Na^+ moves into cell, depolarisation to −70 mV, K^+ moves out of cell, repolarisation to +35 mV
    B. −70 mV, stimulus above the threshold, K^+ moves into cell, depolarisation to −35 mV, Na^+ moves out of cell, repolarisation to −70 mV
    C. +35 mV, stimulus above the threshold, K^+ moves into cell, depolarisation to −70 mV, Na^+ moves out of cell, repolarisation to +35 mV
    D. −70 mV, stimulus above the threshold, Na^+ moves into cell, depolarisation to +35 mV, K^+ moves out of cell, repolarisation to −70 mV

Answer is D: The resting potential is negative, and the first ion movement is sodium ions into of the cell.

56. What term refers to the sudden movement of potassium ions across the cell membrane to the outside of a nerve cell?
    A. Repolarisation
    B. Depolarisation
    C. The action potential

D. The potassium pump

Answer is A: When Na^+ ions move into the axon, it is called depolarisation as the resting potential changes from −70 to +30 mV. Movement out by K^+ ions is called repolarisation as the potential returns to −70 mV. One followed by the other is the action potential.

57. Which of the following is true of anaxonic neurons?
    A. Horizontal and amacrine cells of the retina are anaxonic neurons.
    B. Most sensory neurons of the peripheral nervous system are anaxonic.
    C. All motor neurones that control skeletal muscles are anaxonic.
    D. Anaxonic neurons are found in the retina of the eye.

Answer is A: Choice B would be true of unipolar neurons. Choice C would be true of multipolar neurons. Choice D is true, but A is a better answer as it is more specific.

58. Which of the following is true of bipolar neurons?
    A. Horizontal and amacrine cells of the retina are bipolar neurons.
    B. Most sensory neurons of the peripheral nervous system are bipolar.
    C. All motor neurones that control skeletal muscles are bipolar.
    D. Bipolar neurons are found in the retina of the eye.

Answer is D: The bipolar neurons of the retina are also called bipolar cells. Choice A would be true of anaxonic neurons. Choice B would be true of unipolar neurons. Choice C would be true of multipolar neurons.

59. Which of the following is true of unipolar neurons?
    A. Horizontal and amacrine cells of the retina are unipolar neurons.
    B. Most sensory neurons of the peripheral nervous system are unipolar.
    C. All motor neurones that control skeletal muscles are unipolar.
    D. Unipolar neurons are found in the retina of the eye.

Answer is B: These sensory neurons have their cell bodies in the dorsal root ganglion. Choice A would be true of anaxonic neurons. Choice C would be true of multipolar neurons. Choice D would be true of bipolar neurons.

60. Which of the following is true of multipolar neurons?
    A. Horizontal and amacrine cells of the retina are multipolar neurons.
    B. Most sensory neurons of the peripheral nervous system are multipolar.
    C. All motor neurones that control skeletal muscles are multipolar.
    D. Multipolar neurons are found in the retina of the eye.

Answer is C: Choice A would be true of anaxonic neurons. Choice B would be true of unipolar neurons. Choice D would be true of bipolar neurons.

61. Along what part of the neuron does the action potential travel?
   A. The dendrites
   B. The cell body
   C. The axon
   D. The cytoplasm

Answer is C: The axon. Prior to ravelling along the axon, the potential is referred to as a "graded potential".

62. Which of the following is correct when stated about multipolar neutrons?
   A. All motor neurons are multipolar.
   B. They are found in the retina and olfactory apparatus.
   C. Their cell bodies lie in the dorsal root ganglia.
   D. They have one axon and one fused dendrite.

Answer is A: Choice B refers to bipolar neurons. Choice C refers to sensory neurons. Choice D refers to unipolar neurons.

63. Which neuroglia do **NOT** occur in the central nervous system?
   A. Ependymal cells
   B. Oligodendrocytes
   C. Microglia
   D. Schwann cells

Answer is D: Schwann cells produce the myelin sheath of peripheral neurons. Oligodendrocytes produce the myelin sheath for central neurons.

64. What is a correct statement when applied to the myelin sheath?
   A. It is produced by ependymal cells.
   B. It allows rapid impulse transmission by saltatory conduction.
   C. It insulates the axon by surrounding it with protein.
   D. It surrounds the dendrites of a neuron.

Answer is B: Saltatory conduction refers to the action potential "jumping" from node to node of a myelinated neuron. Myelin is a lipid, not a protein. Myelin surrounds the axon, not the dendrite.

(The peripheral sensory neurons have myelinated dendrites. Short dendrites in the central nervous system are not myelinated.)

65. What is produced when a dendrite of a neuron is stimulated?
   A. An action potential
   B. A graded potential
   C. A threshold potential
   D. A resting potential

Answer is B: A graded potential diminishes in its level of depolarisation of the membrane, as it moves away from the site of stimulation.

66. Which of following events occur when an action potential arrives at the end of the axon's telodendria?
    A. Calcium enters the synaptic knob which causes a neurotransmitter to be released.
    B. Sodium ions rush into the cell through ion channels, then potassium ions rush out.
    C. The resting membrane potential changes from −70 to +30 mV.
    D. The sodium–potassium pump shifts potassium ions into the cell and sodium ions out of the cell.

Answer is A: In order for the action potential to be passed from the neuron to the postsynaptic membrane (of the receiving cell), a neurotransmitter must be released to cross the synaptic cleft.

67. Which part of the brain contains the cells that allow us to consciously control voluntary muscle movements?
    A. Post-central gyrus
    B. Central sulcus
    C. Pre-central gyrus
    D. Prefrontal cortex

Answer is C: The pre-central gyrus of the frontal lobe is anterior to the central sulcus (which divides the frontal lobe from the parietal lobes) and is our primary motor area.

68. What sort of information does an afferent nerve pathway carry?
    A. Information that travels along a corticospinal tract
    B. Sensory information
    C. Information that travels along a descending tract
    D. Motor information

Answer is B: "Afferent" implies information travelling towards the brain, therefore sensory information. A corticospinal tract is descending.

69. Distinguish between a unipolar and a multipolar neurone.
    A. A "multipolar" neurone has two processes extending from the cell body. A "unipolar" neurone has only one process extending from the cell body.
    B. A "multipolar" neurone has more than two processes extending from the cell body. A "unipolar" neurone has only one process extending from the cell body.
    C. A "multipolar" neurone has more than one axon and one dendrite extending from the cell body. A "unipolar" neurone has only one axon and one dendrite extending from the cell body.
    D. Most sensory neurons are multipolar. Most motor neurones are unipolar.

Answer is A: "Multipolar" neurone has more than two processes extending from the cell body: one of them is an axon, and the rest are dendrites (a bipolar neurone has two processes). A "unipolar" neurone has only one process extending from the cell body. The dendrites and axon are continuous, and the cell body lies off to one side. Choice D would be true in the reverse.

70. Distinguish between a "graded potential" and an "action potential".
    A. An "action potential" moves along a dendrite, while a graded potential moves along an axon.
    B. An "action potential" involves multipolar neurones while a graded potential occurs in unipolar and bipolar neurones.
    C. An "action potential" does not diminish as it moves away from its source. A graded potential decreases with distance from the stimulus.
    D. An "action potential" decreases with distance as it moves away from its source. A graded potential does not diminish with distance from the stimulus.

Answer is C: A typical stimulus produces a temporary localised change in the resting membrane potential. The effect, which decreases with distance from the stimulus is called a graded potential. If the graded potential is large enough, it triggers an action potential in the membrane of the axon. An "action potential" is an electrical impulse that is propagated along the surface of an axon and does not diminish as it moves away from its source. This impulse travels along the axon to one or more synapses.

71. Which type of neuroglial cell carries out phagocytosis?
    A. Oligodendrocytes
    B. Astrocytes
    C. Ependymal
    D. Microglia

Answer is D: Phagocytosis is engulfing foreign cells and debris. Microglia do this. Oligodendrocytes produce a myelin sheath around neurones; astrocytes form the blood–brain barrier; ependymal cells help to circulate the CSF.

72. During an action potential, which two membrane ion channels participate?
    A. Potassium and chloride
    B. Potassium and calcium
    C. Sodium and calcium
    D. Sodium and potassium

Answer is D: During an action potential, first the sodium channels open and Na ions rush in. Then they close and potassium channels open and K rushes out.

14.2 Brain and Spinal Cord Anatomy

The brain and spinal cord are enclosed by membranes called meninges: the dura mater, the arachnoid mater and the pia mater. Together the brain and spinal cord are the "central nervous system" (CNS). The peripheral nervous system includes motor nerves that leave the CNS from the brain or the spinal cord and sensory nerves that bring information to the CNS. Sensory nerves are "afferent". That is, carry information from sensory organs to the brain—often via the spinal cord. Motor nerves are "efferent". That is, they carry commands from the brain (usually) to the muscles (usually). Motor and sensory nerve fibres attached to the brain form cranial nerves and those attached to the spinal cord form spinal nerves.

The brain consists of the cerebrum, diencephalon, brainstem and cerebellum while cerebrospinal fluid (CSF) rather than blood circulates through the four ventricles within the brain, through the central canal of the spinal cord and between the arachnoid and pia maters. CSF is formed from blood at the choroid plexuses and returns to the blood in the superior sagittal sinus. The surface of the cerebrum is folded into gyri (ridges) and sulci (valleys) and divided into "lobes": frontal, two parietal, occipital, two temporal and two insula. The central sulcus separates the frontal lobe from the parietal lobes while the precentral gyrus (of the frontal lobe) is noteworthy for being the primary motor area and the post-central gyrus (of the parietal lobe) for being the primary somatosensory area (search for the cortical "homunculus"). A section through the brain displays grey-coloured matter containing cell bodies and white-coloured matter. The whiteness arises from the myelin that wraps around axons as, being a fat, myelin is white in appearance. The diencephalon includes the thalamus, the hypothalamus and the neurohypophysis (of the pituitary) and pineal glands. The brainstem comprises the midbrain, the pons and the medulla oblongata.

In cross-section, the spinal cord displays its characteristic "butterfly-shaped" central grey matter region surrounded by the white matter of myelinated nerves. This white matter is either ascending tracts (carrying sensory information to the brain) or descending tracts (carrying motor instructions to muscles and glands). The axons of sensory neurons within a spinal nerve enter the spinal cord from the dorsal side, while the axons of motor neurons within a spinal nerve exit the spinal cord from the ventral side. This "dorsal root" and the "ventral root" meet and join to form the spinal nerve that then passes through the vertebral foramina.

1. Which of the following lists of structures include all of those contained within the central nervous system?

   A. Cerebellum, cerebrum, spinal cord, diencephalon, brainstem
   B. Midbrain, spinal cord, autonomic nerves, pons, diencephalon
   C. Midbrain, cerebellum, special sense organs, medulla oblongata
   D. Cerebrum, sensory neurons, motor neurons, cerebellum

Answer is A: There are five distinct structures in the CNS. Autonomic nerves, motor and sensory neurones and special sense organs are not part of the CNS.

2. What are the three meninges and three named "potential spaces" that surround the brain, in order from superficial to deep (outermost to innermost)?
   A. Pia, arachnoid, sub-arachnoid, dura, septa, sub-septal
   B. Sub-arachnoid, epidural, dura, pia, arachnoid, sub-pial
   C. Arachnoid, sub-arachnoid, pia, epidural, dura, sub-dural
   D. Epidural, dura, sub-dural, arachnoid, sub-arachnoid, pia

Answer is D: The dura mater, arachnoid mater and pia mater are the three meninges in order.

3. Which of the following is **NOT** composed of "grey matter"?
   A. Spinothalamic tract
   B. Cerebral cortex
   C. Basal nuclei
   D. Post-central gyrus

Answer is A: Grey matter refers to cell bodies of neurons while white matter is aggregations of myelinated axons. The spinothalamic "tract" is a bundle of nerves (axons) carrying information to the brain.

4. Where in the brain is the "primary motor area"?
   A. Midbrain
   B. Thalamus
   C. Basal nuclei
   D. Pre-central gyrus

Answer is D: The pre-central gyrus of the frontal lobe is the primary motor area of the brain and is separated from the primary sensory area (the post-central gyrus), by the central sulcus.

5. The hypothalamus does ALL of the following **EXCEPT** one. Which one?
   A. It is the autonomic control centre.
   B. It directs lower CNS centres to perform actions.
   C. It produces the rigidly programmed, automatic behaviours necessary for survival.
   D. It performs many homeostatic roles.

Answer is C: Rigidly programmed automatic behaviours are controlled by the brainstem (midbrain, pons, medulla oblongata) not the hypothalamus.

6. Which of the following structures together make up the brainstem?
   A. Medulla oblongata, pons, midbrain, cerebellum
   B. Medulla oblongata, pons, midbrain
   C. Medulla oblongata, pons, midbrain, thalamus
   D. Medulla oblongata, pons, midbrain, pineal gland

Answer is B: The cerebellum, thalamus and pineal gland are not part of the brainstem.

7. In which of the following places would you **NOT** find cerebrospinal fluid?
   A. The sub-arachnoid space
   B. The third ventricle of the brain
   C. The epidural space
   D. The central canal of the spinal cord

Answer is C: The epidural space surrounds the dura mater of the spinal cord and so is outside of the meninges and of the CNS. "Epi-" means on top of the dura.

8. What is the name of the lobe of the brain that is immediately superior to the cerebellum?
   A. Dorsal
   B. Occipital
   C. Posterior
   D. Parietal

Answer is B: The cerebellum nestles into the cerebrum immediately inferior to the occipital lobe.

9. Which of the following statements about the blood–brain barrier (BBB) is correct?
   A. The BBB prevents fluctuations of hormone and ion concentrations in blood from affecting the brain.
   B. It is formed by Schwann cells wrapping around capillaries.
   C. The brain is supported by (it floats in) the BBB.
   D. The BBB is formed by the choroid plexus.

Answer is A: Hormones and ions cannot pass through the BBB. (It is astrocytes, not Schwann cells that from part of the BBB; the brain floats on CSF; CSF is formed by the choroid plexus.)

10. In which part of the brain is the thalamus found?
    A. Diencephalon
    B. Cerebrum
    C. Cerebellum
    D. Brainstem

Answer is A: The thalamus, along with the epithalamus, hypothalamus and sub-thalamus form the diencephalon.

11. Where is the autonomic control centre for most of body homeostasis located?
    A. In the limbic system
    B. In the brainstem
    C. In the hypothalamus
    D. In the cerebellum

Answer is C: Hypothalamus is the autonomic centre which is the main visceral control centre of body homeostasis.

12. Which four structures together make up the brain?
    A. Cerebrum, diencephalon, brainstem and cerebellum
    B. Cerebrum, thalamus, brainstem and cerebellum
    C. Cerebrum, diencephalon, meninges and cerebellum
    D. Spinal cord, diencephalon, brainstem and medulla oblongata

Answer is A: The brainstem is at the end of the spinal cord, just inferior to the diencephalon which is surrounded by the cerebrum. The cerebellum is dorsal to the brainstem and inferior to the cerebrum.

13. Which of the following are **NOT** part of the cerebral cortex?
    A. Motor areas, sensory areas and association areas
    B. Pre-central gyrus and post-central gyrus
    C. White matter and cerebellum
    D. The lateral ventricles and the thalamus

Answer is D: The thalamus is part of the diencephalon, not the cerebral cortex (cerebrum).

14. Which parts of the brain allow us to control skilled voluntary muscle movements?
    A. Basal nuclei and sub-thalamic nuclei
    B. Cerebellum and hypothalamus
    C. Pre-central gyrus and cerebellum
    D. Thalamus and post-central gyrus

Answer is C: The pre-central gyrus of the frontal lobe is the primary motor area. It, along with the cerebellum, controls voluntary movements of skeletal muscle.

15. Which of the following roles is **NOT** performed by the hypothalamus?
    A. Autonomic control of heat activity and blood pressure
    B. Relaying visual and auditory information to the cerebral cortex
    C. Production of hormones for the posterior pituitary
    D. Body temperature regulation

Answer is B: The thalamus relays visual and auditory information, not the hypothalamus.

16. Corticospinal pathways crossover from one side of the brain to the other side. Where does this crossover occur?
    A. In the medulla oblongata
    B. In the cerebellum
    C. In the hypothalamus
    D. In the reticular formation

Answer is A: In the medulla oblongata at the “decussation of pyramids”.

17. Which layer of membrane around the brain is the most superficial?
    A. Dura mater
    B. Meningeal mater
    C. Arachnoid mater
    D. Pia mater

Answer is A: The dura mater is in contact with the skull.

18. Which of the following substances is prevented from entering the brain by the blood–brain barrier?
    A. Glucose
    B. Nicotine
    C. Epinephrine (adrenaline)
    D. Ethyl alcohol (in "alcoholic" beverages)

Answer is C: Only trace amounts of circulating neurotransmitters are able to cross the BBB, otherwise they would stimulate neurons in the brain in an uncontrolled way. Glucose is essential to the brain. Nicotine and alcohol, users will tell you, affect the brain, hence can cross the BBB.

19. What part of the brain contains the motor areas and the sensory areas?
    A. Cerebrum
    B. Diencephalon
    C. Brainstem
    D. Cerebellum

Answer is A: The central sulcus of the cerebrum separates the gyri containing the primary motor area in the frontal lobe and the primary sensory areas in the parietal lobes.

20. What part of the brain contains the main visceral control centre of body homeostasis?
    A. Cerebrum
    B. Diencephalon
    C. Brainstem
    D. Cerebellum

Answer is B: The hypothalamus of the diencephalon.

21. What part of the brain subconsciously provides precise timing for the movements of learned skeletal muscle contraction?
    A. Cerebrum
    B. Diencephalon
    C. Brainstem
    D. Cerebellum

Answer is D: Cerebellum subconsciously provides precise timing and appropriate patterns of learned skilled skeletal muscle contraction for smooth coordinated movements, posture and agility.

22. Where does the spinal cord start and finish?
    A. It extends from the foramen magnum to L1–L2.
    B. It extends from the foramen magnum to the sacrum.
    C. It starts at the superior part of the medulla oblongata and extends to the inferior part of the cauda equina.
    D. It extends from C7 to L5.

Answer is A: The foramen magnum is the hole in the occipital bone. Beyond L2, spinal nerves descend within the "lumbar cistern", a space between the arachnoid mater and the pia mater.

23. Where is the cerebral spinal fluid found?
    A. Between the pia mater and the brain
    B. Between the dura mater and the arachnoid mater
    C. Between the dura mater and the pia mater
    D. Between the arachnoid mater and the brain

Answer is D: CSF is located in the sub-arachnoid space and within the ventricles of the brain.

24. Which of the following substances **CANNOT** cross the blood–brain barrier?
    A. Antibodies, proteins, K^+
    B. O_2, CO_2 and H_2O
    C. Fats, fatty acids, fat soluble substances
    D. Alcohol, nicotine, anaesthetics

Answer is A: Large molecules and molecules with an electric charge are prevented from crossing the BBB unless there is a specific transport channel for them. The substances in the other choices can all cross the BBB.

25. What part of the brain contains the midbrain, the pons and the medulla oblongata?
    A. The diencephalon
    B. The cerebrum
    C. The cerebellum
    D. The brainstem

Answer is D: The brainstem consists of the midbrain, the pons and the medulla oblongata.

26. Which of the following statements is **INCORRECT**?
    A. The pituitary gland dangles from the hypothalamus by the infundibulum.
    B. The post-central gyrus houses the primary motor cortex.
    C. The thalamus surrounds the third ventricle.
    D. White matter consists of myelinated axons of neurons.

Answer is B: The post-central gyrus houses the primary sensory cortex (not motor).

27. What is true about the spinothalamic tract?
    A. It is a descending pathway that carries sensory information.
    B. It is a descending pathway that carries motor instructions.
    C. It is an ascending pathway that carries sensory information.
    D. It is an ascending pathway that carries motor instructions.

Answer is C: It is ascending—going from spine to thalamus—and ascending tracts carry sensory information to the brain.

28. What is linked to the posterior grey horn of the spinal cord?
    A. The dorsal root of the spinal nerve that carries motor fibres
    B. The dorsal root of the spinal nerve that carries sensory fibres
    C. The ventral root of the spinal nerve that carries sensory fibres
    D. The ventral root of the spinal nerve that carries motor fibres

Answer is B: Posterior means dorsal. The dorsal root carries axons of sensory neurons.

29. What is the blood–brain barrier?
    A. It is the inner two meninges that surround the brain.
    B. It is formed by the capillaries of choroid plexus and ependymal cells.
    C. It is the endothelial cells of capillaries that supply the brain and their astrocytes.
    D. It is the cerebrospinal fluid that bathes the brain.

Answer is C: The BBB is the endothelial cells of the capillaries (which are joined by tight junctions), their basement membrane and the enveloping astrocytic feet of the astrocytes (one of the cell types called neuroglia).

30. What part of the brain is known as the cerebrum?
    A. It is that part of the diencephalon that surrounds the third ventricle.
    B. It makes up the majority of the brain stem.
    C. The superficial part consisting of sulci and gyri.
    D. The dorsal inferior part adjacent to the occipital bone.

Answer is C: The cerebrum is the cortex (the outer part) of the brain (the left and right cerebral hemispheres) made up of folds called gyri and sulci. That is, the grey matter, the white matter and the basal nuclei.

31. What and where is the pre-central gyrus?
    A. It is the primary somatosensory cortex and is in the frontal lobe.
    B. It is the primary somatosensory cortex and is in the parietal lobe.
    C. It is the primary motor cortex and is in the frontal lobe.
    D. It is the primary motor cortex and is in the parietal lobe.

Answer is C: The pre-central gyrus of the frontal lobe is the primary motor area.

32. What part of the brain receives sensory input before passing it on to another part of the brain for interpretation or action?
    A. Pons
    B. Hypothalamus
    C. Post-central gyrus
    D. Thalamus

Answer is D: The thalamus relays motor and sensory signals to the cerebral cortex. This information is also integrated and modulated. Our awareness, emotional state and arousal is affected by the thalamus.

33. What part of the brain contains the autonomic control centre whose orders regulate food intake, water balance and body temperature?
    A. Hypothalamus
    B. Thalamus
    C. Medulla oblongata
    D. Cerebellum

Answer is A: These homeostatic roles are performed by the hypothalamus.

34. What are the "ascending tracts" of the spinal cord and what do they do?
    A. They are white matter, and they transmit sensory information to the brain.
    B. They are grey matter, and they transmit sensory information to the brain.
    C. They are white matter, and they transmit motor information to the brain.
    D. They are grey matter, and they transmit motor information to the brain.

Answer is A: Tracts means white matter as axons are myelinated which makes them look white. Ascending means that the nerve impulses are going from body to brain, that is they are carrying sensory input to the brain.

35. In what part of the brain is the "decussation of the pyramids" found:
    A. Pons
    B. Medulla oblongata
    C. Midbrain
    D. Hypothalamus

Answer is B: The medulla oblongata is where the pyramidal tracts crossover (decussate).

36. What is the likely result of an injury that severs the spinal cord between C5 and C6?
    A. Respiratory failure and death
    B. Paraplegia
    C. Hemiplegia
    D. Quadriplegia

Answer is D: Spinal nerves below the break cannot pass on impulses from the brain. So quadriplegia results because the spinal nerves that innervate the limbs leave the spinal cord below the level of the spinal cord break (except for some brachial nerves that leave at C5). Fortunately, the phrenic nerve, which innervates the diaphragm, leaves the spinal cord at C3, C4 and C5 so will still carry impulses from the brain. So breathing will continue.

37. To what does the "blood–brain barrier" refer?
    A. The three meninges that surround the brain and spinal cord.
    B. The tight junctions between endothelial cells of the capillaries that serve the brain.
    C. The structures that prevent fat-soluble molecules from entering the brain from the blood.
    D. The structure that produces cerebrospinal fluid from blood.

Answer is B: This answer is not a complete description of the BBB as it does not include astrocytes, but it is the best of the four choices.

38. What functions are controlled from the pre-central gyrus of the frontal lobe?
    A. Automatic visceral functions
    B. Conscious perception of many sensory inputs
    C. Subconscious timing and coordination of skeletal muscle
    D. Voluntary control of skeletal muscle

Answer is D: The pre-central gyrus is the primary motor area of the brain.

39. What is the primary function of the cerebellum?
    A. It regulates such things as body temperature, water balance and emotional responses.
    B. It refines/adjusts learned motor movements so that they are performed smoothly.
    C. It controls our automatic functions such as breathing, digestion and cardiovascular functions.
    D. It is the origin of our conscious thoughts and intellectual functions.

Answer is B: Cerebellum subconsciously provides precise timing and appropriate patterns of learned skilled skeletal muscle contraction for smooth coordinated movements, posture and agility.

40. What do the descending tracts of the spinal cord contain?
    A. White matter and transmit sensory information
    B. White matter and transmit motor commands
    C. Grey matter and transmit sensory information
    D. Grey matter and transmit motor commands

Answer is B: Tracts refers to axons which are encased in myelin which makes them look white. Descending refers to impulses that carry information from the brain to the body (motor commands) usually to muscle.

41. Which of the following lists all of the main sections of the brain?
    A. Cerebrum, brainstem, midbrain, medulla oblongata
    B. Cerebrum, cerebral cortex, cerebellum, mesencephalon
    C. Cerebellum, diencephalon, brainstem, cerebrum
    D. Cerebral cortex, midbrain, diencephalon, cerebellum

Answer is C: Apart from the two "cere" (cerebrum and cerebellum), diencephalon and brainstem must be included.

42. One of the following is not a meninges. Which one?
    A. Pia mater
    B. Alma mater
    C. Arachnoid mater
    D. Dura mater

Answer is B: Alma mater is the university that you attended.

43. Where is that part of the brain that allows us to consciously control voluntary muscle movements located?
    A. Post-central gyrus of parietal lobe
    B. Central sulcus of frontal lobe
    C. Pre-central gyrus of the frontal lobe
    D. Arbour vitae of the cerebellum

Answer is C: The pre-central gyrus is the primary motor area. The "pre-motor cortex" lies just anterior to the precentral gyrus, and the "supplementary motor area" is located on the midline surface of the hemisphere anterior to the primary motor cortex. These two areas allow you make choices about which movement is appropriate and are involved in the initiation of voluntary movements.

44. What part of the brain subconsciously provides the appropriate pattern of smooth coordinated skeletal muscle contraction for movements that we have learned?
    A. The cerebellum
    B. The brainstem
    C. The cerebrum
    D. The diencephalon

Answer is A: Subconscious coordination of learned (and practised) movements is done by the cerebellum.

45. In what respect does "grey matter" differ from "white matter".
    A. Grey matter refers to the CNS while white matter refers to the PNS.
    B. White matter makes up the autonomic nervous system, grey matter does not.
    C. Grey matter is found in the cerebrum, while white matter occurs in the cerebellum and the diencephalon.

D. Grey matter contains the cell bodies of nerve cells white matter contains axons.

Answer is D: The myelin that surrounds the axons of neurons gives them a white appearance.

46. Which of the following is **NOT** a meninges?
    A. Cerebra mater
    B. Pia mater
    C. Dura mater
    D. Arachnoid mater

Answer is A: This choice is nonsense.

47. What is the most superficial part of the brain called?
    A. Diencephalon
    B. Cerebral cortex
    C. Cerebellum
    D. Mesencephalon

Answer is B: Each hemisphere has a superficial layer of "grey matter" called the cerebral cortex, lying over "white matter".

48. What do neurones in the pre-central gyrus do?
    A. They receive information from general sense receptors in muscle and skin.
    B. They communicate with motor, sensory and multi-modal association areas.
    C. They allow conscious control of skilled voluntary muscle movements.
    D. They process and relay auditory and visual input.

Answer is C: The pre-central gyrus is the primary motor area. It allows conscious control over skeletal muscle movement.

49. What is the role of the hypothalamus?
    A. It receives sensory input and relays it to the cerebral cortex.
    B. It is the autonomic control centre which directs the function of the lower CNS.
    C. It uses past experience to analyse and act on sensory input.
    D. It integrates sensory information from association areas and performs abstract intellectual functions.

Answer is B: Hypothalamus is the main visceral control centre of body homeostasis. It is the autonomic control centre from which orders flow to lower CNS centres for execution.

50. With respect to the spinal cord, where is the epidural space?
    A. External to the dura mater
    B. Between the arachnoid and pia maters

C. Between the arachnoid and dura maters
D. Between the two layers of the dura

Answer is A: The epidural space is above or on top of the dura and external to the spinal cord. It is between the spinal cord and the surrounding vertebrae.

51. Which part of the brain controls breathing, heart function, vasoconstriction and swallowing?
    A. Mesencephalon
    B. Cerebellum
    C. Diencephalon
    D. Brainstem

Answer is D: Centres in brainstem produce the rigidly programmed, automatic behaviours necessary for survival, e.g. respiratory centre (breathing rhythm), cardiovascular centre (force and rate of heart contraction, vasoconstriction), vomiting, coughing, hiccoughing, swallowing). The mesencephalon is part of the brainstem.

52. An image of the cross-section of a spinal cord would show "anterior (or ventral) horns". What is in that region?
    A. Spinal nerves
    B. Ascending tracts
    C. Cell bodies of motor neurons
    D. White matter

Answer is C: The "horns" contain grey matter which is the cell bodies of motor neurons. The dorsal horns contain sensory nuclei (cell bodies of neurons).

53. The blood–brain barrier functions to protect the brain from:
    A. Lipid soluble drugs, alcohol and nicotine
    B. Fluctuations in oxygen and carbon dioxide concentrations
    C. Neurotransmitters, bacteria and neurotoxins
    D. Dehydration and fluctuating blood glucose level

Answer is C: These substances would alter the function of, or damage the brain. Glucose, oxygen, carbon dioxide, ethyl alcohol and nicotine can cross the BBB easily.

54. What function does the precentral gyrus of the brain perform?
    A. The primary motor cortex
    B. Our association areas
    C. The primary somatosensory area
    D. Our higher intellectual functions

Answer is A: This gyrus (outfold) is immediately anterior to the central sulcus (infold).

55. In the spinal cord, what do the ascending tracts contain and what do they do?
    A. White matter and they transmit sensory information
    B. White matter and they transmit motor commands
    C. Grey matter and they transmit sensory information
    D. Grey matter and they transmit motor commands

Answer is A: The term "ascending" means travelling upwards to the brain from the body. Hence, this must be sensory information from the body. The tracts are white as they are myelinated which gives the nerves a white appearance.

56. What is found between the arachnoid and pia mater?
    A. Adipose tissue
    B. Venous sinuses
    C. Choroid plexus
    D. Cerebrospinal fluid

Answer is D: Between the arachnoid and pia mater is the sub-arachnoid space, and this is filled with cerebrospinal fluid.

57. Which of the following is found in the epidural space?
    A. Adipose tissue
    B. Venous sinuses
    C. Choroid plexus
    D. Cerebrospinal fluid

Answer is A: The epidural space is on top of (or outside of) the dura mater. It is the space between the spinal cord and the vertebrae. It contains adipose tissue (and the dural sac, spinal nerves, blood vessels and connective tissue).

58. One of the functions of the prefrontal cortex is:
    A. Making conscious decisions
    B. Controlling motor functions
    C. Detecting and integrating sensory information
    D. Enabling word recognition

Answer is A: The pre-frontal cortex (of frontal lobe) performs abstract intellectual functions such as predicting the consequences of actions or events. Hence, it helps us make conscious decisions.

59. What does the term "decussation" mean and where does the "decussation of pyramids" occur?
    A. Decussation = span. It occurs between the superior part of the brain and the inferior part.
    B. Decussation = cross. It occurs between the anterior part of the brain and the posterior part.
    C. Decussation = crossover. It occurs between the pons and the spinal cord.

D. Decussation = associate. It occurs between the sensory area and the sensory association area.

Answer is C: A decussation is a crossing over. Nerves from the left side of the brain crossover to the right side of the spinal cord to innervate the right-hand side of the body (and vice versa).

60. The central sulcus separates gyri involved with which two major functions?
   A. Vision and taste
   B. Vision and hearing
   C. Motor and sensory
   D. Emotion and memory

Answer is C: Motor function resides in the pre-central gyrus, while sensory function resides in the post-central gyrus.

61. What type of nerves are found in the dorsal root?
   A. Only afferent nerves
   B. Only efferent nerves
   C. Both afferent and efferent nerves
   D. Only ganglionic nerves

Answer is A: Afferent (or sensory) nerves/neurons are in the dorsal root of the spinal nerves.

62. In a cross-section view of the spinal cord, there is a butterfly-shaped structure. What would the posterior grey horn of this structure primarily consist of?
   A. The axons of motor neurons
   B. The cell bodies of interneurons
   C. The cell bodies of motor neurons
   D. The cell bodies of sensory neurons

Answer is B: The posterior (or dorsal) grey horn contains interneurons that receive stimuli from sensory axons (whose cell bodies reside in the dorsal root ganglion).

63. An epidural block involves injecting anaesthetic into the epidural space. What is the main reason for this?
   A. Anaesthetic in this space only affects spinal nerves in the immediate vicinity of the injection.
   B. The anaesthetic will be readily distributed along the spinal cord by the cerebral spinal fluid from this space.
   C. There is less chance of damaging the spinal cord when inserting the needle into this space.
   D. The epidural space is highly vascularised and so will the anaesthetic will be quickly absorbed.

Answer is A: The epidural space contains adipose tissue and spinal nerves pass through this space. Injecting anaesthetic into this space anaesthetises (blocks), the nearby spinal nerves.

64. The central sulcus of the brain lies between which two lobes?
    A. Parietal and occipital
    B. Temporal and occipital
    C. Frontal and temporal
    D. Frontal and parietal

Answer is D: The central sulcus separates the frontal lobe from the left and right parietal lobes.

65. If a person suffers a stroke and damage occurs to the occipital lobe of the brain, which function is the most likely to be affected?
    A. The ability to write
    B. Speech
    C. Hearing
    D. Vision

Answer is D: The occipital lobe contains the visual cortex (processes visual information) and the visual association area (interpretation of images). Hence, damage to the occipital lobe would affect vision.

66. If a person had a pre-frontal lobotomy what would be the physiological consequences?
    A. Movement would be impaired.
    B. Sensory function would be impaired.
    C. The ability to assess the consequence of actions would be impaired.
    D. Speech would be impaired.

Answer is C: If the connection between the pre-frontal cortex and the brain is severed, the person would not have anxiety, tension or frustration, or tact and decorum. They would be unaware of the consequences of their actions.

67. What is a function of the thalamus?
    A. Connects two cerebral hemispheres
    B. Connects cerebellum to midbrain
    C. Connects areas within same hemisphere
    D. It is a relay centre

Answer is D: The thalamus is about 13 "nuclei" that receive then distribute (i.e. relays) sensory information from the body to the cerebral cortex and basal nuclei. This information is also integrated and modulated. Our awareness, emotional state and arousal is affected by the thalamus.

68. What is the function of the corpus callosum?
    A. Connects two cerebral hemispheres
    B. Connects cerebellum to midbrain
    C. Connects areas within same hemisphere
    D. It is a relay centre

Answer is A: Commissural fibres pass through the corpus callosum that connect corresponding areas in the two hemispheres.

69. Which one of the following is **NOT** a function of the cerebral spinal fluid?
    A. To produce hormones
    B. To transport nutrients around the brain
    C. To protect the spinal cord
    D. To cushion the brain

Answer is A: The CSF does not produce hormones.

70. Which best describes the function of the association area of the temporal lobe?
    A. It perceives of movement.
    B. It interprets the meaning of sound patterns.
    C. It recognises of geometric shapes and faces.
    D. It perceives meaningful information from different senses.

Answer is B: The temporal lobe contains the auditory cortex. It interprets sound from the hearing mechanism in the adjacent ears.

71. What is the blood–brain barrier? An adaptation of the capillaries serving the brain that:
    A. Prevents fat-soluble molecules from entering the brain.
    B. Inhibits all substances from passing from the blood stream into the brain.
    C. Selectively inhibits many substances from passing from the blood stream into the brain.
    D. Operates from birth to prevent foreign molecules entering the brain.

Answer is C: The BBB is a selectively permeable barrier to many water-soluble particles that is incomplete at birth.

72. What is the purpose of inserting a needle into the epidural space?
    A. It allows access to the cerebrospinal fluid.
    B. It enables the cerebrospinal fluid pressure to be measured.
    C. It enables permanent drainage of cerebrospinal fluid to treat hydrocephalus.
    D. It allows access to administer analgesia and anaesthesia.

Answer is D: The epidural space is on the outside of the spinal cord, hence does not contain CSF. It is into this space that the analgesia for a spinal block is injected.

73. Which of the following statements about the blood-brain barrier is TRUE?
    A. It consists of the meninges which surround the brain.
    B. It is a protective mechanism which limits entry of alcohol into the brain.
    C. It is poorly developed in the newborn who are therefore less sensitive to drugs which act on the brain.
    D. It is unable to prevent entry of lipid-soluble toxins into the brain.

Answer is D: The BBB does not prevent lipid-soluble toxins, such as alcohol and nicotine, from entering the brain.

74. Spinal nerves are formed from a dorsal root and a ventral root. What is true of the ventral root?

    A. They contain sensory neurons carrying afferent impulses.
    B. They contain sensory neurons carrying efferent impulses.
    C. They contain motor neurons carrying afferent impulses.
    D. They contain motor neurons carrying efferent impulses.

Answer is D: The ventral root contains axons of motor neurons. That is, they carry motor information from the brain to the body—this is efferent (outgoing) information.

75. All of the following terms **EXCEPT** for one are applied to an anatomical feature where a crossing over from one side to the other occurs. Which one?

    A. Chiasma
    B. Cruciate
    C. Decussation
    D. Reticular formation

Answer is D: The others refer to the optic chiasma, the cruciate ligament and the decussation of pyramids.

76. Which of the following is **NOT** a meninges that surrounds the brain?

    A. Dura mater
    B. Pia mater
    C. Arachnoid mater
    D. Epidural mater

Answer is D: There are three meninges, but there is no such thing as epidural mater.

77. What is the "blood–brain" barrier?

    A. Blood capillary walls that are selectively permeable
    B. Blood capillaries with impermeable walls
    C. The meninges that surround the brain
    D. The myelin sheaths around central nervous system neurons

Answer is A: The "barrier" is the wall of the capillary that allows some molecules/ions to pass through into the cerebral spinal fluid but prevents the passage of many other molecules.

78. The sections of the cerebrum have their own names. What are these named structures known as?

    A. Sulci
    B. Hemispheres

C. Lobes
D. Cortexes

Answer is C: The cerebral cortex is divided into lobes.

79. Which parts of the brain are contained in the brainstem?
    A. Pons, midbrain, medulla oblongata
    B. Thalamus, hypothalamus, pituitary gland
    C. Cerebellum, cerebrum, limbic system
    D. Basal nuclei, caudate nuclei, metencephalon

Answer is A: The brainstem has these three sections. The three structures of choice B are the diencephalon.

80. Which of the following structures are included in the diencephalon?
    A. Thalamus, hypothalamus
    B. Thalamus, hypothalamus, pineal gland, posterior pituitary gland
    C. Thalamus, hypothalamus, pineal gland, third ventricle, pituitary gland
    D. Thalamus, hypothalamus, pineal gland, third ventricle, mammillary body

Answer is B: The structures within the diencephalon are the thalamus, hypothalamus, pineal and posterior pituitary glands. The anterior pituitary is glandular epithelial tissue rather than nervous tissue so is not part of the brain. The third ventricle is surrounded by the thalamus but contains no brain tissue. The mammillary body is part of the hypothalamus.

81. Which lobes make up each hemisphere of the cerebral cortex?
    A. Central sulcus, frontal, parietal, occipital, temporal
    B. Frontal, cerebellum, parietal, occipital, temporal
    C. Frontal, parietal, occipital, temporal
    D. Frontal, parietal, occipital, temporal, insula

Answer is D: The five lobes of each hemisphere are the frontal, parietal, occipital, temporal and insula. The insula is not visible on the surface of the brain.

82. Which lobe of the brain is **NOT** named after the cranial bone to which it is deep?
    A. Insula
    B. Parietal
    C. Frontal
    D. Temporal

Answer is A: The insula is a deep lobe that is covered by the frontal and temporal lobes, hence is not adjacent to any cranial bone.

83. Which structure(s) contain axons that connect the two hemispheres of the brain?
    A. Projection fibres
    B. Association fibres

C. The corpus callosum
D. The thalamus

Answer is C: The corpus callosum is one of the commissural tracts that connect the left and right hemispheres. Projection fibres enter the hemispheres from lower brain areas or the spinal cord, while association fibres connect different parts of same hemisphere.

84. What is the function of the corpus callosum?

A. It contains axon fibres that connect the left and right hemispheres.
B. It contains fibre bundles that connect the brain stem with the cerebrum.
C. It connects axon fibres that connect the frontal lobes with the occipital lobes.
D. It connects the cerebellum to the cerebrum.

Answer is A: The corpus callosum is a commissural bundle (of myelinated fibres) which means that it connects left and right hemispheres of the cerebrum.

85. What is the role of the lateral geniculate nuclei (or bodies)?

A. They are the part of the visual pathway before the optical radiation.
B. They are the location of the crossover of left and right optic nerves.
C. They are the respiratory centres of the brainstem.
D. The auditory nerves synapse with the temporal lobe in these nuclei.

Answer is A: The optic tracts leaving the optic chiasma synapse with neurons in the lateral geniculate body (of the thalamus). Many axons then continue as the optic radiation of fibres into primary visual cortex.

86. What is the "limbic system" of the brain?

A. It keeps the cortex alert and aroused.
B. It is the emotional or affective part of the brain.
C. It filters out repetitive, familiar or weak sensory signals.
D. It receives sensory information before passing it the cerebral cortex.

Answer is B: The limbic system is also involved in memory and learning. Choices A and C refer to the reticular formation of the brain. Choice D is performed by the thalamus.

87. What is a function of the reticular formation (or reticular activating system) of the brain?

A. It is the emotional or affective part of the brain.
B. It allows emotion to override logic and vice versa.
C. It controls our circadian rhythm.
D. It receives and integrates all incoming sensory input.

Answer is D: The reticular formation is located in the brainstem, and so all sensory information from receptors for pressure touch, pain and limb position pass through here. The RF also regulates reflex activity, muscle tone and postural activity. Vision and sound stimulate the reticular formation. Choices A and B refer to the limbic system. Our circadian rhythm involves the pineal gland.

88. What is the function of the "corpora quadrigemina"?
    A. To relay sensory information to the basal nuclei of the cerebrum
    B. To be sensitive to chemical changes in the blood
    C. To process visual and auditory sensations
    D. To control reflex movements associated with eating

Answer is C: The two superior colliculi receive visual sensation, while the two inferior colliculi receive auditory sensations. Choice A refers to the thalamic nuclei. Choice B refers to the hypothalamus. Choice D refers to the mammillary bodies.

89. Distinguish between the central nervous system (CNS) and the peripheral nervous system (PNS).
    A. CNS includes the brain and spinal cord; PNS includes motor and sensory nerves, receptors and autonomic nervous system nerves.
    B. CNS includes the brain and spinal cord; PNS includes the autonomic NS and the somatic NS.
    C. CNS is seat of higher functions; PNS includes the sympathetic NS and the parasympathetic NS.
    D. CNS includes the autonomic NS and the somatic NS, while the PNS includes the sympathetic NS and the parasympathetic NS.

Answer is A: CNS = brain and spinal cord. PNS = neural tissue outside the CNS. Motor nerves, sensory nerves, receptors, autonomic nervous system nerves. The PNS delivers sensory information to the CNS and carries motor commands from the CNS to peripheral tissues and systems. Choice B is wrong as some of the autonomic and somatic NS are within the CNS. Choice C is wrong as some of the sympathetic NS and the parasympathetic NS is with the CNS.

90. Distinguish between the efferent nervous system and the afferent nervous system.
    A. Efferent system carries nerve impulses away from some effector, while afferent system fibres carry impulses towards peripheral sensory receptors.
    B. Efferent system carries nerve impulses towards peripheral sensory receptors, while afferent system fibres carry impulses away from some effector.
    C. Efferent system carries nerve impulses away from the CNS, while afferent system fibres carry impulses towards the CNS.
    D. Efferent system carries nerve impulses towards the CNS, while afferent system fibres carry impulses away from the CNS.

Answer is C: Efferent system carries nerve impulses away from the CNS and towards some effector, e.g. a muscle (including skeletal, smooth and cardiac muscle) or gland; afferent system fibres carry impulses towards the CNS from peripheral sensory receptors.

91. Distinguish between the autonomic nervous system and the somatic nervous system.
    A. Somatic NS is not under our voluntary control while the autonomic NS is under our conscious control.
    B. Somatic NS sends motor messages from CNS while the autonomic NS sends motor messages from the PNS.
    C. Somatic NS sends motor messages from CNS to cardiac muscle, to smooth muscle and glands while the autonomic NS conducts impulses from CNS to skeletal muscle.
    D. Somatic NS sends motor messages from CNS to skeletal muscle while the autonomic NS conducts impulses from CNS to cardiac muscle, to smooth muscle and glands.

Answer is D: Somatic NS sends motor messages from CNS to skeletal muscle and is under our conscious control. The autonomic NS conducts impulses from CNS to cardiac muscle, to smooth muscle and glands (i.e. our viscera). It is not under our voluntary control.

92. Distinguish between the sympathetic NS and the parasympathetic NS.
    A. The sympathetic division decreases HR, constricts bronchioles and pupils, increases gut motility and secretions, while the parasympathetic division increases HR, dilates bronchioles, redirects blood supply, decreases gut motility, stimulates sweat glands.
    B. The sympathetic division mobilises the body during stressful situations, while the parasympathetic division promotes energy conserving measures.
    C. The sympathetic division innervates heart, bronchioles and pupils, while the parasympathetic division innervates the blood vessels, the gut and sweat glands.
    D. The sympathetic division promotes non-emergency function and conserves energy while the parasympathetic division mobilises the body during emergency situations.

Answer is B: Sympathetic division stirs you up, mobilises the body during emergency (stressful) situations (e.g. increases HR, dilates bronchioles, redirects blood supply, decreases gut motility, stimulates sweat glands). Parasympathetic division promotes non-emergency function; conserves energy (e.g. decreases HR, constricts bronchioles and pupils, increases gut motility and secretions).

93. What are the five main structures of the CNS?
    A. (1) Thalamus, (2) metencephalon, (3) brain stem, (4) cerebellum, (5) spinal cord

B. (1) Cauda equina, (2) diencephalon, (3) autonomic control centre, (4) arbor vitae, (5) mesencephalon
C. (1) Cerebrum, (2) diencephalon, (3) brain stem, (4) cerebellum, (5) spinal cord
D. (1) Cerebrum, (2) midbrain, (3) medulla oblongata, (4) pons, (5) basal nuclei

Answer is C: The brain and the spinal cord make up the CNS. It includes: (1) cerebrum, (2) diencephalon, (3) brain stem, (4) cerebellum, (5) spinal cord. Cerebrum has motor areas, visual association areas, basal nuclei. Brain stem has substantia nigra (midbrain), respiratory centres (medulla oblongata), autonomic control centre (the pons and medulla oblongata). Diencephalon has area for control of body temperature. Cerebellum has arbor vitae. Spinal cord has conus medullaris, cervical enlargement, cauda equina, posterior median sulcus.

94. Which of the following is NOT a lobe of the cerebrum?
A. Subdural
B. Frontal
C. Occipital
D. Temporal

Answer is A: The "subdural" refers to the space between the dura mater and the arachnoid mater.

95. Which of the following is one of the meninges of the brain?
A. Pons
B. Arachnoid
C. Parietal
D. Medulla mater

Answer is B: The arachnoid (or arachnoid mater) is one of the meninges.

96. What is the cauda equina?
A. Spinal nerves that originate from the brainstem
B. That part of the spinal cord located in the sacrum
C. The coccyx or "tail bone"
D. Spinal nerves continuing inferiorly from the spinal cord as individual fibres

Answer is D: The spinal cord ends at about lumbar vertebrae #1. Inferior to L1 spinal nerves continue as an aggregation of individual fibres with the appearance of separate "hairs" like the tail (cauda) of a horse (equina).

97. What is the difference between the spinal cord at vertebra L1 and at vertebra L3?
A. Superior to L1, there is the spinal cord; inferior to L2, instead of a spinal "cord", there are individual spinal nerves.

B. The arachnoid and dura maters that surround the cord end at L1.
C. At L3 the spinal cord expands as the pelvic enlargement.
D. All spinal nerves have exited the spine by L1, and there is no spinal cord inferior to L1.

Answer is A: The solid structure of a cord ends at L1; inferior to this are a bundle of individual fibres still within the vertebral foramina.

98. Which two meninges enclose the cerebrospinal fluid?
    A. The dura mater and arachnoid mater
    B. The arachnoid mater and the pia mater
    C. The dura mater and pia mater
    D. The dura mater and the epidural

Answer is B: The two deeper meninges enclose the CSF.

99. What is the name of the "space" between meninges that contains cerebrospinal fluid?
    A. Subarachnoid space
    B. Subdural space
    C. Epidural space
    D. Central canal

Answer is A: The subarachnoid space is below (sub-) the arachnoid and superficial to the pia mater. Choice D is within the spinal cord.

100. What name is given to the nerves that pass through the epidural space?
    A. The motor nerves
    B. The sensory nerves
    C. The autonomic nerves
    D. The spinal nerves

Answer is D: The spinal nerves formed when the dorsal root of the nerve joins with the ventral root to pass through the epidural space on their way out through the intervertebral foramina. Motor, sensory and autonomic nerves are contained within the spinal nerves.

101. Where are the cell bodies of peripheral **sensory** neurones located?
    A. In the white matter of the spinal cord
    B. In the dorsal root ganglion
    C. In the ventral root ganglion
    D. In the grey matter of the spinal cord

Answer is B: Sensory neurone cell bodies are located outside of the CNS in the dorsal root (of the spinal nerve) ganglion.

102. Where are the cell bodies of peripheral **motor** neurones located?
    A. In the white matter of the spinal cord
    B. In the dorsal root ganglion
    C. In the ventral root ganglion
    D. In the grey matter of the spinal cord

Answer is D: Motor neurone cell bodies are located inside the CNS in the ventral horn of the grey matter of the spinal cord.

103. What is contained in the dorsal root ganglion?
    A. Axons of sympathetic nervous system neurones
    B. Axons of parasympathetic nervous system neurones
    C. The cell bodies of motor neurones
    D. The cell bodies of sensory neurones

Answer is D: The dorsal root ganglion houses the cell bodies of sensory neurones. The cell bodies of motor neurones are in the spinal cord.

104. In which lobe of the brain is the primary visual area found?
    A. Frontal lobe
    B. Parietal lobes
    C. Occipital lobe
    D. Temporal lobes

Answer is C: The occipital lobes has the visual area.

105. Where is the cardiovascular centre and what does it do? It is located in the:
    A. Pons and adjusts heart rate
    B. Pons and adjusts vasoconstriction
    C. Medulla oblongata and adjusts heart rate
    D. Midbrain and adjusts vasoconstriction

Answer is C: The CVC is in the medulla oblongata of the brain stem and controls heart rate and stroke volume. The vasomotor region of the CVC also adjusts blood vessel diameter.

14.3 Autonomic Nervous System, Neurotransmitters, Reflexes

The autonomic nervous system is not under our conscious control. It stimulates cardiac muscle, smooth muscle, the diaphragm and both exocrine and endocrine glands. Hence, it controls heart rate, vasodilation, blood pressure, body temperature, respiration rate, digestive motility, and aspects of urinary function, reproductive function and endocrine function. The ANS has two divisions: the “sympathetic” division (SD) which mobilises the body during exercise, excitement, and emergency situations; and the "parasympathetic" division (PSD) which maintains body activities, e.g. digestion, defecation, diuresis and conserves energy. Some organs (e.g. the heart) are innervated by both divisions.

SD nerve fibres emerge from the spinal cord at the thoracic and lumbar vertebrae; hence, the SD is known as a "thoracolumbar" system. PSD fibres emerge from the brain and sacral spinal cord; hence, the PSD is known as a "craniosacral" system. The vagus nerve (cranial nerve X) carries 90% of PSD impulses. Acetylcholine is a neurotransmitter released by the axon terminals of all PSD neurons. It binds to five types of receptors: one nicotinic and four muscarinic (M1, M2, M3, M4). Norepinephrine is a neurotransmitter released by the axon terminals of most SD post-ganglionic fibres (find out what this means). It binds to five types of receptor: α_1, α_2, β_1, β_2, β_3. It is the type of receptor (not the neurotransmitter molecule) that determines whether the postsynaptic cell responds by being stimulated or being inhibited.

Reflexes are also not under our conscious control despite skeletal muscles being the source of many of the movements. This is because (for spinal reflexes) the sensory neuron carrying the afferent impulse synapses directly with a motor neuron(s) in the spinal cord that carries the efferent impulse to the responding muscles. An interpretation by the brain is not involved.

1. Which of the following sends sensory information to the brain?
   A. The afferent division of the peripheral nervous system.
   B. The efferent division of the peripheral nervous system.
   C. The somatic nervous system.
   D. The autonomic nervous system.

Answer is A: Afferent means "incoming". Incoming information travels from the body's sensors to the brain carrying sensory information.

2. What part of the nervous system prepares the body for action during extreme situations?
   A. The limbic system
   B. The sympathetic division
   C. The efferent system
   D. The parasympathetic division

Answer is B: The sympathetic division (of the autonomic nervous system) "stirs you up". It prepares for action, those body organs that will be required for rapid and vigorous activity.

3. To what does the following description apply? "An unlearned and involuntary but predictable motor response to a stimulus, that is rapid and does not involve any processing by the brain".
   A. Spinal reflex
   B. Autonomic reflex

C. Cranial reflex
D. Learned reflex

Answer is A: Spinal reflexes happen without the involvement of conscious or unconscious decision-making by the brain.

4. Which statement about the sympathetic and/or parasympathetic divisions is correct?
   A. All sympathetic neurons release ACh as a neurotransmitter.
   B. Sympathetic division fibres emerge from brain and sacral spinal cord.
   C. Parasympathetic division stimulates adrenal gland to release norepinephrine and epinephrine.
   D. Some organs are innervated by both sympathetic division and parasympathetic division.

Answer is D: Indeed, some organs (e.g. heart) are innervated by both divisions. In choices A and B, replace "sympathetic" with parasympathetic, and in choice C replace "parasympathetic" with sympathetic, to get a true statement.

5. Which one of the following parts of the nervous system carries impulses towards the brain?
   A. Peripheral nervous system
   B. Somatic nervous system
   C. Autonomic nervous system
   D. Parasympathetic division

Answer is A: The peripheral NS delivers sensory information to CNS as well as carrying motor commands from CNS to peripheral tissues. The somatic NS is part of the efferent division (carrying impulses away from the brain) of the peripheral NS. The autonomic (and parasympathetic division) carry efferent impulses.

6. Which statement about efferent impulses in the spinal cord is correct?
   A. They travel along the spinothalamic and spinocerebellar tracts.
   B. They pass along tracts that are located in the grey matter of the spinal cord.
   C. The nerve cells that they travel through have their cell bodies located in the dorsal root ganglia.
   D. The axons carrying these impulses pass through the ventral root of the spinal nerves.

Answer is D: Efferent impulses are motor impulses so leave the spinal cord via the ventral roots. "Spinothalamic" means from spine to thalamus, i.e. ascending or sensory. Tracts are white matter not grey. Sensory neurons have their cell bodies in the dorsal root ganglia.

7. What statement is true about spinal reflexes?
   A. They cannot be inhibited or reinforced by the brain.
   B. They do not involve processing by the brain.

C. They involve processing by the brainstem.
D. They are all simple monosynaptic pathways.

Answer is B: A spinal reflex does not involve processing by the brain. However, they can be inhibited or reinforced by the brain. Some are polysynaptic.

8. Which one of the statements below about the autonomic nervous system is correct?
   A. The sympathetic division of the autonomic nervous system is part of the somatic nervous system.
   B. The parasympathetic division contains the neurons responsible for spinal reflexes.
   C. The sympathetic division prepares the body for vigorous exercise or emergency situations.
   D. The parasympathetic division uses nerves that emerge from the spinal cord between the T1 and L2 vertebrae.

Answer is C: The sympathetic division “stirs you up”. The ANS and somatic NS are distinct parts of the efferent section of the peripheral nervous system. The parasympathetic division is “craniosacral” not “thoracolumbar”.

9. The nervous system is divided into two divisions. What are they called?
   A. Somatic and autonomic
   B. Central and peripheral
   C. Afferent and efferent
   D. Sympathetic and parasympathetic

Answer is B: The central nervous system and the peripheral nervous system are the two divisions.

10. What do “sympathetic” and “parasympathetic” divisions refer to?
    A. The central nervous system
    B. The efferent neurons of the peripheral nervous system
    C. The autonomic nervous system
    D. The somatic nervous system

Answer is C: The autonomic NS, which in turn is one part (the somatic NS is the other part) of the efferent division of the peripheral NS.

11. Which of the following neural pathways or tracts carry sensory information?
    A. Corticobulbar tracts
    B. Spinothalamic tracts
    C. Corticospinal tracts
    D. Reticulospinal tracts

Answer is B: The first part of the name tells you from where the pathway commences: “spino-”, the second part tells you where the information is going: “thalamic”. If information is going from spinal cord to thalamus (in the brain), it MUST be carrying sensory information.

12. Nerve impulses carried by the parasympathetic division travel along which nerve fibres?
    A. Cranial nerves I and II
    B. The spinocerebellar pathway
    C. The spinal nerves
    D. The vagus nerves

Answer is D: 90% of parasympathetic impulses are carried by the vagus nerves (cranial nerve X).

13. A motor pathway in the autonomic nervous system consists of two neurons. Where does the second neuron start and finish?
    A. Starts in the CNS and runs to the effector organ
    B. Starts in a ganglion and runs to the CNS
    C. Starts in a ganglion and runs to the effector organ
    D. Starts in the CNS and runs to a ganglion

Answer is C: The second neuron runs from ganglion to effector. The first neuron runs from the CNS to a ganglion, then synapses with the cell body of neuron #2.

14. Which of the following is true of parasympathetic neurons?
    A. They emerge from the thoracic and lumbar vertebrae.
    B. They all release ACh as a neurotransmitter.
    C. They all release NE as a neurotransmitter.
    D. They have short pre-ganglionic fibres and long post-ganglionic fibres.

Answer is B: All parasympathetic neurons release ACh as a neurotransmitter. Choices A and D are true of the sympathetic division.

15. Adrenergic receptors are so named because they:
    A. Are located in the adrenal glands
    B. Bind epinephrine and norepinephrine
    C. Are located in the kidneys
    D. Bind acetylcholine

Answer is B: Adrenergic receptors bind adrenalin (and noradrenalin) which is another name for epinephrine (and norepinephrine).

16. What is the purpose of a neurotransmitter?
    A. To pass a nerve impulse along a nerve cell axon
    B. To pass a nerve impulse onto another cell
    C. To pass a nerve impulse onto a muscle cell
    D. To pass a nerve impulse onto another nerve cell

Answer is B: Choices C and D are also correct, but B is the more general answer.

17. What innervates the diaphragm?
    A. The spinal nerves from T6 to T12
    B. The vagus nerve
    C. The phrenic nerve
    D. The sciatic nerve

Answer is C: The phrenic nerve, which originates from spinal nerves leaving the spinal cord at C3, C4 and C5.

18. Consider the following pairs of terms. Which pair has a term that refers to a part of the nervous system that carries sensory information to the brain, and a term that refers to a part that carries motor commands to the peripheral tissues?
    A. Parasympathetic division; sympathetic division
    B. Somatic nervous system; autonomic nervous system
    C. Afferent division; efferent division
    D. Central nervous system; peripheral nervous system

Answer is C: Afferent division carries sensory info to the brain, while efferent refers to motor info leaving the brain.

19. Which statement about the neurotransmitter acetylcholine (ACh) is **NOT** correct?
    A. ACh is released by all autonomic nervous system pre-ganglionic fibres.
    B. ACh is released by all sympathetic division post-ganglionic fibres.
    C. ACh is released by all parasympathetic division post-ganglionic fibres.
    D. ACh is released by all somatic division motor nerve fibres.

Answer is B: In fact norepinephrine (not ACh) is released by most ANS-SD post-ganglionic fibres, i.e. at second synapse. Choices A, C and D are true.

20. Which organisational entity of the brain is divided into an afferent and an efferent division?
    A. The central nervous system
    B. The peripheral nervous system
    C. The somatic nervous system
    D. The autonomic nervous system

Answer is B: The peripheral nervous system has afferent neurons that carry sensory information to the brain and efferent neurons that carry motor information from the brain to the body.

21. Which neurotransmitter do all motor neurons release at their synapses with skeletal muscle cells?
    A. ACh
    B. ATP
    C. GABA
    D. Norepinephrine

Answer is A: Motor neurons release acetylcholine.

22. Which muscle(s) are **NOT** controlled by the autonomic nervous system?
    A. Cardiac muscle
    B. The diaphragm
    C. Skeletal muscle
    D. Smooth muscle

Answer is C: Skeletal muscle is under our conscious control and innervated by the somatic nervous system, which is one part of the efferent division of the peripheral nervous system.

23. Which type of receptor always produces stimulation of the postsynaptic cell when bound by a neurotransmitter?
    A. Nicotinic
    B. Muscarinic
    C. Alpha adrenergic
    D. Beta adrenergic

Answer is A: Nicotinic receptors when bound by ACh are always stimulatory. Muscarinic receptors may be stimulatory or inhibitory.

24. What is one effect that the sympathetic division of the autonomic nervous system have?
    A. Increases gut motility and digestive secretions
    B. Causes bronchioles to constrict
    C. Decreases heart rate
    D. Stimulates sweating from sweat glands

Answer is D: The sympathetic division "stirs you up" to get ready for action. The only choice consistent with vigorous activity is D.

25. To what part of the nervous system does the somatic nervous system belong?
    A. Efferent, central nervous system
    B. Efferent, peripheral nervous system
    C. Afferent, peripheral nervous system
    D. Afferent, central nervous system

Answer is B: The somatic nervous system carries commands from the brain (efferent impulses) to the muscles and glands—which are in the peripheries of the body.

26. What is the parasympathetic division of the autonomic nervous system responsible for?
    A. Rapid predictable motor responses without processing by the brain
    B. Conserving energy and maintaining body activities without conscious brain control
    C. Preparing the body for energetic activity without conscious brain control

D. Gathering sensory information from the viscera that is not interpreted by the brain

Answer is B: The parasympathetic division prepares body for "rest and repose": conserves energy and promotes non-emergency function.

27. The following receptors for neurotransmitters may be stimulatory or inhibitory EXCEPT for one of them. Which One?
    A. Nicotinic receptors
    B. Muscarinic receptors
    C. Adrenergic alpha receptors
    D. Adrenergic beta receptors

Answer is A: Nicotinic acetylcholine receptors bind ACh and are always stimulatory.

28. What do "sympathetic" and "parasympathetic" refer to? Divisions of:
    A. The central nervous system
    B. The efferent neurons of the peripheral nervous system
    C. The autonomic nervous system
    D. The somatic nervous system

Answer is C: The autonomic nervous system is comprised of the sympathetic and the parasympathetic divisions.

29. What structure(s) does a neurotransmitter molecule cross?
    A. Synaptic cleft
    B. Synaptic cleft and the postsynaptic membrane
    C. Presynaptic membrane and the synaptic cleft
    D. Postsynaptic membrane

Answer is C: A neurotransmitter is released from the synaptic knob of the presynaptic neuron, so must get out of the neuron by crossing its plasma membrane, the synaptic cleft, before lodging in the receptor on the postsynaptic membrane.

30. What is a spinal reflex?
    A. It involves rapid processing by the brain and a predictable response.
    B. It involves stimulation of a motor neurone by a sensory neurone without a synapse.
    C. It is a rapid, predictable, learned and involuntary motor response.
    D. It is a predictable, unlearned and involuntary motor response.

Answer is D: Brain processing is not involved but a synapse is. The response is not learned.

31. A "post-ganglionic cholinergic fibre" refers to a neuron that:
    A. Runs from CNS to a ganglion and releases noradrenaline

B. Synapses with an effector cell and whose neurotransmitter stimulates alpha and beta receptors
C. Runs from a ganglion to an effector cell and releases norepinephrine
D. Synapses with an effector cell and releases acetylcholine (ACh)

Answer is D: A "post-ganglionic cholinergic fibre" runs from a ganglion to an effector organ (so choice A is wrong) and releases ACh (so choice C is wrong). Alpha and beta receptors are both adrenergic—that is bind with epinephrine and norepinephrine (= adrenalin and noradrenalin), so choice B is wrong.

32. Which part of the nervous system prepares you for vigorous activity ("to fight or flee")?

A. Sympathetic
B. Parasympathetic
C. Somatic
D. Autonomic

Answer is A: The sympathetic NS "stirs you up" for vigorous activity.

33. What is that part of the nervous system that carries commands to the skeletal muscles called?

A. Somatic nervous system
B. Autonomic nervous system
C. Central nervous system
D. Sympathetic division

Answer is A: The somatic NS carries motor commands to the skeletal muscles (our soma).

34. Which neurotransmitter do all motor neurons release at their synapses?

A. Acetylcholine
B. Norepinephrine
C. Dopamine
D. Adenosine triphosphate

Answer is A: All somatic motor neurons release acetylcholine at their synapses with skeletal muscle fibres.

35. Which neurotransmitter is released by all parasympathetic neurons?

A. Norepinephrine
B. Acetylcholine
C. Nicotine
D. Muscarine

Answer is B: All parasympathetic neurons release ACh as a neurotransmitter. Nicotine and muscarine are not produced by the body.

36. Which choice correctly ends the following sentence? The parasympathetic division is part of the:

    A. Autonomic nervous system
    B. Somatic nervous system
    C. Afferent division of the peripheral nervous system
    D. Central nervous system

Answer is A: The parasympathetic division is part of the autonomic nervous system.

37. What may accurately be said of the postsynaptic membrane?

    A. It is attached to the transmitting axon.
    B. It has receptors for a neurotransmitter.
    C. It is before the synaptic cleft.
    D. It is part of a neurone.

Answer is B: The postsynaptic membrane is stimulated by a neurotransmitter, after it crosses the synaptic cleft, so has receptors for it. The postsynaptic membrane may be part of a muscle cell.

38. Which statement about the vagus nerve is true?

    A. It lies within the cerebrospinal tract.
    B. It arises from the pons.
    C. It has a sensory function in vision and olfaction.
    D. It carries parasympathetic motor impulses.

Answer is D: The vagus nerve carries 90% of parasympathetic motor impulses. It arises mainly from the medulla oblongata.

39. One thing that could **NOT** be correctly said of the sympathetic division is that:

    A. It stimulates the adrenal gland to release adrenaline and noradrenaline.
    B. Its fibres emerge from the spinal cord at the thoracic and lumbar vertebrae.
    C. It promotes the conservation of the body's energy.
    D. It supplies the smooth muscle of blood vessels.

Answer is C: The sympathetic division "stirs you up" rather than promoting the conservation of energy.

40. Choose the true statement.

    A. A spinal reflex does not involve processing in the brain.
    B. Spinal reflexes are rapid predictable learned responses to stimuli.
    C. Spinal reflexes are used to diagnose brain death.
    D. Spinal reflexes are voluntary but unlearned.

Answer is A: Spinal reflexes are not learned, nor used to diagnose brain death. Spinal reflexes are not under voluntary control.

41. If the term "cholinergic" is applied, to a synapse what does it mean?
    A. The target organs are innervated by the sympathetic nervous system.
    B. The receptors are nicotinic.
    C. The result is always stimulatory.
    D. Acetylcholine is released at the synapse.

Answer is D: Cholinergic (or adrenergic) refer to the neurotransmitter acetylcholine (or adrenaline/noradrenalin = epinephrine/norepinephrine) that is released at the synapse.

42. What determines the response of the postsynaptic cell to autonomic impulses?
    A. The neurotransmitter that binds to the cell
    B. The type of receptor on the cell
    C. Whether the innervation is sympathetic or parasympathetic
    D. Whether the fibre is pre-ganglionic or post-ganglionic

Answer is B: It is the type of receptor on the postsynaptic membrane (not the neurotransmitter molecule) that determines how the postsynaptic cell responds.

43. What is the nerve that that carries most of the parasympathetic signals?
    A. Phrenic
    B. Vagus
    C. Sciatic
    D. Trigeminal

Answer is B: The vagus nerves carry 90% of the parasympathetic signals.

44. Which is an example of a cholinergic receptor?
    A. Nicotinic receptor
    B. Adrenergic receptor
    C. Alpha receptor
    D. Beta receptor

Answer is A: A nicotinic receptor binds acetylcholine so is called a "cholinergic" receptor.

45. Which statement about neurotransmitters and/or receptors is correct?
    A. All somatic motor neurons release ACh at their synapse.
    B. Noradrenalin is the major neurotransmitter of the parasympathetic division.
    C. Nicotinic receptors when bound by ACh are always inhibitory.
    D. Noradrenalin binds to nicotinic and muscarinic receptors.

Answer is A: ACh is the major neurotransmitter of the PSD; nicotinic receptors are always stimulatory; ACh binds to nicotinic and muscarinic receptors.

46. The peripheral nervous system is divided into:
    A. Sympathetic division and parasympathetic division
    B. Brain and spinal cord
    C. Somatic system and autonomic system
    D. Motor division and sensory division

Answer is D: The PNS comprises neurons that can transmit efferent motor impulses and neurons that transmit afferent sensory impulses.

47. A neurotransmitter is a molecule that crosses the:
    A. Synaptic cleft
    B. Synaptic cleft and the postsynaptic membrane
    C. Presynaptic membrane and the synaptic cleft
    D. Postsynaptic membrane

Answer is C: A neurotransmitter is released from within one neuron, crosses the cleft to bind with the receptor in the membrane of the next cell.

48. What characterises a spinal reflex?
    A. It involves rapid processing by the brain and a predictable response.
    B. It is a predictable, unlearned and involuntary motor response.
    C. It is a rapid, predictable, learned and involuntary motor response.
    D. It involves stimulation of a motor neurone by a sensory neurone that originates within the CNS.

Answer is B: Peripheral stimulation produces an involuntary, unlearned predictable response that is not initiated from the brain.

49. When their neurotransmitter binds to them, which of the following receptors is always stimulatory?
    A. Muscarinic
    B. Cholinergic
    C. Nicotinic
    D. Adrenergic

Answer is C: Nicotinic receptors when bound by ACh are always stimulatory.

50. Two ways that cells can communicate within the body are by synaptic communication or by endocrine communication. A difference between the two is:
    A. Endocrine communication involves a chemical messenger, whereas synaptic communication does not.
    B. The action caused by synaptic communication may persist for several hours, whereas that caused by endocrine communication persists for several minutes.
    C. Endocrine communication controls cellular activities in distant tissues, whereas synaptic communication affects the adjacent cell.

D. Synaptic communication occurs between adjacent cells, whereas endocrine communication occurs between cells of the same tissue.

Answer is C: A synapse is a communication between one cell and another adjacent cell, separated only by the synaptic cleft.

51. Which of the following is **NOT** indicative of the stimulation of the parasympathetic system?
    A. Constriction of skeletal muscle blood vessels and vasodilation of renal blood vessels
    B. Excitation of the blood flow to the kidneys and smooth muscles of the GI tract
    C. Inhibition of heart rate and smooth muscles of the urinary bladder
    D. Excitation of the smooth muscles of the urinary bladder and GI tract

Answer is A: The parasympathetic division is not involved in vasoconstriction or dilation of skeletal muscle blood vessels.

52. Sensory receptors convert stimuli into what?
    A. Neurotransmitters
    B. Action potentials
    C. Graded potentials
    D. Voltage-gated channels

Answer is C: When a neuron is stimulated by a sensory stimulus, a graded potential is produced.

53. What is the function of a spinal nerve?
    A. Transmit sensory information
    B. Transmit both sensory and motor information
    C. Connect sensory and motor neurons
    D. Transmit autonomic nervous system information

Answer is B: Spinal nerves arise from the combination of a dorsal root that carries sensory information and a ventral root that carries motor information.

54. Sensory receptors convert stimuli that are above threshold into what?
    A. Graded potentials
    B. Neurotransmitters
    C. Action potentials
    D. Motor activity in muscles

Answer is C: Graded potentials that are above the threshold, stimulate an action potential.

55. What is the function of an efferent neuron?
    A. Transmit sensory information
    B. Transmit motor information

C. Connect sensory and motor neurons
D. Transmit both sensory and motor information

Answer is B: Efferent means “outgoing”. Outgoing information travels from the brain to the body carrying motor information to an effector—a muscle or gland.

56. Which autonomic centre controls homeostasis in the viscera?
    A. The hypothalamus
    B. The pons
    C. The thalamus
    D. The limbic system

Answer is A: The hypothalamus is the main visceral control centre of body homeostasis. The thalamus receives sensory information before passing it onto basal nuclei and cerebral cortex.

57. In brain death, why are the corneal and pharyngeal reflexes absent?
    A. In brain death, the brain stem does not respond to stimuli.
    B. The brain is unable to receive nerve impulses from the spinal tracts in brain death.
    C. In brain death, cranial nerves can no longer transmit nerve impulses.
    D. The cortex and diencephalon are unresponsive in brain death.

Answer is A: The corneal (blink) reflex and pharyngeal (gag) reflex are mediated by cranial nerves V and IX, respectively. These nerves are attached to the brain stem. In an unconscious person who is brain dead, even the brain stem is unresponsive.

58. What effect does “brain death” have on the patellar and Achilles reflexes?
    A. There is no effect, these reflexes still work as they are spinal reflexes that do not rely on the brain being alive.
    B. These reflexes are absent as the brain is unable to send nerve impulses to the spinal tracts in brain death.
    C. These reflexes are enhanced as the dead brain no longer exerts a partial suppression to these reflexes.
    D. These spinal reflexes are diminished as in brain death there are no reinforcement impulses from the brain.

Answer is C: Spinal reflexes initiate and complete without the need for nerve impulses to be sent to or from the brain. In fact an active brain will exert some restraint on the muscle activity that is generated. Hence, a dead brain does not exert any restraint so the spinal reflexes are more marked.

59. Which of the following describes the motion that is produced when the patellar reflex is tested by tapping the patellar tendon?
    A. Leg extension
    B. Plantar flexion
    C. Leg flexion

D. Thigh extension

Answer is A: The patellar reflex affects the leg below the knee. The leg (anatomically between the knee and ankle) kicks out. This is extension not flexion.

60. What are the two divisions of the autonomic nervous system?
    A. Central and peripheral
    B. Afferent and efferent
    C. Sympathetic and parasympathetic
    D. Sensory and motor

Answer is C: Choice A refers to the two divisions of the nervous system. Choice B refers to the two divisions of the peripheral nervous system. The terms in choice D are synonyms for those in choice B.

61. Which of the following is a characteristic of inborn spinal reflexes?
    A. They disappear after brain death.
    B. They are rapid but unpredictable responses.
    C. They are used to diagnose damage to cranial nerves.
    D. They do not involve processing by the brain.

Answer is D: Spinal reflexes do not require processing by the brain, hence are not useful for choice C and are present even after brain death.

62. What may be said of the parasympathetic division of the autonomic nervous system?
    A. Its fibres emerge from the thoracic and lumbar regions.
    B. The impulses flow mainly along the vagus nerves.
    C. It stimulates the release of epinephrine and norepinephrine.
    D. Are associated with ganglia that lie close to the spinal cord.

Answer is B: The vagus nerve (CN X) carries about 90% of parasympathetic impulses. Choices A and D are true of the sympathetic division.

63. From what part(s) of the body does the vagus nerve originate?
    A. The spinal cord at the thoracic, lumbar and sacral vertebrae
    B. The spinal cord at the thoracic and lumbar vertebrae
    C. The spinal cord at the thoracic vertebrae
    D. The midbrain

Answer is D: The vagus nerve is cranial nerve X, so originates from the medulla oblongata of the midbrain. It is not a spinal nerve.

64. What type of chemical is stored in the synaptic vesicles of a neurone?
    A. Cerebrospinal fluid
    B. Enzyme
    C. Neurotransmitter

D. Potassium ions

Answer is C: The neurotransmitter, that when released from the neurolemma, crosses the synaptic cleft to transmit the nerve impulse, is stored in the synaptic vesicles.

65. Which of the following is a list of four neurotransmitters?
    A. Globulin, ferratin, albumin, fibrinogen
    B. ATP, ADP, pyruvic acid, glucose
    C. Norepinephrine, glucagon, aldosterone, erythropoietin
    D. Acetylcholine, norepinephrine, dopamine, serotonin

Answer is D: None of the four in choice A is a neurotransmitter. Neither is ADP, pyruvic acid, glucose, glucagon, aldosterone or erythropoietin.

66. In brain death, the corneal and pharyngeal reflexes are absent, whereas the patellar and Achilles reflexes are enhanced. What is the reason for this?
    A. The corneal and pharyngeal reflexes are spinal reflexes, whereas the other two are cranial reflexes, which are absent in brain death.
    B. The corneal and pharyngeal reflexes are cranial nerve reflexes, whereas the other two are spinal reflexes, which are more pronounced in brain death.
    C. The corneal and pharyngeal reflexes are spinal reflexes which are inhibited in brain death, whereas the other two are cranial reflexes.
    D. The corneal and pharyngeal reflexes are cranial nerve reflexes, whereas the other two are spinal reflexes, which are inhibited in brain death.

    Answer is B: Brain death implies that even the brain stem is inactive which means that all of the cranial nerves carry no efferent impulses. The corneal and pharyngeal reflexes occur when cranial nerves respond to a stimulus, but this is not possible in brain death. The patellar and Achilles reflexes are spinal reflexes. This means that the response to the stimulus originates from the spinal cord, not from the brain. The spinal cord can still operate without the brain. In fact the absence of brain activity means that inhibition from the brain is removed which means that spinal reflexes are more pronounced.

67. What is the stretch reflex?
    A. A reflex that involves one of the 12 cranial nerves
    B. The brief contraction of a muscle following the stimulus of a stretch of that muscle
    C. A spinal reflex that does not involve the brain
    D. The contraction of a skeletal muscle causes the antagonist muscle to simultaneously lengthen and relax

Answer is B: Stretch reflex: it is an example of a monosynaptic reflex because there is only one synapse between the single sensory and single motor neurones involved. The patellar reflex is an example of a stretch reflex. The stimulus is a tap on the patellar tendon. This stretches receptors within the quadriceps muscles. The response is a brief contraction of those muscles which produces a noticeable kick. Choice D describes the tendon reflex.

68. Which type (or types) of sympathetic nervous system receptor is found in the lungs?
    A. Beta-2
    B. Nicotinic and beta-2
    C. Nicotinic and muscarinic
    D. Nicotinic, muscarinic and beta-2

Answer is A: Beta-2 receptors are one type of adrenergic receptor and are found in the smooth muscle of the lungs. Nicotinic and muscarinic receptors are in the parasympathetic system.

69. Which type of drug would exhibit parasympathomimetic behaviour?
    A. Adrenergic
    B. Adrenergic and cholinergic
    C. Anticholinergic
    D. Cholinergic

Answer is D: A parasympathomimetic drug, sometimes called a cholinomimetic drug, is a substance that stimulates the parasympathetic nervous system. These chemicals are also called cholinergic drugs because acetylcholine (ACh) is the neurotransmitter used by the parasympathetic NS.

14.4 Special Senses (Eye and Ear)

14.4.1 Eye

Light that strikes the cone cells of the fovea centralis of the macula lutea (of the retina), first passes through the cornea, then the aqueous humour (in the anterior chamber of the eye), the pupil (the central opening in the iris), the lens and through the vitreous humour (in the posterior chamber). Both the cornea and the lens refract (bend) the light that passes through them. However, it is the elasticity of the lens which allows its focal length to be changed and that brings light from any viewed object, regardless of its distance from the eye, to a focus on the retina. This focussing process is called accommodation. It is achieved by the donut-shaped ciliary muscle surrounding the lens and attached to it by the ciliary fibres. By relaxing into a larger diameter hole, the ciliary muscle pulls the ciliary fibres tight, which stretches the biconvex lens to a thinner shape. This thinner shape has a longer focal length which is suited to viewing objects that are distant from the eye. The reverse happens when we focus on nearby objects. Removing the tension from the ciliary fibres allows the elasticity of the lens to regain a thicker biconvex shape.

With age, the lens loses elasticity and so does not ooze back to a more rounded biconvex shape when the ciliary muscle contracts to form a smaller diameter hole, so loosening the ciliary fibres. The more rounded shape is necessary to bring light from nearby objects to focus on the retina. This means that people over about 50 years of age require reading glasses to be able to focus on nearby objects, that is, to read. The loss of lens elasticity and subsequent inability to focus on close objects is called presbyopia.

Away from the fovea centralis, light striking the rod or cone cells of the retina must first pass through the layer of ganglion cells and bipolar cells that transmit the impulses generated by the rods and cones to the optic nerve. The axons of the ganglion and bipolar cells, as well as blood vessels gather together as the optic nerve (cranial nerve II) to pass out of the eyeball at the optic disc. This area (the blind spot) is insensitive to light as there are no rods or cones there. There are three types of cone cells that are maximally sensitive to one of red, green or blue light (they are sensitive also to other wavelengths, just less so). Their combined effect allows us to perceive colour. Rod cells are useful for seeing in conditions where the intensity of light is low such as at night (scotopic conditions), but are unable to distinguish colour.

1. Choose the correct statement about the eye:
   A. Bipolar and ganglion cells occur in the retina except at the fovea.
   B. Rod and cone cells occur in the retina except at the fovea.
   C. Bipolar and ganglion cells occur in the choroid except at the optic disc.
   D. Rod and cone cells occur in the choroid except at the optic disc.

Answer is A: At the fovea, the bipolar and ganglion cells do not cover the cone cells so light can strike the cones directly. Bipolar, ganglion, rod and cone cells occur in the retina, not the choroid.

2. Which part of the retina has the greatest sensitivity to light?
   A. The optic disc
   B. Macula lutea
   C. The choroid
   D. Fovea centralis

Answer is D: The fovea centralis is at the centre of the macula lutea. The optic disc is insensitive to light.

3. The deterioration of sight with age is known by which term?
   A. Protanopia
   B. Presbyopia
   C. Hyperopia
   D. Scotopia

Answer is B: "Presbys" means old (or elder): presbyopia = old age vision; presbycusis = old age hearing.

4. What is the place where the blood vessels and nerve fibres come together and leave the posterior chamber of the eye called?
   A. Macula lutea
   B. Optic disc
   C. Fovea centralis
   D. Choroid

Answer is B: Optic disc is where no rods or cones exists as the nerves and blood vessels pass out through the choroid and sclera here.

5. Accommodation refers to the eye's ability to focus light from objects whatever their distance from the eye. How is this achieved?
   A. By altering the distance between the cornea and the eye's lens
   B. By altering the distance between the lens and the retina
   C. By altering the shape of the eye's lens
   D. By altering the shape of the cornea

Answer is C: The eye lens is flexible. If stretched by the zonules, it adopts a flattened and thin bi-convex shape; if it is allowed to recoil into its relaxed more spherical shape, it can focus on objects very close to the eye (such as a page of writing).

6. What does the ciliary muscle do when accommodation (focussing) in the eye occurs?
   A. It contracts and the tension on the ciliary fibres increases making the eye lens less convex.
   B. It relaxes and the tension on the ciliary fibres decreases allowing the eye lens to become more convex.
   C. It contracts and the pull of the ciliary fibres decreases allowing the lens to become more convex.
   D. It relaxes and the pull of the ciliary fibres increases to make the lens less convex.

Answer is C: The ciliary muscle is doughnut-shaped with the lens in the "hole" and attached to the muscle by ciliary fibres (zonules). When the ciliary muscle contracts, it moves towards the lens which allows the ciliary fibres to reduce their tension, so that the lens is stretched less. Hence, it oozes to a more rounded shape.

7. Which cells of the retina are responsible for detecting light in scotopic (i.e. low light) conditions?
   A. Bipolar cells
   B. Rod cells
   C. Ganglion cells
   D. Cone cells

Answer is B: Rods and cones are the light-sensitive cells. Rods are used for viewing in low light (dim) or nighttime conditions.

8. Accommodation refers to our ability to bring objects at any distance into sharp visual focus by altering which of the following?
   A. The distance between our eye and the object to be viewed
   B. The curvature of our cornea
   C. The curvature of the eye's lens
   D. The distance between the lens and retina

Answer is C: The eye's lens is convex and flexible. So its degree of curvature (convexity) and hence its "focal length" can be altered, so that light from objects at almost any distance can be made to come to a focus on the retina.

9. What is the purpose of the optic chiasma?
   A. To allow images from each eye to crossover to the other side of the brain prior to crossing back at the decussation of pyramids
   B. To allow fibres from the medial aspect of one eye to join fibres from the lateral aspect of the other eye to form an optic tract
   C. To allow the fibres from the lateral aspect of each eye to come together as an optic tract
   D. To allow light entering the left eye to be interpreted by the right hand side of the occipital lobe (and vice versa)

Answer is B: The optic chiasma (crossing) allows the stimulation produced by light striking the rods and cones on the medial aspect of one eye to crossover to the other side of the brain to travel with the stimulation produced in the lateral aspect of the other eye. Hence, some information from each eye is interpreted together in each side of the occipital lobe. In this way, binocular vision and the perception of distance are achieved.

10. What is presbyopia (old-age vision) due to?
    A. The loss of elasticity of the lens of the eye
    B. The change in the curvature of the cornea
    C. The gradual loss of cone cells from the retina
    D. The deviation from a spherical eye-ball shape with aging

Answer is A: Presbyopia refers the progressive loss of the ability to focus on nearby objects (e.g. books) after about age 45–50 years. It is due to the lens becoming stiffer so being less able to ooze into the highly convex shape required to bring light from nearby objects to focus on the retina.

11. In which region of the eye does the most detailed vision occur?
    A. Fovea centralis
    B. Optic disc
    C. Macula lutea
    D. Ciliary body

Answer is A: The fovea centralis is at the centre of the macula lutea. The optic disc is insensitive to light.

12. How does accommodation in the eye occur? When our ciliary muscle:
    A. Contracts and the tension on the ciliary fibres increases allowing a rounder lens
    B. Relaxes and the tension on the ciliary fibres decreases allowing a rounder lens
    C. Relaxes and the pull of our ciliary fibres flattens the lens
    D. Contracts and the pull of our ciliary fibres flattens the lens

Answer is C: The ciliary muscle is doughnut-shaped with the lens in the "hole" and attached to the muscle by ciliary fibres (zonules). When the ciliary muscle relaxes, it moves away from the lens (the diameter of the hole increases) which causes the ciliary fibres to pull on the lens. This tension in the fibres stretches and flattens the lens into a less convex shape.

13. What is the name of the structure that allows nerve fibres from the medial aspect of each retina to join fibres from the lateral aspect of the retina of the other eye?
    A. Optic chiasma
    B. Optic nerve
    C. Optic radiation
    D. Lateral geniculate body

Answer is A: The optic chiasma is where the fibres crossover.

14. What is the light-sensitive cell in the retina that responds to colour called?
    A. Macula
    B. Macula lutea
    C. Cone
    D. Rod

Answer is C: Remember "c" for cone and "c" for colour. Rods are used for conditions where the intensity of light is low (the conditions are "scotopic", for example at night).

15. People over 45 years of age eventually require reading glasses. This condition (known as presbyopia) is the result of which of the following conditions?
    A. The loss of elasticity in the lens
    B. The development of cataracts in the lens
    C. The decrease in the refractive index of the cornea
    D. The degeneration of the cone cells of the retina

Answer is A: The elastic lens loses some of its elasticity with age so is not able to focus light from nearby objects such as books. Hence, "reading" glasses—which have convex lenses—become necessary.

16. Which of the following statements about the structure of the eye is **INCORRECT**?
    A. The retina converts light energy into electrical nerve impulses that are sent to the brain.

B. All detailed vision occurs in a very small area called the fovea centralis.
C. At the blind spot, nerve fibres of the ganglion cells come together and leave the eye as the optic nerve.
D. The black pupil is a sphincter muscle under parasympathetic nerve control.

Answer is D: The black pupil is the central hole in the pupillary muscles of the iris through which light passes to strike the retina. It appears black as normally no light exits the pupil.

17. What is meant by the term "accommodation" when referring to our vision? It refers to the ability of our eye to:
    A. Alter the thickness of the lens to focus on objects whatever their distance from us
    B. Alter the amount of refraction occurring in the cornea
    C. Alter the diameter of the pupil to cope with situations of different light intensity
    D. Use either rods or cones for vision depending on whether we are viewing during daylight or at night

Answer is A: Accommodation refers to the ability of our eye's lens to alter its shape—its thickness or roundness or convexity—to enable it to bring light from objects at any distance to a focus on the retina.

18. Choose the answer that correctly completes the sentence. The "lens" of the human eye:
    A. Is a biconcave lens
    B. Produces more refraction than the cornea
    C. Can have its focal length altered
    D. Consists of rods and cones

Answer is C: The lens, by having its degree of convexity altered, can have its focal length changed. It is in fact biconvex.

19. Which statement about refraction of light in the human eye is correct?
    A. Most of it occurs in the lens.
    B. Its extent is governed by the size of the iris opening to the eye.
    C. Most of it occurs as light enters the cornea from air.
    D. The angle of incidence is equal to the angle of refraction.

Answer is C: The cornea is also a lens, hence light is refracted as it enters the cornea from air. As the difference between air and corneal tissue is greater than the difference between aqueous humour and the lens, more refraction occurs in the cornea than in the "lens".

20. The process of adjusting the eye's lens to view objects at different distances from the eye is called:
    A. Accommodation
    B. Presbyopia
    C. Refraction
    D. Hyperopia

Answer is A: The eye can accommodate its lens to clearly see objects at any distance from the eye.

21. Glaucoma is an eye disease which affects vision. It is caused by:
    A. Blockage of the flow of aqueous humour through the canal of Schlemm and loss of intraocular pressure in the vitreous humour
    B. Detachment of the retina and subsequent loss of vision in this part of the eye
    C. Cataracts that form in the eye's lens which prevent light from reaching the retina
    D. Increased intraocular pressure which collapses the blood capillaries that perfuse the retina so part of it dies

Answer is D: Glaucoma results from higher than healthy intraocular pressure in the eyeball.

22. Which one of the following statements is **INCORRECT**?
    A. A convex lens causes light rays to converge.
    B. The fluid between the cornea and the eye's lens is a lens.
    C. The cornea is a lens.
    D. The eye's lens is a convex lens.

Answer is B: the fluid (aqueous humour) in the anterior chamber is not a lens.

23. In the human eye where does the greatest refraction occur?
    A. In the lens of the eye
    B. At the retina
    C. As light passes from air into the cornea
    D. As light passes from the lens into the vitreous humour

Answer is C: Most refraction occurs as light enters the cornea because the difference between the refractive index of air and that of the cornea is large.

24. In order to focus on objects that are very close to the eye, what must happen to the human lens? It must:
    A. Increase its focal length
    B. Increase the amount of refraction it causes
    C. Be stretched to a thinner shape
    D. Decrease its dioptre value

Answer is B: Increasing the refraction of light will cause light rays from nearby objects to deviate more from their original path and so become focussed on the retina.

25. When the human eye is accommodated to make the image of a distant object focus on the retina, which of the following is true? The ciliary muscles are:
    A. Relaxed, the ciliary fibres are taut and the lens is nearly spherical.
    B. Contracted, the ciliary fibres are slack and the lens is nearly spherical.
    C. Contracted, the ciliary fibres are slack and the lens is thin.
    D. Relaxed, the ciliary fibres are taut and the lens is thin.

Answer is D: To focus on a distant object, the lens must be pulled into a thin profile by the ciliary fibres being taut. This occurs when the ciliary muscles recedes from the lens as it relaxes.

26. Myopia may be corrected with a lens that is
    A. Bifocal
    B. Concave
    C. Cylindrical
    D. Convex

Answer is B: In myopia (near-sightedness), close objects are in focus, but distant ones are not. It may be corrected by using a concave (diverging) lens.

27. What is the change in vision that occurs with ageing called?
    A. Protanopia
    B. Hyperopia
    C. Deuteranopia
    D. Presbyopia

Answer is D: "Presby-" means elderly or old. Choices A and C refer to forms of "colour blindness".

28. As part of the normal ageing process, our eyes deteriorate because:
    A. The ciliary muscles gradually lose their tone.
    B. The lens loses its flexibility.
    C. Parts of the retina detach from the underlying blood vessels.
    D. The distance between the lens and the retina gradually changes.

Answer is B: The lens needs to be elastic in order to become thin (to view distant objects) and to become thick (to view objects that are within 1 m).

29. In which colour ranges do the three pigments in the retina have their major sensitivities?
    A. Red, green and blue
    B. Red, blue and yellow
    C. Green, yellow and red
    D. Green, yellow and blue

Answer is A: Red cones are most sensitive to light with wavelength 560 nm, green cones at 530 nm and blue cones at 420 nm.

30. The human eye is able to alter its focal length in a process called accommodation. Choose the best description of accommodation from the ones below. When the ciliary muscle:

   A. Relaxes, the ciliary fibres are loosened and the eye lens is stretched into a long focal length.
   B. Contracts, the ciliary fibres are pulled taut and the eye lens is stretched into a long focal length.
   C. Relaxes, the ciliary fibres are pulled taut and the eye lens oozes into a short focal length.
   D. Contracts, the ciliary fibres are loosened and the eye lens oozes into a long focal length.

Answer is D: When the ciliary muscle contracts, it forms a smaller diameter circle around the lens causing the fibres that connect it to the lens to slacken, which allows the elastic lens to recoil (ooze) intro a rounder shape. This has a shorter focal length.

31. What is the name of the narrowest aperture of the eye through which light must pass so that we may see?

   A. The cornea
   B. The pupil
   C. The lens
   D. The iris

Answer is B: The pupil is the "black" circle in the middle of the colourful iris. The diameter of the pupil may increase and decrease as the iris dilates and constricts.

32. Visual stimuli travel from the eyes to the part of the brain that interprets them via the "visual pathway". Which of the following list of structures is the pathway in correct order from the eyes?

   A. Optic nerves, optic chiasma, lateral geniculate body, optic radiation, primary visual cortex
   B. Optic radiation, lateral geniculate body, optic nerves, primary visual cortex, optic chiasma
   C. Optic chiasma, optic nerves, primary visual cortex, optic radiation, lateral geniculate body
   D. Primary visual cortex, optic radiation, optic chiasma, lateral geniculate body, optic nerves

Answer is A: The optic nerves leave the eyes and the signal eventually reaches the primary visual cortex in the occipital lobe.

33. What are the names of the two types of photoreceptors in the retina?

   A. Ganglion cells and bipolar cells
   B. Macula lutea and fovea centralis
   C. Photopic cells and scotopic cells
   D. Rods and cones

Answer is D: Cone cells are sensitive to colour while rod cells are more sensitive in scotopic (low light) conditions. Ganglion and bipolar cells are not sensitive to light. The fovea centralis is the central region of the macula lutea of the retina.

34. What term may be applied to the shape of the lens of the eye?
    A. Biconcave
    B. Bipolar
    C. Biconvex
    D. Bifocal

Answer is C: The surface of the eye's lens bulges outwards anteriorly and posteriorly at its centre and is narrower at its edges where it joins to the ciliary fibres. This shape is convex "front and back" so is biconvex. If it bulged inwards, it would be biconcave.

35. What fills the posterior chamber of the eye?
    A. Aqueous humour
    B. Optic chiasma
    C. Vitreous humour
    D. Visual cortex

Answer is C: The vitreous humour (or body) fills the space between the lens and the retina. It is transparent and gel-like, being about twice as viscous as water at 37 °C. "Humour" is an ancient Latin term for liquid or juice from a plant or animal. Aqueous humour fills the eye's anterior chamber.

36. Why does the eye have a blind spot?
    A. Here the layer of ganglion cells and bipolar cells that lie over the retina are absent, hence the rods and cones are insensitive to light.
    B. This may be due to damage to the retina from exposure to extremely bright light.
    C. At this place, the retina is detached from the sclera to allow blood vessels to enter the eye.
    D. Axons from the ganglion cells converge here to leave the eye as the optic nerve; hence, there are no rods or cones.

Answer is D: There is no retina at the blind spot (or optic disc) as it is the exit point for the axons of the ganglion cells. These axons form the optic nerve and the optic disc is also the entry point for the retinal artery and vein. Choice A is incorrectly describing the fovea centralis.

37. One of the following structures is part of a "special sense". That is, it is **NOT** a receptor for the general senses. Which one is it?
    A. Golgi tendon organs
    B. Semicircular canals
    C. Muscle spindles

D. Fingers

Answer is B: Special senses are those that are localised to a particular organ. The semicircular canals of the inner ear are part of our special sense of equilibrium. "Touch" (fingers) is not a special sense as all parts of the integument are sensitive to touch.

38. The visual pathway is best characterised by which of the following neural elements?
    A. Eye lens, retinal ganglion cells, optic radiation, optic chiasma, primary visual cortex
    B. Optic nerves, decussation of pyramids, nucleus pallidus, optic radiation, primary visual cortex
    C. Retina, optic nerves, primary visual cortex, optic radiation, post-central gyrus
    D. Optic nerves, optic chiasma, lateral geniculate nucleus, optic radiation, primary visual cortex

Answer is D: The pathway begins at the photoreceptors, so the eye lens is not included. The decussation of pyramids and post-central gyrus are not part of the visual pathway.

39. Which structure is involved in controlling eye movement?
    A. The corpus callosum
    B. The superior colliculi
    C. The choroid plexus
    D. The inferior colliculi

Answer is B: The superior colliculi receive input from the lateral geniculate nuclei. The inferior colliculi receive auditory sensation.

40. If you have 6/7 vision, you can see clearly at 6 m what most people can see from 7 m away. If you have myopia, what numbers are likely to describe your vision?
    A. 6/6
    B. 6/7
    C. 3/3
    D. 3/6

Answer is D: Myopia is "short-sightedness" meaning that items that are close may be seen clearly but those far away cannot be seen. Myopic people need to be closer in order to see the item clearly. 3/6 vision means that from 3 m away, items may be seen that are normally clear from 6 m away. Choice B, 6/7 is a little myopic, but not as much as 3/6.

41. Hyperopia is a vision condition in which distant objects can be seen clearly, but close ones do not come into proper focus. Farsightedness occurs if the distance between the lens and retina is too short. What type of lens is used to correct this condition?
    A. Bifocal
    B. Contact lenses
    C. Concave
    D. Convex

Answer is D: A convex lens causes light rays to converge. This additional convergence provided by corrective lenses allows light that enters the eye to be focussed on the retina, rather than behind it.

42. If you have 20/20 vision, you can see clearly at 20 ft what should normally be seen at that distance. Given that 1 ft = 0.305 m, what is 20 ft converted to metres?
    A. $2 \times 30.5 = 61$ m
    B. $20 \div 20 = 1$ m
    C. $20 \div 0.305 = 65.6$ m
    D. $20 \times 0.305 = 6.1$ m

Answer is D: If 1 ft is 0.305 m, 10 of them would be 3.05 m and 20 of them would be as shown in choice D.

43. Consider what happens when the beam of a small torch is shone into the cornea of the left eye (but not the right eye) to test the pupillary light reflex. Which of the following happens?
    A. The left eye pupil diameter contracts, while the right eye pupil does not.
    B. The pupils of both eyes contract by the same amount.
    C. The left eye pupil diameter contracts, while the right eye pupil contracts but not as much.
    D. The right eye's pupil diameter contracts, while left eye pupil dilates.

Answer is B: Unless there is some damage to the nerves to the right eye, both eyes will display the same response—the pupil diameter will decrease.

44. What is the role of the ciliary body and the suspensory ligaments in the process of accommodation?
    A. The ciliary body is a muscle that is attached to the eye lens by the suspensory ligaments. When it relaxes, the lens can focus on distant objects.
    B. When the ciliary body contracts, the suspensory ligaments tighten, causing the eye lens to stretch.
    C. The contractile suspensory ligaments are attached to the ciliary body, which in turn is attached to the lens. When the ligaments are relaxed, the lens can focus on distant objects.

D. When the ciliary body contracts, the suspensory ligaments loosen, causing the eye lens to stretch.

Answer is A: The ciliary body is a muscular ring that surrounds the eye lens and is attached to the eye lens by the suspensory ligaments. When the ciliary body relaxes, it forms a larger diameter ring around the lens. This pulls on the suspensory ligaments which in turn stretch the elastic eye lens into a thinner shape which is able to focus light from distant objects onto the retina.

45. Which three pieces of information are required in order to explain why "colour blind" people are more likely to be male (XY) than female (XX)?
    A. The "colour blindness gene" is carried on the X chromosome. The colour blindness gene is "dominant". Females have 2 × X chromosomes while males have one X and a (short) Y.
    B. Females have 2 × X chromosomes while males have one X and a (short) Y. The "colour blindness gene" is carried on the Y chromosome. The colour blindness gene is "recessive".
    C. The colour blindness gene is "recessive". Females have 2 × X chromosomes while males have one X and a (short) Y. The "colour blindness gene" is carried on the X chromosome.
    D. The "colour blindness gene" is carried on the Y chromosome. Females have 2 × X chromosomes while males have one Y and a (short) X. The colour blindness gene is "dominant".

Answer is C: The "colour blindness gene" is a recessive gene that is carried on a part of the X chromosome that has no matching part on the short Y chromosome which males possess. This means that males have only one allele of the "colour blindness gene". If it is the recessive form, they will have a colour vision deficiency. If they have the dominant allele, they will have normal colour vision. Females have two copies of the X chromosome. The colour blindness gene is "recessive" which means that females who have the dominant allele on their other X chromosome cannot be colour blind. (The term "red-green colour-blind" is not accurate. There are degrees of anomalous colour vision so we refer to "colour vision deficiency".)

46. Visual acuity may be tested by determining the smallest size of letter which may be read from a chart placed a certain distance away from the reader. "6/6" vision means that from a distance of 6 m, you can read what the average person can read from 6 m. What does 6/9 vision imply?
    A. Normal vision. From a distance of 6 m, you can read what the average person can read from 9 m.
    B. Better than average vision. From 9 m away you can read letters that the average person needs to be at 6 m to read.
    C. Worse than average vision. The average person can read from 9 m away, the letters that you can read from 6 m away.
    D. Average vision. From a distance of 9 m, you can read what the average person can read from 9 m.

Answer is C: 6/9 vision means that the average person can read from 9 m away, the letters that you can read from 6 m away. So your vision is not as good as that of the average person. Normal/average vision is 6/6 not 6/9. 6/9 is not better than average. (20/20 vision is "normal" but expressed in units of feet rather than metres. 20 ft = 6.01 m.)

47. Why is the **optic disc** of the retina also known as the "blind spot"?
    A. Rod cells are absent at the disc, so it is insensitive to light.
    B. Only cone cells are present at the optic disc whereas rods are also required for vision.
    C. The optic disc is not part of the retina, so is not a light-sensitive structure.
    D. There are no rod or cone cells in the disc, so it is insensitive to light.

Answer is D: There are no rod or cone cells in the optic disc because it is the entry/exit point for the blood vessels and nerve fibres that penetrate the eyeball's wall. Only rods and cones have the ability to detect light when it strikes them.

48. What type of lens has the same curvature as your eye's lens?
    A. Biconvex
    B. Biconcave
    C. Planoconvex
    D. Planoconcave

Answer is A: The eye's lens in convex on both sides—it bulges in the centre and is thinner at the edges. This allows it to converge light onto the retina. A planoconvex lens is flat on one surface while convex on the other.

49. What is the function of the iris of the eye?
    A. Increases (and decreases) in diameter to allow more (and less) light to pass through the pupil
    B. Suspends the lens within the eye and pulls on it in order to alter its curvature
    C. Transmits electrical impulses to the brain's optical centre to be interpreted as sight
    D. Contracts or relaxes to accommodate the lens for viewing objects at different distances from the eye

Answer is A: The iris is the colourful muscle that determines the size of the pupil. Choice B refers to the ciliary fibres; C to the optic nerve; D to the ciliary muscle.

50. What is the problem with a myopic eye AND how it is compensated for?
    A. The distance between the lens and the retina is too short. A converging lens is used in eyeglasses to compensate.
    B. The distance between the lens and the retina is too long. A diverging lens is used in eyeglasses to compensate.
    C. The distance between the lens and the retina is too short. A diverging lens is used in eyeglasses to compensate.

D. The distance between the lens and the retina is too long. A converging lens is used in eyeglasses to compensate.

Answer is B: A myopic eye is "short sighted" and sees distant objects as blurry because the eye has too much refraction or is too long. A concave (diverging) lens is used to reposition the focal point back onto the retina.

51. What is the problem with a hyperopic eye AND how it is compensated for?
    A. The distance between the lens and the retina is too short. A converging lens is used in eyeglasses to compensate.
    B. The distance between the lens and the retina is too long. A diverging lens is used in eyeglasses to compensate.
    C. The distance between the lens and the retina is too short. A diverging lens is used in eyeglasses to compensate.
    D. The distance between the lens and the retina is too long. A converging lens is used in eyeglasses to compensate.

Answer is C: A hyperopic eye is "long sighted" and sees close objects as blurry because the eye has too little refraction or is too short. A convex (converging) lens is used to reposition the focal point forward onto the retina, that is, to add extra refraction.

52. What is the function of the cornea?
    A. It allows the eye to accommodate to focus on close or on distant objects.
    B. It is the light-sensitive part of the eye that produces the nerve impulse to the brain.
    C. It is the white of the eye that forms the tough protective outer layer.
    D. It allows light to enter the eye and produces most of the refraction of the eye.

Answer is D: The cornea is the transparent part of the sclera that is superficial to the iris. Hence, light may enter the eye. Being a curved surface and have a "refractive index" very different from air, it produces most of the refraction (and therefore focussing) of the eye.

14.4.2 Ear

Adults can hear sounds with frequencies between about 60 and 15,000 Hz. However, teenagers and children can hear frequencies beyond these values. Sound is directed into the ear canal by the pinna (ear lobe) where the air pressure variations cause the tympanic membrane to bow in and out. As the malleus rests on the tympanic membrane, this movement is passed onto the three connected ossicles (malleus incus and stapes) that are contained in the middle ear. They behave like a system of levers whose movement causes the stapes to vibrate against part of the cochlea called the "oval window". In this way, variations in air pressure (i.e. sound waves) are transformed into vibrations of the liquid (perilymph) with the cochlea. A physicist would say that the outer and inner ear is an impedance matching device between air and endolymph. Without this mechanism, sound waves would reflect from, rather than enter through, the oval window.

The middle ear contains air whose pressure returns to atmospheric pressure whenever the Eustachian tube is opened. This 3–4 cm long tube connecting the mouth to the inner ear is usually squashed flat but is opened by yawning or swallowing (etc.).

The inner ear contains the snail-shaped organ called the cochlea—imagine a long thin cylinder that has been coiled into a helix whose radius decreases. The cylinder is divided into three longitudinal chambers (scalae) by the basilar membrane and the vestibular membrane. Formed between these membranes is the cochlear duct (or scala media) containing endolymph and the organ of Corti that rests along the length of the basilar membrane. Above the scala media is the scala vestibuli and below is the scala tympani. Both of these contain perilymph. When the stapes pushes against the oval window, a wave of pressure moves through the perilymph. This causes the basilar membrane and organ of Corti to "rise and fall" which in turn causes the hair cells of the organ of Corti to rub against the tectorial membrane that is located just overhead. The hair cells are stimulated by this contact and pass on the stimulus to a branch of the auditory nerve (cranial nerve VIII).

One decibel (dB) is a tenth of a bel. It is a unit of "level" and is the logarithm of the ratio of two sounds—one being the reference sound, which for the case of sound level measurement is "silence" (= 0 dB). Silence does not mean that sound energy is absent, rather it means that the human ear cannot hear it.

1. By which of the following pathways does sound entering the ear reach the organ of Corti?

   A. Basilar membrane, middle ear, oval window, endolymph
   B. Tympanic membrane, ossicles, oval window, cochlear fluid
   C. Tectorial membrane, Eustachian tube, ossicles, cochlear fluid
   D. Oval window, ear canal, auditory tube, endolymph

Answer is B: Sound impinges on the tympanic membrane first. The Eustachian tube and the auditory tube are the same thing and do not conduct sound—they connect the middle ear to the mouth.

2. Which part of the ear contains the apparatus that we use to distinguish between different frequencies of sound?

   A. The cochlea
   B. The Eustachian (or auditory) tube
   C. The tensor tympani
   D. The auditory meatus

Answer is A: The cochlea contains the organ of Corti (the spiral organ) and its basilar membrane.

3. What is the range of frequencies that the human ear is most sensitive to?
   A. 50–500 Hz
   B. 12,000–20,000 Hz
   C. 500–6000 Hz
   D. 20–20,000 Hz

Answer is C: "Most sensitive" means the sound is loudest to us. Some people can hear sound with frequency as low as 20 Hz or as high as 20,000 Hz, but the sound will be very faint. Sounds with frequency between 500 and 6000 Hz seem to be loudest to the human ear.

4. Which one of the following statements is **WRONG**?
   A. The middle ear lies between the tympanic membrane and the oval and round windows.
   B. The outer ear is vented by the Eustachian (or auditory) tube.
   C. The stapes is located in the middle ear.
   D. The tectorial membrane and the basilar membrane are located in the inner ear.

Answer is B: The middle (not outer) ear is filled with air and is vented to the mouth via the Eustachian tube.

5. The ossicles of the ear pass on sound vibrations to the fluid in the inner ear. In what structure is this fluid located?
   A. The organ of Corti
   B. The cochlea
   C. The Eustachian tube
   D. The saccule and utricle

Answer is B: The cochlea contains the scala vestibuli and the scala tympani, both of which are filled with perilymph. It also contains the cochlear duct filled with endolymph.

6. Sound produces vibrations in the cochlear fluid of the inner ear. The movement of the fluid then produces motion in which of the following?
   A. Tectorial membrane
   B. Basilar membrane
   C. Otolithic membrane
   D. Crista ampullaris

Answer is B: When the cochlear fluid moves, it displaces the basilar membrane which pushes the hair cells within the organ of Corti against the tectorial membrane.

7. What is the function of the middle ear ossicles?
   A. To protect the cochlea from excessively loud noises
   B. To increase the sound intensity by resonating for sounds of frequencies near 3000 Hz

C. To amplify the sound intensity that reaches the tympanic membrane
D. To cause the sound energy of waves in air to be transmitted into the cochlear fluid

Answer is D: The ossicles are a mechanical lever system that are attached to the tympanic membrane at one end and to the oval window at the other. They transmit the movements in the tympanic membrane across the middle ear and to the perilymph behind the oval "window".

8. Sound waves are conducted from the air outside the ear to the inner ear by the processes of:
   A. Absorption, transmission and refraction
   B. Reflection, transmission and scattering
   C. Resonance, leverage and amplification
   D. Resonance, diffraction and refraction

Answer is C: The air in the external ear canal resonates (at about 3000 Hz), the ossicles are a lever system that amplifies the movement of the tympanic membrane, the oval window, being about one twentieth the size of the tympanic membrane also amplifies the pressure applied to the cochlear fluid compared to that applied to the tympanic membrane.

9. Sound level (measured in decibels, dB) is a subjective measure of the loudness of a sound. A sound of 90 dB:
   A. Will produce hearing damage if the ear is subjected to it chronically
   B. Is beyond the normal audible frequency range of human hearing
   C. Cannot be heard by the human ear even though it carries energy
   D. Will be perceived to be the same loudness at all audible frequencies

Answer is A: This sound pressure level (or perhaps even 85 dB) will produce hearing damage with long-term exposure. The decibel is not a statement about frequency. 0 dB is the threshold of human hearing. Different frequencies if all played at 90 dB would be perceived as different in loudness.

10. Of the following lists of four anatomical features, which one has them in the correct order of the path taken by sound energy as it is transmitted through the ear?
    A. Tectorial membrane, malleus, oval window, cochlear fluid
    B. Tectorial membrane, incus, round window, organ of Corti
    C. Tympanic membrane, malleus, oval window, cochlear fluid
    D. Tympanic membrane, stapes, round window, organ of Corti

Answer is C: The tympanic membrane (ear drum) is the first structure. The ossicles pass on the vibrations to the oval (not round) window of the cochlea.

11. Which of the following is **NOT** a small bone involved in hearing?
    A. Meatus

B. Malleus
C. Stapes
D. Incus

Answer is A: Meatus is a canal-like tube. For example external acoustic meatus (of the temporal bone), urethral meatus (of the penis), the nasal meatus.

12. Which one of the following lists of anatomical features is in the correct order of the path taken by sound energy as it is transmitted through the ear?
    A. Tympanic membrane, malleus, oval window, cochlear fluid
    B. Basilar membrane, incus, round window, fluid of Corti
    C. Tympanic membrane, stapedius, round window, fluid of Corti
    D. Basilar membrane, malleus, oval window, cochlear fluid

Answer is A: Tympanic membrane comes first. Then any of the ossicles, then the oval window and cochlear fluid.

13. What is the function of the tensor tympani and stapedius muscles in the middle ear?
    A. To protect the tympanic membrane and ossicles from excessively loud noises
    B. To increase sound intensity by resonating for sounds with frequencies near 3000 Hz
    C. To amplify the sound intensity that reaches the oval window
    D. To cause the sound energy of waves in air to be transmitted into the cochlear fluid

Answer is A: The tensor tympani limits the amount of movement in the tympanic membrane. The stapedius reduces the movement of the stapes against the oval window.

14. The loudness of a sound wave as perceived by the human ear depends on which of the following pairs of wave properties?
    A. Speed and frequency
    B. Intensity and frequency
    C. Amplitude and phase
    D. Speed and intensity

Answer is B: Intensity and amplitude are related. Our ear perceives some frequencies to be louder than others.

15. If the frequency of a sound wave is increased from 50 to 3000 Hz, its loudness also increases. This occurs because:
    A. The ear is more sensitive to 3000 Hz than to 50 Hz.
    B. Sounds of higher frequency carry higher energy.
    C. As frequency increases, the sound intensity also increases.
    D. Loudness is proportional to frequency.

Answer is A: The ear as a hearing mechanism is able to better able to respond (detect) frequencies near 3000 Hz, whereas frequencies near 50 Hz are almost inaudible.

16. Our ears are most sensitive to sounds with frequencies that lie between about 3000 and 3500 Hz. The reason for this is that:
    A. Sounds with these frequencies have the largest decibel rating.
    B. The majority of human speech sounds are composed of frequencies that lie in this range.
    C. The external ear canal has dimensions that allow it to resonate with a frequency that is in this range.
    D. The largest part of the basilar membrane is receptive to this range of frequencies.

Answer is C: The ear canal is about 2.7 cm long and a tube of this length, closed at one end, resonates at about 3000 Hz (resonates means that a "standing wave" is set up).

17. What can be said about noise-induced hearing loss?
    A. It affects sound frequencies near 4000 Hz most.
    B. It is also called presbycusis.
    C. It is likely to be caused by sounds above 65 dB.
    D. It is due to otosclerosis.

Answer is A: The "notch" in someone's measured audiogram at about 4000 Hz is characteristic of all noise-induced hearing loss.

18. For which condition is a hearing aid most successful at treating?
    A. Conductive hearing loss
    B. Perceptive deafness
    C. Nerve deafness
    D. Sensorineural hearing loss

Answer is A: Conductive loss is used to describe the situation where sound is not conducted well to the inner ear, which is otherwise healthy. The other three choices are different names for the same thing which is damage to the cochlea or auditory nerve.

19. What is the purpose of the diaphragm on the bell of a stethoscope?
    A. To prevent external sounds from interfering with auscultation
    B. To eliminate any air gap between the skin and stethoscope
    C. To resonate with the sound being listened to
    D. To transmit the body sounds to the earpieces.

Answer is C: The bell matches the impedance of the air and the skin by resonating with the sounds coming from the body.

20. Which structure lies on the boundary between the middle and inner ear and has the stapes bound to it?
    A. Ampulla
    B. Oval window
    C. Round window
    D. Tympanic membrane

Answer is B: The stapes presses onto the oval window to transmit the vibration of the tympanic membrane to the cochlear fluid.

21. Which membrane lies over the hair cells found in the organ of Corti?
    A. Basilar
    B. Tectorial
    C. Vestibular
    D. Cochlear

Answer is B: The tectorial membrane lies over the hair cells of the organ of Corti (which sits on the basilar membrane).

22. The inner ear (or internal ear) maybe described as a series of tubes. What are the tubes filled with?
    A. Air
    B. Perilymph
    C. Endolymph
    D. Perilymph and endolymph

Answer is D: Both perilymph and endolymph are contained in the scalae of the cochlea and the semicircular canals of the labyrinth. The middle ear contains air.

23. What are the cells found in the maculae of the utricle and saccule that are responsible for our sense of equilibrium?
    A. Supporting cells
    B. Otoliths
    C. Hair cells
    D. Epithelial cells

Answer is C: The hairs (stereocilia) of the hair cells are stimulated when our head tilts. Hair cells are embedded within the supporting cells. Otoliths are not cells.

24. Which of the following auditory structures are filled with fluid?
    A. The inner ear
    B. The middle ear
    C. The external meatus
    D. The Eustachian tube

Answer is A: The cochlea and the three semicircular canals of the inner ear are filled with fluid.

25. To which membrane is the ear's organ of Corti attached?
    A. The basilar membrane
    B. The tympanic membrane
    C. The tectorial membrane
    D. The Eustachian membrane

Answer is A: The basilar membrane is located in the inner ear and separates the scala tympani from the scala vestibuli. The tectorial membrane arches over the organ of Corti, while the tympanic membrane is the ear drum.

26. There are three ossicles in the middle ear. What are their names?
    A. Malleus, stapedius, otolith
    B. Cochlea, stapes, incus
    C. Stapedius, ampulla, cochlea
    D. Malleus, stapes, incus

Answer is D: Malleus, stapes, incus (aka: hammer, stirrup, anvil) are in the middle ear. The stapedius is a muscle, while the cochlea is the inner ear.

27. Which chamber or section of the ear is **NOT** filled with air?
    A. The cochlea
    B. The middle ear
    C. The outer ear
    D. The Eustachian tube

Answer is A: The cochlea contains liquids known as perilymph and endolymph. The Eustachian tube is usually squashed flat, but when it is opened, it contains air that is entering or leaving the middle ear.

28. The unit of sound level is decibel (dB), and the unit of sound frequency is hertz (Hz). Which sounds below, characterised by their sound level and frequency, would sound loudest to the human ear?
    A. 100 dB and 20 Hz
    B. 60 dB and 1000 Hz
    C. 10 dB and 3000 Hz
    D. 90 dB and 21,000 Hz

Answer is B: 60 dB is approximately the sound level of normal conversation, while the human ear is very sensitive to a frequency of 1000 Hz. 20 Hz is below our hearing range while 21,000 Hz is beyond our range. 10 dB is very quiet compared to 60 dB.

29. What is the purpose of the muscles known as the stapedius and the tensor tympani?
    A. To hold the three ossicles together, so that movement in the tympanic membrane is transmitted to the oval window.

B. They move the organ of Corti against the tectorial membrane to stimulate its hair cells.
C. To reduce the amount of ossicle movement that is produced by loud sounds.
D. To hold the malleus against the ear drum and the stapes against the oval window.

Answer is C: The tensor tympani pulls the malleus and stiffens the ear drum, while the stapedius pulls the stapes to reduce its movement against the oval window. In this way, the damage that loud sounds may do to the hearing mechanism is reduced.

30. Which of the following is an example of a "special sense"?
A. Hearing
B. Proprioception
C. Touch
D. The stretch reflex

Answer is A: A special sense is one that is localised into a specific organ such as the ear. Sight is also a special sense being localised in the eye. The sense of touch is not localised as the entire skin is sensitive to touch.

31. Through which structure must sound pass in order to enter the middle ear?
A. The tympanic membrane
B. The tectorial membrane
C. The Eustachian tube
D. The oval window

Answer is A: Sound travels from the air in the outer ear to the middle ear by vibrating the tympanic membrane (the ear drum). Sound enters the cochlea (inner ear) via the oval window.

32. How does the ear distinguish between sounds of different frequency?
A. By their different intensities (loudness): Soft sounds are detected closer to the oval window, while louder sounds can travel deeper into the cochlea.
B. Different regions of the basilar membrane are displaced by different frequencies.
C. Sounds of low frequency are more energetic than sounds of higher frequency and so displace a greater area of the basilar membrane than less energetic sounds.
D. High frequencies displace the basilar membrane close to the oval window, low frequencies displace more distal parts.

Answer is D: Sounds of the highest frequencies travelling through the endolymph displace that part of the basilar membrane that is close to the oval window. As frequency gets lower, progressively more distant sections of the basilar membrane are displaced. Choice B is correct but not as good an answer as choice D.

33. Which structure(s) control reflex movement of the head and neck in response to auditory stimuli?
    A. The superior colliculi
    B. The arbor vitae
    C. The inferior colliculi
    D. The medulla oblongata

Answer is C: The inferior colliculi receive auditory stimuli via the medulla oblongata. The superior colliculi receive visual sensation.

34. Which of the following situations constitutes sensorineural hearing loss rather than conductive hearing loss?
    A. Impacted ear wax in the ear canal
    B. Damaged hair cells in the organ of Corti
    C. The use of ear plugs while sleeping
    D. A perforated eardrum

Answer is B: The organ of Corti in the inner ear generates auditory nerve impulses when the hair cells are stimulated. If some of the hair cells are damaged, they cannot produce impulses in the auditory nerve.

35. When you rise in altitude rapidly, why do your internal ears feel uncomfortable until they "pop"?
    A. Atmospheric air pushes the tympanic membrane inwards.
    B. The Eustachian tubes are squashed by the increase in pressure.
    C. Air in the middle ear expands which pushes on the tympanic membrane.
    D. The Eustachian tubes are squashed by the decrease in pressure.

Answer is C: Atmospheric air pressure decreases as altitude increases. The air trapped in the middle ear remains at atmospheric pressure if we rapidly gain altitude, so this trapped air volume in the middle ear expands (Boyle's law). This causes the ear drum to be uncomfortably deformed outwards. When the Eustachian tube (which is normally squashed flat) opens to allow some air to escape from the middle ear compartment, comfort is restored.

36. What is the function of the pinna of the ear?
    A. It is the tube that allows air to enter the middle ear.
    B. It contains endolymph to transmit vibration to the basilar membrane.
    C. It funnels sound onto the tympanic membrane.
    D. It is an ossicle that transfers sound vibrations to the cochlea.

Answer is C: The pinna is the "external ear". It acts like a sound horn (a funnel) to direct sound waves into the ear canal. Choice A refers to the Eustachian tube; B to the cochlea.

37. What is the function of the ossicles of the ear?
    A. They contain otoliths that detect acceleration of the head.

B. They are three mutually perpendicular semi-circular tubes that provide our sense of balance.
C. They lie on the basilar membrane and rub against the tectorial membrane to distinguish sound frequencies.
D. They transfer a vibration of the tympanic membrane to the cochlear liquid.

Answer is D: The ossicles are three small bones that are attached laterally to the tympanic membrane and medially to the oval “window” of the cochlea. Their movement generates movement of the fluid in the cochlea. Choices A and B refer to the semicircular ducts that attach to the utricle; choice C to the organ of Corti.

38. What is the function of the Eustachian tubes of the ear?

A. They allow the air space of middle air to be vented to the atmosphere.
B. They detect rotational motion of the head.
C. They contain endolymph and perilymph and allow detection of sounds of different frequency.
D. They contain the vestibulocochlear nerves that transmit sensory stimulation to the brain.

Answer is A: The Eustachian tubes (auditory tubes) are normally squashed flat. When they are opened, the air pressure within the middle ear can equalise with the atmospheric pressure. Choice B refers to the semicircular ducts; choice C to the cochlea.

39. What is the function of the cochlea of the ear?

A. It transfers a vibration of the tympanic membrane to the cochlear liquid.
B. It contains tubes and the organ of Corti that are able to detect sounds of different frequency.
C. It contains three mutually perpendicular semicircular tubes that provide our sense of balance.
D. It is made up of the saccule and utricle and detects rotational motion of the head.

Answer is B: The cochlea consists of the scala tympani (tympanic duct), scala vestibuli (vestibular duct) filled with perilymph, and the cochlear duct (filled with endolymph). The cochlear duct houses the organ of Corti. It is the motion of the round window that generates a wave in the perilymph, which displaces a particular part of the basilar membrane to move the organ of Corti, which in turn rubs against the tectorial membrane to generate a nerve impulse. The brain interprets stimulation from a particular part of the basilar membrane as a sound of a particular frequency. Choice A refers to the ossicles.

40. What is the healthy range of sound frequencies that may be heard by an average 18-year-old person?

A. 50–19,000 Hz
B. 3000–18,000 Hz

C. 50–14,000 Hz
D. 20–15,000 Hz

Answer is A: Very few people if any, can hear frequencies below 50 Hz or above 19,000 Hz. Older people lose the ability to hear frequencies above 14,000 Hz.

41. Sam is a person with normal hearing but cannot hear sounds with frequency above 14,000 Hz. How old is Sam likely to be?
    A. Less than 10 years old
    B. Less than 20 years old
    C. More than 50 years old
    D. More than 70 years old

Answer is C: The ability to hear high frequencies decreases progressively with age. Everyone less than age 20 years will be able to hear 14 kHz, but very few older than 50 years will hear that sound.

Chapter 15
Reproductive System

The purpose of sexual reproduction is to produce genetic variability in individuals, that is, babies that will develop to be different from their mother and father and from their siblings. This is achieved through the processes of gamete production (meiosis) and the fusion of a sperm with an ovum. Males result when fertilisation produces a zygote with an X and a Y chromosome, while a female results from a zygote with two X chromosomes. The organs of the reproductive tract that are present at birth are called the primary sex characteristics, and you should know them. The physical features that develop after puberty are known as the secondary sex characteristics.

In females, the hypothalamus releases GnRH, which causes the anterior pituitary to release FSH (and LH). FSH acts on an ovary to stimulate the development of some primordial follicles. They produce oestrogens which cause the rate of production of GnRH to increase which in turn causes the anterior pituitary to release LH. LH promotes the development of one follicle and causes its ovum to be ovulated. This releases the ovum into the abdominal cavity from where it enters the fallopian tube in which it may be fertilised. The remnant granulosa cells of the follicle develop into the corpus luteum which produces and releases progesterone. Progesterone prepares the uterus (its endometrium thickens) to receive a fertilised egg. The progesterone also decreases the rate of GnRH and FSH release so that additional primordial follicles are not stimulated to continue their development. If pregnancy does not occur, the corpus luteum deteriorates and oestrogen and progesterone levels drop. This drop allows the rate of GnRH release to increase and so the next cycle begins.

In males, the hypothalamus releases GnRH, which causes the anterior pituitary to release FSH (and LH). LH stimulates the interstitial (Leydig) cells of testes to produce testosterone. FSH stimulates the nurse (sustentacular/Sertoli) cells in testes which, in the presence of testosterone, promotes spermiogenesis. Sperm are continually produced in the seminiferous tubules and stored in the epididymis. The vas (ductus) deferens transports sperm by peristaltic contractions to the ejaculatory

M. Caon, *Examination Questions and Answers in Basic Anatomy and Physiology*, https://doi.org/10.1007/978-3-030-47314-3_15

ducts. Here the sperm mix with semen from the seminal vesicles (which capacitates them) and prostate fluid before being ejaculated through the urethra.

15.1 Reproductive Anatomy

1. Where are the male ejaculatory ducts?
   A. In the testicles before the epididymis
   B. In the penis
   C. Between the bulbourethral glands and the urethra
   D. At the end of the vas deferens (ductus deferens)

   Answer is D: The ejaculatory ducts commence where the ducts of the seminal glands join the vas deferens and end where the vas deferens join with the urethra.

2. In which of the following lists are the structures of the male reproductive tract listed in correct order from testes to urethra?
   A. Ejaculatory ducts, seminiferous tubules, epididymis, vas deferens
   B. Seminiferous tubules, epididymis, vas deferens, ejaculatory ducts
   C. Epididymis, ejaculatory ducts, seminiferous tubules, vas deferens
   D. Vas deferens, seminiferous tubules, epididymis, ejaculatory ducts

   Answer is B: Seminiferous tubules are in the testes so should be before the epididymis and vas deferens; ejaculatory ducts must be after the vas deferens.

3. Which one of the following is **NOT** a secondary sex characteristic?
   A. Adult male body shape
   B. Thicker vocal cords of a male
   C. Pubic hair
   D. Penis

   Answer is D: Secondary sex characteristics are those that develop after puberty, that is, are not present at birth.

4. Where should fertilisation of the egg by a sperm occur?
   A. In the cervix
   B. In the uterus
   C. In the fallopian tube
   D. In the abdominal cavity between ovary and fallopian tube

   Answer is C: Fluid from the "peg" cells of fallopian tube assists in the capacitation of sperm allowing them to fertilise the egg.

5. Which of the listed structures does the male reproductive tract pass through?
   A. Prostate
   B. Bulbourethral gland

C. Seminiferous vesicles
D. Bladder

Answer is A: The ducts from the seminal glands join with the vas deferens to become the ejaculatory ducts and enter the prostate. There they both join the urethra which then exits the prostate.

6. What is the name of the tube that carries sperm from the testes to the prostate gland?
   A. Vas deferens
   B. Ejaculatory duct
   C. Seminiferous tubule
   D. Urethra

Answer is A: The vas deferens (= ductus deferens) begins at the epididymis and ends at the ejaculatory duct in the prostate.

7. Where are the ejaculatory ducts?
   A. In the testes
   B. Between the testes and the prostate
   C. Within the prostate
   D. Between the prostate and the external urethral meatus

Answer is C: Each ejaculatory duct is formed by the union of the vas deferens with the duct of the seminal vesicle. They pass through the prostate and open into the urethra.

8. Where does fertilisation of the ovum normally occur?
   A. In the cervical canal
   B. In the ovary
   C. In the uterus
   D. In the fallopian tube

Answer is D: The sperm meet the ovum in the ampulla of the fallopian tube (= uterine tube) where fertilisation takes place, facilitated by the secretions of the peg cells.

9. Which list of structures in the male reproductive tract has them in correct sequence?
   A. Seminiferous tubules, epididymis, vas deferens, urethra, ejaculatory duct
   B. Epididymis, seminiferous tubules, vas deferens, ejaculatory duct, urethra
   C. Seminiferous tubules, epididymis, vas deferens, ejaculatory duct, urethra
   D. Epididymis, seminiferous tubules, vas deferens, urethra, ejaculatory duct

Answer is C: Seminiferous tubules must precede the epididymis; and the ejaculatory duct must precede the urethra.

10. Male sterilisation (vasectomy) involves the cutting of which tube?
    A. Ejaculatory duct
    B. Epididymis
    C. Urethra
    D. Ductus deferens

Answer is D: Vasectomy involves removing a piece of the ductus deferens (vas deferens) from a site within the scrotum.

11. Which cells develop into the corpus luteum?
    A. Granulosa cells
    B. Interstitial cells
    C. Cells of the antrum
    D. Thecal cells

Answer is A: Granulosa cells invade the ruptured antrum and proliferate to become the CC.

12. From which source does the majority of the volume of a male ejaculation come?
    A. Epididymis
    B. Seminiferous tubules
    C. Seminal vesicles
    D. Prostate gland

Answer is C: The seminal vesicles provide about 60% of ejaculate volume.

13. What is a fertilised egg known as?
    A. Ovum
    B. Zygote
    C. Embryo
    D. Blastocyst

Answer is B: The fertilised egg, before it begins to divide, is known as a zygote.

14. What is the section of the male reproductive tract within which sperm are produced called?
    A. Urethra
    B. Epididymis
    C. Vas deferens
    D. Seminiferous tubules

Answer is D: The seminiferous tubules of the testes produce sperm.

15. In the male reproductive tract, where are sperm produced?
    A. Seminiferous tubules
    B. Epididymis
    C. Sertoli cells of the testes

D. Leydig cells of the testes

Answer is A: In the seminiferous tubules, stem cells called spermatogonia undergo meiosis to become sperm cells.

16. What is the “external urethral meatus” another name for?

A. Shaft of the penis
B. The opening of the tube at the end of the penis
C. The prostate gland
D. The scrotum

Answer is B: The external urethral meatus, also known as the external urethral orifice, is the opening or meatus of the urethra. It is the point where urine exits the urethra in males.

17. After ejaculation, sperm travel through the structures of the female reproductive tract in which order?

A. Vagina, uterus, fallopian tube, ovary
B. Cervix, vagina, uterus, fallopian tube
C. Vagina, cervix, uterus, fallopian tube
D. Cervix, urethra, uterus, fallopian tube

Answer is C: Sperm eventually reach the fallopian tube after gaining entry to the uterus via the cervix. The urethra is not part of the female reproductive tract.

18. The hormone progesterone is released by which structure?

A. Anterior pituitary
B. Corpus luteum
C. Hypothalamus
D. The adrenal glands

Answer is B: If progesterone was produced by any of the other structures, then men would have it too.

19. What is the function of follicle-stimulating hormone in the male? To stimulate the:

A. Production of sperm
B. Release of testosterone
C. Release of gonadotropin-releasing hormone
D. Release of luteinising hormone

Answer is A: FSH stimulates the production or maturation of gametes. In the male, the production of sperm.

20. Sperm gain motility as they pass through which structure?

A. Lumen of the seminiferous tubule
B. Prostatic part of the urethra

C. Ductus deferens
D. Epididymis

Answer is B: In the prostate, sperm are mixed with seminal fluid from the seminal vesicles. This fluid mobilises the flagellum of a sperm cell.

21. Which of the following lists the structures of the female perineal area in the correct order?

    A. Clitoris, vaginal opening, urethral opening, anus
    B. Clitoris, urethral opening, vaginal opening, anus
    C. Urethral opening, clitoris, vagina, cervix
    D. Anus, clitoris, urethral opening, vaginal opening

Answer is B: Clitoris is the most ventral structure, while the anus is the most dorsal. The urethra is between the clitoris and vagina.

22. Which of the following produces male sex hormones?

    A. Interstitial cells
    B. Corpus luteum
    C. Anterior lobe of the pituitary gland
    D. Seminal vesicles

Answer is A: The interstitial or Leydig cells of the testes produce testosterone.

23. Which organ does **NOT** add a secretion to semen?

    A. Testes
    B. Prostate gland
    C. Penis
    D. Seminal vesicles

Answer is C: There is no addition to the semen as it passes through the penis.

24. The male reproductive tract has two of each of the following structures **EXCEPT** for one of them. Which structure is present as a single structure?

    A. Prostate gland
    B. Seminal vesicle
    C. Vas deferens
    D. Ejaculatory duct

Answer is A: There is only one prostate gland at which place the two ejaculatory ducts join the urethra.

25. Which cells are targeted by follicle-stimulating hormone (FSH) inducing the production of which of the listed products?

    A. Leydig cells to promote spermiogenesis
    B. Thecal cells to produce androgens
    C. Nurse/Sertoli cells to produce testosterone

D. Granulosa cells to proliferate around the ovum

Answer is D: FSH also targets nurse cells which promote spermiogenesis (they do not produce testosterone). Thecal cells do produce androgens, but that comes after the granulosa proliferate.

26. Which cells are targeted by luteinising hormone (LH) inducing the production of which of the listed products?
    A. Nurse/Sertoli cells to promote spermiogenesis
    B. Granulosa cells to proliferate around the ovum
    C. Leydig cells to produce testosterone
    D. Uterine epithelial cells to proliferate

Answer is C: Choices A and B are caused by FSH, while progesterone causes the uterus epithelium to proliferate.

27. Where do FSH and LH come from?
    A. The posterior pituitary gland
    B. The anterior pituitary gland
    C. The hypothalamus
    D. The adrenal cortex

Answer is B: GnRH is released from the hypothalamus to stimulate the anterior pituitary gland to release FSH and LH.

28. From the list below, what do MEN have as part of their reproductive system that women do not?
    A. Androgens
    B. Follicle-stimulating hormone
    C. Tunica vaginalis
    D. Gametes

Answer is C: The remarkably named tunica vaginalis is the membrane that lines the scrotal cavity in men. Both women and men produce gametes, androgens and FSH.

29. From the list below, what do WOMEN have as part of their reproductive system that men do not?
    A. An X chromosome
    B. Bartholin's (vestibular) glands
    C. Luteinising hormone
    D. An inguinal canal

Answer is B: Bartholin's glands (or greater vestibular glands) are possessed by women. The male homologue is Cowper's gland (bulbourethral glands). The inguinal canal is present in women where it attaches to the round ligament of the uterus, as well as in men.

30. Which of the following is **NOT** a secondary sex characteristic?

    A. A penis
    B. Axillary hair
    C. Female hips being wider than male hips
    D. Developed mammary glands

Answer is A: The penis is present at birth in male babies, so is a primary sex characteristic. The other choices all develop after puberty.

31. Spermatozoa are capacitated by mixing with the secretions of "peg" cells. Where are these cells located?

    A. In the seminal vesicles
    B. In the prostate gland
    C. In the epididymis
    D. In the fallopian tubes

Answer is D: Peg cells are non-ciliated epithelial cells within the fallopian tube. Their cellular secretions capacitate spermatozoa by removing glycoproteins and other molecules from their cell membranes.

32. Which organ produces and releases progesterone?

    A. Adrenal cortex
    B. Corpus luteum
    C. Ovaries
    D. Anterior pituitary

Answer is B: The corpus luteum (left behind when the ovum is released) produces most of the progesterone.

33. Which list presents the structures of the male reproductive tract in the order that a spermatozoa would pass through them?

    A. Epididymis, ductus (vas) deferens, ejaculatory duct, urethra
    B. Seminiferous tubule, ductus (vas) deferens, urethra, ejaculatory duct
    C. Epididymis, ductus (vas) deferens, external urethral meatus, penis
    D. Seminiferous tubule, external urethral meatus, ejaculatory duct, penis

Answer is A: Testis, epididymis, ductus (vas) deferens, ejaculatory duct, urethra, then through penis and to the outside via external urethral meatus.

34. Where do spermatozoa become "capacitated"?

    A. In the epididymis
    B. In the prostate gland
    C. In the fallopian tube
    D. In the prostate gland and in the fallopian tube

Answer is D: Spermatozoa become capacitated in a two-stage process: when mixed with seminal vesicle fluid as they pass through the prostate, their flagellum is activated; when in the fallopian tube of the female reproductive tract, fluid from "peg" epithelial cells of fallopian.

35. What are the male secondary sex characteristics?

    A. Penis, testes and reproductive tract
    B. Reproductive tract excluding external genitalia
    C. Hair distribution, muscle mass, body size, fat deposits, thicker vocal cords, Adam's apple
    D. The Y chromosome, testosterone, spermatozoa

Answer is C: The secondary sex characteristics are those that appear after puberty. These include hair distribution, muscle mass, body size, fat deposits, thicker vocal cords, Adam's apple, thicker skin, body shape, larger hands and feet, and male behavioural effects.

36. What are the female primary sex characteristics?

    A. Breasts, uterus and vagina
    B. Hair distribution, breasts, location of adipose tissue, voice, wide pelvis
    C. The X chromosome, Breasts, ovaries, uterus and vagina
    D. Reproductive tract, ovaries and uterus

Answer is D: Primary sex characteristics are those that a girl is born with. Ovaries, reproductive tract and external genitalia. Choice B lists the secondary sex characteristics.

37. What part of the uterine tube is closest to the ovary?

    A. Ampulla
    B. Isthmus
    C. Infundibulum
    D. Fimbriae

Answer is D: In order from closest to furthest from the ovary the structures are: fimbriae, infundibulum, ampulla, isthmus.

38. Where are the seminal glands?

    A. Immediately inferior to the prostate
    B. In the testes
    C. In the epididymis
    D. Immediately superior to the prostate.

Answer is D: The two seminal glands and the two vas deferens tubes enter the prostate together and form the prostatic urethra.

39. What is the name given to the passage through which the spermatic cord enters/leaves the abdomen?

    A. The ejaculatory duct
    B. The umbilicus

C. The inguinal canal
D. The ductus deferens

Answer is C: The inguinal canal is thus a weak spot in the male abdominal muscle wall and is the site where an inguinal hernia may occur.

15.2 Reproductive Physiology

1. What is the function of luteinising hormone?
   A. It stimulates the interstitial (Leydig) cells to produce testosterone.
   B. It stimulates sustentacular (Sertoli) cells to produce sperm.
   C. It stimulates the anterior pituitary to release follicle-stimulating hormone.
   D. It stimulates the ovary to develop follicles.

Answer is A: Luteinising hormone targets the interstitial cells of the testes inducing them to secrete testosterone. (In women a surge in LH triggers the completion of meiosis I, ovulation and the formation of corpus luteum.)

2. What does gonadotropin-releasing hormone (GnRH) do?
   A. Stimulates the anterior pituitary to release LH
   B. Stimulates the anterior pituitary to release both LH and FSH
   C. Stimulates the anterior pituitary to release FSH
   D. Stimulates the corpus luteum to release progesterone

Answer is B: GnRH stimulates the ant pit to release both LH and FSH, which are two gonadotropins. They act on the gonads and control gamete and sex hormone production.

3. Which of the following is/are **NOT** associated with the male reproductive system?
   A. Oestrogens
   B. Androgens
   C. FSH and LH
   D. Tunica vaginalis

Answer is A: Oestrogens are female sex hormones. Tunica vaginalis is the serous membrane around the scrotal cavity.

4. Where is the hormone progesterone produced?
   A. By the thecal cells that surround the follicle
   B. In the anterior pituitary
   C. In the corpus luteum
   D. By the developing follicle

Answer is C: The corpus luteum (which develops from the follicle after ovulation) produces progesterone.

5. Below is a list of structures and (in brackets) a hormone that they might produce. Which list is correct?
   A. Hypothalamus (FSH); anterior pituitary (GnRH); follicle (oestrogens); corpus luteum (LH)
   B. Hypothalamus (GnRH); anterior pituitary (oestradiol); follicle (LH); corpus luteum (progesterone)
   C. Hypothalamus (oestrogens); anterior pituitary (FSH); follicle (progesterone); corpus luteum (oestrogens)
   D. Hypothalamus (GnRH); anterior pituitary (FSH and LH); follicle (oestrogens); corpus luteum (progesterone)

Answer is D: The anterior pituitary produces and releases FSH and LH.

6. What effect does luteinising hormone have?
   A. It stimulates the growth of a few follicles each month.
   B. It stimulates ovulation and maintains the corpus luteum.
   C. It prepares the uterus for pregnancy.
   D. It establishes and maintains the secondary sex characteristics.

Answer is B: Luteinising hormone stimulates ovulation and the formation of the corpus luteum.

7. Which of the following do the testes produce?
   A. Capacitated spermatozoa
   B. About 60% of the ejaculate
   C. Slightly acidic fluid
   D. Physically mature spermatozoa

Answer is D: The sperm are physically mature when they enter the epididymis, but need to be capacitated by fluid from the seminal glands and the fallopian tube's peg cells. Most of the volume of the ejaculate comes from the seminal glands.

8. What do the thecal cells that surround the follicle produce?
   A. Mucus
   B. Luteinising hormone
   C. Androgens
   D. Oestrogens

Answer is C: The thecal cells produce androgens (androstenedione) which diffuse to the granulosa cells which convert androgens to oestrogens.

9. In the sequence of events known as the ovarian cycle, which of the following does **NOT** occur?
   A. The anterior pituitary releases FSH and LH.
   B. FSH stimulates a follicle to develop.
   C. The hypothalamus releases GnRH.
   D. The developing follicle produces progesterone.

Answer is D: The corpus luteum (not the follicle) produces progesterone.

10. What hormone is released by the corpus luteum in the greatest quantity?
    A. Progesterone
    B. Oestrogens
    C. Luteinising hormone
    D. Follicle-stimulating hormone

Answer is A: The corpus luteum releases progesterone which prepares the uterus for pregnancy (and causes the number of peg cells to increase).

11. Choose the correct statement about LH or FSH.
    A. LH targets Leydig cells of the testes which produce testosterone.
    B. LH targets Sertoli cells of the testes which promotes spermiogenesis.
    C. FSH targets Leydig cells of the testes which promotes spermiogenesis.
    D. FSH targets Sertoli cells of the testes which produce testosterone.

Answer is A: LH targets Leydig (interstitial) cells to produce testosterone. (FSH targets Sertoli (nurse/sustentacular) cells which (in the presence of testosterone) promote spermiogenesis.)

12. Which cells produce the majority of oestrogens?
    A. The cells of the corpus luteum
    B. The cells of anterior pituitary
    C. Endometrial cells
    D. Granulosa cells of the follicle

Answer is D: The thecal cells of the follicles produce androstenedione which is absorbed by the granulosa cells and used to produce oestrogens. (The corpus luteum does secrete some oestrogens too.)

13. What is the function of progesterone?
    A. To stimulate the development of follicles
    B. To maintain the corpus luteum
    C. Prepare and maintain the uterus for pregnancy
    D. To stimulate ovulation

Answer is C: Progesterone which prepares the uterus lining for pregnancy (and causes the number of peg cells in the fallopian tube to increase).

14. Which of the following could be accurately said of oogenesis and spermiogenesis?
    A. They both occur after puberty.
    B. They both cease after menopause.
    C. The former occurs before birth while the latter continues from puberty to death.
    D. The former is promoted by FSH while the latter is promoted by LH.

Answer is C: Oogenesis occurs before birth while spermiogenesis continues from puberty to death. Both are promoted by FSH.

15. What is the function of the hormone progesterone?

   A. To maintain secondary sex characteristics
   B. To pause meiosis until the ovum is fertilised
   C. To stimulate oestrogen production
   D. To prepare the uterus for pregnancy

Answer is D: Progesterone causes the lining of the uterus to thicken in preparation for implantation.

16. What term is applied to the second 2 weeks of the menstrual (uterine) cycle?

   A. Menses
   B. Secretory phase
   C. Luteal phase
   D. Proliferation

Answer is B: The secretory phase begins at ovulation and continues while corpus luteum is intact. The endometrial glands of the uterus secrete glycogen—a nutrition source for embryo. (The second phase of the ovarian cycle is called the luteal phase.)

17. Both testosterone and FSH contribute to which of the following effects in the male?

   A. The stimulation of spermiogenesis.
   B. Metabolic rate stimulation.
   C. They establish and maintain the secondary sex characteristics.
   D. The stimulation of muscle and bone growth.

Answer is A: FSH in the presence of testosterone stimulates spermiogenesis.

18. What is the role of progesterone?

   A. To stimulate follicle development
   B. To stimulate the maturation of the uterine lining
   C. To stimulate the oocyte to complete meiosis I
   D. To stimulate the release of FSH

Answer is B: Progesterone stimulates the maturation of the uterine lining and the secretions of the uterine glands.

19. Which statement about the "granulosa cells" is **NOT** correct?

   A. They produce oestrogens and inhibin
   B. They form a single layer around the primary follicle
   C. They form the corpus luteum
   D. One of them will develop into the ovum

Answer is D: Granulosa cells surround the ovum that was formed before birth.

20. What is the role of luteinising hormone (LH) in the female?

   A. It causes release of FSH from the anterior pituitary.
   B. It maintains the secondary sex characteristics.

C. It allows the frequency of pulses of GnRH to increase.
D. It causes ovulation.

Answer is D: A surge of LH release from the anterior pituitary gland causes the mature follicle to release its ovum. GnRH stimulates the release of FSH, while the secondary sex characteristics are maintained by oestrogens.

21. Which one of the following statements is **INCORRECT**?
    A. Corpus luteum releases progesterone.
    B. Hypothalamus releases GnRH.
    C. Anterior pituitary releases FSH.
    D. Granulosa cells produce androgens.

Answer is D: Granulosa cells produce oestrogens—female sex hormones. Androgens are male sex hormones.

22. Which one of the following is true about the secretory phase of the menstrual cycle?
    A. It occurs as the uterine epithelium regrows under the stimulation of oestrogens.
    B. It begins at ovulation and continues while the corpus luteum is intact.
    C. It refers to the release of progesterone by the corpus luteum.
    D. During this phase, the follicle develops prior to ovulation.

Answer is B: The secretion alluded to is glycogen secreted by endometrial glands of the uterus. It begins at ovulation and continues while the corpus luteum is active.

23. What is the function of the epididymis?
    A. Production of sperm
    B. Stores sperm and facilitates their maturation
    C. Stores sperm and produces seminal fluid
    D. Carries semen out through the penis

Answer is B: The epididymis store sperm for 2–3 months during which time they mature.

24. In males, what is the function of luteinising hormone (LH)?
    A. It stimulates interstitial cells in the testes to produce testosterone.
    B. It stimulates sustentacular cells in the testes to produce sperm.
    C. It promotes the maturation of spermatozoa.
    D. It stimulates the anterior pituitary to release FSH.

Answer is A: LH targets interstitial (Leydig) cells to produce testosterone. GnRH stimulates the ant pit to secrete FSH and LH.

25. Which one of the following is **NOT** a function of testosterone?
    A. Stimulates spermiogenesis
    B. Maintains secondary sex characteristics

C. Maintains glands of the reproductive tract
D. Stimulates anterior pituitary to release FSH and LH

Answer is D: GnRH stimulates the ant pit to secrete FSH and LH.

26. What is the name given to a young woman's first menstrual period?
    A. Menarche
    B. Menses
    C. Eclampsia
    D. Amenorrhea

Answer is A: Menarche is the first menstrual cycle, or first menstrual bleeding, in female humans.

27. What is the result of meiosis in males? The production of:
    A. One spermatid and three polar bodies
    B. One spermatid and two polar bodies
    C. Two primary spermatocytes
    D. Four spermatids

Answer is D: Meiosis results in four haploid cells being produced. Sperm in the male.

28. What are the phases of the ovarian cycle?
    A. Menarche, menstrual cycle and menopause
    B. Luteal phase, menses and proliferative phase
    C. Follicular phase and the luteal phase
    D. Menses, proliferative phase and secretory phase

Answer is C: The ovarian cycle is separated into follicular phase and luteal phase by ovulation.

29. Which of the following is **NOT** a function of oestrogens?
    A. Stimulating bone and muscle growth
    B. Maintain female secondary sex characteristics
    C. Maintaining the corpus luteum
    D. Initiating growth and repair of the endometrium

Answer is C: The corpus luteum is maintained by luteinising hormone.

30. Which statement is **NOT** correct?
    A. The ova in the ovary are continually forming from mitosis.
    B. The ovarian follicles are embedded in the ovary cortex.
    C. Ovulation is the ejection of an oocyte from the ovary.
    D. The corpus luteum develops from the ruptured follicle.

Answer is A: Ova are formed before birth in baby girls.

31. Which statement is **NOT** correct?
    A. Testosterone is required to maintain the adult male secondary sex characteristics.
    B. Testosterone increases the sex drive in both sexes.
    C. Testosterone inhibits the closure of the epiphyseal plate in the long bones.
    D. Testosterone stimulates the growth and maturation of the male genitalia.

Answer is C: In fact, the closure of the epiphyseal plate is caused by testosterone. While testosterone does increase libido in females, oestrogen normally performs this function.

32. What does gonadotrophin-releasing hormone stimulate?
    A. Hypothalamus
    B. Anterior lobe of pituitary gland
    C. Spermatogenesis or oogenesis
    D. Production of testosterone or oestrogen

Answer is B: GnRH stimulates the ant pit to release FSH and LH.

33. Select one **INCORRECT** statement from the following:
    A. Meiosis results in the reduction of chromosome numbers in cells from $2n$ to n.
    B. The primordial follicle cells are present in females at birth.
    C. Spermatogenesis and oogenesis result in the production of either four sperm or four ova.
    D. Puberty is the time of life when the reproductive organs begin to mature.

Answer is C: Spermatogenesis results in the production of four sperm, but oogenesis results in the production of one ovum.

34. The menstrual cycle can be divided into three phases. Starting from day 1 of menstruation, which of the following orders are the phases in?
    A. Menstrual, secretory, proliferative
    B. Proliferative, secretory, menstrual
    C. Menstrual, proliferative, secretory
    D. Secretory, proliferative, menstrual

Answer is C: Starts with menstruation and ends with secretory when the endometrium is mature.

35. After passing through puberty, what ability does the human possess?
    A. The ability to produce primary sex characteristics.
    B. The ability to produce gametes capable of being fertilised.
    C. The ability to grow body hair and breasts.
    D. Eccrine (merocrine) sudiferous glands begin to secrete.

Answer is B: After puberty, humans are sexually mature, meaning that they can produce viable sex cells (gametes). Secondary (not primary) sex characteristics develop during puberty, while apocrine sweat glands (not eccrine) become functional after puberty.

36. Which of the following is a female primary sex characteristic?
   A. Uterus
   B. Menstruation
   C. Spermiogenesis
   D. Pubic hair

Answer is A: Female babies are born with a uterus, so this is the primary sex characteristic.

37. Which one of the following is a primary sex characteristic of a male human?
   A. Spermiogenesis
   B. Prostate
   C. Comparatively deep voice
   D. Body hair

Answer is B: Male babies are born with a prostate gland. All the other listed characteristics develop during puberty.

38. Ovulation in the female follows which of the listed hormonal stimuli?
   A. A surge in rate of release of luteinising hormone
   B. A rise in blood oestrogen level
   C. The release of follicle-stimulating hormone
   D. The release of progesterone

Answer is A: When the release of LH from anterior pituitary gland rises suddenly to three times its previous level (in response to a gradual rise in GnRH concentration to a maximum), ovulation is induced.

39. What is the function of testosterone in the male?
   A. It establishes and maintains primary sex characteristics.
   B. It stimulates spermiogenesis.
   C. Along with fluid from the seminal vesicles, it capacitates spermatozoa.
   D. It oversees the autonomic control of ejaculation.

Answer is B: In conjunction with the FSH, testosterone stimulate spermiogenesis. Choice A is wrong as primary sex characteristics are present at birth.

40. What happens in the ejaculatory duct?
   A. Sperm are stored in a regulated fluid environment to facilitate their maturation.
   B. Here sperm are mixed with fluid from the seminal vesicles.
   C. In the female, it receives the ejaculate as it leaves the male.

D. Sperm pass through this immediately prior to being expelled from the penis.

Answer is B: Ejaculatory ducts (~2 cm long) begin at the end of vas deferens. Here seminal vesicles empty seminal fluid into tract, and it mixes with and dilutes sperm. The two ejaculator ducts enter prostate and join the urethra (distal to internal urethral sphincter) which carries semen (or urine) out through meatus of penis. Choice A describes what happens in the eipdidymus.

41. What is the composition of semen?

  A. Plasma proteins, leucocytes, spermatozoa, protein, fructose, vitamin C
  B. Leucocytes, spermatozoa, water, mucus, enzymes, electrolytes
  C. Spermatozoa, enzymes, electrolytes, protein, fructose, vitamin C
  D. Urea, spermatozoa, enzymes, electrolytes, prostaglandins, water

Answer is C: Ejaculated semen contains spermatozoa, enzymes, electrolytes, protein, fructose, lipids, vitamin C, carnitene, prostaglandins, water, mucus. Semen does not contain leucocytes, plasma proteins or urea.

42. What is a function of oestradiol (an oestrogen)?

  A. Maintains accessory repro glands/organs.
  B. It acts on the ovary to stimulate follicle development.
  C. It maintains the corpus luteum.
  D. It causes FSH (and LH) release from anterior pituitary.

Answer is A: In addition, oestradiol maintains secondary sex characteristics, affects CNS, initiates growth and repair of endometrium and promotes development of breasts. Choice B refers to FSH, choice C refers to LH, choice D refers to GnRH.

43. What is meant by menopause?

  A. It is the first menstrual bleeding cycle in female humans.
  B. The time of life at which the last menstrual cycle occurs.
  C. It is the time interval between successive menstrual cycles.
  D. The time between the release of the ovum from the ovary and the commencement of menstrual bleeding.

Answer is B: Usually occurs in women at about 45–55 years of age. Choice A is menarche.

15.3 Chromosomes and Genetics

1. Which statement is true?

  A. Males have two "X" chromosomes.
  B. Females do not produce any primordial follicles after they are born.
  C. Females have one "X" chromosome.
  D. Fertilisation occurs in the pelvic cavity before the start of the fallopian tube.

Answer is B: Female babies are born with all of their primordial follicles (about two million) paused at the prophase stage of meiosis I. Males are XY, females are XX, not the reverse.

2. How many chromosomes does a human gamete have?
   A. 46
   B. 46 pairs
   C. 23
   D. 23 pairs

Answer is C: A gamete (a sperm or egg) has 23 chromosomes, so when they fuse to form the fertilised egg, there is a total of 46 (= 23 pairs).

3. If a cell is said to be "haploid", what does it mean?
   A. It has 23 chromosomes.
   B. It has chromosomes that all consist of one chromatid.
   C. It has the "2*n*" number of chromosomes.
   D. It is NOT a gamete (or sex cell).

Answer is A: Haploid is the term used when a cell has half the usual number of chromosomes. Diploid means the cell has "2*n*" chromosomes, that is, two of each chromosome.

4. Which of the following statements is correct?
   A. Ova all contain a Y chromosome.
   B. Half of the ova carry an X chromosome and half carry a Y chromosome.
   C. Half of the sperm cells carry an X chromosome and half carry a Y chromosome.
   D. Sperm all carry an X chromosome.

Answer is C: The somatic cells of the male each carry an X and a Y chromosome. When sperm are produced, they have 23 chromosomes (rather than 46), thus they have equal chance of having an X or a Y chromosome. All ova have an X chromosome.

5. What is the chromosome complement of a sperm?
   A. 46 chromosomes (23 pairs including a pair of Y chromosomes)
   B. 46 chromosomes (22 pairs plus one X chromosome and one Y)
   C. 23 chromosomes including one X chromosome OR one Y
   D. 23 chromosomes including one X chromosome AND one Y.

Answer is C: Sperm are haploid (have 23 chromosomes) so that when they fuse with the egg (also with 23 chromosomes), the zygote has 46 chromosomes. The X or Y chromosome of the sperm, in conjunction the X of the egg, then determine the sex of the baby.

6. How do sperm cells differ from other cells in the male body?
   A. They contain 23 chromosomes.
   B. They all contain an X chromosome.
   C. They all contain a Y chromosome.
   D. They undergo mitosis.

Answer is A: Somatic cells contain 46 chromosomes, whereas a sperm cell has 23. It has either an X or a Y chromosome.

7. Which statement is true of the 23 chromosomes within a sperm?
   A. 23 chromosomes is the diploid number.
   B. 11 chromosomes came from the father, 11 chromosomes came from the mother, while one of either the Y or the X came from the father or mother, respectively.
   C. Some of the 23 came from the father and the rest came from the mother.
   D. 11 chromosomes and the Y came from the father, while 11 chromosomes came from the mother.

Answer is C: A sperm has 23 chromosomes which is the haploid number. The male that produced the sperm has 46 chromosomes (23 pairs) in their somatic cells—23 each from the man's father and mother. When sperm are produced, the 46 chromosomes assemble and pair up. The man's X and Y chromosomes pair up. Each pair of chromosomes then separates, so that only one of each pair moves into a new sperm. Which one of the pair ends up in which sperm is a random process? It is possible that an individual sperm has any number of chromosomes between 0 and 23 that originated from the father.

8. Which one of the following statements is correct?
   A. The zygote has the haploid number of chromosomes.
   B. Spermatids have the diploid number of chromosomes.
   C. The primary oocyte has the diploid number of chromosomes.
   D. The fertilised egg has the diploid number of chromosomes.

Answer is D: The sperm and the oocyte each have the haploid number of chromosomes (23), so that when fertilisation occurs, the fertilised egg (the zygote) has the diploid number (46).

9. A male with normal colour vision (X Y) has children with a female who carries the "colour-blind" allele on one of her X chromosomes (X_c X). What proportion of their children will have anomalous colour vision?
   A. Half of the males and none of the females
   B. All of the males and female children
   C. All of the male children and none of the females
   D. Half of the females and none of the males

Answer is A: Half of the mother's eggs will inherit the X chromosome with the normal colour vision allele and half will have the chromosome with X_C. Hence half of the males will be X Y and half will be X_C Y. The males with X_C Y will have anomalous colour vision (be "colour-blind"). Half the female children will all be X_C X and half will be X X, and all will have normal colour vision as the X_C is recessive. However, the female children with X_C X will be carriers.

10. A male with anomalous colour vision (X_C Y) has children with a female who carries the "colour-blind" allele on one of her X chromosomes (X_C X). What proportion of their children will have anomalous colour vision?
    A. Half of the males and none of the females
    B. Half of the males and all of the females
    C. None of the males and half of the females
    D. Half of the males and half of the females.

Answer is D: Half of the mother's eggs will inherit the X chromosome with the normal colour vision allele and half will have the chromosome with X_C. Hence, half of the males will be XY and half will be X_CY. The males with X_C Y will have anomalous colour vision (be "colour-blind"). Half of the female children will inherit X_C X_C and half will inherit X_C X. The girls with X_C X_C will be "colour-blind" while the latter will have normal colour vision. However, they will be carriers of the colour-blind allele.

11. A female with normal colour vision (X X) and who does not carry the colour-blind allele of the colour vision gene has children with a male who has anomalous colour vision (X_C Y). What proportion of their children will be "colour-blind"?
    A. Half of the girls will be colour-blind but none of the boys.
    B. Half of the boys will be colour-blind but none of the girls.
    C. All of the girls but none of the boys will be colour-blind.
    D. None of the children will be colour-blind.

Answer is D: All of the mother's eggs will have the normal colour vision gene (X) while all of the father's daughters will inherit his colour-blind allele (X_C). Hence, all boys will be XY and have normal colour vision. Half of the girls will be X X and half will be X X_C and this latter half will be carriers. As the "colour-blind gene" is recessive, and girls have two copies of the X chromosome, they will all have normal colour vision.

Chapter 16
Waves, Light Waves, Sound Waves, Ultrasound (the Physics of)

16.1 Waves

Mechanical waves (sound waves, waves on water) are a mechanism for transferring energy through a medium (the air or water) without transferring matter. Another definition is: A periodic disturbance in some property of the medium, the medium itself remaining relatively at rest. Waves have the following measurable properties:

1. Wavelength (symbol λ) is the distance between two successive crests (in metres, m). A typical value is ~500 nm for light and ~20 cm for sound.
2. Frequency (f) is the number of λ that pass by in 1 s (in hertz, Hz). Typical values are 500 THz for light, 500 Hz for sound. Frequency is related to pitch (for sound) and colour (for visible light).
3. Period (T) is the time it takes for one λ to pass by (in seconds, s).
4. Speed (v) is how fast a wave is moving in the direction of propagation (in metres per second, m/s). The speed of light travelling through air is 3×10^8 m/s, while for sound, speed in air is about 330 m/s. In tissue, sound moves faster, at about 1560 m/s.
5. Amplitude (A) is the *maximum* displacement from the mean (or rest) position. For example the vertical distance between a trough and a crest of a wave in water is two times the amplitude. Amplitude (or intensity) is related to loudness of sound and brightness of light and to the amount of energy being carried by the wave.
6. Phase refers to how far out of step the oscillation of one part of a wave is when compared with another part. A phase of 0° or 360° means that the two parts are in step, while a phase difference of 180° means that the two points are completely out of step. Differences in phase between the sounds entering each ear allow us to localise the source of a sound.

Some of these quantities are related to each other by the following formulae:

$$v = f\lambda; f = 1/T; E = h \times f.$$

M. Caon, *Examination Questions and Answers in Basic Anatomy and Physiology*, https://doi.org/10.1007/978-3-030-47314-3_16

Electromagnetic waves include (in increasing order of frequency), microwaves, infrared waves, visible red light, visible blue light, ultraviolet, X-rays and gamma rays. Low-frequency electromagnetic waves do not require a medium to travel through, but are able to travel through transparent media. They can also travel through "empty" space. They are not mechanical waves, they are oscillations in the magnitude and direction of electric and magnetic fields. They do not cause the medium to vibrate and are called "non-ionising radiation". They cause the electrons in the atoms of the medium to vibrate in step with the electric field. Lowfrequency em waves include microwaves, and em waves used in radio, TV, cell phone, remote control devices and radar.

Our ears can detect audible sound, while infrasound and ultrasound are both inaudible. Our eye can see visible light, while infrared and ultraviolet are both invisible.

1. The amplitude of a wave is related to which of the following?
   A. The distance between two successive crests
   B. The number of wavelengths that pass by per second
   C. The speed of the wave's travel
   D. The amount of energy it carries

Answer is D: The greater the amplitude, the greater the energy carried.

2. Which is the **BEST** definition of a wave?
   A. The method by which the energy carried by visible light is propagated
   B. Travelling oscillations in the magnitude and direction of electric and magnetic fields that do not require a material medium
   C. A periodic disturbance in some property of the medium, the medium itself remains (relatively) at rest
   D. A mechanism for the transfer of energy without the transfer of matter

Answer is D: The other choices also describe a wave but not in the most general terms.

3. What does the term "wavelength" mean when applied to a wave?
   A. The number of complete cycles that pass by in 1 s
   B. The distance between two successive crests (or compressions)
   C. The time it takes for one wavelength to pass by
   D. How fast a wave is moving in its direction of propagation

Answer is B: Wavelength is the shortest distance between two points that are displaying the same displacement.

4. The wave equation may be written in symbols as $v = f\lambda$ where f stands for frequency, v stands for velocity and λ stands for wavelength. If a wave has a speed of 3×10^8 m/s, what is its frequency and wavelength?
   A. $f = 10{,}000$ Hz and $\lambda = 0.0003$ m
   B. $f = 5 \times 10^{14}$ Hz and $\lambda = 6 \times 10^{-7}$ m

C. $f = 6 \times 10^4$ Hz and $\lambda = 0.5 \times 10^{-4}$ m
D. $f = 2$ MHz and $\lambda = 3 \times 10^5$ m

Answer is B: Which combination of frequency and wavelength will give the speed 3×10^8 m/s?

$v = f\lambda = 5 \times 10^{14}$ Hz $\times 6 \times 10^{-7}$ m $= 30 \times 10^{14-7} = 30 \times 10^7 = 3 \times 10^8$.

5. The wave equation may be written in symbols as $v = f\lambda$ where f stands for frequency, v stands for velocity and λ stands for wavelength. If a wave has a speed of 2×10^8 m/s, what is its frequency and wavelength?
   A. $f = 4 \times 10^4$ Hz and $\lambda = 0.5 \times 10^{-4}$ m
   B. $f = 5 \times 10^{14}$ Hz and $\lambda = 4 \times 10^{-7}$ m
   C. $f = 10{,}000$ Hz and $\lambda = 0.0002$ m
   D. $f = 2$ MHz and $\lambda = 1 \times 10^5$ m

Answer is B: Which combination of frequency and wavelength will give the speed 2×10^8 m/s?

$v = f\lambda = 5 \times 10^{14}$ Hz $\times 4 \times 10^{-7}$ m $= 20 \times 10^{14-7} = 20 \times 10^7 = 2 \times 10^8$.

6. What name is given to the change in the observed frequency of a wave (or its reflection) because of the motion of the source (or the reflecting object) or of the observer?
   A. Beat frequency
   B. Red shift
   C. Phase inversion
   D. Doppler effect

Answer is D: The change in frequency when the source of the (say) sound is in motion relative to the observer is called the Doppler effect. (The red shift in astronomy is due to the Doppler effect.)

7. Which of the statements about sound and light is **INCORRECT**?
   A. Sound is a mechanical wave while light does not require a medium to travel in.
   B. Light is a transverse wave phenomenon while sound is a longitudinal wave phenomenon.
   C. The speed of light is much greater than the speed of sound.
   D. Ultrasound and ultraviolet light have frequencies less than infrasound and infrared light.

Answer is D: In fact both ultrasound and ultraviolet light have frequencies GREATER than infrasound and infrared light respectively.

8. A wave may be defined as which of the following?
   A. The oscillation of a particle of the medium
   B. A series of crests and compressions that propagate through space
   C. A mechanism for the transfer of energy without transferring matter

D. The transport of the medium due to the oscillation of its particles

Answer is C: Essentially, a wave transfers energy without any matter moving away from its location.

9. Which of the following definitions of a wave is the best?
   A. A wave is an event or disturbance that is localised at a particular location.
   B. A mechanical wave is a periodic disturbance in a material medium.
   C. Waves are a means of transferring energy without transferring matter.
   D. Waves are a phenomenon characterised by their wavelength and their displacement of the medium from the mean position.

Answer is C: This is the classic definition of a wave. Choices B and D are also true statements.

10. Which quantity most closely describes the amount of energy that is transported by a wave?
    A. Frequency
    B. Amplitude
    C. Wavelength
    D. Velocity

Answer is B: Energy and amplitude are directly related.

11. What is the amplitude of a wave?
    A. The maximum displacement from the rest position
    B. The distance between adjacent troughs
    C. The energy carried by the wave
    D. The frequency multiplied by the wavelength

Answer is A: As a wave passes through a medium, the particles of the medium are shifted (displaced) from the position occupied before they were disturbed by the wave. The amplitude of the wave is the maximum displacement.

12. Waves have properties (such as wavelength, frequency, period, speed, amplitude, intensity, direction and phase) which can be measured. Some of these properties are related. That is, if one of a pair of related properties is known, then the other can be worked out. Which of the following lists contain properties that are all **unrelated** to each other?
    A. Amplitude, period, intensity, phase
    B. Frequency, amplitude, period, wavelength
    C. Direction, speed, amplitude, phase
    D. Phase, wavelength, speed, frequency

Answer is C: Frequency and wavelength are related, as are amplitude and intensity.

13. The wave equation for electromagnetic radiation may be stated (in words) as "the product of wavelength and frequency is a constant called the speed of light". The speed of light is 3×10^8 m/s. Use this information to determine the frequency of an electromagnetic radiation with a wavelength of 600 nm.
    A. 5×10^{14} Hz
    B. 2×10^{-17} Hz
    C. 1.8×10^{14} Hz
    D. 1.8×10^{11} Hz

Answer is C: $v = f\lambda$ so $f = v/\lambda = 3 \times 10^8$ m/s ÷ 600 nm = 3×10^8 m/s ÷ 600×10^{-9} m = $1800 \times 10^{8-9} = 1800 \times 10^{17} = 1.8 \times 10^{14}$ Hz.

14. Given the formulae: speed = frequency × wavelength ($v = f\lambda$) and frequency is one over the period ($f = 1/T$), what is the wavelength of a light wave (speed = 3×10^8 m/s) whose period is 2×10^{-15} s?
    A. 1.5×10^{-7} m
    B. 6×10^{-7} m
    C. 1.67×10^6 m
    D. 1.5×10^{23} m.

Answer is B: First determine the frequency: $f = 1/T = 1/(2 \times 10^{-15}) = 0.5 \times 10^{15}$ Hz.
Then rearrange the wave equation for wavelength: $v = f\lambda$ so $\lambda = v/f$.
Then substitute the numbers: $\lambda = v/f = 3 \times 10^8$ m/s ÷ 0.5×10^{15} Hz = 6×10^{-7} m.

15. What is the quantity that is most characteristic of an electromagnetic wave?
    A. Amplitude.
    B. Wavelength.
    C. Frequency.
    D. Velocity.

Answer is C: Frequency of an em wave does not change as it travels through different media, while the other three quantities do.

16. In a wave, what is the distance between two adjacent troughs called?
    A. Period
    B. Displacement
    C. Amplitude
    D. Wavelength

Answer is D: The length of a wave is the distance between the two nearest points that are undergoing the same movement. This is happening at two adjacent troughs (or two adjacent crests too).

17. The wave equation for electromagnetic radiation may be stated as "the product of wavelength and frequency is a constant called the speed of light". The speed of light is 3×10^8 m/s. Use this information to determine the frequency of an electromagnetic radiation with a wavelength of 600 nm.

    A. 6×10^{14} Hz
    B. 1.5×10^{14} Hz
    C. 1.67×10^{-15} Hz
    D. 0.6×10^{10} Hz.

Answer is B: $v = f\lambda$ so $f = v/\lambda = 3 \times 10^8$ m/s ÷ 500 nm = 3×10^8 m/s ÷ 500×10^{-9} m = $1500 \times 10^{8-9} = 1500 \times 10^{17} = 1.5 \times 10^{14}$ Hz.

16.2 Light Waves

Electromagnetic radiation that stimulates the sensation of vision in our eyes is called visible light. This radiation is also "non-ionising" in that the energy it carries is insufficient to produce ions in the medium through which it passes. Light that passes through the curved surfaces of our cornea and the eye's lens has its direction of travel changed. That is, the direction of travel of the light ray is changed or refracted. In this way, our eye can focus the light coming from an object at any distance away. Light or "visible" light has a range of frequencies from red light (3.9×10^{14} Hz) to blue (7.5×10^{14} Hz). Our eye is able to distinguish between the different frequencies and perceives them as different colours. Our eye can see visible light, while infrared and ultraviolet are both invisible.

A LASER is a source of light that is medically useful because the light it produces is monochromatic (all of one wavelength) and "coherent" (is oscillating in phase), and the beam is very parallel (it does not diverge). This means the energy in the beam can be focussed to a tiny spot, so that it is intense enough to vaporise the tissue at that spot. Hence a laser may be used as a surgical tool to cut through tissue.

1. Visible light waves are examples of

    A. Electromagnetic waves.
    B. Mechanical waves.
    C. Longitudinal waves.
    D. Compressional waves.

Answer is A: Visible light is electromagnetic waves that are detectable by the human eye.

2. Which phrase would best describe waves of dim violet light?

    A. High frequency and high amplitude.
    B. Low frequency and high amplitude.
    C. High frequency and low amplitude.
    D. Low frequency and low amplitude.

Answer is C: The violet end of the spectrum has waves of higher frequency than the red end; dim light refers to low amplitude light.

3. The phenomenon of refraction is due to
   A. Light rays bending when they enter a different medium.
   B. The decrease in speed when a ray enters a less dense medium.
   C. The difference in refractive indices of two media.
   D. The different speeds with which different frequencies of light travel through media.

Answer is C: In choice D, the different speeds with which different frequencies of light travel through media are due to the different refractive indices of the different frequencies.

4. Consider the situation where a light wave travelling in air strikes a glass surface with an angle of incidence of 20°. Which of the following statements is true?
   A. The angle formed by the incident ray and the normal will be 20°.
   B. The angle of refraction will be greater than 20°.
   C. The angle formed by the incident ray and the glass surface is 20°.
   D. There will be no refracted ray. That is, total internal reflection will occur.

Answer is A: The "angle of incidence" is defined as angle formed by the incident ray and a line perpendicular with the surface (i.e. the "normal"). Total internal reflection may occur when light passes from glass to air, not air to glass.

5. The term "refraction" may be applied to which of the following?
   A. The effect on light rays as they enter the cornea of the eye.
   B. Light rays from an object meeting at a point to form a sharp image.
   C. A light ray travelling away from a surface at the same angle that the light ray approached the surface.
   D. A light ray travelling through glass with a faster speed than it would have in air.

Answer is A: The cornea bends (refracts) light that enters the eye.

6. What is a convex lens? One that:
   A. Can accommodate to different focal lengths.
   B. Will cause light rays to diverge.
   C. Is thicker in the middle than at the edge.
   D. Will correct for myopia (short sightedness).

Answer is C: A convex or converging lens is "fat" in the middle.

7. In the electromagnetic spectrum, the frequencies that we call visible light have values that lie above
   A. Ultraviolet and below infrared.
   B. Infrared and below microwaves.

C. Microwaves but below ultraviolet.
D. Infrared but below radio range.

Answer is C: Microwaves are low-frequency em radiation while UV radiation is relatively high. The frequencies of visible light are between the two.

8. In which of the following sequences are the types of electromagnetic radiation listed in correct order of energy with lowest energy first and highest energy last?
   A. Visible, ultraviolet C, ultraviolet A, X-rays.
   B. Microwaves, infrared, ultraviolet, gamma rays.
   C. X-rays, ultraviolet, visible, microwaves.
   D. Infrared, microwaves, X-rays, gamma rays.

Answer is B: Choice C is in reverse order; in choice A, UVC has higher frequency than UVA; it is possible for an X-ray frequency (e.g. from a hospital linear accelerator) to be higher than a particular gamma ray (e.g. from radioactive technetium).

9. The energy carried by a photon of electromagnetic radiation is proportional to its
   A. Frequency.
   B. Speed.
   C. Wavelength.
   D. Amplitude.

Answer is A: When referring to the energy in a photon (a packet) of radiation, the equation $E = h \times f$ relates the energy E, of a photon to its frequency, f.

10. Ultraviolet radiation is damaging to the eye because
    A. The heat produced as it is absorbed distorts the cornea.
    B. It causes an increase in the pressure in the eyeball which results in glaucoma.
    C. The energy of the radiation destroys the cones in the fovea.
    D. The energy of ultraviolet radiation is mainly absorbed in the lens which harms the cells.

Answer is D: UV radiation is absorbed by the eye lens, and the energy of the absorbed photons can break the bonds between atoms.

11. Which of the following types of electromagnetic radiation is most likely to be responsible for the incidence of skin cancers?
    A. Infrared.
    B. Ultraviolet-A (UVA).
    C. Ultraviolet-B (UVB).
    D. Ultraviolet-C (UVC).

Answer is C: While UVC has higher energy than UVB and therefore is potentially more damaging; however, UVC does not reach the ground as it does not penetrate the earth's atmosphere.

12. To what radiation does the term "non-ionising radiation" refer? To radiation that consists of photons with energy that is:
    A. Not high enough to knock electrons out of atoms but is high enough to produce ions.
    B. Sufficient to generate ions in the material that the photons enter.
    C. Not sufficient to generate ions in the material that the photons enter.
    D. High enough to knock electrons out of atoms but not high enough to produce ions.

Answer is C: The term "non-ionising" implies that ions are not produced. Ions are produced when one or more electrons are removed from an atom.

13. An oscillating magnetic field will be produced by
    A. An electric field of constant magnitude.
    B. A permanent magnet moving at constant speed.
    C. The direct current that powers portable radios.
    D. The alternating current in household electrical appliances.

Answer is D: An alternating current (one whose direction reverses the direction frequently in a second) will produce a changing (oscillating) magnetic field. Whereas a steady current will produce a steady magnetic field.

14. What can be said about the collimated beam of electromagnetic energy produced by an Nd-YAG laser?
    A. It has a higher frequency than visible light energy.
    B. It stimulates the production of light as it passes through air.
    C. It is produced when electrons change energy levels.
    D. It consists of waves of two or more frequencies.

Answer is C: The em energy produced by a laser occurs when electron drop to a lower energy level from their "excited" state. An Nd-YAG laser produces one frequency of near infrared light (which is of lower frequency than visible light).

15. Material that is unsuitable for use in a LASER lacks which of the following properties?
    A. A metastable excited state.
    B. A short-lived ground state.
    C. The ability to spontaneously emit radiation.
    D. A long-lived ground state.

Answer is A: A metastable excited state—that is, one that persists for longer than an excited state normally exists, is required for lasing action.

16. Consider an electron that is in its ground state in an atom. When this electron moves to another state in the atom, it is said to have
    A. Gained energy and to be in an excited state.
    B. Gained energy and to be in a metastable state.
    C. Lost energy and undergone stimulated emission.
    D. Lost energy and to be part of a population inversion.

Answer is A: "Ground state" is the state of lowest energy; hence, it can only move to a higher energy state. In this case, it is said to be excited.

17. What property of a LASER makes it suitable for surgical procedures? A LASER:
    A. Beam's energy can be focussed onto a very small spot to vaporise tissue
    B. Can be passed down an optical fibre in an endoscope
    C. Beam produces monochromatic (all of one wavelength) photons
    D. Emits photons which all have the same energy

Answer is A: The ability to vaporise tissue means a laser can "cut" through tissue at the point of vaporisation.

18. What is a concave lens? One that:
    A. Can accommodate to different focal lengths
    B. Will cause light rays to converge
    C. Is thicker in the middle than at the edge
    D. Will correct for myopia (short sightedness)

Answer is D: Concave lens will cause light rays to diverge. So can be used to correct for short-sightedness as they shift the focal point further away from the eye's lens.

19. Which of the following conditions may be corrected by a diverging (concave lens)?
    A. Hyperopia
    B. Myopia
    C. Presbyopia
    D. Red minus dichroma

Answer is B: In myopia, the image forms ventral (in front of) retina. A diverging lens will extend the focal length of a myopic eye, so that the image falls onto the retina.

20. In what way is solar ultraviolet (UV) radiation potentially capable of causing damage to biological systems?
    A. Solar UV generates heat which can disrupt chemical bonds within molecules.
    B. Absorption of solar UV can raise the temperature of muscle sufficiently to denature the muscle proteins.

C. Solar UV is absorbed by the eye lens which can cause it to become opaque (produce cataracts).
D. Solar UV can disrupt the bonds in the DNA molecule to produce lesions which may be improperly repaired.

Answer is D: Solar UV can damage DNA. Despite the repair mechanisms that exist, some lesions escape without or with faulty repair. If so, a cancerous tumour may result. Cataracts will occur in the lens too, but D is the better answer.

21. In what way is LASER radiation potentially capable of causing damage to biological systems?
    A. A focussed LASER beam can vaporise tissue.
    B. LASER light can disrupt the bonds in the DNA molecule to produce lesions which may be improperly repaired.
    C. A LASER beam that enters the eye will damage the retina before the blink reflex can operate.
    D. Absorption of LASER energy can raise the temperature of muscle sufficiently to denature the muscle proteins.

Answer is C: A LASER beam that is inadvertently reflected from a shiny surface and into an eye will be focussed by the eye lens onto the retina. This will damage the retina before the eyelid has the chance to blink. Choice A is true of course but is unlikely to cause unintended damage as the focal distance for surgical LASERS is short.

16.3 Sound

Sound waves require a solid, liquid or gas (the medium) to travel through. They are mechanical waves. They cause particles (atoms and molecules) of the medium to move (vibrate). The young healthy ear has an audible range of ~20 to ~20,000 Hz. However, sounds of different frequencies stimulate our hearing by different amounts. Our ear is most "sensitive" between 500 and 6000 Hz. It is not very sensitive below 200 Hz or above 12,000 Hz. A sound's intensity (in watts per square metre, W/m^2) is the objectively measurable amount of *sound energy* carried by a sound. A value as low as 10^{-12} W/m^2 is at our threshold of hearing and may be called silence. The human perception of the loudness of a sound is called "sound level" and is expressed in decibels (dB). A sound level of 0 dB is perceived as silence. This does not mean that the amount of sound energy is zero, just that we cannot hear it! We can also hear a sound energy of 1 W/m^2 as a painful (and ear damaging) sound level of 120 dB.

Two sounds that differ by 0.1 dB can barely be perceived as different in loudness while a sound that is 3 dB louder than another, sounds twice as loud. Continuous noise at 90 dB will, over time, produce hearing damage without causing pain.

1. Which of the following statements about sound and light is **INCORRECT**?
   A. They both extend over a range of frequencies.
   B. Sound is a longitudinal wave and light is a transverse wave.
   C. The speed of travel of light is fast while sound travels relatively slowly.
   D. Sound will travel through a vacuum while light requires a transparent medium.

Answer is D: Sound cannot travel through a vacuum, but all em radiation including light can.

2. What is the measure known as "sound level"?
   A. A subjective measure of the perceived loudness of a sound (in dB).
   B. A graph of the levels of sound of different frequencies that we perceive as equal in loudness (in phone).
   C. The frequency of the sound that produces the loudest response in a healthy hearing mechanism (in Hz).
   D. The objectively measured amount of sound energy carried by a sound wave (in W/m^2).

Answer is A: Sound level is subjective, in that it is perceived by the hearer and depends on the frequency/ies of the sound. The unit is the decibel.

3. What is the speed of sound in water at 20°C?
   A. 330 m/s
   B. 1480 m/s
   C. 2540 m/s
   D. 3400 m/s

Answer is B: If you have an idea of the approximate value of the speed of sound in air, choices A, C and D can be easily rejected.

4. How may sound waves be characterised? As:
   A. Longitudinal waves because the oscillations that they cause are along a line at right angles to the direction of propagation.
   B. Mechanical waves because they can only travel through a material medium.
   C. Waves because they allow for the oscillation of a medium without the transfer of energy.
   D. Compressions and rarefactions because the nature of the oscillation of their electric and magnetic fields.

Answer is B: Sound waves require a medium. Longitudinal waves oscillate parallel to their direction of propagation. Electric and magnetic fields are not involved. Choice C is correct but not specific enough.

5. If the frequency of a sound is 1000 Hz, what will its period will be?
   A. 1 s
   B. 0.1 s

C. 0.001 s
D. 0.0001 s

Answer is C: Period = 1/frequency, hence = 1/1000 = 0.001 s.

6. A sound wave of frequency 1000 Hz is travelling through air with a speed of 330 m/s. What is its wavelength?
   A. 33 cm
   B. 330 cm
   C. 33 m
   D. 330 m

Answer is A: Rearrange the wave equation for wavelength: $v = f\lambda$ so $\lambda = v/f$. Then substitute: $\lambda = 330\ \text{m/s} \div 1000\ \text{Hz} = 0.33\ \text{m} = 33\ \text{cm}$.

7. As the frequency of a sound in air is made to decrease, which of the following will happen?
   A. The period will increase.
   B. The wavelength will decrease.
   C. The amplitude will decrease.
   D. The velocity will decrease.

Answer is A: Frequency and period are inversely related, so as frequency decreases, the period of the wave will increase.

8. What is the unit called the decibel (dB) used in the measurement of?
   A. Sound frequency (pitch)
   B. Sound intensity (energy)
   C. Sound pressure
   D. Sound level (loudness)

Answer is D: Sound level uses the unit dB. Frequency uses hertz, intensity uses watts per square metre and pressure uses pascal.

9. An increase in sound level of 20 dB represents an increase in intensity (W/m^2) of:
   A. 10,000
   B. 1000
   C. 200
   D. 100

Answer is D: If the sound level difference between two sounds whose intensities are I_1 and I_2 is 20 dB, $20 = 10 \times \log\left(\frac{I2}{I1}\right)$, so a sound level of 20 dB means that the log term must be 2. Now the log(100) = 2, hence $I_2/I_1 = 100$. Thus one sound is 100 times the intensity of the other.

10. When a sound with a particular frequency is played at a loudness of 60 dB and compared to the same sound played at 10 dB, what can be said about the intensity (in W/m²) of the 60 dB sound?
    A. 50 times more intense than the 10 dB sound
    B. 5000 times more intense than the 10 dB sound
    C. 50,000 times more intense than the 10 dB sound
    D. 100,000 times more intense than the 10 dB sound

Answer is D: If a sound is 10 times more intense (10 × more W/m²) than another sound of the same frequency, then it is 10 dB (1 bel) louder. If it is 100 times more intense, it sounds 20 dB louder. Hence, remembering the logarithmic nature of the loudness scale, a sound that is 50 dB (5 bel) louder than another is also 100,000 times (10^5) more intense.

11. How much louder does a sound of 30 dB sound when compared to the same sound played at 10 dB?
    A. 2 times louder
    B. 4 times louder
    C. 20 times louder
    D. 100 times louder

Answer is D: 1 bel = 10 dB. Each time sound intensity increases by a factor of 10, sound level increases by 1 bel. Thus if a sound is 2 bel (20 dB) louder, it is 10 × 10 = 100 times louder.

12. Which of the following sound levels, if prolonged, will damage the hearing mechanism of the ear without being perceived as painful?
    A. 50 dB
    B. 70 dB
    C. 90 dB
    D. 120 dB

Answer is C: Prolonged exposure to 90 dB will result in hearing loss. So will 120 dB but that will be noticeably painful.

13. Decibels (dB) measure the ratio of a given intensity I to the threshold of hearing intensity (10^{-12}). The ratio is defined by the formula: $dB = 10\log\left(\frac{I}{10^{-12}}\right)$, where I is the intensity (in W/m²) of the sound in question. How many decibels separate a sound of intensity 10^{-10} W/m² and another which carries 100 times this intensity (i.e. 10^{-8} W/m²)?
    A. 10 dB
    B. 20 dB
    C. 30 dB
    D. 40 dB

Answer is B: Substituting 10^{-10} for I in the formula gives: dB = 10 × log(100) = 10 × 2 = 20. Substituting 10^{-8} for I in the equation gives: dB = 10 × log(10,000) = 10 × 4 = 40. Hence the difference is 20 dB.

16.4 Ultrasound

Ultrasound is inaudible sound above 20,000 Hz in frequency. Medical ultrasound is high-frequency longitudinal waves of frequency 2 MHz or 4 MHz or thereabouts. Ultrasound is a mechanical wave that must travel in a medium. It is not an electromagnetic radiation so is non-ionising. It is fundamentally different from electromagnetic radiation. Ultrasound is produced in the transducer (a piezoelectric crystal that can vibrate very rapidly when electrified with high frequency AC voltage). The transducer is coupled to the skin with acoustic gel. This excludes air from between the transducer and the skin which would cause ultrasound to reflect from it.

Diagnostic imaging with ultrasound projects ultrasound into the body and detects the ultrasound that is reflected from the boundaries between organs that differ in their density and elasticity. Ultrasound travels at about 1540 m/s in tissue and is reflected at (i.e. echoes from) the interface between one tissue and the next as long as their impedance differs (impedance = density × speed of sound). Echoes are "loudest" from bone-tissue and air-tissue interfaces. From the time taken for the reflections to return and the intensity of the echo, an ultrasound image is reconstructed from the boundaries that produced the reflections. That is, from the boundaries between organs or tissues.

Ultrasound is not used to image lung (or bowel with gas) as the very different densities of gas and tissue cause all the ultrasound to be reflected. Ultrasound can be made to deliver energy to tissues deep to the skin, that is, to warm tissues. That is, ultrasound can be used in physiotherapy as well as for medical imaging. The amount of energy delivered is not sufficient to ionise atoms or molecules so ultrasound is non-ionising radiation.

If ultrasound reflects from a moving object such as the beating walls of the heart or the red blood cells flowing in a blood vessel, the ultrasound is reflected with a changed frequency. This change-of-frequency phenomenon is known as the Doppler effect, so the examination modality is known as Doppler ultrasound. Quantitative measurements such as speed of blood flow, rate of heart beat and volume of blood remaining in a ventricle after systole can be made with this modality.

1. What frequencies are attributed to the inaudible sound known as ultrasound?

   A. Greater than 2 kHz
   B. Greater than 2 MHz
   C. Greater than 20,000 Hz or less than 20 Hz
   D. Less than 20 Hz

Answer is B: Frequencies less than 20 Hz are called infrasound. Frequencies greater than 2 kHz (2000 Hz) are still audible. The best answer is 2 MHz. While this is a frequency typical of medical ultrasound, any frequency greater than 20,000 Hz is inaudible and called ultrasound.

2. Prior to an imaging examination using ultrasound, the skin surface is coated with a gel substance known as a "coupling agent". What is the purpose of the gel?
   A. To reduce the friction between the skin and the ultrasound transducer.
   B. To eliminate air, which would reflect the ultrasound, between the skin and transducer.
   C. Since ultrasound travels faster through denser materials, the time delay caused by travelling through air is avoided by using gel.
   D. To avoid an unpleasantly cold sensation that would otherwise be produced by the ultrasound transducer.

Answer is B: The impedance difference between air and skin is large, so most of the ultrasound would reflect from the skin, rather than enter the body. Using gel avoids the ultrasound passing through air.

3. Which statement about a Doppler ultrasound stethoscope is true?
   A. They amplify the echo produced when ultrasound strikes a boundary between two tissues of different impedance.
   B. Their operation depends on the reflected ultrasound being at a different frequency to the emitted ultrasound.
   C. They emit ionising radiation.
   D. The depth of penetration of ultrasound into tissue increases as the frequency of ultrasound increases.

Answer is B: A Doppler "stethoscope" produces ultrasound and detects the reflected ultrasound from moving red blood cells (or the moving heart wall). The motion of the object reflecting the ultrasound causes the reflected ultrasound to be shifted in frequency. Its depth of penetration in human tissue decreases as it frequency increases.

4. Which of the following is considered to be the diagnostic ultrasound frequency range?
   A. 1–15 mHz
   B. 10–20 Hz
   C. 20 Hz to 20 kHz
   D. 1–20 MHz

Answer is D: Ultrasound has frequencies greater than 20 kHz. In medical applications, the frequencies used are greater than 1 MHz.

5. What will cause a large percentage of the ultrasound energy to be reflected from the interface between two media? This will happen if:
   A. The acoustic impedance of each medium is the same.
   B. The speed of ultrasound in the two media differs greatly.
   C. The acoustic impedance of each medium is very different.
   D. The densities of the media are different.

Answer is C: A large difference in acoustic impedance will cause a large reflection. Impedance depends on both speed of sound in the medium and the density of the medium.

6. Ultrasound radiation may be characterised as which one of the following?
   A. Ionising radiation
   B. Audible radiation
   C. Longitudinal waves
   D. Electromagnetic waves

Answer is C: Ultrasound, and audible sound, is a longitudinal wave. It is neither ionising nor electromagnetic.

7. What is ultrasound?
   A. Frequencies less than 50 Hz and more than 20,000 Hz
   B. A mechanical wave (it requires a medium)
   C. A form of ionising radiation
   D. More penetrating (in human tissues) as its frequency increases

Answer is B: Ultrasound is a mechanical wave. Its depth of penetration in human tissue decreases as it frequency increases.

8. What will a sound of frequency 250,000 Hz be?
   A. Audible
   B. Painful to listen to
   C. Ultrasonic
   D. Close to the threshold of hearing

Answer is C: 250 kHz is way above the upper limit of human hearing so is ultrasound.

9. What is ultrasound least useful for examining?
   A. Heart
   B. Lungs
   C. Kidneys
   D. Uterus

Answer is B: The lungs contain air which reflect all of the ultrasound reaching it, making the lungs opaque to ultrasound.

10. The change in pitch (frequency) of a sound when there is relative motion between the sound source and the human observer is known as the:

    A. Resonance effect
    B. Interference effect
    C. Diffraction effect
    D. Doppler effect

Answer is D: This change of frequency effect is apparent when listening to the sound of a motor vehicle engine as it approaches you and then as it continues past you.

11. Basically, what does a Doppler ultrasound examination for peripheral vascular disease involve?

    A. The measurement of blood speed
    B. Listening to the reflected ultrasound frequencies
    C. Using ultrasound to produce an image on a screen
    D. Increasing blood flow by warming the deep tissues

Answer is A: The faster the red blood cells are moving, the greater will be the difference in frequency between the incoming and reflected ultrasound frequency. RBCs move faster as they pass through a narrowing in a blood vessel. Ultrasound is inaudible, so you cannot hear the reflected ultrasound.

12. A Doppler ultrasound device produces ultrasound of frequency 5 MHz and the echo from red blood cells has a frequency of 5,002,500 Hz. The beat frequency produced will be:

    A. Inaudible
    B. 250 Hz
    C. 2500 Hz
    D. 0.25 MHz

Answer is C: Beat frequency is the difference between the two frequencies. 5,000,000 – 5,002,500 = 2500 Hz. The frequency of the beats is within the audible range so we can hear them.

13. When ultrasound strikes the boundary between two different body tissues, the amount of reflection that will occur is proportional to the difference in what?

    A. The speed of sound in the tissues
    B. The density of the two tissues
    C. Elasticity in the tissues
    D. Impedance between the two tissues

Answer is D: The impedance depends on the speed and density. Speed depends on the elasticity of the medium. So impedance is the more complete answer.

14. When a Doppler ultrasound device is used to investigate the blood flow in an artery, what happens to the ultrasound when it reflects from a red blood cell?
    A. The frequency of the ultrasound is increased.
    B. The frequency of the ultrasound is changed.
    C. The frequency of the ultrasound is decreased.
    D. The frequency of the ultrasound is unchanged.

Answer is B: When ultrasound reflects from a moving object (e.g. a RBC), its frequency is changed. This is a manifestation of the "Doppler effect". The frequency will increase if the RBC is travelling towards the ultrasound transducer. The frequency will decrease if the RBC is travelling away from the transducer. As the question does not state the direction of travel of the RBC, one cannot choose either of A or C.

15. A typical sound produced by a Doppler device when used to assess the speed of blood flow in the brachial artery is tri-phasic. That is there are three sounds. Why is this?
    A. Blood flow through elastic arteries is pulsatile.
    B. The three sounds correspond to the pumping action of the left ventricle, the right ventricle and the left atrium.
    C. Blood flow is affected by the actions of the left ventricle, and by lung inhalation and exhalation.
    D. The sounds are the original ultrasound frequency, the changed reflected frequency and the frequency of the beats between the two frequencies.

Answer is A: Blood flow is fastest as the left ventricle contracts. The velocity of blood flow decreases between contractions. In fact flow is momentarily retrograde as the aortic valve fills and closes. Then forward flow continues as the elastic artery walls recoil from being distended. The other answers are nonsense.

16. When using a Doppler ultrasound "stethoscope" at the radial pulse and an inflatable arm cuff to determine systolic blood pressure, why does the apparatus produce no Doppler sounds when the blood pressure cuff is fully inflated?
    A. Ultrasound is not audible to the human hearing apparatus.
    B. An inflated cuff squashes the artery and so blood flow is stopped.
    C. Blood flows too slowly in the radial artery to produce a velocity signal.
    D. Doppler sounds are produced by turbulent blood flow. This type of flow is not present in the radial artery.

Answer is B: Doppler sounds are produced when the incident ultrasound frequency is different from the frequency of the beam reflected by the moving red blood cells. An inflated cuff completely squashes (occludes) the artery, so that no blood flow is present. Hence, the RBCs are not moving. This means that the two ultrasound frequencies (the incident one and the frequency of the beam reflected from the RBC) are the same.

17. What is the medically useful frequency range of diagnostic ultrasound considered to be?
    A. 1–15 mHz
    B. 10–20 Hz
    C. 20 Hz to 20 kHz
    D. 1–20 MHz

Answer is D: Ultrasound is beyond the range of human hearing so is greater than 20 kHz. However, diagnostic US is in the megahertz range.

18. When will a large percentage of the ultrasound energy reflect from the interface between two media? If:
    A. The acoustic impedance of each medium is the same.
    B. The speed of ultrasound in the two media differ greatly.
    C. The acoustic impedance of each medium is very different.
    D. The densities of the media are different.

Answer is C: The "acoustic impedance" of a medium is the product of density of the medium and speed of sound in the medium. Hence, choices B and D are incomplete.

19. What is the speed of sound in water at 20°C?
    A. 330 m/s
    B. 1480 m/s
    C. 1540 m/s
    D. 3400 m/s

Answer is B: Sound travels faster at higher temperature. Choice A is the speed of sound in air.

20. If an interface lies 15.4 cm below the patient's skin, what is the time interval between emitting the ultrasound pulse and detection of its echo? (speed of sound is 1540 m s^{-1}).
    A. 2.0 ms
    B. 1.0 ms
    C. 0.2 ms
    D. 0.1 ms

Answer is C: Sound travels at 1540 m s^{-1} = 154,000 cm s^{-1}. Hence, it would take 0.0001 s to travel 15.4 cm from skin to interface and another 0.0001 s for the echo to make the return journey. Hence 0.0002 s or 0.2 ms.

21. What property of tissue does the product $\rho \times \nu$ (density of tissue multiplied by the speed of sound in the tissue) determine?
    A. The elasticity of the tissue
    B. The acoustic impedance
    C. The mechanical index

D. The spatial-peak, temporal-average intensity

Answer is B: This is a formula for the acoustic impedance of a material which determines the strength of reflection from a boundary between two tissues.

22. How does pulse waved (PW) Doppler differ from continuous wave (CW) Doppler? In the former:
    A. Ultrasound is used to produce colour flow images.
    B. The Doppler shift is displayed in colour, superimposed on the 2D grey scale image.
    C. Uses the beat frequency to determine the Doppler shift.
    D. The ultrasound intensity is measured as "spatial average intensity".

Answer is A: CW Doppler cannot be used to produce colour flow images.

23. The piezoelectric effect may be described as:
    A. The deformation of a crystal material when placed in a magnetic field
    B. The change in frequency when a wave is reflected from a moving source
    C. The production of an electric field in a crystal when it is deformed
    D. The transmission of ultrasound through an interface when the two materials are matched in impedance

Answer is C: A piezoelectric crystal material suffers a deformation when placed in an electric field. The reverse is also true: When a piezoelectric crystal material is deformed, it generates an electric field.

24. In what way is pulsed Doppler ultrasound radiation potentially capable of causing damage to biological systems?
    A. It can disrupt the bonds in the DNA molecule to produce lesions which may be improperly repaired.
    B. High-power pulsed Doppler US can potentially produce a blood thrombosis in a vein.
    C. It can produce ions in insonated tissue which can go on to damage biological molecules.
    D. Pulsed Doppler US uses high power which can potentially raise the temperature of imaged tissue.

Answer is D: The power level of pulsed Doppler US exposure can cause temperature rises in tissue, which if greater than 4 °C for over 5 min should be considered potentially hazardous. Choices A and C are wrong as US is not ionising radiation.

Chapter 17
Ionising Radiation

17.1 Medical Imaging with X-Radiation

X-ray photons directed at a patient do one of three things:

1. Pass straight through the patient (who is mostly empty space) without deviating from their trajectory, to strike the detector and form the medical image.
2. Are stopped within the patient by ionising some of the patient's atoms, leaving their energy in the patient's tissues. This absorbed energy contributes to the patient's "absorbed dose" of radiation.
3. Are deflected from their path so do not reach the detector. These "scattered" photons do not contribute to the image. They may increase the radiation dose to the patient and will irradiate people nearby. Hence radiographers wear protective "lead" aprons which absorb this (relatively low energy) scattered X-radiation.

A diagnostic procedure uses the X-rays that pass, undeflected, through the body to be detected by a device on which produces a "shadow picture" of internal structure. Bones, being dense tissue (containing calcium Ca—element number 20, and phosphorus P—element number 15), absorb most of the X-rays that fall on them so cast a shadow on the detector. Soft tissue is easily penetrated by X-rays, so they strike the measuring surface. Soft tissue contains mostly hydrogen, carbon and oxygen atoms which are less dense than Ca and P. The higher is the energy of the X-ray photon, the more penetrating it is. Low-energy photons (less than 20 keV) will be stopped—even by soft tissue—before they get out of the body. These photons will never influence the radiograph so should not be allowed into the patient in the first place. They are filtered out of the X-ray beam by passing the beam through a sheet of aluminium or copper before the X-rays enter the patient.

X-radiation is a form of electromagnetic radiation that is produced by an electrical machine. That is, when the machine is switched off, no radiation is present. X-ray photons have enough energy to remove electrons from their atoms, that is, to create ions. Some X-rays have sufficient energy to break the chemical bonds within molecules. This breaks the molecule into fragments that may undergo unintended

M. Caon, *Examination Questions and Answers in Basic Anatomy and Physiology*, https://doi.org/10.1007/978-3-030-47314-3_17

chemical reactions. This ability to deposit energy within a living cell can cause harm to that cell. On the other hand, because atoms are mostly empty space, the majority of X-ray photons pass through human tissue without colliding with anything. The ability of most X-rays to pass through tissue and on emerging, to be detected by some device, while some X-rays are absorbed within the tissue, is the basis for the medical imaging modalities known as mammography, X-ray pictures (radiographs) and CT scanning. Mammography uses X-rays below about 30 keV of energy, while CT scanning uses X-rays with energies up to 140 keV. While ionising radiation has the potential to cause damage to living cells, cells have mechanisms to repair the damage. The radiation dose that is received by someone undergoing an imaging procedure with X-rays (less than 10 mSv for CT) is medically acceptable considering the diagnostic information it provides and the benefit of this for the patient.

The ability of X-radiation to damage living cells is utilised when a hospital linear accelerator is used to produce the very high energy X-rays (4–20 MeV) that are used to kill cancerous cells. These high energy X-rays are more penetrating than those with lower energy, so can be used to target tumours deep within the body.

1. Which statement about the differences between medical imaging using X-rays and a nuclear medicine scan using gamma rays is correct?
   A. An X-ray procedure leaves the patient with residual radioactivity while nuclear medicine does not.
   B. A gamma ray source can be switched off after which no gamma radiation is produced while an X-ray source will continue to produce radiation until the source decays.
   C. X-rays produce an image of internal anatomy while a nuclear medicine scan provides information about the functioning of an organ or tissue.
   D. A beam of gamma rays is fired at the patient and detected on the other side, while X-rays are produced by the nucleus of a radionuclide incorporated in the patient's body.

Answer is C: X-rays that pass through the body without being absorbed by the body are used to produce an image of internal structure. In nuclear medicine, a radioactive material is incorporated into the body, travels to certain organs from where a gamma ray is emitted. If physiology is altered sufficiently to affect the way the radioactive material moves about the body, the resulting image provides information about how much alteration there has been.

2. Conventional radiography—such as a chest X-ray (CXR)—differs from computed tomography (CT) in what respect?
   A. CT produces an image of all internal anatomy while in CXR, overlying anatomical structures obscure the view of underlying structures.
   B. In CT the patient is left with some residual radioactivity, but not with CXR.
   C. CT produces a lower absorbed dose of radiation to the patient than does a CXR.

D. CT involves the use of ultrasound while a CXR results from X-rays.

Answer is A: CT mages are unobstructed by the "shadow" of overlying structures.

3. Which of the following imaging modalities does **NOT** involve the use of "ionising radiation"?
   A. Mammography
   B. Ultrasound
   C. A scintigram using technetium
   D. A chest X-ray

Answer is B: Ultrasound produces an oscillation in the particles of the body, but does not use ionising electromagnetic radiation.

4. When inspecting an X-ray image, the order of densities from blackest to whitest is:
   A. Bone, water, fat, air
   B. Air, fat, water, bone
   C. Air, water, fat, bone
   D. Bone, air, water, fat

Answer is B: The blackest part on an X-ray image is air, while the whitest is bone.

5. Which of the following imaging modalities uses X-rays?
   A. Computed tomography (CT)
   B. Single-photon emission computed tomography (SPECT)
   C. Positron emission tomography (PET)
   D. Nuclear medicine scan (scintigram)

Answer is A: CT machines generate X-rays. The others are all nuclear medicine procedures.

6. Radiation which is "ionising" includes which of the following?
   A. X-rays and gamma rays
   B. Infrared radiation
   C. Radiation emitted by mobile phones
   D. Microwaves

Answer is A: The other three choices, while also being forms of electromagnetic radiation, do not utilise energies that are sufficient to remove electrons from their atoms.

7. The lead aprons that are used for the protection of staff in diagnostic radiography procedures do not provide protection against the ionising radiation used in nuclear medicine or radiotherapy. Why is this? Because:
   A. Charged particles are much easier to stop (are less penetrating) than photons are.

B. Gamma rays are more penetrating than x-rays even if both have the same energy.
C. Such aprons do not cover the arms, feet, head and neck.
D. The shielding provided by aprons is not sufficient to stop photons with energies above 100 keV.

Answer is D: Diagnostic radiography uses X-rays with an average energy of 70 keV or less, which can be stopped by relatively thin amounts of lead (or lead-like) material. In order to stop the gamma rays used in nuclear medicine imaging, the thickness of the "aprons" would make them prohibitively heavy to wear.

8. When compared to visible light, which is not very penetrating, why can radiation such as X-rays and gamma rays pass right through the human body? Because:

   A. The density of the human body is relatively low.
   B. They have no mass and no charge.
   C. Atoms in the body are mostly empty space.
   D. They have very high energy.

Answer is D: X- and gamma rays are high energy radiation, which means high frequency, which means short wavelength, which means that many of the photons will pass through the body without interacting with any atoms. The other three choices do not distinguish between different types of radiation.

9. Which of the following is true?

   A. A patient exposed to diagnostic X-rays will emit X-rays for a short time after the procedure.
   B. A cancer patient treated with a megavoltage beam of X-rays will emit X-rays for a short time after the treatment.
   C. For a short time after having a bone scan using the radionuclide technetium-99m, the patient will emit gamma rays.
   D. The human body does not contain any radioactive material unless it has been exposed to man-made radioactive material.

Answer is C: A bone scan (any nuclear medicine scan) involves taking some radioactive materials into the body which then emits gamma rays to be detected outside the body. The body takes some time to excrete the material (and much of it decays) so until that happens, the body is more radioactive than is usually the case.

10. What does the term *ionising radiation* refer to?

   A. The radiation that is emitted by ionised atoms
   B. That part of the electromagnetic spectrum with wavelengths less than 300 nm which has enough energy to produce ions
   C. Alpha, beta and gamma rays spontaneously emitted from radionuclides
   D. Radiation with enough energy to produce ionisation in the material which absorbs it

Answer is D: If the radiation produces ions when it interacts with any substance, then it is ionising radiation. Choice B is true but does not include particulate radiation. Choice C is true but does not include X-rays or cosmic rays.

11. Which of the following is a correct use of the unit known as the "electron volt" (eV)?
    A. One electron volt is the amount of radioactivity that results in one disintegration per second.
    B. Radiopharmaceuticals contain gamma photon emitting radionuclides whose energy is usually in the range 100–250 keV.
    C. One electron volt is equal to 1.9×10^{16} J of energy.
    D. A photon of visible light has energy of about 1.5 MeV.

Answer is B: eV is a unit of energy (not radioactivity). It is equal to 1.9×10^{-16} J of energy. Visible photons have energy of 1.5 eV or less.

12. Which one of the statements about the penetrating ability of radiation is true?
    A. 750 keV gamma rays are more penetrating than 750 keV X-rays.
    B. 140 keV gamma rays are more penetrating than 60 keV X-rays.
    C. 2 MeV beta rays (electrons) are more penetrating than 1 MeV gamma rays.
    D. 1 MeV gamma rays are more penetrating than 2 MeV X-rays.

Answer is B: Gamma rays and X-rays are indistinguishable, once they have travelled away from their site of production. So a 140 keV gamma or X-ray is more penetrating than a lower energy one.

13. What may the term "ionising radiation" be applied to?
    A. All electromagnetic radiation
    B. Radiation that produces ions when it interacts with matter
    C. Infrared radiation
    D. Radiation that is emitted by ions

Answer is B: The production of ions (i.e. removal of electrons from an atom) is the sign of ionising radiation. Such a change in the medium through which radiation passes is significant.

14. What does it mean when an X-ray tube is operated at an accelerating voltage of 120 kV?
    A. The maximum energy that an X-ray photon can have will be 120 keV.
    B. The characteristic X-rays will have energy 120 keV.
    C. All of the X-ray photons will have an energy of 120 keV.
    D. The X-ray beam will contain photons with every energy from 0 keV up to 120 keV.

Answer is A: X-rays are emitted from the tube with a range of energies (a spectrum) which will range from mid-teens to low 20s of keV (depending on the amount of "filtration" that the photons pass through) up to the maximum value which will equal the accelerating voltage.

15. Why do some X-ray photons pass through the human body without deflection? Because:
    A. Carbon, hydrogen and oxygen atoms are transparent to X-rays.
    B. The energy of diagnostic X-rays is too low to produce interactions
    C. The wavelength of X-rays is too long to interact with an object with the dimensions of the human body.
    D. The interior of atoms is mostly empty space.

Answer is D: Apart from the positions where electrons and the nuclei are located, the rest of the body is empty space through which photons can travel unimpeded. They will interact only if they "hit" these subatomic particles. The other answers contain errors.

16. How will increasing the filtration of an X-ray beam reduce the intensity of the X-ray spectrum?
    A. Equally at all frequencies
    B. More at lower frequencies than at higher frequencies
    C. More at higher frequencies than at lower frequencies
    D. Only at lower frequencies

Answer is B: Lower frequency radiation is less penetrating than higher frequency radiation, so the low frequency are preferentially absorbed compared to high frequency radiation.

17. What is the difference between X-rays and gamma rays?
    A. X-rays are ionising radiation and gamma rays are not.
    B. Gamma rays have higher energies than X-rays.
    C. Gamma rays can be turned off by switching the power supply off.
    D. X-rays are produced in an electrical machine whereas gamma rays emerge from an atomic nucleus.

Answer is D: Apart from the difference in how they are produced, X-rays and gamma rays are the same phenomenon.

18. To what energy can an X-ray tube that is operated at 110 kV accelerate electrons?
    A. 110 eV
    B. 110 keV
    C. 110 J
    D. 110,000 keV

Answer is B: An electron volt is the amount of energy gained by an electron when it is accelerated by a potential difference of 1 V. 110 keV is the amount of energy gained by an electron when it is accelerated by a potential difference of 110 kV.

19. What is the purpose of adding filtration to an X-ray beam? To:
    A. Prevent high-energy photons entering the patient

B. Increase the mean energy of the beam
C. Decrease the scattered radiation
D. Increase the ratio of low energy photons to high energy photons

Answer is B: X-rays are produced in an X-ray tube with a range of energies. Filtration will absorb the lowest energy photons more than higher energy ones. This will result in the average energy of the remaining spectrum increasing.

20. Consider the "inverse square law". Compared to being 1 m from a point X-ray source, what will be the photon flux at a distance of 4 m from the source?

A. Sixteen times as great
B. Eight times as great
C. One eighth as great
D. One sixteenth as great

Answer is D: "Inverse square" means at double the distance, the flux will be one half squared of the original flux $(½)^2$ = ¼. At 4× the distance, the flux will be $(¼)^2$ = 1/16.

21. What is the purpose of an intensifying screen?

A. It converts a small number of X-ray photons into a large number of visible light photons.
B. It converts low energy X-ray photons into high energy visible light photons.
C. It improves the absorption efficiency of X-rays.
D. It protects the radiologist's eyes from the damage that would be caused by X-rays.

Answer is A: It converts X-rays (which have high energies but are invisible) into visible light photons. Because visible light photons have low energy, an X-ray photon can be made to produce lots of them.

22. One difference between the X-radiation in the primary beam and the scattered radiation is that:

A. Photons in the primary beam degrade contrast in radiographic images.
B. Scattered radiation is more penetrating than the primary beam.
C. Scattered radiation may be absorbed in the imager.
D. Scattered radiation is travelling at an angle to the main beam.

Answer is D: "Scattered" radiation has interacted with an atom in the medium so is diverted from the direction of the primary beam.

23. The advantage of computed tomography (CT) over conventional radiography is:

A. CT delivers lower doses than conventional radiography.
B. CT images are faster to acquire than conventional radiographs.

C. CT produces a cross-sectional image that is not obscured by overlying anatomical structures.
D. CT projects a 3D structure onto a 2D image.

Answer is C: It is the ability to image internal structures without having them partially obscured by the tissue on either side as is the case for "conventional" radiography that makes CT such a useful diagnostic tool. (Also there is no need to cut the person open to see what is inside!)

24. The contrast of a CT image displayed on a monitor may be increased by:
    A. Increasing the window width
    B. Increasing the window level
    C. Decreasing the window width
    D. Decreasing the window level

Answer is C: Decreasing the width of the "window" means restricting the range of grey scale displayed on the monitor screen (0–255), to a more limited range of the available Hounsfield numbers (−1000 for air to 3000 for dense bone). A lung window may be −1250 to +250; a soft tissue window may be −160 to +240; a bone window may be −650 to +1350.

25. Which one of the following beams is the most penetrating?
    A. 2 MeV gamma rays
    B. 2 MeV X-rays
    C. 4 MeV X-rays
    D. 8 MeV X-rays

Answer is D: Penetrating ability of electromagnetic radiation increases as energy does.

26. Exposing a foetus or young baby to X-rays should be avoided. What is the cause of the danger most likely to be due to?
    A. Denaturing of cells due to the increase in temperature in cells absorbing radiation.
    B. Damage to a cell's DNA.
    C. The baby becoming radioactive.
    D. The formation of a blood clot.

Answer is B: X-rays are ionising radiation. This means that ions may be formed from the atoms that make up DNA. In this case, the chemical bonds between atoms will be broken and the molecule will be changed.

27. Which of the following is **NOT** a feature of mammography?
    A. A low accelerating voltage is used for the X-ray tube.
    B. Non-ionising radiation is used.
    C. The radiation dose is small (<1 mSv).
    D. The X-ray tube utilises a molybdenum target.

Answer is B: Mammography uses X-rays which are ionising radiation, albeit giving the patient a small dose of radiation.

28. Which anatomical plane is usually displayed in a CT scan?
    A. Transverse
    B. Sagittal
    C. Coronal
    D. Longitudinal section

Answer is A: A cross-section or transverse slice is the usual image viewed. CT images are displayed as if viewed from the direction of the feet.

29. Why does an interventional cardiac angiography procedure have the potential to deliver a high dose of radiation to the patient?
    A. A radioactive tracer is injected into the patient.
    B. An extensive region of the torso is irradiated.
    C. Cardiac angiography employs ionising radiation.
    D. The X-ray generator is switched on for 10s to 100s of seconds.

Answer is D: The long exposure time is the potential problem. All X-ray imaging uses ionising radiation. A radioactive material is not used in cardiac angiography.

30. What does the term "ionising radiation" refer to?
    A. Any long wavelength electromagnetic radiation
    B. Radiation that is emitted by ions
    C. Ultrasound radiation
    D. Radiation that produces ions when it interacts with matter

Answer is D: If ions are produced when radiation passes through a material, it is termed ionising radiation. It also means that energy is deposited within that material.

31. When a mammography X-ray tube with an Mo target (characteristic X-ray at 19.6 keV) with an Mo filter is operated at an accelerating voltage of 28 kV, what may be said of the resulting spectrum?
    A. The maximum energy that an X-ray photon can have will be 28 keV.
    B. Most of the X-rays will have an energy of 19.6 keV.
    C. All of the X-ray photons will have an energy of 28 keV.
    D. The X-ray beam will contain photons with every energy from 0 keV up to 28 keV.

Answer is B: The Mo filter has the effect of stopping photons with energy between 20 and 28 keV. Characteristic X-rays are present with much greater intensity than photons in the rest of the spectrum. So the majority of photons will have energy of about 19 keV. This is the best energy for imaging the compressed breast.

32. Why will many X-ray photons pass through the human body without deflection?
    A. The carbon, hydrogen and oxygen atoms are transparent to X-rays.
    B. The Compton interaction does not occur at diagnostic X-ray energies.
    C. The wavelength of X-rays is too long to interact with an object with the dimensions of the human body.
    D. The wavelength of X-rays is too short to interact with many atoms.

Answer is D: X-ray photons are very small and atoms are mostly empty space. Furthermore they are uncharged so are not attracted to or repelled from electrons and protons. To interact, an X-ray photon must pass extremely close to the nucleus an electron. Hence most pass straight through.

33. If the filtration of an X-ray beam is increased from 2 mm aluminium to 5 mm aluminium, how will the intensity of the X-ray spectrum change?
    A. It will decrease equally at all frequencies.
    B. It will decrease more at lower frequencies than at higher frequencies.
    C. It will decrease more at higher frequencies than at lower frequencies.
    D. It will decrease only at lower frequencies.

Answer is B: Intensity will certainly decrease. The lower frequencies are "less penetrating" than higher frequencies and so are more easily stopped than the higher frequencies.

34. What is a difference between X-rays and gamma rays?
    A. X-rays emerge from a radioactive atomic nucleus whereas gamma rays are produced in an electrical machine.
    B. Gamma rays can be turned off by switching the power supply off.
    C. X-rays are produced in an electrical machine, whereas gamma rays emerge from an atomic nucleus.
    D. X-rays can penetrate deeper into solid material than can gamma rays.

Answer is C: When the electrical supply is turned off, no more X-rays are produced. Gamma rays continue to emanate naturally from a radioactive substance until all the atoms have decayed.

35. What is the maximum energy that an X-ray tube energised to 140 kV will accelerate electrons to?
    A. 140 eV
    B. 140 keV
    C. 140 J
    D. 140,000 keV

Answer is B: The unit "electron volt" is the amount of energy gained by an electron when it is accelerated by a potential difference of 1 V. 140 keV is the amount of energy gained by an electron when it is accelerated by a potential difference of 140 kV.

36. Why is an X-ray beam passed through a metal foil (i.e. filtered) before irradiation a patient? To:
    A. Remove electrons from the X-ray beam
    B. Prevent high-energy photons entering the patient
    C. Increase the mean energy of the beam
    D. Increase the ratio of low energy photons to high energy photons

Answer is C: The filter material will remove the lowest energy X-ray photons from the beam. As a result, it will also increase the mean energy of the beam.

37. What is the purpose of an intensifying screen when used to view an X-ray image?
    A. It converts a small number of X-ray photons into a large number of visible light photons.
    B. It converts low-energy X-ray photons into high-energy visible light photons.
    C. It improves the absorption efficiency of X-rays.
    D. It improves the conversion efficiency of the detector.

Answer is A: An X-ray photon has much higher energy than a visible photon; hence, one X-ray photon has sufficient energy to produce a great many visible photons. Furthermore, visible light can be seen by human eyes but X-rays cannot. Such screens mean that a lower radiation dose can be given to the patient while still producing a useful image.

38. Which element is NOT used in the construction of the anode of an X-ray tube?
    A. Rhenium
    B. Tungsten
    C. Aluminium
    D. Molybdenum

Answer is C: Aluminium (atomic number 13) is too light to be used in an X-ray tube anode. It is however used as a filter to remove low-frequency X-rays from the beam.

39. In which one of the following situations are characteristic X-rays produced?
    A. When an outer shell electron is knocked out of the atom and the vacancy is filled by another electron.
    B. When electrons bombarding the atom have less than 20 keV of energy.
    C. When the bombarding electrons have LESS energy than the binding energy of the K shell.
    D. When a K shell electron is knocked out of a target atom and the vacancy is filled by an outer electron.

Answer is D: Electron shells are labelled K, L, M, etc. from innermost to outermost shell. Electrons in the K shell, being closer, are most tightly bound to the nucleus and require more energy to be knocked out of the atom. When the hole left behind is filled by another electron, it gives up some of its energy as a characteristic X-ray photon.

40. The binding energy of an electron in the K shell of a molybdenum atom is 20.0 keV. Given that one of the characteristic X-rays of molybdenum atom is 17.8 keV, what is the binding energy of the other electron energy level?
    A. 37.9 keV
    B. 20.0 keV
    C. 17.8 keV
    D. 2.2 keV

Answer is D: Electrons in the K shell are the most tightly bound and so have the most binding energy. That is, 20 keV is the most energetic characteristic X-ray for Mo. If 17.8 keV is a characteristic X-ray, then there must be an energy difference of 17.8 keV between the 20.0 keV level and the one we want. 20–17.8 = 2.2 keV.

41. Why is the anode surface of an X-ray tube is inclined at an angle to the bombarding electron beam? So that:
    A. The X-ray beam is directed towards the exit window.
    B. The effective focal spot is smaller than the actual focal spot.
    C. A diverging X-ray beam is produced.
    D. The emerging beam is parallel.

Answer is B: A small focal spot will produce a sharper X-ray image. A small focal spot will also mean a large amount of heat is deposited in that part of the anode. Inclining the anode angle to about 7° will spread the bombarding electrons onto a larger area, while making the size of the emerging X-ray beam narrower.

42. The bremsstrahlung interaction produces all of the following except one. Which one?
    A. Characteristic radiation
    B. Heat in the anode
    C. A bombarding electron with less kinetic energy
    D. Photons with a range of energies

Answer is A: Bremsstrahlung radiation is emitted when a fast-moving electron is slowed down by passing close to a nucleus in the target anode. Characteristic radiation is emitted when an electron bombarding the anode collides with a K shell electron and knocks it out of the atom.

43. The benefit of having a small focal spot on the anode of an X-ray tube is that:
    A. Scattered radiation is minimised.
    B. Heat production in the anode is minimised.
    C. The resulting image will be sharply defined.
    D. The anodes may be made smaller.

Answer is C: A small focal spot ensures that the produced X-rays come from close to a "point" source. Such a source will produce an X-ray image with very little "penumbra", that is, the image will be sharper.

44. Given that the absorption of photons by the photoelectric effect increases with the Z of the absorbing atom and decreases with energy of the photon, which of the following is likely to be true?
    A. The photoelectric effect is of greater importance in soft tissue than in bone.
    B. The photoelectric effect is of greater importance in bone than in soft tissue.
    C. There is no marked difference in the occurrence of the photoelectric effect in soft tissue than in bone.
    D. Photons with energy greater than 60 keV are more likely to undergo the photoelectric effect than photons of energy less than 60 keV.

Answer is B: Bone contains Ca and P which have atomic numbers of 20 and 15, respectively. The elements C, H and O which predominate in soft tissue have the lower Z values of 6, 1 and 8, respectively. Hence, the photoelectric interaction is more common in bone.

45. What is the "heel effect" that is displayed in an X-ray spectrum is due to?
    A. Poor collimation
    B. Absorption of X-rays as they pass out of the anode
    C. The use of wedge filters in the X-ray beam
    D. Operating the X-ray tube with the beam horizontal

Answer is B: X-ray photons are produced at different depths within the anode. Hence, they traverse different distances through the anode before they emerge. This ensures that those traversing the greatest distance suffer the greater absorbance. So a noticeable modification to the spectrum (a "heel") with angle from the anode is discernible.

46. The photoelectric effect produces all of the following except one. Which one?
    A. Characteristic radiation (a secondary X-ray photon)
    B. A scattered X-ray photon
    C. An energetic photoelectron
    D. A dose of radiation to tissue

Answer is B: An incoming photon with sufficient energy is totally absorbed by an electron which is then ejected from the atom as a "photoelectron". As the energy of the incoming photon is all absorbed, there is not scattered photon.

47. Consider 100 X-ray photons (all of the same energy) that are fired at a 4 mm thick sheet of aluminium and that 80 pass through without interacting. What is the linear attenuation coefficient (in cm^{-1}) for Al at this energy?
    A. 0.2 cm^{-1}
    B. 0.5 cm^{-1}
    C. 2.0 cm^{-1}
    D. 5.0 cm^{-1}

Answer is B: 80 out of 100 photons (80%) pass through. Thus, 20% are stopped. Hence, 20% per 4 mm, or 5% per 1 mm is = 0.05 mm^{-1} = 0.5 cm^{-1} as there are 10 mm per cm.

48. An X-ray beam with a high HVL (half-value layer) is:
    A. Said to be "softer" than a beam of low HVL
    B. Less penetrating than a beam of low HVL
    C. Likely to produce a larger absorbed dose to the skin
    D. More penetrating than a beam of low HVL

Answer is D: Having a HVL of "x" mmAl means that half of the photons in the beam are able to penetrate "x" mmAl. Hence, the higher is the HVL, the more penetrating is the beam. Also it will be said to be "harder" and to produce a lower absorbed dose to tissue.

49. Which statement about the Compton effect is NOT correct?
    A. A Compton interaction is more likely in bone than in muscle tissue.
    B. The deflected photon has less energy than the original photon.
    C. An electron is ejected from the atom.
    D. Compton events contribute to scattered radiation.

Answer is A: A Compton interaction occurs when a photon uses some of its energy to eject an electron, while the remaining energy becomes a scattered photon. The probability that an X-ray photon will undergo a Compton interaction depends on the density of the issue and the energy of the photon. Tissues such as muscle, blood and solid organs create the greatest amount of scatter radiation in the body. Bone tends to absorb more of X-rays.

50. Use the "inverse square law" to compare to being 1 m from an X-ray source, with the photon flux at a distance of 3 m from the source. The flux will be:
    A. Nine times as great
    B. Six times as great
    C. One sixth as great
    D. One ninth as great

Answer is D: "Inverse square" means at double the distance, the flux will be reduced to one half squared of the original flux $(½)^2 = ¼$. At 3× the distance, the flux will be $(1/3)^2 = 1/9$.

51. Imagine that the X-radiation from a diagnostic machine was measured by a dose-area-product meter that could count individual photons. For a particular exposure, 8×10^{12} photons were collimated to pass through a 10 cm × 10 cm aperture at 1 m from the X-ray source. What would be the number of photons passing through a 20 cm × 20 cm aperture at 2 m from the X-ray source?
    A. 8×10^{12} photons
    B. 2×10^{12} photons
    C. 16×10^{12} photons

D. 4×10^{12} photons

Answer is A: By the inverse square law, a 10 cm × 10 cm beam at 1 m from the source would diverge to a 20 cm × 20 cm beam at 2 m from the source (that is to four times the area). However, the same number of photons would be passing through this wider area.

52. Why are iodine and barium ideal for use as contrast agents in imaging of soft tissue?
    A. They are high *Z* atoms so the probability of a Compton scatter is high.
    B. The binding energy of their K shell is in the middle of the diagnostic energy range.
    C. The energy of their characteristic X-rays is in the middle of the diagnostic energy range.
    D. They are high *Z* atoms so the probability of a photoelectric interaction is high.

Answer is B: As the binding energy of their K shell is in the middle of the diagnostic energy range, many X-ray photons will be absorbed by I and Ba. Hence, organs containing these elements will appear white on the image and so will contrast with the grey shades of the surrounding soft tissue.

53. One difference between the X-radiation in the primary beam and the scattered radiation is that:
    A. Scattered radiation contributes dose to the surroundings, primary beam does not.
    B. Scattered radiation is more penetrating than the primary beam.
    C. Scattered radiation is not absorbed by the patient while photons in the primary beam are.
    D. Primary X-ray photons that have passed through the patient contribute to patient dose.

Answer is A: Scattered radiation has had its direction of travel altered and so may irradiate people standing nearby. Scattered radiation is less penetrating than the primary beam and may be absorbed by the patient and so contribute to their radiation dose. Choice D is wrong as emerging photons do not contribute to dose.

54. The detectors in a CT scanner measure which one of the following?
    A. The linear attenuation coefficient of each pixel
    B. The grey scale value of the image
    C. The Hounsfield unit of each pixel
    D. The average linear attenuation coefficient

Answer is D: The average linear attenuation coefficient along the path from where the beam enters the patient's body to where it emerges. This number is then attributed to every pixel along the path. A variety of techniques is then used to create a cross-sectional image of anatomy and to assign a "Hounsfield unit" to each pixel of the image.

55. If a CT (Hounsfield) number associated with a given tissue is close to zero, then what is the tissue likely to be?
    A. Mostly water
    B. Predominantly lung
    C. Mainly muscle
    D. It probably contains bone

Answer is A: The numerator of Hounsfield number is calculated by subtracting the "linear attenuation coefficient" of water from the attenuation coefficient of the tissue. If the tissue coefficient is close to that of water, the result will be close to zero.

17.2 Radioactivity

Radioactive atoms are unstable and will spontaneously emit some form of ionising radiation eventually. In doing so, they become an atom of a different element (if they emit α or β particles) or they lose energy without changing to a different element (if they emit a γ photon). Radioactivity is a characteristic of many naturally occurring isotopes such as ^{3}H, ^{14}C, ^{40}K and all elements that have between 84 and 92 (inclusive) protons in their nucleus. These isotopes and elements are either being produced continuously by the cosmic rays that impinge on our atmosphere or have always been present on earth (primordial radionuclides) and because their "half-lives" are very long, are yet to emit their radiation. They contribute to the background radiation that continuously bathes all life on earth.

The level of radioactive behaviour of an element is described by its half-life. That is, the amount of time that will elapse while half of all atoms present in a sample of the element emit their radiation. Whether a particular atom will "decay"—that is, emit its radiation—is a random phenomenon. It may occur almost immediately, or the particular atom may wait for thousands of years without emitting anything. Hence technetium-99m, with a half-life of about 6 h, is highly radioactive. However, after 48 h, almost all of its radiation has been emitted, and it is considered safe. On the other hand, plutonium-239 has a half-life of 24,100 years which means a sample of ^{239}Pu is not emitting radiation (in the form of alpha particles) very rapidly and so is not very radioactive. However, it will continue to be radioactive for a very very long time.

The amount of radioactivity of a sample of material is characterised by the unit becquerel (Bq), where 1 Bq = 1 emission per second. A 1 g sample of technetium-99m contains about 6.17×10^{21} atoms and immediately after its preparation would have a radioactivity of 1.43×10^{17} Bq. After 6 h, approximately 3.085×10^{21} atoms will have emitted their gamma photon. This is a half of all of the atoms present. A 1 g sample of plutonium-239 contains about 2.47×10^{21} atoms whose radioactivity is about 1.62×10^{9} Bq. This make ^{99m}Tc 100 million times more radioactive than plutonium (for a short time!) After 6 h, 3.5×10^{13} plutonium atoms will have emitted their alpha particle. This is a very small fraction of the number of plutonium atoms in the 1 g sample.

1. Which of the following statements about radioactivity is **NOT** correct?
   A. Some of the atoms in our body are radioactive.
   B. Radioactivity occurs naturally in the environment.
   C. Radioactivity is associated with the nucleus of an atom.
   D. Radioactivity is involved in diagnostic X-rays.

Answer is D: Diagnostic X-rays are not associated with radioactivity.

2. If a patient who has been administered a radiopharmaceutical containing technetium-99m (half-life = 6 h) is allowed to go home 36 h after the administration, what fraction of the original radioactivity is left in the patient (ignoring excretion from the body)?
   A. 1/216
   B. 1/64
   C. 1/36
   D. 1/6

Answer is B: After every 6 h, the amount of technetium-99m that remains is one half of the amount that was present 6 h ago. 36/6 = 6, so six half-lives will elapse over 36 h, thus the amount remaining is: $\frac{1}{2} \times \frac{1}{2} \times \frac{1}{2} \times \frac{1}{2} \times \frac{1}{2} \times \frac{1}{2} = (\frac{1}{2})^6 = 1/64$.

3. Which of the following statements about radioactivity is correct?
   A. X-rays can be produced by radioactivity.
   B. The spontaneous emission of microwaves from the nucleus of an atom is one form of radioactivity.
   C. Radioactivity is the spontaneous emission of particles or photons from the nucleus of an atom.
   D. Radioactivity is involved in the FM radio frequency band but not the AM radio band.

Answer is C: Radioactivity involves the nucleus of an atom. Microwaves are not produced by radioactivity. While radiation from a nucleus could remove an inner shell electron from another atom, and when this electron is replaced by another "falling" into its place while emitting an X-ray photon, this is not classed as radioactivity.

4. Given that a radiopharmaceutical containing the radionuclide technetium-99m ($^{90m}_{43}Tc$) has an activity of 4 MBq immediately prior to injection into a patient. About 12 h later, the patient is measured to have an activity of about 1 MBq. How long is the half-life of technetium-99m?
   A. 6 h
   B. 12 h
   C. 24 h
   D. 48 h

Answer is A: After one half-life, half of the radioactive atoms have decayed, so only half remain. One MBq which is one quarter of 4 MBq (= ½ × ½) remains after 12 h, so two half-lives have elapsed in the 12-h period. 12/2 = 6 h.

5. Which of the following pairs of nuclides are isotopes?
   A. ${}^{60}_{27}\text{Co}$ and ${}^{60}_{28}\text{Ni}$
   B. ${}^{99}_{43}\text{Tc}$ and ${}^{98}_{42}\text{Mo}$
   C. ${}^{40}_{19}\text{K}$ and ${}^{19}_{40}\text{K}$
   D. ${}^{40}_{19}\text{K}$ and ${}^{39}_{19}\text{K}$

Answer is D: Isotopes are two forms of the same element that have different numbers of neutrons in the nucleus. This means same symbol, and same atomic number (subscript), but different mass number (superscript).

6. What may the term "radioactive" be correctly used to describe?
   A. A diagnostic X-ray machine used to produce radiographs
   B. A linear accelerator used to produce X-rays for radiotherapy
   C. A patient undergoing a nuclear medicine scan
   D. A patient undergoing a CT (computed tomography) examination

Answer is C: A patient who has been prepared for a NM scan has had a radioisotope incorporated into their body, so are themselves (temporarily) radioactive.

7. What may the term *radioactivity* correctly used to refer to?
   A. The spontaneous emission of electromagnetic radiation from the nucleus of an atom
   B. The particles or photons emitted from an unstable nucleus
   C. The emission of particulate radiation from a radionuclide
   D. The alpha, beta or X-radiation which emanates from some atomic nuclei

Answer is B: Unstable nucleus = radioactive. Choices A and C also describe radioactivity but not in as general terms as choice B.

8. What does the physical *half-life* of a pure sample of radioactive material refer to?
   A. The amount of time taken for half of the radioactive atoms to decay.
   B. Half of the time that it would take for all of the radioactive atoms to decay.
   C. The average time taken for any particular radioactive atom to decay.
   D. The time it takes for half of a sample of ingested radioisotope to be cleared from the body.

Answer is A: The half-life of a particular radionuclide is the time it takes for half of the remaining atoms of that particular radionuclide to decay. Choice B is not the same thing.

9. The radionuclide technetium-99m (${}^{90m}_{43}\text{Tc}$) is often incorporated into radiopharmaceuticals. When it decays by gamma emission, the daughter nucleus

may be represented as ${}^{90}_{43}X$ (X is **not** the *real* chemical symbol for the daughter nucleus!). Which of the following statements is correct?

A. The daughter nucleus displays chemical behaviour identical to ${}^{90m}_{43}Tc$.
B. *X* is the symbol of a chemical element different from technetium.
C. The daughter nucleus detaches from the radiopharmaceutical and is excreted.
D. The daughter nuclide is not radioactive.

Answer is A: "X" is still technetium as the atomic number (43) has not changed. Consequently, the chemical behaviour of the two nuclides is the same. A daughter nucleus may or may not be radioactive (in this case it is).

10. If the half-life of technetium-99m is 6 h. After five half-lives, what fraction of the original amount of technetium would remain?

A. $\frac{1}{5}$
B. $\frac{1}{10}$
C. $\frac{1}{30}$
D. $\frac{1}{32}$

Answer is D: The half-life of a particular radionuclide is the time it takes for half of the remaining atoms of that particular radionuclide to decay. $(½)^5 = 1/32$.

11. If the half-life of technetium-99m is 6 h. After seven half-lives, what fraction of the original amount of technetium would remain?

A. $\frac{1}{7}$
B. $\frac{1}{14}$
C. $\frac{1}{42}$
D. $\frac{1}{128}$

Answer is D: The half-life of a particular radionuclide is the time it takes for half of the remaining atoms of that particular radionuclide to decay. $(½)^7 = 1/128$. This means that after seven half-lives, more than 99.2% of the radionuclide has decayed.

12. Which statement about the atom and its nucleus is correct?
    A. The nucleus contains neutrons with a positive charge and protons with no charge.
    B. Most of the volume of the atom is occupied by the nucleus.
    C. The majority of the atom's mass is due to the electrons
    D. The nucleus is one ten thousandth times the diameter of the atom.

Answer is D: The nucleus is tiny compared to the atomic diameter. Choice A is wrong as protons have the positive charge and neutrons have no charge.

13. The term "radioactive" when applied to the nucleus of an atom refers to which one of the following phenomena?
    A. The spontaneous emission of a particle or of electromagnetic radiation.
    B. The emission of X-rays, gamma rays, alpha particles or beta particles.
    C. The formation of a daughter nucleus by the decay of a parent nucleus.
    D. The spontaneous emission from a nucleus which has the optimum ratio of neutrons to protons.

Answer is A: Choice B is wrong as X-rays don't come from a nucleus. Choice C is wrong as sometimes a nucleus is in an "excited" state and returns to ground state by emitting a gamma ray while the nuclear structure is unchanged. Choice D is wrong as a nucleus which has the optimum ratio of neutrons to protons is not unstable.

14. Which statement is correct? Radioactivity is the emission from the nucleus of an atom of:
    A. Electrons or beta particles or gamma rays.
    B. Alpha or beta particles or X rays.
    C. Electrons or beta particles or X rays.
    D. Alpha or beta particles or gamma rays.

Answer is D: X-rays are not involved in radioactivity. Electrons are beta particles.

15. What are isotopes? Atoms of the one element whose nuclei:
    A. Contain the same number of protons and neutrons.
    B. Have different numbers of protons.
    C. Have different numbers of neutrons.
    D. Do not have the optimum neutron to proton ratio.

Answer is C: The number of neutrons does not affect the nature of the element.

16. Which one of the following statements best describes what "ionising radiation" is?
    A. High-frequency electromagnetic radiation.
    B. Charged particles that are emitted from radionuclides.
    C. Radiation that is emitted from a radioactive nucleus.
    D. Radiation that can remove electrons from matter.

Answer is D: To produce an ion, one or more electrons must be removed from an atom.

17. Which of the following is not a form of ionising radiation?
    A. Beta particles.
    B. Gamma rays.
    C. Infrared rays.
    D. X-rays.

Answer is C: Every object emits infrared rays in proportion to its temperature. IR rays have too little energy to ionise the medium that they strike.

18. Alpha and beta rays share several properties. Which of the following is one? That they:
    A. Are electrically charged.
    B. Both contain protons.
    C. Are forms of electromagnetic radiation.
    D. Have the same mass.

Answer is A: Alpha particles are He^{++} nuclei, while beta particles are electrons (negative charge) or positrons (positive charge).

19. Which of the statements about the following three-step radioactive decay series is **TRUE**?

    Molybdenum99 → β + technetium-99m
    Technetium-99m → γ + technetium-99
    Technetium-99 → β + ruthenium-99

    A. Technetium-99m is the parent of ruthenium-99.
    B. Molybdenum-99 is the parent of technetium-99.
    C. Ruthenium-99 is the daughter of molybdenum-99.
    D. Technetium-99 is the grand-daughter of molybdenum-99.

Answer is D: Technetium-99 results after two steps, from molybdenum-99. But "grand-daughter" may not be true technical terminology.

20. In the following nuclear reaction, ${}^{238}_{92}\text{U} \rightarrow \text{X} + {}^{4}_{2}\text{He}$, one of the products (${}^{4}_{2}\text{He}$) is an alpha particle. What is the other product (X)?
    A. ${}^{234}_{90}\text{Th}$
    B. ${}^{234}_{91}\text{Pa}$
    C. ${}^{236}_{21}\text{Pa}$
    D. ${}^{236}_{90}\text{Th}$

Answer is A: Consider the mass numbers: 238 – 4 = 234. Then the atomic numbers: 92 – 2 = 90. Hence the nuclide X must be thorium 234.

21. Which nuclide does the symbol X in the nuclear reaction ${}^{99m}_{43}\text{Tc} \rightarrow \text{X} + {}^{0}_{0}\gamma$ refer to?
    A. ${}^{99m}_{43}\text{Tc}$
    B. ${}^{99}_{43}\text{Tc}$
    C. ${}^{99}_{44}\text{Ru}$

D. $^{99}_{42}Mo$

Answer is B: As a gamma photon has been emitted, nuclide X must have the same atomic and mass numbers as technetium 99m. What has happened is that the metastable (hence "m" in the symbol) state of technetium has de-excited to technetium.

22. In the nuclear reaction: $^{60}_{27}Co \rightarrow {}^{0}_{0}\gamma + X$, what does the symbol "X" represent?
    A. $^{60}_{28}Co$
    B. $^{59}_{27}Co$
    C. $^{60}_{27}Co$
    D. $^{60}_{26}Co$

Answer is C: As a gamma photon has been emitted, nuclide X must have the same atomic and mass numbers as cobalt 60. Choices A and B are incorrect as cobalt has an atomic number of 27.

23. Which of the following is the principal unit used when measuring the energy of ionising radiation?
    A. Joule.
    B. Electron-volt.
    C. Grey.
    D. Sievert.

Answer is B: The electron-volt or keV or MeV is preferred to the joule as the energy of a photon is a tiny fraction of 1 J.

24. What time does the half-life of a radioactive sample refer to? The time for the:
    A. Activity to halve.
    B. Count rate to double.
    C. Parent nuclei to decay.
    D. Number of nuclei to halve.

Answer is A: This is the best answer, although the daughter nuclei may themselves also be radioactive and so contribute to the radioactivity of the sample. When a nucleus decays, it (or the daughter nucleus) is still there, so choice D is wrong.

25. "Half-life" when applied to atoms of a radioactive isotope refers to the:
    A. Midpoint of the time span for which the isotope will emit its radiation.
    B. Effective time for which the isotope is considered to be dangerous.
    C. Length of time taken for half of the isotope to emit its radiation.
    D. Time after which the radioactivity of the sample is half of its original value.

Answer is C: This is the best answer. Half-life refers to the atoms of an isotope of the one element. The daughter nuclei may themselves also be radioactive and so contribute to the radioactivity of the sample now containing a mixture of original isotope atoms and daughter atoms.

26. In an experiment to determine the half-life of a particular radionuclide, measurements of the mass of that nuclide are made. If masses of 8 and 2 μg are recorded at intervals 8 days apart, what is the half-life of the nuclide?
    A. 8 days
    B. 6 days
    C. 4 days
    D. 2 days

Answer is C: If we start with 8 μg of a radionuclide and when it decays it becomes another nuclide, then after one half-life there will be 4 μg remaining and after another half-life, there will be 2 μg. Hence two half-lives have elapsed during the 8-day period. This means one half-life is $8 \div 2 = 4$ days.

27. A particular radionuclide has a half-life of 6 h. When first measured, its activity is 10,000 Bq. After 12 h, what will its activity be?
    A. 0 Bq
    B. 1250 Bq
    C. 2500 Bq
    D. 5000 Bq

Answer is C: After one half-life the starting activity of 10,000 Bq will be halved to 5000 Bq. After another half-life, the activity will be halved again from 5000 to 2500 Bq.

28. What is a radionuclide with a short half-life (say 5 h) said to be?
    A. Highly radioactive.
    B. Weakly radioactive.
    C. Of high penetrating ability.
    D. Of low penetrating ability.

Answer is A: A radionuclide with a short half-life must be emitting a lot of radiation in that time as half of the nuclei present will have emitted their radiation. Hence, it may be termed "highly radioactive", albeit for a short time.

29. Which of the following does **NOT** contribute to the background radiation?
    A. Cosmic rays.
    B. Fluorescent lights.
    C. Radon gas.
    D. Uranium.

Answer is B: Fluorescent lights contain mercury vapour at low pressure which emits ultraviolet light when excited by the passage of electrons through the tube. The UV photons are absorbed by the phosphor that coats the inside of the tube which re-emits the energy as visible light. No ionising radiation is emitted.

30. Which one of the following sources contributes to the background radiation?
    A. Medical X-rays of bones.
    B. Potassium 40 and carbon 14 in our bodies.
    C. Ultraviolet B radiation from the sun.
    D. Microwave radiation.

Answer is B: Background radiation is that radiation that is impossible to avoid as it is part of our natural environment. Potassium 40 and carbon 14 are two naturally occurring radionuclides that form part of the background radiation.

31. Which one of the following radiations, all of 1 MeV energy, is the least penetrating?
    A. Alpha rays.
    B. Beta rays.
    C. Gamma rays.
    D. X-rays.

Answer is A: Alpha rays are doubly ionised helium nuclei, which, because of their large mass and double charge interact strongly with any matter through which they pass. They are able to travel a mere few centimetres in air and a fraction of a millimetre in solid matter.

32. Which is the most penetrating nuclear radiation?
    A. Alpha particles of energy 7 MeV.
    B. Beta particles of 0.5 MeV energy.
    C. X-rays with energy of 5 MeV.
    D. Gamma rays with 140 keV of energy.

Answer is C: X-rays are photons so have no mass or charge. Hence they interact only weakly with the matter through which they pass. In addition, 5 MeV are very energetic X-rays so are more penetrating than gamma ray photons with less energy.

33. Why are gamma rays able to penetrate "solid" walls but alpha or beta rays cannot? Because:
    A. The speed of gamma rays (3×10^8 m/s) is very much faster than either alpha or beta radiation.
    B. Gamma rays do not have an electric charge whereas alpha is charged +2 and beta −1.
    C. To a gamma ray, the atoms of the wall appear to be mostly empty space.
    D. Gamma rays possess much more energy than either alpha or beta rays.

Answer is B: It is the absence of an electrical charge that allows photons to pass by electrons and atomic nuclei with being attracted or repelled by them. This means that photons are undeflected unless they suffer a direct hit.

34. In the nuclear reaction $^{99}_{42}\mathrm{Mo} \rightarrow \mathrm{X} + {}^{99m}_{43}\mathrm{Tc}$, what does the symbol X represent?
    A. An electron.

B. A positron.
C. An X-ray.
D. A gamma ray.

Answer is A: Mass number must be conserved, hence $99 = s + 99$, this means $s = 0$. Atomic number must be conserved, hence $42 = a + 43$, this means $a = -1$. The particle X with zero mass and charge of negative one is an electron (a beta negative particle).

35. The "half-value layer" for gamma rays of 0.5 MeV is 0.42 cm of lead. What percentage of the original ray would penetrate four half-value layers?
    A. 50%
    B. 25%
    C. 12.5%
    D. 6.25%.

Answer is D: Four half-value layers would absorb all but ½ × ½ × ½ × ½ = 1/16 (6.25%) of the photons.

36. Given that the half-value layer of lead for a gamma photon of energy 0.5 MeV is 0.42 cm, which of the situations listed below would result in the **LEAST** exposure to radiation? Staying in the same room as the gamma source for:
    A. 5 min at a distance of 1 m from the source with 0.42 cm of lead shielding
    B. 10 min at 0.5 m with 0.84 cm of lead as shielding
    C. 15 min at 2 m from the source using no shielding
    D. 40 min at a distance of 1 m using 0.84 cm of lead as shielding.

Answer is A: Choice B would result in 2× the dose (due to doubling the time), multiplied by 4× the dose (due to halving the distance) multiplied by ½× the dose due to the extra half-value layer = 4× the dose of choice A.

Choice C would result in 3× the dose multiplied by 1/4× multiplied by 2× the dose = 1½× the dose of choice A.

Choice D would result in 8× the dose multiplied by 1× multiplied by ½× the dose = 4× the dose of choice A.

37. Why are photons of high-frequency electromagnetic radiation, such as X and gamma radiation, more penetrating than particulate radiation such as alpha or beta radiation?
    A. Atoms are mostly empty space which makes it unlikely that photons will collide with a nucleus.
    B. Photons carry no electrical charge so will not be attracted to the electrically charged particles within an atom.
    C. High-frequency em radiation has a wavelength longer than the dimensions of an atom so will not "see" sub-atomic particles.
    D. Alpha and beta particles are electrically repelled by the charged particles within an atom, hence are deflected from their paths.

Answer is B: It is the absence of an electrical charge that allows photons to pass by electrons and atomic nuclei with being attracted or repelled by them. This means that photons are undeflected unless they score a "direct hit" on an electron or the nucleus. Choice C is wrong as high-frequency radiation has a short wavelength. Choice D is wrong as alpha and beta particles will experience attraction to electrons and the nucleus, respectively.

17.3 Radiotherapy, Nuclear Medicine, Radiation Safety

Radiotherapy can utilise high energy x-radiation from a Linac, or the radiation emitted from radioactive atoms to kill the cancerous cells in a tumour. In the case of the latter, radioactive sources are placed within an applicator, and the applicator is inserted into or adjacent to the tumour to be treated. This is called brachytherapy. The amount of energy used to treat a tumour is about 60 Gy (grey), where 1 Gy = 1 J of energy deposited per 1 kg of tissue.

Radioactive atoms that emit low-energy (100–200 keV) gamma radiation can be attached to special molecules that are given intravenously to a patient. These carrier molecules are designed, so that they transport the attached radioactive atoms through the blood stream to aggregate in the tissues of interest. Large concentrations of radioactivity within a tissue show up as a hot spot while anatomical regions that fail to accumulate radioactivity (perhaps because their function is impaired) are "cold". The gamma radiation emitted from particular tissues can be detected as it emerges from the body to provide an image of the tissue and of how it is working. The image is a "scintigram" and is acquired by a nuclear medicine procedure. Such a patient is "radioactive" until most of the radioactive atoms have decayed. The most commonly used isotope is technetium-99m which has a half-life of about 6 h and emits gamma radiation as photons of energy 140 keV. This means that after 6 h, half of the radioactive atoms have decayed and that after 12 h, only one quarter of the radioactive technetium atoms remain.

Patients who undergo a nuclear medicine procedure will receive a dose of radiation which is less than 10 mSv. Such a dose is regarded as "small" and is considered to be medically justified given the diagnostic information produced and the benefit of this for the patient. For the period that a patient is significantly radioactive, they are cautioned to minimise the time that they spend close to another person. This is because if you halve the time spent near someone, they will receive half as much dose. Furthermore, if they double their distance from the other person, the other person will receive one quarter the dose that would have been received at the closer distance.

1. In radiotherapy, why is the patient's irradiation treatment "fractionated", that is, consist of (say) 20 sessions and spread over (say) 4 weeks—rather than given all at one session?
   A. Fractionation allows time for the normal healthy tissue that is also irradiated, to recover in between irradiations.

B. Extremely high-energy electrons bombard the target of a linear accelerator, fractionation is necessary to allow the X-ray target to cool.
C. In order to irradiate the tumour over the period of time that it is growing.
D. Irradiating in a single session takes too long, people cannot remain immobile for the time it would require.

Answer is A: Radiotherapy also damages the healthy tissue through which it passes. Healthy, well-oxygenated tissue recovers from radiation damage more rapidly than does tumour tissue. By fractionating, the time between fractions allows the healthy tissue to recover from radiation damage more rapidly than the tumour cells can recover. This leads to a better therapeutic outcome.

2. In radiotherapy for a deep tumour, why does the patient's irradiation treatment utilise incident therapy beams that converge on the tumour from different directions, rather than just a single beam from one direction:
   A. Multiple beams deliver a more uniform dose to the tumour.
   B. Extremely high-energy electrons bombard the target of a radiotherapy linear accelerator, shifting beam position is necessary to allow the X-ray target to cool.
   C. Beams from different direction allows dose to radiosensitive tissue to be minimised.
   D. To maximise the dose to the tumour while minimising dose to the overlying healthy tissue.

Answer is D: Radiotherapy also damages the healthy tissue through which it passes. Choosing a different path for the radiotherapy beam through the body for the next "fraction", while at the same time ensuring that the tumour is always in the beam, means that the tumour is delivered a higher dose (the sum of the dose from several beams) than the adjacent healthy tissue (which only get dose from one beam). Choices A and C both have some merit.

3. Why is radiotherapy using high-energy X-rays, an effective way of treating some cancer? Because:
   A. If enough energy is deposited in tumour cells, they can be killed.
   B. Most of the X-rays pass through the body without harming healthy tissue.
   C. The dose to the skin surface is lower than with low-energy X-rays.
   D. High-energy X-rays do not kill healthy cells.

Answer is A: The ability of X-rays to be switched off and on and to deposit enough energy within a cell to kill it, makes it suitable for treatment of cancer tumours.

4. X-rays for radiation therapy are produced by:
   A. A linear accelerator.
   B. Cobalt 60.
   C. An after-loading brachytherapy device.
   D. Technetium-99m.

Answer is A: A compact medical linear accelerator is used to produce high-energy X-rays for radiotherapy. The other choices all involve gamma ray producing radionuclides.

5. What advantage does brachytherapy as a form of radiation therapy have over external beam therapy (also called teletherapy)?
   A. In brachytherapy, the radiation does not have to pass through healthy tissue to reach the tumour.
   B. In brachytherapy, there is the choice of using electrons as well as gamma rays.
   C. In teletherapy, the energy of the radiation is limited to the energies available from the radionuclide.
   D. In teletherapy, the total radiation dose must be given all at the one session.

Answer is A: In brachytherapy, the radioactive isotope is placed adjacent to or within the structure to be irradiated, so healthy tissue receives less dose.

6. In a radiotherapy treatment plan for a deep tumour, which of the following is part of the treatment plan?
   A. Saturating the healthy tissue with oxygen to minimise its sensitivity to radiation.
   B. Using a single large dose to destroy all the tumour cells quickly.
   C. Using low-energy alpha particles in order to minimise dose to healthy tissue.
   D. Splitting of the total dose into a number of smaller doses given daily.

Answer is D: "Fractionation" of the dose allows the healthy tissue some time to recover between irradiations. It also allows the tumour to be targeted from another direction by passing the beam through a different part of the body.

7. How has remote-controlled after-loading reduced the dose to staff involved in brachytherapy?
   A. Patient exposure to radiation is reduced.
   B. The radionuclides are only in place in an operating theatre.
   C. Staff members are never exposed to radiation during the treatment.
   D. Only one staff member at a time is required in the brachytherapy room.

Answer is C: The radioactive sources are loaded into their applicators automatically and remotely when staff have left the room. They are automatically retrieved and returned to a shielded safe when staff enter the room.

8. Why are radioisotopes that emit low energy (100–250 keV) gamma rays preferred for the diagnostic procedures of nuclear medicine? Because:
   A. High-energy gamma rays are too easily stopped by body tissue.
   B. Radioisotopes that emit gammas within this energy range have a half-life that is ideal for diagnostic procedures.

C. Photons of this energy are sufficiently penetrating to escape from the body but are able to be detected.
D. Charged particles are too difficult to shield against.

Answer is C: There is a trade-off where the photons should be penetrating enough so that they are not stopped within the body, but not so penetrating that too few are stopped by the detector placed next to the body. This energy range satisfies both needs.

9. A radiopharmaceutical is comprised of two components. These are:
   A. A gamma emitter and a beta emitter.
   B. Technetium-99 and a non-radioactive carrier.
   C. A non-radioactive carrier and a radionuclide.
   D. A radionuclide and a radioactive carrier.

Answer is C: A radiopharmaceutical requires a gamma-emitting nuclide and a non-radioactive carrier to transport the radionuclide to the organ(s) of interest. Most often Tc99m (not Tc99) is used.

10. Which one of the following statements about the radionuclide selected for inclusion in radiopharmaceuticals is **NOT** correct?
    A. It should have a short half-life.
    B. It should emit alpha or beta rays.
    C. It should emit gamma rays in the range 100–250 keV.
    D. All the atoms should be radioactive.

Answer is B: Particulate radiation (alpha and beta rays) is not penetrating enough to pass out of the body so cannot be used in radiopharmaceuticals.

11. Which of the statements about radionuclides selected for inclusion in radiopharmaceuticals used in medical imaging is correct? They should:
    A. Have a long half-life.
    B. Emit alpha or beta particles.
    C. Be insoluble in water.
    D. Emit low-energy gamma rays.

Answer is D: Low-energy gamma rays are sufficiently penetrating so that most will pass out of the body and be available for detection. But not so penetrating that they will pass through a "gamma camera" without being detected.

12. Why are radionuclides that emit low-energy gamma radiation preferred to other radionuclides for in vivo diagnosis using a nuclear medicine technique? Because:
    A. Other forms of radiation are emitted with too much energy.
    B. High-energy gamma radiation is not penetrating enough.
    C. Most of the radiation will emerge from the patient's body.

D. Sources of X-rays require more extensive technical support than gamma sources.

Answer is C: If most gamma photons emerge from the body, the maximum amount of information will be available and the minimum amount of radiation dose will be deposited in the patient.

13. Why is a radionuclide with a short half-life and which emits low-energy gamma radiation preferred for in vivo diagnosis? Because they:
    A. Have a low activity.
    B. Are highly penetrating.
    C. Emit their radiation in a short time span.
    D. Are very damaging to cancerous tissue.

Answer is C: Having a short half-life means that lot of radiation is emitted in a short time period. If the radiation was emitted over a long time, the patient would be required to lie motionless for that long time while the image (the scintigram) was produced.

14. Why is a 150 keV gamma emitter more useful for nuclear medicine than an 800 keV gamma emitter? Because:
    A. More 150 keV photons are stopped in the detecting crystal of the gamma camera.
    B. 800 keV photons cannot penetrate the body in sufficient quantities to produce an image.
    C. A 150 keV photon results in a lower dose to the patient than an 800 keV photon.
    D. 800 keV photons require more shielding for the staff.

Answer is A: 800 keV gamma photons are more penetrating than 150 keV photons, so fewer of them will be detected by the "camera". Hence longer exposure, or a greater dose of radiopharmaceutical would be required to produce an image.

15. Why are radioactive isotopes of the stable elements that occur normally in the body useful for tracing metabolic pathways? Because:
    A. Such isotopes undergo chemical reactions that are identical to those of stable isotopes.
    B. Radioisotopes are indistinguishable from non-radioactive isotopes of the same element.
    C. Nuclear radiation can be detected outside the body and be used to produce an image of internal structures.
    D. Radioactive forms of elements that exist naturally in the body do not produce toxic effects when used as a radiopharmaceutical.

Answer is A: Radioactive forms of elements that occur naturally in the body will be handled by the body in exactly the same way as the non-radioactive isotope. Hence their location within the body can be traced by the gamma radiation they emit.

16. Why is it possible to trace metabolic pathways in the body using radioactive isotopes of the naturally occurring elements in the body? Because, radioactive isotopes:
    A. Are used in such small quantities that they produce no toxic effects.
    B. Have a very short half-life so soon they decay to safe levels.
    C. Are chemically identical to non-radioactive isotopes of the same element.
    D. Are physically identical to non-radioactive isotopes of the same element.

Answer is C: Radioactive forms of elements that occur naturally in the body will be chemically handled by the body in exactly the same way as the non-radioactive isotope. Hence their location within the body can be traced by the gamma radiation they emit.

17. Which anatomical plane is usually displayed in a nuclear medicine bone scan (a scintigram)?
    A. Transverse
    B. Sagittal
    C. Coronal
    D. Longitudinal section

Answer is A: A coronal view or a view of the anterior anatomical position is the most common.

18. Given the statement: "Your exposure to radiation varies inversely with the square of your distance from the source of radiation", what is the correct way to finish the sentence? "If you increased the distance between you and a patient with a radioactive implant, from 1 to 3 m, your exposure would:
    A. Increase by a factor of 9
    B. Decrease by a factor of 1/9
    C. Decrease by a factor of 1/3
    D. Increase by a factor of 3

Answer is B: If you triple your distance, your exposure would be one third of a third of your exposure while closer. $(1/3)^2 = 1/9$.

19. Why is increasing the distance between yourself and a source of radiation is an effective way of reducing your exposure? Because:
    A. The amount of exposure to radiation is inversely proportional to distance.
    B. The intensity of radiation decreases as the square of the distance.
    C. The intensity of radiation decreases exponentially with distance.
    D. An expanse of air is an effective shield for gamma and X radiation.

Answer is B: As radiation leaves a point source, it diverges and spreads equally in all directions over an ever-expanding spherical surface. A spherical surface area is proportional to the radius squared, so exposure to radiation diminishes as it spreads more thinly in space.

20. Suppose a gamma source is placed 10 cm from a radiation detector and in succession, a 1-cm-thick slab of each of the following materials is used to shield the source. For which one would the count rate be lowest?
    A. Lead.
    B. Wood.
    C. Aluminium.
    D. Cardboard.

Answer is A: Shielding material with the highest density is able to absorb the greatest amount of gamma photons. Lead has the highest density so would reduce the count rate the most.

21. Why is a shield made of lead effective in reducing the exposure to gamma rays? Because the:
    A. Lead shield absorbs all of the gamma rays.
    B. Number of gamma rays absorbed increases exponentially with the thickness of the shield.
    C. Number of gamma rays absorbed increases in an inverse square fashion with thickness.
    D. Lead shield reflects all the gamma rays.

Answer is B: The number of gamma photons that can pass through a medium is described by an exponential relationship. Their number decreases exponentially as the thickness increases.

22. What does the unit the Sievert describe?
    A. Radioactivity.
    B. Absorbed radiation dose.
    C. Absorbed radiation equivalent dose.
    D. Energy of radiation.

Answer is C: Absorbed radiation dose is measured in "grey" (= J/kg). When the different types of radiation are weighted by their biological effect (so that a gamma ray is "equivalent" to an alpha particle say), the grey becomes the Sievert.

23. Why is keeping one's distance from a source of radiation effective in minimising exposure? Because:
    A. Exposure is inversely proportional to distance.
    B. Exposure decreases as the inverse square of distance.
    C. Exposure decreases exponentially with distance.
    D. Electromagnetic radiation is absorbed by air.

Answer is B: If you multiply your distance from the source by "r", then your exposure is multiplied by $1/r^2$ (that is decreased).

24. Why is an absorbed radiation dose of 10 Gy, absorbed over 1 day (i.e. acutely) more damaging to living tissue than the same dose received as 1000 exposures of 0.01 Gy over 10 years (i.e. chronically)?
    A. 0.01 Gy is below the threshold dose that is known to damage cells.
    B. Rapidly dividing cells are more susceptible to damage from acute doses of radiation.
    C. Healthy cells can recover from low levels of radiation if the whole body is not exposed.
    D. Cells can repair radiation damage if given time between exposures to do so.

Answer is D: There are repair mechanisms within a cell that can repair radiation damage, given time and damage that is not too severe.

25. What does the "maximum permissible dose" of radiation (100 mSv/5-year period) refer to?
    A. The average dose for the general population as a whole.
    B. The dose for an individual not exposed to radiation through their work.
    C. The dose allowed to people exposed to radiation through their work.
    D. The dose above which radiation is likely to cause harm to humans.

Answer is C: 100 mSv/5 years is the permitted dose to workers who because they are employed derive the benefits of employment which outweighs the small theoretical detriment they suffer due to their exposure.

26. What does the "relative biological effectiveness" (RBE) of radiation depends on?
    A. The recommended maximum permissible dose of radiation.
    B. The amount of radiation absorbed by the whole body.
    C. The dose of radiation that is actually absorbed in the tissue.
    D. The energy deposited in the tissue per millimetre of distance travelled through the tissue.

Answer is D: If radiation deposits energy of 1 MeV along its path of 1 mm, more damage would be produced (the radiation would be more "effective") than if the 1 MeV of energy deposited was spread out along a path length of 10 cm.

27. Which of the following body parts is the *least* sensitive to radiation?
    A. Lens of the eye.
    B. Red bone marrow.
    C. Gonads.
    D. Hands.

Answer is D: The other three body parts are all named by the ICRP and have been assigned a "radiation weighting factor" determined by their known sensitivity to radiation.

28. The intensity of gamma radiation decreases as the inverse square of the distance. What does this statement mean? If the distance doubles, then the intensity:
    A. Doubles.
    B. Is halved.
    C. Is four times as great.
    D. Is one quarter as great.

Answer is D: If the distance doubles, the inverse is ½. The square of the inverse is $(½)^2 = ¼$.

29. At a distance of 1 m, exposure to a particular radionuclide is found to result in an absorbed dose of 12 mGy. If the time of exposure remains constant, what will the absorbed dose be at a distance of 2 m?
    A. 3 mGy
    B. 6 mGy
    C. 12 mGy
    D. 24 mGy

Answer is A: If the distance doubles, the dose would be expected to decrease according to the inverse square of distance: $(½)^2 = ¼$. One quarter of 12 is 3.

30. Which of the following absorbed radiation dose equivalents would cause the most harm to your body?
    A. 0.05 Sv per year to the whole body for the human lifespan (75 years) = 3.75 Sv
    B. 1 Sv (acutely) to the gonads only
    C. 40 Sv (acutely) to the heart only
    D. 7 Sv (acutely) to the whole body

Answer is D: This dose given acutely to the entire body will probably be lethal. Higher doses can be withstood if only a limited part of the body is exposed, or if the dose is accumulated over a long time.

31. Of the following, which one would be considered the lowest level for a fatal whole body dose of radiation if received as an acute dose?
    A. 100 R
    B. 10 Sv
    C. 100 mSv
    D. 10 mSv.

Answer is B: Ten Sievert acutely is a lethal dose. 100 R is close to 1 Sv dose to soft tissue. Both 10 mSv and probably also 100 mSv would produce no noticeable effect on the body.

32. A caesium 137 source produces an absorbed radiation dose of 400 mGy/h at a distance of 1 m. What would be the dose received at a distance of 5 m in 2.5 h? (consider the effect of distance first)
    A. 16 mGy

B. 20 mGy
C. 40 mGy
D. 100 mGy.

Answer is C: At a distance of 5 m, the dose would be $(1/5)^2 = 1/25$ of that at a distance of 1 m. Hence dose would be 400/25 = 16 mGy/h. If exposed for 2.5 h, dose would be 2.5 × 16 = 40 mGy.

33. If a thickness of 7.35 mm of lead can absorb half of the 1 MeV gamma photons that enter the lead, then what fraction of gamma rays will be absorbed by twice this thickness (i.e. 14.7 mm)?
    A. Three quarters.
    B. Seven eighths.
    C. Fifteen sixteenths.
    D. Sixteen sixteenths.

Answer is A: Since half of the gamma photons are absorbed by 7.35 mm Pb, then that is the "half-value layer". A second half-value layer placed after the first one would again absorb half of the photons that emerge from the first slab of lead. This would leave half of a half, that is one quarter emerging from the second half value layer. Hence ¾ have been absorbed.

34. Which of the following is **NOT** a principle used to set radiation safety standards?
    A. There is no completely safe dose.
    B. Any dose given must show a positive net benefit.
    C. Any dose received should be as low as reasonably achievable.
    D. Any dose received should not exceed that due to natural background radiation.

Answer is D: This is often not achievable.

35. Why should the use of staff of child-bearing age to nurse patients with implanted radioactive sources be avoided?
    A. Because such sources are highly radioactive.
    B. Because there is a risk of damage to the gametes of the staff.
    C. Because younger people are more at risk of developing radiation induced cancers.
    D. Because implanted sources cannot be adequately shielded.

Answer is B: If it is possible to reduce risk to staff by assigning them to other duties, then it should be done.

Chapter 18
Electricity

Electrical devices perform the tasks that they are designed for thanks to the energy supplied to them by "electricity". AC generators at a power station provide the electromotive force that causes electrons to oscillate within conductors. The energy of oscillating electrons is transformed into useful work within the electrical device. The human body, inside the skin, can also conduct electricity because its solutions contain many dissolved ions (electrolytes). If, due to a fault, electric current flows through the skin into a person from a domestic electrical device, operating on a 50Hz, 240 V circuit capable of delivering 5 A (say), the energy so deposited in the person will cause harm and perhaps death. This is called macro-electrocution as the current required to produce a significant shock is greater than about 5 mA. There are safety precautions in place to prevent this from happening.

In a hospital situation, a patient may be connected to intravenous fluids, to a cardiac pacemaker, to indwelling catheters, etc. That is, they are connected to electrical conductors and electrical devices which penetrate the skin and make contact with the electrically conducting solutions within the body. The skin, being "dead" and dry, usually presents a relatively high resistance to the flow of electric current. If the skin is penetrated, then this natural protection is by-passed and a much smaller current can cause significant damage. This is called micro-electrocution as the current likely to cause death is probably less than 100 μA. There are special safety precautions in place in treatment rooms to prevent this from happening.

The body is a generator of electricity. The distribution of ions within the body and their movement from one side of the cell plasma membrane to the other produces local electrical currents. This polarisation and depolarisation of membranes then produces contraction of muscle fibres and the generation of nerve impulses. These endogenously generated electric currents can be measured for the heart (the measurement is called the ECG) for skeletal muscle (called EMG) and for the brain (called EEG) and provide considerable diagnostic information.

M. Caon, *Examination Questions and Answers in Basic Anatomy and Physiology*, https://doi.org/10.1007/978-3-030-47314-3_18

1. Which is the correct statement about the behaviour of electrical charges?
   A. An electron and a proton will repel each other.
   B. Two like charges would repel each other.
   C. Like charges would attract each other.
   D. Unlike charges repel each other.

Answer is B: Two positive charges would repel each other. Two negative charges will repel each other. So "like" charges repel.

2. If an ion has a positive charge, then what do we know about it?
   A. It will attract another ion with a positive charge.
   B. It has gained some protons.
   C. It has lost some electrons.
   D. It has more electrons than protons.

Answer is C: An ion will have a positive charge when it has lost one or more electrons. Hence, it will have more protons in its nucleus than electrons in the orbitals.

3. Choose the alternative which correctly completes the following sentence: "There are two types of electric charge called:
   A. Protons and electrons and they attract each other".
   B. Positive and negative and they attract each other".
   C. Anions and cations and they repel each other".
   D. Electrons and ions and they repel each other".

Answer is B: The two types of charges are positive and negative which may exist on a variety of particles such as those named in the other choices.

4. In the fluids of the human body, what are the carriers of charge that move through fluids called?
   A. Cations
   B. Ions
   C. Anions
   D. Electrons

Answer is B: The charge carriers are ions (sometimes called electrolytes); they may have a positive or a negative charge.

5. Why is the human body (inside the skin) a conductor of electricity?
   A. It contains nerves and blood vessels which behave like electrical wires.
   B. Its solutions contain ions (electrolytes) which allow the body to conduct electricity.
   C. Muscle cells generate a voltage across their cell membrane, which produces current flow.
   D. Neurons generate a voltage across their cell membrane, which produces current flow.

Answer is B: Na^+ and Cl^- are present in the blood, inside cells and in the interstitial fluid—they are the most common ions in the body. These and other ions/electrolytes, being charged particles, allow the body to conduct electricity. Choices C and D are not as good as B as muscle cells and neurons produce a current flow when the cells depolarize. That is, when the voltage across their cell membrane changes.

6. What is one difference between static electricity and current electricity?
   A. Static electricity flows in the human body, while current electricity flows in electrical appliances.
   B. No useful purpose has been found for static electricity.
   C. In current electricity, charges are moving whereas in static electricity, charges do not move.
   D. Direct current involves static electricity, while alternating current involves current electricity.

Answer is C: The word "current" means flowing while "static" means not moving.

7. Why does a plastic rod after being rubbed vigorously with rabbit fur attract and deflect a thin stream of water from a tap, but not a thin stream of oil?
   A. Water molecules have the opposite electrical charge to that on the rod, while oil molecules have no charge.
   B. Water molecules have the same electrical charge to that on the rod, while oil molecules have no charge.
   C. Water molecules are polar, while oil molecules are non-polar.
   D. Water molecules are non-polar, while oil molecules are polar.

Answer is C: A plastic rod after being rubbed with fur will have a charge of static electricity. Water molecules have a +ve end and a −ve end (are polar). The electrically charged rod attracts the end of the water molecules that have the opposite charge to the rod. This attraction is seen as the stream of water molecules is deflected towards (bends towards) the rod. The hydrocarbon chains of the triglyceride molecules in oil are very non-polar (carbon and hydrogen have similar electronegativity) and so have no charged end to be attracted to the oppositely charged rod, so no deflection. (Water molecules have no net charge, so choices A and B are wrong.)

8. What is one difference between direct current (DC) and alternating current (AC)?
   A. DC can produce a fatal shock, whereas AC cannot.
   B. AC can supply power to portable devices, but DC cannot.
   C. AC can be transmitted over long distances more cheaply than but DC.
   D. DC can be easily transformed to a different voltage, but AC cannot.

Answer is C: (I may be picking a fight here) Transformers (which change voltage of electrical power) and circuit breakers are dramatically less expensive for AC than for DC.

9. Which one of the following statements is true?
   A. An electric current exists when electrons are moving through insulators.
   B. Electrical resistance is a measure of the difficulty with which electrons move through a conductor.
   C. Electrons are the carriers of electrical charge within our bodies.
   D. Static electricity results from a build-up of charge on conductors.

Answer is B: The higher is the resistance of a conductor, the smaller will be the current that flows in it (for a given potential difference).

10. Consider a "conductivity kit" (a pair of electrodes connected to a 1.5 V torch globe and a 6 V power supply). When the electrodes are immersed in pure water contained in a beaker, no current flows and the globe does not light up. What happens when 30 g of Na^+Cl^- is dissolved in water?
    A. No current flows and the globe does not light up.
    B. No current flows and the globe does light up.
    C. Current flows and the globe does not light up.
    D. Current flows and the globe does light up.

Answer is D: Salt is an electrolyte. When added to water, it dissolves the Na^+ ions separate from the Cl^- ions. These ions can diffuse freely throughout the solution and, being charged particles, will conduct electricity. The Na^+ ions will be attracted to the –ve electrode, while the Cl^- ions will be attracted to the +ve electrode. In this way, an electric current will flow that is sufficient to cause the globe to light.

11. The unit of potential difference is volt. What does the number of volts tell us?
    A. How much energy 1 C of electrons loses as it moves through the potential difference.
    B. How much difficulty 1 C of electrons has in moving through a circuit.
    C. How many coulombs of electrons are moving per second.
    D. How much energy per second electrons lose as they move through the circuit.

Answer is A: Choice B refers to electrical resistance. Choice C refers to the current in amperes (1 A = 1 C/s). Choice D refers to power (P = A/s).

12. Which three electrical quantities does Ohm's law relate to each other?
    A. Current, resistance and potential difference.
    B. Potential difference, current and voltage.
    C. Resistance, charge and ohms.
    D. Charge, current and potential difference.

Answer is A: Current (symbol I), resistance(R), potential difference (V). In choice B, voltage is another name for potential difference. In choice C, ohm is the unit of resistance. In D, current is the amount of charge that is flowing.

13. How much current (I) will flow through someone who touches a 240 V active wire with one hand and the ground with the other, given that their hand to hand resistance (R) is 2400 Ω?
    (Use the equation $V = I \times R$)
    A. 100 μA
    B. 10 mA
    C. 100 mA
    D. 10 A

Answer is C: $V = I \times R$, so $I = V \div R$; substituting the values in the equation gives: $I = 240 \div 2400 = 0.1$ A = 100 mA.

14. How much current (I) will flow through someone who touches a 240 V active wire with one hand and ground with the other, given that their hand to hand resistance (R) is 24,000 Ω? ($V = I \times R$)
    A. 100 μA
    B. 10 mA
    C. 100 mA
    D. 10 A

Answer is B: $V = I \times R$, so $I = V \div R$; substituting the values in the equation gives: $I = 240 \div 24{,}000 = 0.01$ A = 10 mA.

15. Given Ohm's law, i.e. potential difference (V) is the product of current (I) and resistance (R), which of the following statements is true? If potential difference is fixed (at say, 240 V):
    A. A high resistance means that a low current will flow.
    B. A high resistance means that a high current will flow.
    C. A low resistance means that a low current will flow.
    D. Current will be fixed whatever the resistance.

Answer is A: As $240 = I \times R$, if the value of R is large, then the value of I must be small enough so that their product is 240.

16. A household light globe with a power rating of 60 W operates at a potential difference of 240 V and has a resistance of 960 Ω. Using this information and Ohm's law, the current (I) may be calculated to be:
    A. $I = V \div R = 240 \div 960 = 0.25$ A
    B. $I = R \div V = 960 \div 240 = 4\ A$
    C. $I = P \div V = 60 \div 240 = 0.25$ A
    D. $I = V \div P = 240 \div 60 = 4$ A

Answer is A: Ohm's law: $V = I \times R$, so $I = V \div R$. Choice C also produces the correct answer for current, but does not use Ohm's law.

17. A person, holding to the edge of the bath, steps out onto an exposed wire in a frayed electrical cord attached to a domestic (240 V) radiant heater on the floor. Consequently a fatal current of 100 mA flows between their hand and foot. What must be the electrical resistance between these two points?

    A. $240 \times 100 = 2.4\ \Omega$
    B. $240 \times 0.1 = 24\ \Omega$
    C. $240 \times 0.1 = 2400\ \Omega$
    D. $240 \times 100 = 24{,}000\ \Omega$

Answer is C: Using Ohm's law: $V = I \times R$, so $R = V \div I$ and realising that 100 mA = 0.1 A, substituting the values gives: $R = 240 \times 0.1 = 2400\ \Omega$.

18. Ohm's law may be stated in the form: "the current flowing in a circuit is the potential difference divided by the electrical resistance in the circuit". If a person whose hand to hand resistance is 80,000 Ω touches a live wire (attached to 240 V mains supply) and a wall at the same time, what is the magnitude of the current that will flow through the chest?

    A. 3 mA
    B. 30 mA
    C. 300 mA
    D. 3000 mA

Answer is A: Ohm's law as stated is: $I = V \div R$; substituting the values gives:
$I = V \div R = 240 \div 80{,}000 = 0.003$ A = 3 mA.

19. Which of the following is an equivalent and correct way of writing Ohm's law ($V = I \times R$) where V stands for voltage, I for current and R for resistance?

    A. $I = R \div V$
    B. $I = V \times R$
    C. $R = V \times I$
    D. $R = V \div I$

Answer is D: Given that $V = I \times R$. When we solve for I we get: $I = V \div R$; hence, choices A and B are wrong. When we solve for R, we get $R = V \div I$.

20. In an Australian domestic AC electrical circuit, what is the potential difference (voltage) in the three wires?

    A. Active wire oscillates between +240 V and −240 V, while neutral and earth are both at 0 V.
    B. Neutral wire at −240 V, active at +240 V, earth at 0 V.
    C. Active wire at +240 V, while neutral and earth are both at 0 V.
    D. Neutral wire and active wire both at +240 V, earth at −240 V.

Answer is A: In a domestic supply, the potential of the active wire oscillates (alternates) from about +270 V through 0 V to −270 V (the rms values are ±240 V), while the neutral wire is at 0 V. The earth wire is for safety purposes, kept at 0 V to provide a low resistance path to the ground in the case of a fault.

21. Inside a typical household electrical appliance that has a three prong plug, what is the earth wire connected to?
   A. The fuse
   B. The neutral wire
   C. The metal casing
   D. The on/off switch

Answer is C: The earth wire in an appliance with a metal case is connected to the inside of the case so that if a fault occurs that causes the metal case to be "live", current can flow down the earth wire, rather than into your hand.

22. In order for the "on-off" switch and the fuse to operate as intended in AC circuits, which wires must they be placed on?
   A. Switch on active wire, fuse on earth wire
   B. Switch on earth wire, fuse on neutral wire
   C. Both switch and fuse on active wire
   D. Both switch and fuse on earth wire

Answer is C: The active wire is connected to the electricity grid and to the electrical device. So the on switch must be able to interrupt this supply. The fuse is also on the active wire so that it "melts" if the current passing through it is above its set value.

23. The purpose of a fuse (or circuit breaker) is to prevent any flow of current. How is this done?
   A. By connecting the active wire to earth
   B. By connecting the earth wire to the supply cables
   C. By disconnecting the neutral wire from the supply cables
   D. By disconnecting the active wire from the supply cables

Answer is D: The fuse is on the active wire so when it "blows" the wire has a gap in it.

24. When electrical equipment becomes faulty, the casing (if made of metal) may become live. In this case what is the function of the earth wire on the equipment?
   A. To provide a path to ground that has high electrical resistance
   B. To provide a path to ground that has low electrical resistance
   C. To provide a path to ground for excess heat to flow through
   D. To melt and thus break the circuit so that current can no longer flow

Answer is B: The earth wire allows current to flow to earth rather than through a person touching the case. In addition, the fuse should melt almost immediately.

25. What are the three wires connected to household "three-pin plugs" on electrical cords called?
   A. Earth, active and fuse
   B. Live, return and fuse

C. Active, neutral and earth
D. Active, neutral and live

Answer is C: Active (or live), neutral (or return) and earth.

26. Suppose that while crawling around inside the roof space of your house, you touch a bare wire with your bare hand while touching the brick wall with the other bare hand. Which of the following wires would be most likely to deliver a fatal shock?
    A. The neutral wire
    B. The active wire
    C. The telephone wire
    D. The earth wire

Answer is B: The active wire is live in the roof space between your power board (fuse box) and the power points in the walls. The telephone wire is live too, but is at low voltage.

27. In an electrical device, a correctly installed earth wire provides protection against what?
    A. The electrical device receiving too high a current.
    B. It protects the fuse or circuit breaker from too much current.
    C. Electrocution of a person touching the faulty electrical device.
    D. It protects against stray currents that may exist in the ground (the soil).

Answer is C: The earth wire will only carry current if the electrical device is faulty, and the metal casing becomes live. It prevents a shock being received by anyone touching the device.

28. Why does a correctly installed earth wire provides protection against electrocution by a faulty electrical device? Because:
    A. It is made of thick copper wire so presents almost zero resistance.
    B. A fuse is located on the earth wire which will "blow" if excessive current flows.
    C. The earth wire is connected to ground and to the metal case of the appliance.
    D. The on/off switch is on the earth wire, so that power can be turned off.

Answer is C: A faulty appliance (with a metal case) may become "live". If it does, a large current may flow through the earth wire, rather than through a person touching the case. The large current will blow the fuse which is located on the active wire (as is the on/off switch).

29. When can electrical microshock occur? When:
    A. Contact between the skin and the 240 V domestic supply results in a current greater than 100 mA.
    B. A conductor carrying a current greater than 1 μA enters the body through the skin.

C. A conductor carrying a current greater than 100 μA enters the body through the skin.
D. A current of 2–4 A passes through the heart forcing it to clamp shut.

Answer is C: Microshock refers to electrocution by a current above about 100 μA if delivered by a conductor that has penetrated the skin. Currents less than this are probably not fatal. If the current greater than 100 mA is applied to the exterior of the skin, it is called microshock. Choice D is referring to external defibrillation.

30. The effects on a human that result from an electric shock depend upon all of the following EXCEPT one. Which one?
    A. Amount of current flowing
    B. Whether current is AC or DC
    C. The path through the body taken by the current
    D. Time for which current flows

Answer is B: Both AC and DC can electrocute a human, with DC requiring higher current, probably more than 300 mA.

31. The electrical resistance of the body, measured from hand to hand, will be different on different occasions. Why is this? Because the resistance of the skin:
    A. Increases as the skin gets drier
    B. Increases as the skin gets damper
    C. Decreases as the skin gets drier
    D. Decreases as the hands are brought closer together

Answer is A: Dry human skin presents greater resistance to current flow than damp skin.

32. What could be a correct definition of macro-electrocution? "That phenomenon which results from a prolonged macro-shock produced by:
    A. An electric current flowing directly to the heart without having to cross the skin"
    B. The contact of bare skin to alternating voltages of over 100,000 V″
    C. A fatal current in direct contact with unprotected skin"
    D. Switching on a faulty device that was earthed and had a fuse on the active wire"

Answer is C: Macro-electrocution refers to electrocution by a relatively large current that is applied to the skin surface.

33. Choose the correct alternative to make the following sentence correct. "If a person is in contact with the active wire of a domestic 240 V electricity supply, a macroshock will occur if the current that flows is 'above/below' 1 mA and there is a second connection between the person and the 'active/ground'".
    A. Below → active wire
    B. Below → ground

C. Above → active wire
D. Above → ground

Answer is D: A macroshock will be noticed if the current is above 1 mA, and the person is electrically connected simultaneously to the active wire and to the ground.

34. Complete the sentence: "Defibrillation" is the process where for a few milliseconds, a direct current of about:
    A. 6 A is applied directly to the heart through a conducting path that bypasses the skin.
    B. 100 mA is applied to the chest wall through two "paddles".
    C. 6 A is applied the chest wall through two "paddles".
    D. 100 mA is applied directly to the heart through a conducting path that bypasses the skin.

Answer is C: The current is large and is applied to the outside of body to bare skin.

35. What is meant when a medical procedure room is electrically wired as a "body protected area"? Patients that are connected to an electrical device are:
    A. Protected from macro-electrocution
    B. Protected from micro-electrocution
    C. Protected from macro- and micro-electrocution
    D. Unlikely to experience ventricular fibrillation

Answer is A: Procedures conducted in body protected areas do not involve a direct electrical connection to the heart. Hence only faults that allow relatively large currents can produce electrocution (macro-electrocution).

36. What is meant when a medical procedure room is electrically wired as a "cardiac protected area"? Patients who are connected to an electrical device are:
    A. Protected from macro-electrocution
    B. Protected from micro-electrocution
    C. Only permitted to be connected via electrodes that are placed on their skin
    D. Unlikely to experience ventricular fibrillation

Answer is B: Procedures conducted in cardiac protected areas may involve a direct electrical connection to the heart from a device that is connected to the electrical mains. Such a connection bypasses the relatively high insulation provided by dry skin. Hence electrical devices must be prevented from allowing even very small faulty electrical currents to flow into the patient (micro-electrocution).

37. If a patient is connected to an ECG machine to obtain an electrocardiogram, to what standard must the electrical supply to the room be installed?
    A. Cardiac-protected
    B. Domestic standard
    C. Body-protected

D. Double insulated

Answer is C: ECG electrodes are applied to the patient's skin, so do not bypass the barrier to electric current flow that is the skin. Consequently, the protection must be against macro-electrocution. A body-protected area is required. "Double insulated" is a term that is applied to domestic appliances with an external plastic structure.

38. For patients in a medical treatment room that are receiving an intravenous infusion driven by an electrical pump, what term is used to describe the electrical supply to the room?
    A. Cardiac protected
    B. Body protected
    C. Equipotentially earthed
    D. Double insulated

Answer is D: The catheter that delivers the IV infusion passes through the skin to enter a vein. This makes a direct electrical connection to the heart (via the blood). Consequently the patient must be protected against micro-electrocution and be treated in a cardiac-protected room. An equipotential earthing system is a part of the installation of a cardiac-protected area.

Chapter 19
Biomechanics

The study of mechanics involves the appreciation of the mass of a body and its inertia and of the interaction of mass and gravity to produce the object's weight. Newton's laws describe forces and their interaction with masses to alter the motion of the masses. Examples of unbalanced forces (those that will change an object's motion) are gravity, friction and the contraction of skeletal muscle.

The bones of the skeleton may be thought of as levers that turn around the synovial joints (the fulcra) when the contraction of skeletal muscles provide the effort force. The load force is the weight of the limb being moved along with whatever is being held by or being made to move by the limb. Many configurations of bone, muscle and joint that occur in the body can be described as "third class" lever systems, and hence as "inefficient". That is, the effort force produced by contracting muscles is greater than the load force to be shifted.

Correct and safe patient-handling procedures (manual handling) involves minimising the amount of lifting by asking the patient to move themselves—and instructing them how to. If the carer is required to shift a patient, manual handling also involves minimising the use of weak muscles and bones, such as those of the back. Instead, the strong muscles of the legs are used; the length of the load arms are minimised (by getting your centre of gravity close to the person being moved—the load), and friction is minimised by using a slide sheet or board. Furthermore correct technique requires that stability is maintained (by keeping your centre of gravity over your base of support) and that the help of gravity, and of other health care workers, is enlisted.

Sometimes a fractured bone needs to be held in the correct anatomical position while healing progresses, by applying traction. This uses the force applied by suspended masses that pull on cords that may be directed by pulleys and are ultimately firmly attached to the patient, to counteract the force of the patient's muscles that would otherwise hold the ends of the fractured bones away from each other. Traction systems can be understood in terms of the vector addition of forces so that the traction force is applied in the appropriate direction.

M. Caon, *Examination Questions and Answers in Basic Anatomy and Physiology*, https://doi.org/10.1007/978-3-030-47314-3_19

19.1 Force, Vectors and Levers

1. Choose the **INCORRECT** statement from the four below.
   A. Mass is the amount of matter contained in an object.
   B. Gravity is the name of the force that acts between any objects with mass.
   C. Weight is the pressure with which earth's gravity acts on an object.
   D. For a standing person, base of support is the area bounded by their feet.

Answer is C: Weight is a force while pressure is a force divided by the area upon which it is acting.

2. If a traction hanging mass is 3 kg, use Newton's second law (or otherwise) to determine the gravitational force acting on the mass.
   A. 0.3 N
   B. 3 N
   C. 30 N
   D. 3 kg

Answer is C: Newton's second law: $F = m \times a$; substituting in the numbers gives: $F = 3 \times 10 = 30$ N ($a = 9.8$ ms^{-2} but 10 is close enough).

3. A person whose mass is 65 kg would have a weight closest to:
   A. 65 kg because mass and weight are directly proportional to each other (on earth)
   B. 650 kg because weight = mass × 10 (approximately)
   C. 65 N because $F1 = -F2$ (Newton's third law)
   D. 650 N because $F = m \times a$ (Newton's second law)

Answer is D: Choice B is almost correct, but the unit of force is the newton while the kilogram is the unit of mass.

4. A person whose mass is 50 kg would have a weight closest to:
   A. 50 kg because mass and weight are directly proportional to each other (on earth)
   B. 50 N because $F1 = -F2$ (Newton's third law)
   C. 500 kg because weight = mass × 10 (approximately)
   D. 500 N because $F = m \times a$ (Newton's second law)

Answer is D: Choice C is almost correct, but the unit of force is the newton while kilogram is the unit of mass.

5. In which of the following situations is an unbalanced force acting on a patient's body? When the patient is:
   A. Lying stationary and supine on the bed
   B. Lying stationary and supine on the bed while the bed is being pushed a constant speed along a straight corridor

C. Being assisted to sit still and upright on the edge of the bed
D. Lying stationary and supine on the bed while the bed is being pushed around a corner in a corridor

Answer is D: Changing the direction of motion by turning a corner (even if the speed does not change) requires an unbalanced force. Being still or moving at constant speed indicates that there are no unbalanced forces acting.

6. Which statement concerning friction is **INCORRECT**?
   A. Friction within liquids is greater than that between dry solids.
   B. Within the human body friction is reduced by fluids such as saliva, serous fluid, mucus.
   C. Friction exists whenever two surfaces are in contact.
   D. Sliding friction is less than static friction.

Answer is A: Frictional forces within liquids are lower than within solids. Sliding friction is less than static friction.

7. Which one of the following statements is **WRONG**?
   A. A patient's inertia may be decreased by using several people to assist in their transfer.
   B. The position of a person's centre of gravity may be altered by altering the position of their arms and legs.
   C. Base of support may be increased by adopting a wide stance with your feet.
   D. Friction between a patient and the bed may be reduced by using a slide sheet.

Answer is A: Inertia is a fixed attribute of an object that cannot be changed without changing its mass.

8. What is the weight of a nurse if her mass is 65 kg? (Use acceleration due to gravity of 10 m s^{-2})
   A. 65 kg
   B. 650 kg
   C. 65 N
   D. 650 N

Answer is D: Weight is the force with which gravity attracts you to the earth. Use Newton's second law:
Weight, $F = m \times a = 70 \times 10 = 700$ N. The unit of force is the newton, N.

9. Choose the one correct statement.
   A. The unit of weight is the kilogram.
   B. The unit of mass is the newton.
   C. Weight is a force.
   D. Inertia is mass multiplied by 9.8.

Answer is C: Weight is the force with which gravity attracts you to the earth. Your weight would be different on the moon (say), but your mass would be the same.

10. Which of the following is not an example of a force?
    A. Tension.
    B. Friction.
    C. Inertia.
    D. Weight.

Answer is C: Inertia is not a force, it is a measure of mass. A measure of inertia of an object is the difficulty of setting the object in motion.

11. What does the term 'friction' refer to?
    A. The tension force generated when a muscle contracts.
    B. A force that acts in the opposite direction to a motion.
    C. The resistance force that is overcome by an effort force.
    D. The force of gravity that causes an object to fall.

Answer is B: Friction is a force that causes a moving object to slow in speed. That is, it produces a negative acceleration—one in the opposite direction to the object's velocity.

12. Which one of the following describes what could happen to an object when a balanced force is acting on it? The object:
    A. Starts to move.
    B. Changes its direction but not its speed.
    C. Object changes its shape.
    D. Nothing happens.

Answer is D: As the force is balanced, there is no net force acting in any direction, so there is no change in motion.

13. Which one describes when an unbalanced force is acting?
    A. An object's centre of gravity is above its base of support.
    B. A moving object maintains a constant speed and direction.
    C. A muscle contracts and causes a limb to move.
    D. A soldier is standing rigidly to attention.

Answer is C: An unbalanced force will result in some movement. The movement of the limb indicates that an unbalanced (muscle) force was acting.

14. In cars, the aim of safety features such as seatbelts, padding on the dashboard, collapsible steering columns, airbags and body panels that crumple progressively is to minimise the unbalanced force on occupants during a crash. Bearing Newton's second law in mind. How do such features achieve this? They:
    A. Minimise the occupant's deceleration.
    B. Maximise the occupant's deceleration.
    C. Prevent whiplash injuries.
    D. Convert an unbalanced force into a net force.

Answer is A: Say the occupant has a mass of "*m*". Newton's second law states: $F = m \times a$ to make the unbalanced force F as small as possible, then the acceleration "*a*" must be made small. Acceleration may be positive or negative, the latter is often called a "deceleration".

15. Which of the following is **NOT** consistent with Newton's second law?
    A. Weight = mass × 9.8.
    B. Acceleration = weight ÷ mass.
    C. Mass = acceleration ÷ weight.
    D. Force = mass × acceleration.

Answer is C: Newton's second law states: $F = m \times a$. This may be rearranged to $m = F \div a$ or in words: mass = weight ÷ acceleration.

16. What is the best definition of the WEIGHT of an object? Weight is the:
    A. Force of attraction between the earth and the object.
    B. Tendency of a body to maintain its state of motion.
    C. Amount of matter contained in the body.
    D. Mass of the object multiplied by its acceleration.

Answer is A: Weight is a force. It is the attractive force between the earth's mass and the object's mass.

Vectors

17. When two forces of magnitude 6 and 10 N are added vectorially, what value could the resulting vector **NOT** have?
    A. 4 N
    B. 6 N
    C. 10 N
    D. 18 N

Answer is D: The maximum value that can be obtained by adding two vectors of length 6 and 10 N is 16 N when the vectors are in the same direction.

18. When two forces of magnitude 8 and 12 N are added vectorially, what value could the resulting vector **NOT** have?
    A. 4 N
    B. 10 N
    C. 20 N
    D. 25 N

Answer is D: The maximum value that can be obtained by adding two vectors of length 8 and 12 N is 20 N when the vectors are in the same direction. Hence Choices A, B and C are all possible.

19. Which of the following pairs of quantities does **NOT** contain a scalar quantity and a vector quantity?
    A. Mass and weight.
    B. Traction and counter-traction.
    C. Speed and velocity.
    D. Distance and displacement.

Answer is B: Traction and counter-traction are both forces so are both vector quantities.

20. What may be said about the measurable quantities which are referred to as vectors?
    A. They are arrow-headed line segments.
    B. They include time, mass, pressure and energy.
    C. They need a magnitude to be completely defined.
    D. They include force, velocity, acceleration and electric field strength.

Answer is D: These four quantities are all vectors. Vectors need a direction as well as a magnitude. Mass is not a vector.

Levers

21. Consider the action of moving from having both feet flat on the ground to standing on "tip toes". Which of the following correctly identifies the fulcrum, the load force, the effort force and the lever?
    A. Fulcrum is the heel, load is the body's weight, effort is the pull of the hamstrings, lever is the tarsal bones of the foot.
    B. Fulcrum is the ball of the foot, load is the body's weight, effort is the pull of the calf muscles, lever is the tarsal and metatarsal bones of the foot.
    C. Fulcrum is the ankle, load is the weight of the foot, effort is the pull of the hamstrings, lever is the tibia and fibula bones.
    D. Fulcrum is the knee, load is the weight of the foot, effort is the pull of the calf muscles, lever is the tibia and fibula bones.

Answer is B: Fulcrum has to be the "ball" of the foot (the rounded area superficial to the joint between metatarsal and proximal phalanx of the hallux) and load must be the weight of the body.

22. Most of our bones that articulate at freely movable joints can be described as third class levers. What does this mean?
    A. They are "efficient" levers.
    B. The "effort arm" is longer than the "load arm".
    C. The muscle's tendon is inserted between the load and the joint.
    D. The muscle's tendon is inserted close to the joint

Answer is C: In third-class levers, the muscle force (effort) is between the load and the fulcrum (the joint). Choice D can be satisfied by first and third class levers.

23. Which of the following statements is characteristic of "third-class levers"?
    A. The fulcrum lies between the effort and the load.
    B. The muscular effort involved in shifting them exceeds the load that is shifted.
    C. They are efficient levers.
    D. The effort and the load are equally distant from the fulcrum.

Answer is B: Third class-levers are "inefficient" in the sense that the muscular force that needs to be exerted will exceed the load force to be shifted. Nevertheless human muscles are able to generate the forces required for human activities, and we have the benefit of a very large range of movement for our limbs.

24. Levers are acted upon by forces known as the load, effort and fulcrum. "Third-class" levers are characterised by having the:
    A. Effort located between the other two forces
    B. Load located between the other two forces
    C. Fulcrum located between the other two forces
    D. Resistance located between the other two forces

Answer is A: The effort, that is the insertion of the muscle tendon, is close to the fulcrum (the joint), but between the joint and the load (i.e. the rest of the limb) being moved. The load and resistance are the same force.

25. Levers are acted upon by forces known as the load, effort and fulcrum. "First-class" levers are characterised by having the:
    A. Effort located between the other two forces
    B. Load located between the other two forces
    C. Fulcrum located between the other two forces
    D. The effort and the load on opposite sides of the fulcrum

Answer is C: For example the atlas vertebra is the fulcrum for the lever action of nodding yes. The sternocleidomastoid and the trapezius muscles provide the effort forces.

26. What is the reason that a third-class lever is inefficient?
    A. The muscle's effort force is applied closer to the fulcrum than the load force.
    B. The load force is applied closer to the fulcrum than the muscle's effort force.
    C. The fulcrum separates the effort from the load.
    D. The load force is greater than the muscle's effort force.

Answer is A: Consider flexing the forearm using the biceps brachii. Its tendon is inserted on the radius bone very close to the elbow joint, while the load (the forearm) is further away. Such an arrangement allows a short contraction of the biceps brachii to be converted to a large sweep of motion of the hand.

27. Third-class lever systems are always inefficient for which one of the following reasons?

    A. The effort arm is longer than the load arm.
    B. The effort force required is less than the load force.
    C. The fulcrum lies between the effort and the load.
    D. The load arm is longer than the effort arm.

Answer is D: By definition of third-class levers, the load arm is always longer than the effort arm. A spanner tightening a nut has a long effort arm and is a mixture of first- and second-class lever.

28. In the schematic diagram of an arm being a lever, rod 1 represents the humerus and rod 2 the radius and ulna. M is a block being supported by the 'hand'. What are located at the positions indicated by p, q, r and s?

    A. Effort (p), centre of mass of the radius and ulna (q), fulcrum (r) and load (s).
    B. Fulcrum (p), effort (q), centre of mass of radius and ulna (r) and load (s).
    C. Centre of mass of radius and ulna (p), effort (q), load (r) and fulcrum (s).
    D. Fulcrum (p), centre of mass of radius and ulna (q), effort (r) and load (s).

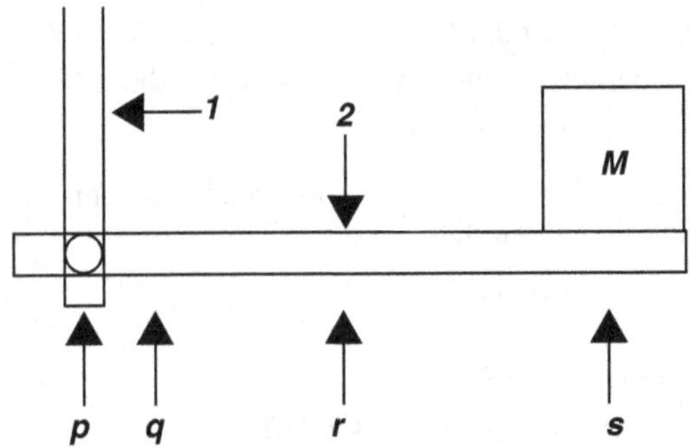

Answer is B: p is the fulcrum (the elbow joint); q is the effort (the insertion point of the biceps brachii tendon); r is halfway along the rod 2, so is the centre of mass of the radius/ulna.

29. When the biceps brachii muscle flexes the forearm, what is the forearm an example of?

    A. A first-class lever.
    B. A second-class lever.
    C. A third-class lever.
    D. An efficient lever.

Answer is C: Third-class levers have the effort force between the fulcrum and the load.

30. Why are the third-class lever systems of the human musculoskeletal system inefficient? Because:

    A. Third-class lever systems are the least efficient.
    B. The muscle insertion is closer to the joint than the load is.

C. Muscles can contract only by about 20% of their relaxed length.
D. The force of muscle tension is less than the weight of the load.

Answer is B: When the muscle insertion is closer to the joint than is the load, the effort arm is short. This means that the effort force must be large to provide the same torque exerted by the load acting on its longer load arm.

31. Many of the muscle-bone combinations in the body behave as third-class levers. Why do these require a muscle tension (the effort force) much greater than the load's force?

A. Because the effort force is applied further from fulcrum than is the load force.
B. Because the effort force is applied closer to the fulcrum than is the load force.
C. Because the fulcrum lies between the load and the effort.
D. Because the fulcrum is further from the effort force than is the load force.

Answer is B: The closer is the effort to the fulcrum, the greater is the effort it must exert to shift the load.

Stability

32. A person who is standing on both of their feet is "stable" when:

A. Their base of support is as wide as is comfortably possible.
B. Their centre of gravity is above their base of support.
C. Their centre of gravity is close to their base of support.
D. The position of their centre of gravity and base of support coincide.

Answer is B: When the imaginary line joining the earth's centre with the person's centre of gravity passes through the area of a person's base of support they are stable. They is, they will not overbalance and tip over.

33. If the imaginary line joining a person's centre of gravity to the centre of the earth passes through the person's base of support, what can we say about that person? They are:

A. Unstable.
B. Balanced.
C. Using their back as a lever.
D. Not doing any work.

Answer is B: They are balanced or stable. That is, they will not need to shift their feet to prevent them from "overbalancing".

34. A person (assumed to be healthy, awake and on their feet) is stable—that is, will not overbalance and fall—as long as:

A. They have a large base of support.
B. Their centre of gravity is close to the ground.

C. Their centre of gravity remains within their body.
D. Their centre of gravity is above their base of support.

Answer is D: As long as the centre of gravity is above the base of support, there will be no unbalanced force tending to tip you over. Try standing with your back against the wall, then bend over to touch your toes.

19.2 Manual Handling and Traction

1. Which statement is consistent with "good lifting technique"?
   A. Keep your centre of gravity and that of the object being lifted as close as possible
   B. Maximise the length of the effort arms of the body levers you use to lift the load
   C. Use the relatively strong back muscles rather than the relatively weak thigh muscles
   D. Minimise your base of support and bend your legs at the knees before you lean over

Answer is A: This strategy will keep the load arms of the levers, as short as possible. In addition adopt a wide base of support, bend your knees and use your leg muscles to lift.

2. Which one of the following is a nurse affecting, by positioning her body close to that of the patient while executing a patient handling procedure?
   A. The mass of the patient to be shifted
   B. The maximum effort that her muscles are able to produce
   C. The distance between the position of muscle insertion and the fulcra of her limbs
   D. The distance between the centre of gravity of the patient and that of the nurse

Answer is D: Minimising the separation between the centres of gravity also minimises the load arms of the levers involved. This will decrease the muscular effort required.

3. What does a nurse attempt to minimise during a patient manual handling procedure?
   A. The patient's centre of gravity
   B. Friction between patient and bed
   C. The patient's inertia
   D. The nurse's base of support

Answer is B: Minimising friction will both minimise the muscular effort (force) required and also the shearing force on the patient's skin.

4. What is a nurse attempting to minimise by getting close to a patient during a manual handling procedure?
   A. The load arm
   B. The size of the fulcrum
   C. The effort arm
   D. The patient's weight

Answer is A: The shorter is the load arm, the smaller is the muscular effort that is required to move the lever and load.

5. Good manual handling technique on an unconscious patient involves which of the following?
   A. Using the muscles of the arms
   B. Instructing the patient on how to shift themselves
   C. Using the muscles of the legs
   D. Using the muscles of the back

Answer is C: The leg muscles are the largest (and therefore the strongest) in the body. An unconscious patient cannot take instruction.

6. Good manual handling technique generally requires extensive use of which muscles?
   A. Biceps brachii and triceps brachii
   B. Erector spinae and abdominal muscles
   C. Gluteus maximus and rectus abdominus
   D. Quadriceps and hamstrings

Answer is D: The muscles of the thighs are the strongest and should be used in manual handling tasks.

7. What is the purpose of using of a slide sheet in patient manual handling?
   A. To extend the patient's base of support
   B. To facilitate raising the patient's centre of gravity
   C. To minimise friction between the patient and the bed
   D. To minimise the patient's inertia

Answer is C: A slide sheet reduces friction, making it easier to slide the patient. It also provides comfortable hand holds and prevents fragile skin being torn as it slides along the bed.

8. What is the purpose of "counter-balancing" in manual handling? To:
   A. Ensure that your large leg muscles are used to shift a patient
   B. Ensure a firm grip on the slide sheet
   C. Increase the size of your base of support
   D. Use gravity, acting on your weight, to shift a patient

Answer is D: Counter-balancing involves placing the centre of gravity to one side of the base of support, so that the gravitational pull on your body assists in moving the patient.

9. Most manual handling manoeuvres require bending at the knees. This is so that
   A. The strong thigh muscles are used.
   B. A wide base of support can be adopted.
   C. A stable body position is achieved.
   D. The back can be used as a third-class lever.

Answer is A: Bending the legs at the knees (and hips) allows the strong thigh muscles to be used to extend the legs while rising to a standing position with the load.

10. Which of the following is **NOT** a reason for using a slide sheet during manual handling?
    A. To provide the handler with convenient hand holds while shifting the patient
    B. To increase the patient's base of support
    C. To reduce the friction between the patient and their bed
    D. To reduce the risk of damaging fragile skin

Answer is B: The patient's base of support (their area of contact with the bed) is not changed by a slide sheet.

11. What is the aim of 'correct lifting technique'? To:
    A. Maintain balance by keeping the centre of gravity over the base of support.
    B. Avoid working with heavy loads that are on the ground.
    C. Use the bones and muscles of the leg.
    D. Keep the back straight while using it as a lever.

Answer is C: The leg bones and muscles are the strongest in the body so are most suited to coping with large forces.

12. Which of the following is **NOT** considered part of good lifting technique?
    A. Keeping the feet close together.
    B. Keeping the back virtually straight.
    C. Standing close to the object to be lifted.
    D. Bending the knees.

Answer is A: It is best to keep the feet separated to widen the base of support. A wide stance also makes it easier to get close to the object to be lifted.

Traction

13. Which of the following is correct?
    A. Fixed traction employs a moveable pulley to provide mechanical advantage.
    B. In suspended traction, the traction force is equal to the tension in the cord.
    C. In suspended traction, counter-traction is provided by the friction between the patient and the bed.

D. The traction force must be equal in magnitude but opposite in direction to the counter-traction force.

Answer is D: This a statement of Newton's third law. Fixed traction does not employ a moveable pulley. In suspended traction, the traction force is greater that the tension in the cord. Counter-traction is supplied by gravity as well as friction.

14. In a Hamilton-Russell traction system, a moveable pulley provides a mechanical advantage. What does this mean?
    A. Weight of the hanging mass is greater than the traction force.
    B. Weight of the hanging mass is equal to the traction force.
    C. Traction force is greater than the weight of the hanging mass.
    D. Traction force is greater than the counter-traction force

Answer is C: When the effort force is less than the load force, there is a mechanical advantage. Traction and counter traction force must be equal.

15. In straight leg traction (Buck's extension), the force of counter-traction is supplied by friction (between the patient and the bed) and also by:
    A. The component of the patient's weight that is perpendicular to the bed
    B. The component of the patient's weight that is parallel to the bed
    C. Gravity acting on the hanging mass
    D. Using cords to attach the patient to the head of the bed

Answer is B: The bed is tilted head down so that patient is tending to side away from the traction mass.

16. A requirement in traction is that the traction force is equal in magnitude but in the opposite direction to the counter traction force. This requirement is really a statement of:
    A. Ohm's law.
    B. Pascal's principle.
    C. Newton's first law.
    D. Newton's third law.

Answer is D: Newton's third law: $F_1 = -F_2$. When an object (the hanging mass) exerts a force on another object (the patient in traction), the second object exerts an equal but opposite force on the first object.

17. Traction forces may be represented by vectors. In Hamilton-Russell traction, which of the following vectors is equal in magnitude to the traction force?
    A. The component of the patient's weight that is perpendicular to the bed.
    B. The component of the patient's weight that is parallel to the bed.
    C. The vector in A subtracted from the patient's weight.
    D. The resulting vector when the vectors in A and B are added.

Answer is B: By elevating the foot of the bed, the patient tends to slide towards the head of the bed which is opposite to the direction of traction pull.

18. Fixed traction may be applied by a device such as a 'Thomas splint'. In this case, the counter-traction force is supplied by:
    A. The push of the appliance on a fixed point on the body (such as the ischial tuberosity).
    B. An adhesive bandage wrapped around the lower leg.
    C. The pull of a wire that is made taut by turning a wingnut attached to the patient's foot.
    D. The weight of the patient's leg and the friction between the patient and the bed.

Answer is A: Fixed traction does not involve cords and hanging masses. The Thomas splint involves "stretching" the leg between the foot and the pelvis.

19. The counter-traction force on a patient's leg is often increased by tilting the bed so that the patient's head is lower than their feet. Why is this done? Because tilting the bed increases the:
    A. Component of the patient's weight that is parallel to the bed.
    B. Traction force that is exerted by the hanging masses.
    C. Force of gravity that is acting on the patient's body.
    D. Component of the patient's weight that is perpendicular to the bed.

Answer is A: Unless the bed is tilted, there is no component of the patient's weight that is parallel to the bed. The force of gravity and the force exerted by the hanging masses do not change.

20. The magnitude of the traction force in a Hamilton-Russell traction is determined by:
    A. The vector addition of the forces in the cords.
    B. The hanging mass multiplied by 9.8.
    C. The component of the patient's weight that is perpendicular to the bed.
    D. The number of pulleys in the system.

Answer is A: The moveable pulley allows the one cord to exert more than one force on the leg. So the magnitude of the force is determined by adding the forces vectorially.

21. In a Hamilton-Russell traction system, the traction force is greater than the weight of the hanging mass because:
    A. Three cords are attached to the patient's leg.
    B. There is a moveable pulley attached to the patients foot.
    C. Four pulleys are in the system.
    D. The traction force is parallel to the femur.

Answer is B: The moveable pulley provides a mechanical advantage. A fixed pulley merely redirects the direction of traction without altering its magnitude.

22. The figure below whose corners are labelled ABCD could represent the vector diagram for a Hamilton-Russell traction system if the lines were the correct length and an arrow pointing in the correct direction was drawn on them. Which of the following modifications to the diagram would make it more closely resemble the correct diagram?
    A. The three lines D to C, C to B and B to A should be the same length and an arrow pointing from D to A should be drawn on the fourth line.
    B. Arrows should be drawn so that their directions describe an anticlockwise circuit and the line A to D should be the longest.
    C. The resultant traction force is represented by the vector from C to B and should be longer than the other three lines.
    D. All four lines should be the same length and the vector directions are from D to C, from C to B, from B to A, and from A to D.

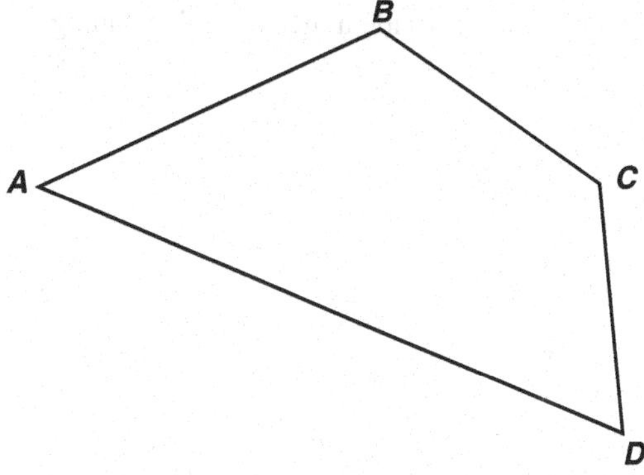

Answer is A: The three lines between D–C, C–B, B–A should be the same length as they are all the same force in three cords. They add to form the resultant (traction) vector from D to A.

23. Which of the following statements about the pulleys used in traction systems is correct?
    A. A moveable pulley changes the direction of the traction force.
    B. A moveable pulley changes the direction of the traction force and provides a mechanical advantage.
    C. A fixed pulley changes the direction of the traction and provides a mechanical advantage.
    D. A fixed pulley is one that is attached to the limb undergoing traction.

Answer is B: The provision of a mechanical advantage is the important feature of a moveable pulley. Traction direction is also changed.

24. What is one of the roles of the pulley in a traction system?
    A. To supply the counter traction.
    B. To enable the vector addition of forces.

C. To change the direction of the traction force.
D. To prevent the hanging masses from resting on the floor.

Answer is C: Pulleys do change the direction of the traction force so that the downward pull of the hanging mass can be redirected to a direction parallel (say) to a leg.

25. What are the two purposes served by pulleys used in traction systems? These are to:

A. Reduce friction and provide support for the limbs in traction.
B. Supply a mechanical advantage and attachment points to the bed frame.
C. Allow the hanging masses to exert a traction force greater than their weight and redirect the direction of the effort force.
D. Change the direction of the cords and suspend the limb from the bed.

Answer is C: Pulleys change the direction of the force provided by the hanging masses and (if the pulley is a moveable one) provide a mechanical advantage.

Chapter 20
Work, Energy, Body Temperature and Heat Loss

Energy has the units of joule or kilojoule for food energy values. When energy is transferred from one object to another we say "work has been done". How fast the energy is transferred (the rate of change of energy) is known as the power which has units of watts. Heat is the name given to the energy that is transferred from one object to another object that is at a lower temperature. The temperature of an object is a measure of the kinetic energy of its particles—roughly speaking, how fast they are moving. The units of temperature are degrees Celsius (or more correctly kelvin, where Celsius degrees = kelvin − 273).

Humans maintain their internal body temperature at close to 37 °C. A temperature above 38 °C is hyperthermia, while temperatures below 35 °C are hypothermic. Humans can gain or lose heat by the physical processes of convection, conduction and radiation. The physiological process of sweating will cool the body, while physical activity increases the amount of energy liberated from ATP and hence our metabolic rate which will cause an increase in body temperature. In addition vasoconstriction and vasodilation will redistribute heat within the body and its surface as the blood is redistributed. This redistribution of blood to the skin facilitates the loss of heat from the body by increasing the rate of radiation and heat loss by sweating. Furthermore, ingesting hot or cold food and liquid will alter the body's heat content as will eliminating and urinating.

Mitochondria within cells liberate the energy stored in the chemical bonds of the small organic molecules (e.g. glucose) which result from the hydrolysis (digestion) within the digestive tract of the large molecules ingested as food. This liberated energy is stored in the chemical bonds of a different molecule—the third phosphate bond of ATP. ATP is then available to the body's cells when required. If we ingest energy that is additional to that required to support our physical and metabolic activities, it is stored as fat in adipocytes and our body mass increases.

M. Caon, *Examination Questions and Answers in Basic Anatomy and Physiology*, https://doi.org/10.1007/978-3-030-47314-3_20

20.1 Work and Energy

1. What is the meaning of "work" in the scientific sense?
   A. An artist's completed painting.
   B. The amount of energy transferred between objects.
   C. The functions performed during the course of paid employment.
   D. Sustained physical or mental activity.

Answer is B: The amount of work done on an object is the amount of energy transferred to that object.

2. In which of the following cases is the greatest amount of work being done on the object that is experiencing the force?
   A. 10,000 N is exerted on a wall for 30 s.
   B. 2000 N is used to raise an object by a height of 10 m.
   C. 5000 N is used to push an object a distance of 4 m.
   D. 1000 N is used to pull an object over 50 m.

Answer is D: Work = force × distance, so 1000 N × 50 m = 50,000 Nm is the greatest amount.

3. Energy may be defined from the concepts of:
   A. Force and heat.
   B. Heat and joules.
   C. Force and work.
   D. Work and temperature.

Answer is C: When a force causes an object to move, work (in joules) has been done to the object. The work done changes the amount of energy (in joules) stored in an object.

4. Which of the following statements about work is **NOT** correct?
   A. More work occurs when a force acts over a small distance ($W = F \div s$).
   B. Simple machines allow us to perform work more easily.
   C. Work (done on an object) is the amount of energy that transfers to that object.
   D. It takes energy to perform work.

Answer is A: The reverse is true as: $W = F \times s$.

5. Work is done when an object is moved by a force. What is energy?
   A. The capacity to do work.
   B. The rate at which work is done.
   C. A force that results in no movement.
   D. The transformation of work from one form to another.

Answer is A: Choice B refers to power. Choice C refers to a balanced force.

6. The Système International (Standard International) unit of power is watt. This unit is the same as which one of the following?
   A. Electronvolt.
   B. Joule.
   C. Joule per second.
   D. Kilojoule.

Answer is C: Power is the rate of doing work or of converting energy from one form to another. The unit of power = J/s is renamed the watt in honour of James Watt. All the other choices are units of energy.

7. Which of the following is the most acceptable definition of the term "power"?
   A. The rate of doing work
   B. That which is stored and can be fully recovered and converted to kinetic energy
   C. The concept applied to that which gives an object the ability to do work
   D. The sum of an object's potential and kinetic energies

Answer is A: Power refers to how fast work is done. That is, the time taken for the energy in joules to be used. The unit of power is "watt" which is the same as joules per second.

8. When a particular energy value is ascribed to a food, what type of energy is being referred to? Its:
   A. Translational kinetic energy.
   B. Gravitational potential energy.
   C. Chemical potential energy.
   D. Average kinetic energy per molecule.

Answer is C: The energy value of food is stored in the chemical bonds within the food molecules.

9. Which one of the following statements best describes what is meant by the "principle of conservation of energy"?
   A. Internal energy is the sum of kinetic energy, thermal energy and potential energy.
   B. Energy may be transformed from one form into another, but it cannot be created or destroyed.
   C. Kinetic energy is gradually transformed into potential energy and vice versa.
   D. The earth has a finite amount of energy and modern society must learn to use less of it and to use it more efficiently.

Answer is B: Energy is noticed when it is transformed from one form to another. However, the total amount when all of its forms are accounted for does not change.

10. A ball initially with 20 J of potential energy has rolled halfway down a slope and is still moving. Which of the following values of kinetic energy (KE) and gravitational potential energy (PE) is it likely to have?
    A. 0 J of KE and 0 J of PE
    B. 0 J of KE and 10 J of PE
    C. 10 J of KE and 10 J of PE
    D. 10 J of KE and 0 J of PE.

Answer is C: The total amount of PE + KE must still be 20 J. As the ball has rolled halfway down the slope, half of its gravitational potential energy has been transformed into kinetic energy.

11. Which one of the following statements does **NOT** accurately describe energy?
    A. Energy can be created but not destroyed.
    B. The total amount of all types of energy remains constant.
    C. The energy gained by an object is the amount of work done on the object.
    D. Energy is the heat given out when oxidation occurs.

Answer is A: Energy can be neither created nor destroyed.

12. What is a person's metabolic rate defined as? The rate:
    A. Of energy utilisation by their body
    B. Of energy utilisation by their body during "absolute rest".
    C. At which they consume oxygen
    D. At which they produce heat

Answer is A: Choice B refers to basal metabolic rate, which may be measured by oxygen consumption while at rest.

13. In order to lose body fat, diet and exercise must be organised so that the energy value of the food intake is:
    A. Less than the energy used daily.
    B. More than the energy used daily.
    C. Equal to the daily energy use.
    D. Greater than the daily exercise.

Answer is A: If more energy is expended than is taken in as food, the body will use some of its energy stored as fat, to make up the difference. Hence, the percentage of body mass that is fat will decrease.

14. What may be said about the amount of energy in the form of infrared electromagnetic waves that is radiated from our bodies?
    A. It depends on our body's temperature.
    B. It would be greater if our layer of subcutaneous fat was thicker.
    C. It may be increased by contact between the body and an object with temperature lower than the body's temperature.

D. It may be increased by exposing more bare skin.

Answer is A: The amount of energy radiated as infrared depends on the body temperature ($E \propto T^4$).

15. Which of the following correctly states the principle of conservation of energy in terms of the human body? (Assume no foodstuffs are consumed and no urine or faeces are excreted.)
    A. $Q = s \times m \times \Delta T$ (Q = energy, s = specific heat of tissue, m = body mass, T = body temperature).
    B. The energy stored in the human body is equal to the energy lost from the body plus the work done by the body.
    C. The energy value of the food we eat must exceed the energy value of the muscular activity we perform.
    D. The change in the energy stored in the body is equal to the heat lost from the body plus the work done by the body.

Answer is D: Energy that leaves the body (without mass) must be in the form of heat loss or work done by body movements. The formula in A may be used to determine the energy value of a food burned in a calorimeter.

16. Which of the statements is a description of basal metabolic rate in a human?
    A. The sum total of the energy released per minute by all of the chemical reactions that occur in the body
    B. The rate of energy utilisation during "absolute rest"
    C. The power generated by the body's activities
    D. The oxygen consumption (in L/min) of an individual

Answer is B: "Basal" metabolic rate refers to the minimum amount. This occurs when there is no movement (apart from breathing and cardiovascular movements) and little mental activity.

20.2 Heat Transfer

1. A block of wood (a poor conductor of heat) whose temperature is 15 °C is placed in contact with a block of steel (a good conductor of heat) of the same size but whose temperature is 20 °C. Both are touched with a hand whose skin temperature is 28 °C. Which of the following is true?
    A. The steel block will feel colder than the wood.
    B. The wood block will withdraw more heat from the hand than will the steel block.
    C. Heat will flow from the steel block to the wood block.
    D. Heat will flow from the steel block to the hand.

Answer is A: "Feeling colder" is somewhat subjective. Placing a warm hand against a steel block at a lower temperature will make the hand feel cold as heat flows from the hand to all parts the steel block. Wood does not produce this feeling as it conducts heat poorly so the hand will soon warm up the wood in contact with it. Choice C will happen only slightly as the wood block is a poor conductor of heat.

2. What is heat? It is a:
   A. Measure of the temperature of an object.
   B. Transfer of energy by convection currents.
   C. Measure of the average translational kinetic energy of the particles.
   D. Form of energy transfer along a temperature gradient.

Answer is D: Heat is the energy that is transferred between two objects due to their difference in temperature. Choice C refers to the temperature of an object.

3. Which of the statements about heat is true?
   A. Heat is one of the forms of infrared radiation.
   B. Heat is transmitted through a solid object by convection currents.
   C. Heat is a measure of the temperature of an object.
   D. Heat is the flow of energy from one body to another at a lower temperature.

Answer is D: Heat is the energy that is transferred between two objects due to their difference in temperature. Infrared radiation is one form of heat (but not the reverse).

4. What does kinetic theory allows us to understand that the temperature of an object measures?
   A. Heat that it contains.
   B. Average kinetic energy of its particles.
   C. Hotness or coldness of it.
   D. Number of degree kelvin it is.

Answer is B: All particles of an object move or vibrate. Temperature is the average kinetic energy of the particles of an object.

5. What may the "thermal energy" of the particles in an object be defined as?
   A. The amount of heat that is contained in the object.
   B. Sum of the random translational, rotational and vibrational kinetic energies.
   C. Sum of the random translational, rotational and vibrational kinetic energies and the work done to overcome the intermolecular forces.
   D. Average random kinetic energy of the particles.

Answer is B: In gases, thermal energy is energy stored as kinetic energy, rotational motion, vibrational motion and associated potential energies. In other substances, in cases where some of the thermal energy is stored in atomic vibration or by increased separation of particles having mutual forces of attraction, the thermal energy is partitioned between potential energy and kinetic energy.

6. Which of the following factors contributes LEAST to the human sensation of "hot" or "cold" when an object is touched?
   A. The amount of heat in the object.
   B. The temperature of the object being touched.
   C. The thermal conductivity of the object.
   D. The local skin temperature.

Answer is A: The amount of heat in an object depends on the size of the object. The sensation of hot or cold depends on the other three things.

7. Two identical beakers hold 100 ml and 200 ml of water at 20 °C. They are heated for 1 min each using the same Bunsen burner flame and their temperature is measured. What will their temperatures be?
   A. The same but the heat transferred to them will be different.
   B. The same and so will be the heat transferred to them.
   C. Different but the heat transferred to them will be the same.
   D. Different and so will be the heat transferred to them.

Answer is C: The heat transferred to them will be the same as the Bunsen burner flame was the same as was the time of heating. Their temperatures will be different as the greater volume of 200 ml of water requires more heat to reach the same temperature as the smaller volume.

8. On a normal winter day, by what means does the human body lose most of its heat?
   A. Convection.
   B. Conduction.
   C. Radiation.
   D. Evaporation.

Answer is C: Heat loss through radiation (of infrared radiation) occurs continuously and despite being clothed. Heat loss by conduction is low as clothes are insulators and also decrease convection losses. Sweating will be low on a cool winter day so evaporation will not be a major heat loss avenue.

9. What factor does **NOT** affect the amount of heat lost from the human body by radiation?
   A. The area of skin facing the external environment.
   B. The temperature difference between the skin and the surroundings.
   C. The surface area of the body.
   D. The mass of the body.

Answer is D: Body mass produces heat, but radiative losses are affected by the other three factors.

10. By what means does a person sitting in the shade of a tree on a very hot dry summer day lose most of their heat?
    A. Radiation.
    B. Conduction.
    C. Convection.
    D. Evaporation.

Answer is D: On a hot day, sweating will be occurring and its evaporation into a dry atmosphere would produce cooling. As the surroundings and air are hot, not much heat will be lost by radiation, convection or conduction.

11. Why is heat loss from a hot object prevented when a material that is a poor conductor of heat, is wrapped around a hot object? Because:
    A. The particles of the poor conductor easily transfer the kinetic energy of their vibrations to their neighbours.
    B. Water vapour is prevented from escaping to the air.
    C. It reflects radiated heat back into the hot object.
    D. Air trapped within the poor conductor prevents convection currents from occurring.

Answer is D: In this case, "clothing" the object in an insulator will prevent air convection from carrying away the warm air in contact with the hot object. Hence cool air is prevented from replacing the heated air.

12. Why does the evaporation of perspiration from our bodies cool our skin? Because the evaporated water carries with it the:
    A. Latent heat of vaporisation.
    B. Heat lost by radiation.
    C. Latent heat of fusion.
    D. Heat lost by convection.

Answer is A: Radiation and convection are not involved. Latent heat of fusion refers to the heat required to melt a solid, latent heat of vaporization refers to the heat that leaves the body with the evaporated sweat.

13. Which of the following forms of heat therapy relies mainly on conduction?
    A. Heat lamp.
    B. Microwave diathermy.
    C. Hot water bottle.
    D. Ultrasound waves.

Answer is C: A hot water bottle will transfer heat to objects in contact with it. The other three rely on a form of radiation.

14. What is the most effective way to deposit heat in bones and joints?
    A. Ultrasound.
    B. Infrared radiation.

C. Conductive heating (heat packs).
D. Diathermy.

Answer is A: Ultrasound is the most effective form of diathermy as bones absorb ultrasound more effectively than soft tissue.

15. Which of the following is a biological mechanism of preventing heat loss?
    A. The production and evaporation of sweat
    B. Increasing muscular activity
    C. Vasoconstriction of superficial blood vessels
    D. Seeking a warm environment

Answer is C: Vasoconstricting blood vessel close to the body surface will withdraw blood from the surface and allow the skin temperature to cool.

16. Which of the following would be the main mechanism of heat loss for a student sitting in a lecture theatre while attending a lecture?
    A. Conduction
    B. Radiation
    C. Convection
    D. Evaporation of perspiration

Answer is B: Heat loss through radiation (of infrared radiation) occurs continuously and despite being clothed. Heat loss by conduction is low as clothes are insulators and also decrease convection losses. Sweating will be absent in an air-conditioned lecture theatre.

17. Which of the following statements is correct? As a means of losing heat from the body, evaporation of sweat works:
    A. Provided that the surrounding air is not saturated with water vapour.
    B. Provided that the surrounding environment is at a lower temperature that the body.
    C. Because sweat is at a lower temperature than blood.
    D. Because a film of sweat acts as an absorber of infrared radiation from the surroundings.

Answer is A: Sweat will evaporate and cooling will be achieved as long as the air is not already saturated with water vapour.

18. When the vibrating atoms of an object (at a temperature of 40 °C) pass on energy to the more slowly vibrating atoms in an adjacent object with which it is in contact (and which is at a lower temperature), what is this energy transfer known as?
    A. Insulation
    B. Convection
    C. Radiation
    D. Conduction

Answer is D: Conduction is the transfer of heat between contacting objects at different temperatures.

19. In what situation will the ability to lose heat by evaporation of sweat be diminished?
    A. When the body is dehydrated.
    B. When the ambient temperature of the surrounding environment is significantly higher than body temperature.
    C. When the relative humidity of the surrounding air is very low.
    D. When very little bare skin is exposed.

Answer is A: If the body is dehydrated, its ability to secrete sweat is diminished. The amount of bare skin does not affect sweating.

20. The amount of heat lost as radiation depends on the following things except one. Which one?
    A. The area of bare skin
    B. The degree of vasodilation of blood vessels in the dermis
    C. The temperature difference between the skin and the surrounding objects
    D. The surface area of the body

Answer is A: Radiation will occur whether the skin is covered by clothing or not.

21. Which of the listed factors does **NOT** affect the amount of heat that the human body loses by radiation?
    A. The temperature difference between the skin and the surroundings
    B. An individual's behaviour
    C. Being wrapped in a "space blanket" with a silver foil lining
    D. The area of uncovered skin

Answer is D: Heat will be lost by radiation even through clothing. A person may seek a cool environment to maximize their heat loss to the environment, so their behaviour may influence heat loss. A silver foil lining will reflect infrared radiation back towards the body.

22. In which situation will the evaporation of sweat from the skin be ineffective as a heat loss method?
    A. Air temperature is greater than human body temperature.
    B. Air temperature is less than human body temperature.
    C. Surrounding air is saturated with water vapour.
    D. Human body is immersed in the water of a swimming pool.

Answer is C: Sweat will evaporate only if the surrounding air is not already saturated with water vapour. High humidity decreases the effectiveness of sweating as a heat loss mechanism.

23. Which mechanism of heat loss from the human body is minimised by wearing clothes?
    A. Convection of air
    B. Radiation to the environment
    C. Warming of inhaled air to body temperature before exhaling
    D. Evaporation of sweat

Answer is A: Clothes trap a layer of air close to the skin. This air is warmed to skin temperature and is prevented from blowing away by being contained within the clothing.

24. Evaporation of sweat cools our body because the evaporating water molecules:
    A. Have a high heat capacity
    B. Transfer kinetic energy away from us
    C. Radiate heat away from us
    D. Remove heat by conduction

Answer is B: The water molecules in sweat that have the highest kinetic energy are the ones that evaporate first. This leaves the slower moving molecules behind, and they have a lower temperature. These slower molecules will gain energy from the body heat and eventually have enough kinetic energy to evaporate, again cooling the body as more kinetic energy is transferred away from the body.

25. What causes the cooling effect that we experience when sweating?
    A. Emission of infrared radiation
    B. Conduction of heat to the surrounding air then convection
    C. Dripping off of warm sweat from our skin
    D. Evaporating water molecules taking their kinetic energy with them

Answer is D: Evaporating water molecules take with them more kinetic energy than the average amount. This decreases the average kinetic energy of the remaining water molecules, hence leaving them at a lower temperature.

26. How does the evaporation of sweat operate as a heat loss mechanism?
    A. Sweat is produced at a lower temperature than skin so cools the body by conduction.
    B. Sweat promotes vasodilation which promotes heat loss by infra-red radiation.
    C. Water molecules with the greatest energy evaporate, leaving the remaining ones at a lower temperature.
    D. Sweat flows across the skin surface so promotes heat loss by convection.

Answer is C: If the most energetic water molecules depart, the remainder will have a lower average energy than before the fastest ones evaporated. This means that the temperature of the remaining sweat will be lower.

27. Which heat loss avenues are reduced by wearing clothes?
    A. Radiation and evaporation
    B. Conduction and convection
    C. Evaporation and convection
    D. Radiation and conduction

Answer is B: Clothes trap air between the clothes and skin, minimising convection. Clothes lying between skin and a contacting surface insulate against conducting losses. Radiation is not prevented by clothing.

28. Except for one situation described below, water molecules changing state from liquid to gas are involved in the cooling effect. Which one?
    A. A cold wind blowing against your skin
    B. Drying off after a swim
    C. Evaporation of sweat
    D. Exhalation of breath from the lungs

Answer is A: A cold wind cools us by convection. It continually removes the air that has been warmed by contact with the skin.

29. How does subcutaneous adipose tissue assist the body to regulate its temperature?
    A. It conducts heat more readily than lean tissue so promotes heat loss.
    B. It stores heat energy so acts like a "heat sink".
    C. It produces sweat for secretion via sweat glands.
    D. It conducts heat less readily than lean tissue so insulates the body.

Answer is D: Adipose tissue (fat) conducts heat less readily than does lean tissue. Hence, a layer of fat insulates the body against heat loss.

30. When is radiation an effective form of heat loss from the body?
    A. When we expose a greater amount of bare skin.
    B. When our body temperature is greater than that of our surroundings.
    C. When blood vessels close to the body surface are vasoconstricted.
    D. When our body temperature is less than that of our surroundings.

Answer is B: All objects radiate IR rays. If our body temperature is greater than that of our surroundings, we will radiate more than we absorb from the surroundings. It is body surface area, not amount of bare skin, that determines the amount of energy radiated.

31. Which choice explains how the evaporation of sweat "cools" our body?
    A. Evaporating water molecules carry with them more than the average amount of kinetic energy which leaves the remaining molecules with a lower average kinetic energy.
    B. The body loses more heat through the infrared radiation emitted by sweat than it gains from the infrared radiation emitted by the surroundings.

C. Sweat is at a lower temperature than our core body temperature so sweat on our skin cools us by conduction.
D. The water molecules in sweat are at a higher temperature than our core body temperature so losing sweat leaves us cooler due to the mass of water lost.

Answer is A: Water molecules have a range of KE. The water molecules with the most KE are those that vaporise first. Sweat will be at the same temperature as the skin it sits on.

32. Why does an ice pack applied to a bruise reduce swelling?
    A. Less fluid leaks from the bruise due to the diminished nerve impulses.
    B. It causes vasoconstriction.
    C. It reduces the metabolic rate at the local site.
    D. It increases the viscosity of blood below the ice pack.

Answer is B: If blood vessels constrict, less blood can flow out of them. Choices C and D are true statements but are beside the point.

33. The skin is usually at a lower temperature than the body's core temperature (37 °C). What is the reason for this?
    A. The layer of adipose tissue in the hypodermis insulates the skin from the core temperature.
    B. The sweating mechanism is able to lower the skin's temperature.
    C. Vasoconstriction restricts the amount of blood that is brought close to the skin surface.
    D. Heat loss through conduction, convection and radiation keeps the skin at a lower temperature.

Answer is C: Vasoconstriction limits the amount of blood (at 37 °C) that flows near the skin which allows the skin temperature to approach that of the surroundings.

34. A person lightly dressed in shorts, shoes and a shirt is sitting on a cushioned chair in their shady backyard where the air temperature is 12 °C. There is no wind yet they feel uncomfortably cool. What is their major avenue of heat loss?
    A. Conduction
    B. Evaporation of sweat
    C. Radiation
    D. Convection

Answer is C: The lack of wind minimises convective loss, while the light clothing prevents conductive loss to the chair and ground. The low temperature means that sweating is not occurring. Radiative loss will occur as the body temperature is greater than that of the surroundings.

35. Why will the metal bell of a stethoscope that is at the room's air temperature produce the sensation of cold when placed on the patient's skin?
    A. The skin is at a lower temperature than the bell.
    B. The stethoscope bell is a good conductor of heat.
    C. The stethoscope bell is a poor conductor of heat.
    D. Sweat evaporating from under the bell cools the skin.

Answer is B: Being a good conductor of heat, the metal bell will gain heat from the skin until all its mass reaches skin temperature. A non-conductor would only gain heat until the surface in contact reached skin temperature.

36. Why does perspiring cause heat energy to be lost from the body? Because:
    A. Being at a lower temperature than the skin, sweat cools skin by conduction.
    B. Evaporating water molecules remove heat in the form of their own kinetic energy.
    C. The presence of sweat on the skin prevents infrared radiation being absorbed.
    D. Sweat on the skin allows heat to be lost to the air by conduction.

Answer is B: Vaporising water molecules carry away more than the average kinetic energy. Temperature of an object is the average kinetic energy of its particles. The skin surface is left at a lower temperature.

37. In which situation will the skin lose heat by conduction to an object that is in contact with it? When the:
    A. Object is a good conductor of heat.
    B. Skin is not covered by clothing.
    C. Object is a poor conductor of heat.
    D. Object is at a lower temperature than the skin.

Answer is D: For the skin to lose (rather than gain) heat, the object in contact must be at a lower temperature than the skin, regardless of whether it is a good or poor conductor.

38. Which explanation of the cooling effect produced by an ice pack when applied to the skin is the best one?
    A. Cold applied to the skin causes vasodilation, thus allowing more blood to pass through the tissue adjacent to the ice pack and be cooled.
    B. Heat withdrawn from the body is used to provide the ice with the latent heat of vaporisation it requires to melt, until melting is complete, the temperature of the ice pack remains constant.
    C. Cold transfers along its temperature gradient—from ice pack to skin—the melting ice and water mixture ensures good contact with the skin.

D. Heat withdrawn from the body (at 37 °C) is used to provide the ice (at 0 °C) with the latent heat of fusion it requires to melt, until melting is complete, the temperature of the ice pack does not rise appreciably.

Answer is D: Cold causes vasoconstriction no vasodilation. Body heat provides the ice pack with latent heat of fusion (melting) not vaporisation. Heat transfers, not "cold".

39. A cold pack, applied to reduce swelling, is more effective if it contains melting ice at 0 °C rather than water at 0 °C. Why is this?

    A. Because ice cools by conduction whereas water cools by convection.
    B. Because initially the melting ice is colder than the cold water.
    C. Because ice has a higher latent heat of vaporisation than water.
    D. Because melting ice remains at 0 °C until it has all melted.

Answer is D: The heat that transfers from the bruised part is used to break the bonds between adjacent water molecules to melt the ice, rather than to increase the kinetic energy of the molecules. Temperature measures the average kinetic energy of the molecules. Hence, temperature does not rise until all the ice has melted.

40. One of the following mechanisms for losing body heat does **NOT** increase heat loss as we increase the amount of bare skin that we expose. Which one?

    A. Respiratory heat loss
    B. Heat loss by conduction
    C. The evaporation of sweat
    D. Heat lost through convection

Answer is A: Respiratory heat loss occurs because the air we inhale is usually at less than 37 °C while it is warmed to 37 °C before it is exhaled. (Also exhaled air is saturated with water vapour which carries with it the heat of vaporisation.) These mechanisms are not affected by the amount of skin we expose.

41. Consider a human who is inside in the shade during a heat wave where the temperature of the air and objects within the room is 40 °C. Assume a relative humidity of 30%. What are the means by which the body can lose heat?

    A. A. Conduction and drinking cold water.
    B. B. Sweating, fanning and drinking cold water.
    C. C. Convection, fanning and urination.
    D. D. Radiation and minimising muscular exertion.

Answer is B: As relative humidity is low enough to allow sweat to evaporate, evaporation of sweat will cool the body, as long as water is ingested to replace that sweated out. Conduction, convection and radiation will not work as the surroundings (40 °C) are at a higher temperature than the body (37 °C). Turning on the air-conditioner will work but is outside the scope of the question.

42. Which of the following ideas is **incorrect** and so could not be used in an explanation of how sweating allows the body to lose heat?
    A. A. The water molecules on the skin are continually gaining energy from blood near the skin surface.
    B. B. The body is cooled when sweat (at 37 °C) drips off the body or is wiped off.
    C. C. Evaporating water molecules leave with more than the average kinetic energy of all water molecules.
    D. D. The average kinetic energy of water molecules that have not evaporated is lower (than those that do) so their temperature is lower.

Answer is B: Sweat that drips off or is wiped off has not evaporated so has not contributed to a reduction in temperature of the skin from whence it came. The average kinetic energy of the evaporating water molecules in sweat is higher than the average KE of those remaining, hence those remaining have a lower temperature. This contributes to cooling the body.

43. Which statement distinguishes between convection and conduction as two means of transferring heat?
    A. Convection is the movement of heat due to the movement of a fluid, while conduction is the transfer of energy through a stationary object.
    B. Convection is the movement of heat due to the movement of a fluid, while conduction is the transfer of energy via an infrared wave.
    C. Convection is the transfer of heat due to the evaporation of water, while conduction is the transfer of energy through a stationary object.
    D. Convection is the transfer of heat via infrared radiation, while conduction is the transfer of energy by the vibration of the constituent particles.

Answer is A: A convection current exists when warm air (or liquid) moves away from a place and is replaced by cooler air (or liquid). Conduction of heat occurs between two touching objects that are at different temperatures. The energy transfer occurs due to vibrating particles in one object bumping into the particles of the adjacent object.

20.3 Body Temperature

1. How does a clinical (or fever) thermometer differ from a standard thermometer?
    A. It contains mercury.
    B. It is a maximum reading thermometer.
    C. It measures temperature in kelvin.
    D. It contains a capillary tube.

Answer is B: A clinical thermometer should maintain its reading (be maximum reading), so that the value does not change when it is removed from the measuring site.

2. To convert 30 °C into degree kelvin, what must be done?
   A. Add 212.
   B. Subtract 212.
   C. Add 273.
   D. Subtract 273.

Answer is C: Degree kelvin = degrees Celsius + 273.

3. The thermodynamic temperature scale has an "absolute zero" at −273 °C. In what sense is the zero absolute?
   A. It is 0 K.
   B. It is exactly 0.00000000....
   C. Particle motion has ceased at this temperature.
   D. It is impossible to reach this temperature.

Answer is C: As temperature is a measure of the kinetic energy of particles, when all particle motion has stopped, temperature is at its lowest point and cannot decrease further.

4. What is the reason that the temperature of a substance does not rise while it is in the act of melting, even if heat is added to it?
   A. Melting will not occur until the latent heat of vaporisation has been supplied, this prevents a rise in temperature until all the solid has melted.
   B. The added heat energy is used to increase the vibrational and rotational kinetic energies rather than the temperature (i.e. translational kinetic energy) of the particles.
   C. The presence of colder unmelted particles alongside particles that are in liquid form causes their temperature to remain low.
   D. Any energy added to the substance is used to break the bonds that hold the particles into the solid form so does not contribute to a temperature rise.

Answer is D: Once the bonds between the adjacent particles of a solid have been broken, they are free to move about, and the particles can increase their kinetic energy, that is, rise in temperature.

5. How does the body attempt to cope with hyperthermia?
   A. By increasing muscular activity
   B. With peripheral vasodilation and sweating
   C. With peripheral vasoconstriction and shivering
   D. By reducing heat loss by radiation and convection

Answer is B: Hyperthermia is a body temperature that is too high. Hence the body must lose heat by evaporating sweat. Vasodilation brings more blood closer to the surface allowing radiation loss to increase.

6. What does "hypothermia" refer to?
    A. Body temperature below 41 °C
    B. Body temperature below 38 °C
    C. Body temperature below 35 °C
    D. Body temperature above 38 °C

Answer is C: Hypothermia is a body temperature that is below the healthy range. That is, below 35 °C.

7. What is the definition of the temperature of an object?
    A. Temperature is a measure of the average kinetic energy of the particles of the object.
    B. Temperature is the amount of heat energy contained within the object.
    C. Temperature is the amount of energy that flows from the object being measured to the thermometer.
    D. Temperature is a measure of the hotness (or coldness) of the object.

Answer is A: The faster the particles of an object are vibrating, the higher is its temperature. Temperature is the average kinetic energy of its particles.

8. Human body core temperature is usually maintained within which of the following ranges?
    A. 35.0–37.5 °C
    B. 36.5–38.5 °C
    C. 36.5–37.5 °C
    D. 35.0–38.5 °C

Answer is C: This is considered the usual healthy range. Above 38 °C is a fever (except if as a result of vigorous exercise), while below 35 °C is hypothermia.

9. Hyperthermia is a core body temperature *greater* than which of the following temperatures?
    A. 36 °C
    B. 37 °C
    C. 38 °C
    D. 39 °C

Answer is C: 38 °C is the lower limit of what is considered to be a fever in a resting adult.

10. What is the temperature (in °C) below which the body is said to be hypothermic?
    A. 38
    B. 37
    C. 36
    D. 35

Answer is D: 35 °C is the upper limit of what is considered to be hypothermia in an adult.

11. What is the healthy human core body temperature?
    A. Below 38 °C except during fever.
    B. It lies between 36.5 and 37.5 °C.
    C. It lies between 36 and 37 °C.
    D. It lies between 37 and 38 °C.

Answer is B: Healthy human body temperature is close to 37 °C. It can be elevated to 38 °C (and more) by vigorous activity.

12. Which is the correct distinction between temperature and heat?
    A. Temperature is a measure of the amount of heat in an object and heat is the energy that flows between objects which have different temperatures.
    B. Heat is a measure of the energy contained within an object and temperature is the objective measurement of heat.
    C. Temperature is a measure of the average kinetic energy of particles and heat is the energy that flows between objects as a result of a temperature difference.
    D. Heat is the energy that flows from a cold object to a hotter object and temperature measures the average random kinetic energy of the particles.

Answer is C: Heat flows from the object at higher temperature to the object at lower temperature. Choices A and B are nonsensical in their reference to temperature.

13. How does an "infrared thermometer" determine body temperature?
    A. It detects radiation emitted from the body and converts that to a skin temperature.
    B. It directs an infrared beam at the skin and from the reflected energy determines skin temperature.
    C. It detects radiation from the body surface and converts that to a core temperature.
    D. It directs a laser beam at the skin and from the amount absorbed, determines a core temperature.

Answer is C: An IR "thermometer" has a sensor that detects emitted IR radiation. The radiated energy is proportional to the fourth power of temperature of the radiating body (Stefan–Boltzmann law), so the detected IR is converted to a core temperature by the device electronics (surface temp is not of interest). Choice D is wrong, while choice B is nonsense.

14. A person is measured to have an oral (sublingual) temperature of 36.8 °C, an axillary temperature of 36.5 °C, forehead temperature (measured with infrared thermometer) of 37.1 °C and a tympanic temperature of 35.9 °C. Which is the correct temperature?
    A. Oral
    B. Axillary
    C. Tympanic

D. All are correct

Answer is D: A physiological variable like temperature varies depending on when, where and with what it is measured. All the temperature values are accurate measurements of the body site used. Just be sure that you measure the same site with the same instrument if you want to compare the value with a previously measured value.

15. Which of the following is the best definition of "heat"? Heat is the energy that is transferred by:
    A. A. Conduction and ingesting hot liquids.
    B. B. Conduction, convection, ingesting hot liquids and urinating.
    C. C. Conduction, convection, radiation, ingesting hot liquids, defecating and urinating.
    D. D. Conduction, convection, radiation and evaporation of water and hydrolysis of molecules within the mitochondria.

Answer is D: The energy that is **transferred** between two objects (or locations) is known as heat. There are a variety of ways that energy can be passed along. Choice D is the most comprehensive list for humans.

16. Choose the best definition of what "temperature" is:
    A. A. Temperature is a measure of the speed of vibration of the atoms.
    B. B. Temperature is that quantity measured by a thermometer.
    C. C. Temperature is the thermal energy within an object.
    D. D. Temperature measures the heat within an object.

Answer is A: All matter is made of atoms. Atoms in a solid are always vibrating "on the spot" to some extent. The more energy they have, the more rapidly they vibrate (the greater is their vibration velocity v). The more rapid the vibration, the greater is the kinetic energy ($KE = \frac{1}{2}\, m \times v^2$) they possess. This vibration energy is what temperature is proportional to.

17. In what sense is the temperature 0 K an "absolute zero"?
    A. Temperatures below 0 K have never been achieved.
    B. At this temperature, all particle motion has ceased, hence lower temperature is not possible.
    C. Temperatures below 0 K do not exist, hence 0 K is the lowest possible temperature.
    D. Zero centigrade is the temperature at which water freezes so is relative to water, 0 K does not rely on a reference temperature.

Answer is B: As temperature is a measure of the kinetic energy of particles, when all particle motion has stopped KE is zero, so temperature is at its lowest point and cannot decrease further. Temperature less than 0 K do not exist. Choices C and D are true but do not give an explanation.

18. How is temperature understood in terms of the kinetic molecular theory of gases? Temperature is a measure of the:
    A. Average speed of the gas molecules.
    B. Average of the squared speed of the gas molecules.
    C. Average kinetic energy of the gas molecules.
    D. Average momentum of the gas molecules.

Answer is C: Kinetic energy = $\frac{1}{2} \times m \times v^2$. Thus both the mass of the molecules and their average speed squared determine the KE, and temperature is: $T = \frac{2}{3}$ KE ÷ ($N \times k_b$).

19. Given that 0 °C = 273 K, what is the result when 15°C is converted to kelvin.
    A. 15 K
    B. 268 K
    C. 288 K
    D. 18.2 K

Answer is C: When 273 is added to degree centigrade, the measurement is converted into kelvin. Given that 0 °C = 273 K, then 15 °C is 273 + 15 = 288 K.

Chapter 21
Sixty-Four Essay Topics for a Written Assignment Assessment in Anatomy and Physiology

21.1 Introduction

A different form of assessment in anatomy and physiology that allows students to display skills different from answering MCQ is a written assignment (an essay). This affords the students the opportunity to research the published literature about a topic and to write about it logically, clearly and succinctly while learning to cite the scientific literature that is referred to. Answering scientific essay questions by writing 1500–2000 words of prose is challenging. But then, studying a university course in anatomy and physiology should be appropriately challenging. The resources available to the student to help them rise to the challenge are extensive. The exercise is not time limited in the way that a sit-down exam in an examination venue is; hence, the essay may be drafted, reviewed and revised. The writer may use authoritative sources from textbooks and other published work. They may discuss their work with their peers, their tutors and others. They may revise their work based on the comments of the markers and gain additional credit (see below for a suggested approach to marking an essay).

The essay topics given below are intended to be descriptive rather than prescriptive. That is, students are given the freedom to take the essay topic in almost any direction. The statements that follow the essay topic are intended to guide students in deciding what might be included in the essay. Additional material should be included if it is appropriate for the essay. Students are advised to use a variety of published textbooks such as those on anatomy and physiology, physiology, pathophysiology, physical science texts, medical texts, as well as some journal articles. The essay should show evidence of careful library research from published sources. But not from website pages. Wikipedia and similar sites are not allowed nor are personal/commercial/institutional web pages as these are not peer-reviewed and the information they contain will change. Nor are lecture handouts (as these are not "published"), encyclopaedias or dictionaries. However, e-copies of articles from published journals or books are allowed. One purpose of the essay is to research the

M. Caon, *Examination Questions and Answers in Basic Anatomy and Physiology*, https://doi.org/10.1007/978-3-030-47314-3_21

published scientific literature that is available to anyone, that has undergone the peer scrutiny that commercial publishing requires and that will be available, unchanged, on the public record "forever".

Another purpose of a written assignment is to provide the opportunity to learn to write a good essay. The researched information must be paraphrased and organised into a logical structure that tells a story. Hence, the essay will have an Introduction that will introduce the content of the essay topic by giving some relevant background to set the scene and to explicitly tell the reader what theme or approach to the topic the essay will follow, and what will and will not be discussed. Additionally there should be several descriptive subheadings between the Introduction and Summary that are pertinent to the content of the essay. The Summary briefly reiterates the major points that were made in the essay without introducing any new material.

The three-step marking procedure that I use for essays is:

1. Mark a draft (say for a nominal 5 marks). This is intended to stimulate students into beginning their essay in a timely manner. Not all students choose to submit a draft.
2. Mark the essay (say out of 40 marks) and provide extensive critical comments for the student. The intention is: to identify any shortcomings; to provide guidance on how to improve the essay and; to grade the student relative to their peers. Students are required to revise their essays on the basis of the comments provided.
3. Mark the revised essay (say out of 20 marks). All students are required to revise their essay while addressing the marker's comments. The second marking is a brief marking, merely checking that all the marker's comments were addressed; looking at what the students modified for themselves and scoring the summary.

In step 2, I usually contact students whose essay is scored at less than half marks (because it is really bad) and tell them what is wrong with it and that it is so bad that it is not worth marking. By not marking under-prepared essays, a "fail" is avoided. I ask them to try again. In this way, students are not permitted to hand in rubbish. Such students are awarded 50% of marks on satisfactory resubmission. In addition, markers keep a histogram of their marks which I use to equilibrate marking between different markers. If deemed necessary, I add or subtract marks from essays marked by different markers so that their mean marks awarded more or less coincide.

Revision after critical comment (step 3) is intended to mimic the process of submitting a manuscript to a journal for peer review. The intention is to force students to improve their work by considering its weaknesses (as identified by the marker) and to allow the students to show some initiative by identifying for themselves other ways to improve their work. This also allows students a chance to improve their grade. Hence, in theory, after revision all students end up with a reasonable essay.

I use an essay or written assignment as a standard part of my assessment regime each year. To prevent a trade in essays from previous years, the essay topics are changed each year (that is how the list of topics below came about). To discourage students from buying an essay from a website, show them the websites that advertise "we write your essay for a fee", discuss academic dishonesty, plagiarism and their reason for being at university. And be vigilant.

21.2 Advice for the Students When Writing Scientific Essays

Essays that are about 1500 words in length (not counting words of one or two letters, or words in the subheadings, reference list or citations) will be sufficient to address the listed topics. The ability to write succinctly and unambiguously is valued in scientific writing. Hence, your essay should be substantially free of spelling errors, serious grammatical errors and examples of incomplete or awkwardly expressed sentences.

You should use a variety of published textbooks such as anatomy and physiology texts, physiology texts, pathophysiology books, physical science texts, medical texts and journal articles. Be sure to cite the publications that you draw your essay content from. For example by using the "author–date" (i.e. the Harvard) system.

Your essay should show evidence of careful library research from published sources DO NOT cite WIKIPEDIA or similar sites or personal/commercial/institutional web pages. However, digital "pdf" copies of articles from published journals or books ARE allowed, but do not give the web address or "viewed on" date. This is because published articles with a DOI number do not change with time. Do not cite your lecture handouts as these are not "published". Do not cite encyclopaedias or dictionaries as these are generalist publications that often do not treat anatomical and physiological concepts in correct context.

Direct quotations from published sources are almost never used in science essays. If they are, quotations should only be used where there is something very special about the expression used in the quoted passage which would be lost if it was expressed differently. The essay should then go on to explain the significance of the expression used. Paraphrasing of your selected sources will be necessary, but this must involve the expression of ideas in your own words and not just a modification of a set of words already used by someone else. This will show your mastery of the content that has been discussed.

Your essay assignment should have the following subheadings within the text:

Introduction: This will introduce the content of the essay topic by giving some relevant background and explicitly tell the reader what theme or approach to the topic, the essay will follow, what will and will not be discussed (your essay will not cover everything). It will not restate the question.

Several descriptive subheadings between the Introduction and Summary, there should be several subheadings that are pertinent to the content within that subsection of the essay and appropriate to your particular approach to the essay topic.

Summary: A brief summary of the major points that you have made in the essay. New material should not be introduced in the Summary.

References (not a bibliography): For example listed alphabetically in the Harvard style. This has the author names first, followed by the year of publication, then the title of the work. Then for books, state the publisher and perhaps the ISBN or for journal articles, state the journal title, the volume and issue and the first and last pages and perhaps the DOI number.

21.3 An Example of an Essay Marking Rubric

Student's name: Marks

Scope of research:

Was a variety of textbook references used? (between 8 and 14 refs are sufficient).

Were relevant journal articles used? (2 or 3 are sufficient) /6

Construction of essay:

Does the Introduction make it clear what ideas or what focus or what unifying concepts the author is going to use/emphasise in their essay?

Does the Introduction give some background to the essay topic?

Is the essay thoughtfully put together so that a *theme* is evident?

Has the topic been *analysed* thoughtfully?

Does the author comment on the materials cited?

Does the author make insightful comments about the essay content?

Has an effort been made to relate the material presented to the rest of the essay?

Have (appropriate) references been used well in essay construction? /7

Referencing and citation:

Are refs cited often enough/cited correctly in essay, and presented Harvard style in ref. list?

Were web pages used? (they should NOT be).

Are page numbers cited for text books?

Are sources paraphrased rather than quoted?

Arealllistedreferencescited(andviceversa!) /4

Essay layout:

Is the essay organised by *descriptive* headings within the text?

Is similar content gathered together under the appropriate heading?

If diagrams were used, were they referred to in the text?

Does the discussion progress logically through the content that is included?

Is repetition avoided? /4

Scientific content:

Is there *enough* basic science in the explanation/description in the essay?

Has the appropriate content been included?

Is the science content that is presented **correct**?

Are scientific principles explained sufficiently to demonstrate the author's understanding?

Are examples given to support the discussion?

Is the *level* of science in the essay appropriate for a first year essay?

Is the *depth* of discussion of the science content appropriate? /15

Spelling and sentence/paragraph construction:

Are spelling mistakes few or unimportant?

Is the discussion written in sentences? (*Lists or dot points should **NOT EVER** be used!*)

Are sentences short, constructed grammatically and clear in their meaning? /4

First marking /40

General comments:

Second, Revised Submission Mark Sheet.
(attach this to your second submission)
Revision due: Fri **** 20**, 3.00 pm.

1. As well as the revised and modified version of your essay, **hand up your original essay and mark sheet again attached to the revised version** (so we can compare the two).

2. In your revision, underline the changes that have been made so that the marker can easily see where modifications have been made (and indicate where deletions have been made). So that we can identify the changes!

☐ Deduct a mark if this mark sheet is not attached.

☐ Deduct a mark if the changes are not underlined.

☐ Deduct a mark if the original marked essay (with its mark sheet) is not attached.

Revisions to essay.

In the completed and revised essay, has the student adequately addressed the marker's comments from the first marking (*by clarifying descriptions, adding scientific explanations and examples*)?

Has the essay been modified at all places where the marker has made explicit comments?

/8

Has the student SHOWN INITIATIVE and used the general comments of the marker to ***identify for themselves*** other places in their essay (which were not explicitly identified by the marker) where improvements could be made (and made them)?

/6

Essay Summary
Does the Summary summarise the main points of the essay?
Does the Summary avoid vacuous generalisations?
New material should NOT be presented in the summary. /6
Second marking /20

21.4 Essay Topics

The essay topics below are grouped together into broad subject areas covering tissues, solution chemistry, homeostasis, and the body systems in the same order as the preceding MCQs. However, they are difficult to assign to just one content area as many of them were purposely constructed to cross over organ system boundaries and to incorporate both physical and biological science.

1. **Write an essay that discusses epithelial and connective tissues.**
   Your essay should contain some physiology, and use many examples of different structures in the body that are composed of these tissues to describe what the tissues do. Your essay should discuss the cells that occur in the tissue as well as the non-cellular parts of the tissue. You may wish to discuss the occurrence of a type of tissue across a variety of organs (or organ systems) in the body. Or you may wish to investigate epithelial and connective tissue within a single-organ system in the body. (Do not write about the detailed classification of epithelial or connective tissue.)
2. **Discuss the roles that calcium ions perform in the body**.
   Calcium is required for bones, skeletal muscle contraction, nerve signal transmission at a synapse, blood clotting, enzyme activation via calmodulin, cardiac muscle contraction, for the amoeboid movement of white blood cells and acts as a "second messenger" for some protein hormones. You may choose to discuss some or all of these roles. You could discuss the regulation of calcium concentration in the blood by hormones. You could include a discussion of how calcium is absorbed from the gut or how calcium is stored and released from the skeleton.
3. **Discuss the roles that sodium ions perform in the body**.
   Sodium is the most common cation in the body. Its concentration is much greater in the extracellular fluid than in the intracellular fluid. You could discuss the maintenance of this difference by the "sodium/potassium pump". You should describe the normal concentrations of sodium in the body. You could discuss how the concentration of sodium in the blood is regulated by aldosterone and ANP and is monitored by osmo-receptors in the hypothalamus. You may choose to discuss sodium's involvement in the transmission of an action potential, in the reabsorption of water from the kidney filtrate, or sodium's role in fluid and electrolyte balance or in hypertension.
4. **Write an essay about the sodium–potassium "pump"**.
   In the plasma membrane of all cells is a type of solute pump called the "sodium–potassium ATPase enzyme" that pumps sodium ions out of cells while

pumping potassium ions into cells. Both ions are moved against their concentration gradients so this requires that energy be expended. You should discuss why the Na-K pump exists and its mechanism of action. You could discuss how ATP is converted to ADP to release energy and how this energy is used. You may choose to discuss secondary active transport and/or the resting membrane potential in muscle or nerve cells. You could discuss the qualities of intracellular fluid versus those of extracellular fluid.

5. **Discuss body fluids.**

   There is a great variety of body liquids (fluids), and they all have a particular function. Discuss the composition, concentration, formation, function and physiological significance of some body liquids. You may choose from bile, blood, menstrual fluid, mucus, pus, semen, saliva, sweat, tears, urine, gastric juice, synovial fluid, cerebrospinal fluid, breast milk, pancreatic juice. You may wish to describe the various ways by which solution concentration may be stated, for example: density (g/ml), specific gravity (density of substance/density of water); % concentration (g/100 ml), osmotic pressure, partial pressure (for dissolved gases), molarity (mol/L), osmolarity (osmol/L), osmolality (osmol/kg).

6. **Write an essay about the properties of the solution known as plasma.**

   Plasma is 55% by volume of blood, and 90% of this is water (the remaining 10% is dissolved substances). Your essay should include the relevant definitions, a general discussion about solutes, solvents, solution concentration (and "tonicity") and the properties of blood plasma. You may wish to discuss the solutions added to blood by means of intravenous therapy and the effects they have. You may wish to compare plasma to some other solutions within the body such as lymph and intercellular fluid.

7. **Write an essay on organic compounds as medicines.**

   Your essay should have two examples of spectacularly successful drugs and explain chemically how they work. You should also include two examples of disasters in chemical therapy. Medicines you may consider include anaesthetics, drugs for pain relief, antibiotics, as well as specific drugs such as digoxin, angiotensin, the pill and anticoagulants. You will need to detail the relevant chemical basis for your essay such as chemical structures, functional groups, doses and how the body treats the chemical.

8. **Discuss how drugs (medicines) can affect the normal functioning of the cell.**

   Drugs are transported throughout the body in the blood stream. Your essay should explain how drugs interact with the cell membrane and the membrane's structural characteristics relevant to drug interaction. You may wish to discuss how drugs move from the blood stream into the extracellular fluid (noting the blood–brain barrier), some or all of the ways in which drugs act, the way they are eliminated from the cell. (Do not discuss the various routes of drug administration.)

9. **Write an essay about the integumentary system.**

   Your essay should describe the cell and tissue types in the skin, the role of the skin in homeostasis and its function in the immune system. You may wish to discuss the functions of the skin (for example its role in thermoregulation) and how its structure allows it to perform these functions. You could discuss the problem presented to the body when the skin is burnt.

10. **Discuss the role of the integument in protecting the body**.

    The skin encases the human body. The environment inside the skin is very different from the external environment. Discuss how the integument protects the body. You may choose to limit your description by focusing on one (or several) of protection from dehydration, from bacteria, from abrasion, from overheating and from excessive heat loss. You may wish to discuss the secretions produced by the glands of the skin or how your skin gathers sensory information about the immediate surroundings. You could discuss the roles of the different types of cells in the epidermis. (Do not merely describe the anatomy of the integument in the manner found in your textbook.)

11. **Write an essay about body temperature**.

    You should describe what a healthy human core body temperature is and why is it needed to be at this value. You could discuss how the body temperature is measured, how the body produces heat—both mechanically and by chemical means and how body temperature is maintained at the desired level. That is, the mechanisms by which heat is gained and lost (sweating, vasodilation, radiation, conduction, convection). You might discuss the body's homeostatic control of body temperature. You could discuss the consequences of hypothermia and or hyperthermia, or even frostbite.

12. **Write an essay about homeostasis in the human body**.

    The body is able to allow its internal conditions to vary while maintaining them within relatively narrow limits. The term coined to name this ability is "homeostasis".

    Your essay should include the relevant definitions, a general discussion about homeostasis and control mechanisms (negative and positive feedback). You should also choose an organ (or organ system) and its control mechanism(s) and use it to illustrate your discussion of the process of homeostasis. You may wish to discuss homeostasis on a cellular level as well.

13. **Discuss the roles that the skeleton performs**.

    You may choose to discuss some or all of the skeleton's biomechanical function, its role in haemopoiesis and its role in calcium metabolism. You could discuss the microscopic structure of bone and how it is remodelled. You could include a discussion of the cells in bone and their function and/or a brief discussion of some bone pathologies. (Do not merely describe the anatomy of the skeletal system in the manner found in your textbook.)

14. **Discuss the inter-relationship of the skeletal muscular system to the skeleton and how they function together biomechanically**.

    You may approach the topic from an anatomical perspective or from a biomechanical perspective (in either case your essay should contain some anatomy and some physical science). Your essay should explain how the anatomical structures are related to each other (attached to each other) and are well suited to perform the functions that they do. Your essay could include the physiology of muscle contraction and/or bone remodelling or ossification. It may also discuss one or two defects of the musculoskeletal systems and their explanation.

15. **Discuss the inter-relationship of the skeletal muscular system to the skeleton.**

    Your essay should contain some physiology and some physical science. It should explain how the anatomical structures are related to each other (attached to each other) and are well suited to perform the functions that they do. Your essay could include the physiology of muscle contraction and/or bone remodelling or ossification. It may also discuss one or two defects of the musculoskeletal systems and their explanation.
16. **Discuss the statement that "the role of the musculoskeletal system is to provide locomotion".**

    You may choose to discuss both muscles and bones together or one of the systems separately. You may wish to discuss the role of muscle and bones in producing locomotion. You should explain why maintaining a healthy musculoskeletal system is important. Your essay should contain some physiology, and (where appropriate) some physical science and chemistry. You could include (among other things) appropriate discussions about other roles performed by muscles and the skeleton and why muscles and bones have the structure they have. (Do not merely describe the anatomy of the musculoskeletal system in the manner found in your textbook.)
17. **Discuss the metabolism within the skeletal muscular system.**

    Skeletal muscle makes up about 40% of male body mass (~29% for females). Discuss the contribution that skeletal muscle makes to metabolism. You may wish to discuss the role of mitochondria in ATP production, the role of muscle in oxygen consumption and carbon dioxide production (and what muscle produces and what it consumes). You could discuss heat production in muscle (and heat removal), what stimulates them to function, or how they interact with other body systems. (Do not merely describe the anatomy of muscles in the manner found in your textbook.)
18. **Write an essay that discusses the digestion of foodstuffs as a chemical process.**

    You may approach the topic from an anatomical perspective or from a chemical perspective (in either case your essay should contain some anatomy and some physical science). Your essay should explain how the anatomical structures are well suited to digesting food and producing digestive enzymes. Your essay should also contain the chemical basis for the breakdown of food products into the end products which are suitable for absorption. You may wish to discuss the role of glucose, gluconeogenesis and glucogenolysis. (Do not write about nutrition.)
19. **Write an essay about the digestive system.**

    Your essay should contain some physiology and some appropriate chemistry. It should explain how the anatomical structures of the digestive system are well suited to perform the functions that they do. Your essay could include a discussion of the role of the accessory organs in digestion. Your essay could discuss the chemical reactions that constitute the breakdown of complex molecules into simple molecules. It could follow in detail how the digestion and absorption of protein or carbohydrate or lipid occurs (or all three in less detail). It may also discuss how the processes of digestion are controlled. (Do not merely describe the anatomy of the digestive system or write about teeth and mouth.)

20. **Discuss the proposition that "the contents of your gut are outside your body".**

    You may choose to discuss both how material enters your gut and how it leaves or is absorbed from the gut. You should describe what the gut contains and how it is modified as it moves along the gut. Your essay should contain some physiology, and (where appropriate) some physical science and chemistry. You may wish to discuss the roles of an organ or several organs in digestion. You could include (among other things) appropriate discussions about what happens in the gut and why the gut has the structure it has. (Do not merely describe the anatomy of the gastro-intestinal system in the manner found in your textbook.)

21. **Write an essay about the role of the liver in metabolism, digestion and in dealing with blood.**

    Your essay should describe the function behind the structure of the lobule (the kidney's functional unit). You may wish to write the essay from a pharmacological point of view and discuss the liver's role in the metabolism of drugs (but this is not a mandatory approach). You may wish to discuss the first pass effect and half-lives of drugs. You could also discuss the manufacture of bile, the storage of vitamins and the metabolism of amino acids, fatty acids and glucose. It is not necessary that your essay be an exhaustive discussion about everything that the liver does. Consequently, your Introduction should inform the reader on which aspects of the liver the essay will address and in what detail.

22. **Discuss the role that the liver has in the physiology of the body.**

    The liver performs a great many functions—decide which ones you will include in your essay. You may want to discuss the metabolism of carbohydrates, fats and proteins by the liver. You should explain the production of bile. You may wish to discuss the liver's ability to construct molecules needed by the body or its ability to change molecules into different molecules. You could include (among other things) appropriate discussions about the liver's role in excretion or blood clotting. You may choose to include a discussion about pathology of the liver. (Do not merely describe the anatomy of the liver in the manner found in a textbook.)

23. **Discuss the role that the pancreas has in the physiology of the body.**

    The pancreas is both an endocrine organ and an exocrine organ. Your essay should contain some discussion of both the functions. You should discuss what substances the pancreas produces, what they do and where they go. Your essay could include a discussion of how pancreatic secretions are controlled. You may choose to include a discussion about pathology of the pancreas. (Do not merely describe the anatomy of the pancreas in the manner found in a textbook.)

24. **Discuss the need for, the safety of and the perceived risks of adding chemicals to preserved food and grocery items.**
 In packaged and processed foods, there is usually a need to include chemicals (called food additives) as well as the foodstuff in order to give the grocery item the desired nutritional quality, maintain or improve keeping quality or make the food's taste, colour or consistency more attractive. On the other hand, some grocery items proudly proclaim the absence of these additives. In addition, sometimes the removal of substances such as fat, sugar and "salt" is announced.
25. **Write an essay about the place of alcoholic drinks in the diet including a discussion of their benefits (if any) and their deleterious effects.**
 Much has been written about the place of alcohol in the diet and its abuse. In certain quantities, some alcoholic drinks may be beneficial to well-being; however, its deleterious effect on the body's physiology is more often portrayed. You might discuss how the liver detoxifies alcohol. You may choose to discuss how alcohol is absorbed from the gut and how it affects the nervous system. You could discuss the energy released by the alcohol molecule when metabolised by the mitochondria.
26. **Write an essay about the pancreas, insulin and type II diabetes.**
 Type II diabetes is a disease of the pancreas or insulin resistance. Discuss insulin and the role of the pancreas in its production and what goes wrong when type II diabetes develops. You may choose to discuss what insulin is and does, and how it affects cells and blood sugar. You may wish to discuss how physiology and metabolism are different (or changes) when type II (and/or gestational) diabetes develops. You could discuss (scientifically) how T2D is diagnosed and what can be done about it. (Do not merely describe the anatomy of the pancreas in the manner found in your textbook.)
27. **Write an essay about the endocrine system.**
 Your essay should contain some physiology, and some reference to homeostasis and feedback. Your essay should discuss (at least) one endocrine gland and its hormones in detail. You should discuss the role of the hypothalamus in the way your chosen gland(s) function. You may wish to discuss some or all of hormone receptors, the chemical structure of hormones and the method of hormone action at your chosen level of detail. (Do not write about the anatomy of the endocrine system.)
28. **Compare and contrast the roles that the endocrine system and the autonomic nervous system have in communicating with and controlling the body.**
 You will need to decide what to include and what to leave out of your essay, there is no need to discuss the entire endocrine and nervous system. You should describe how the two systems get their messages to their target(s) in the body, the nature of the message carriers and how the systems are controlled. Your essay should contain some physiology and (where appropriate) some physical science and chemistry (of for example hormones and neurotransmitters). You may wish to discuss the roles of a particular organ or several organs as an

example of how the systems work. You could include (among other things) appropriate discussions about the effects that a hormone(s) or stimulation by the ANS have on the body. (Do not merely describe the anatomy of the endocrine or nervous system in the manner found in your textbook.)

29. **Discuss the role of the hypothalamus and pituitary gland in controlling the body's functions.**

    Your essay should contain some discussion of how the two structures are linked and interact. You should explain how they maintain control over other organs in the endocrine system. Your essay could include a discussion of the effect of the endocrine system on the nervous system. Your essay could discuss hormones and their chemical nature. You may choose to include in your discussion all of the organs influenced by the pituitary or you may wish to concentrate on one or a few. (Do not merely describe the anatomy of the hypothalamus and pituitary gland in the manner found in your textbook.)

30. **Discuss hormones in the body.**

    You may choose to discuss circulating hormones (those produced by the endocrine system), local hormones (i.e. paracrines), hormones produced by the gut, or recently discovered hormones such as ghrelin, obestatin and leptin. You could discuss their structure, their effects and/or how they exert their effects on their target organs and tissues. You could concentrate on one (or a few) hormones or organs, or you could be more exhaustive in your coverage. (Do not merely describe the anatomy of the structures in the endocrine system in the manner found in your textbook.)

31. **Discuss the role of the kidneys in controlling blood pressure and blood pH.**

    You may choose to discuss both blood pressure and pH, or just one of them. You should explain why maintaining an appropriate blood pressure and/or pH is important. Your essay should contain some physiology and (where appropriate) some physical science and chemistry. You may wish to discuss the roles of blood volume, renin, aldosterone, ADH and the production of bicarbonate ions. You could include (among other things) appropriate discussions about pressure, vasoconstriction/dilation, and/or the carbonic acid/bicarbonate buffer. (Do not merely describe the anatomy of the kidney in the manner found in your textbook.)

32. **Discuss the role that the renal system has in controlling blood pressure.**

    You may choose to discuss the kidney's control of blood volume or the renin–angiotensin mechanism (or both). You should explain why maintaining an appropriate blood pressure is important. Your essay should contain some physiology and (where appropriate) some physical science and chemistry. You may wish to discuss the roles of blood volume, renin, aldosterone, ADH and urine concentration. You could include (among other things) appropriate discussions about pressure, vasoconstriction/dilation. (Do not merely describe the anatomy of the kidney in the manner found in your textbook.)

33. **Discuss the roles that blood performs.**

    You may choose to discuss how the blood interacts with the digestive or respiratory or urinary systems. You should introduce all the functions of blood and then probably narrow your focus to one or two. You could discuss the role of capillaries. You might wish to include a discussion of the cells found in blood and what they do. (Do not merely describe the types of blood cells or the anatomy of the blood vessels the manner found in your textbook.)

34. **Discuss how the body keeps blood pH within the healthy range.**

The blood's pH changes very little despite the acidic substances produced by the body's metabolism and despite the acidic substances we eat. Discuss how and why the blood pH is so regulated. You should discuss the blood's buffer systems and how they are maintained. You may choose to discuss how breathing out CO_2 and producing urine are ways of excreting acid. You could discuss what happens if some pathology causes blood pH to stray outside the healthy range. (Do not merely describe pH, blood buffers, the anatomy of the lungs or kidneys in the manner found in a textbook.)

35. **Discuss the heart and coronary artery disease (CAD).**

You may choose to discuss heart and how it works or the histology of the heart and/or the electrical activity. You should discuss the heart's own blood supply and how the structure of arteries change in disease. You could discuss the ECG and how it is used to diagnose heart disease. You might wish to include a discussion of the causes of CAD and how it might be avoided. (Do not merely describe the anatomy of the structures in the heart in the manner found in your textbook.)

36. **Discuss the function of capillaries in the body and how their structure facilitates their functions.**

You may choose to discuss how material enters and leaves capillaries by crossing the walls. You should introduce the different types of capillaries. Then you may choose to elaborate on the capillaries that are found in the liver, or the kidney, or the brain, or the alveoli, or the small intestine (or several of them). You should discuss how the structure of each type of capillary that you discuss facilitates the function that they perform. (Do not merely describe the types of capillaries and their anatomy in the manner found in your textbook.)

37. **Describe thrombolytic therapy and its efficacy in the treatment of heart attack patients.**

Heart attacks are the result of coronary artery disease. These days, clot dissolving drugs such as streptokinase are routinely administered to patients who have suffered a heart attack (this is called thrombolytic therapy). You may wish to discuss the coagulation of blood. You could discuss the two ways clot-busting agents can be given: through a peripheral IV or through a catheter that has been navigated to the site of the clot. You may discuss the risks associated with the therapy. You may wish to discuss thrombolytic therapy for conditions other than a heart attack.

38. **Discuss the role of pressure in the cardiovascular system.**

You may approach the topic from an anatomical perspective or from a biomechanical perspective (in either case your essay should contain some anatomy and some physical science). Your essay should explain how the anatomical structures produce pressure, cope with pressure and regulate pressure.

Your essay could include a discussion of why auscultatory blood pressure is taken and what it means. You may wish to discuss why the right and left ventricles are different, why veins have valves and why blood flow in the capillaries is slow. (Do not write about the ECG.)

39. **Discuss how the brain and cardiovascular system work together to control blood pressure.**

    Your essay should contain some discussion of how the brain affects the way the heart works and affects the blood vessels. You should explain why maintaining an appropriate blood pressure is important. You should also discuss how changes in the CVS are monitored/detected by the brain. Your essay could include a discussion of the effect of heart rate, stroke volume, body position or exercise on blood pressure. Your essay could discuss baroreceptors and chemoreceptors. You may choose to include a discussion about the electrical conduction system of the heart and/or the ECG in your response to the essay topic. (Do not merely describe the anatomy of the brain and CVS system in the manner found in your textbook.)

40. **Discuss how the cardiovascular system works at controlling blood pressure.**

    Your essay should contain some discussion of heart rate, stroke volume, vasoconstriction and vasodilation. You should explain why maintaining an appropriate blood pressure is important. You should also discuss how changes in the CVS are monitored/detected by the brain. Your essay could include a discussion of the effect of heart rate, stroke volume, body position or exercise on blood pressure. Your essay could discuss baroreceptors and chemoreceptors. You may choose to include a discussion about the electrical conduction system of the heart and/or the ECG in your response to the essay topic. (Do not merely describe the anatomy of the CVS system in the manner found in your textbook.)

41. **Discuss cholesterol and steroids and cardiovascular disease.**

    Cholesterol is one of the steroids found in the body. Discuss cholesterol (and/or other steroids) and its (their) roles in the body and what might result from an excess of cholesterol. You may choose to discuss what steroids are and what they do and how they change the body. You may wish to discuss cholesterol's influence on cardiovascular disease. You could discuss what can be done to reduce blood cholesterol level. (Do not merely describe the anatomy of the plasma membrane in the manner found in your textbook.)

42. **Write an essay on negative pressures in the body.**

    In your discussion you may want to refer to breathing, wound drainage, suctioning, low pressure devices, lavage, osmotic pressure, entrainment and taking blood. However, your essay need not talk about all (or even most) of these. Your essay should discuss at an appropriate level the relevant physical science (such as atmospheric pressure, pressure units and pressure gradient).

43. **Write an essay about pressure and the control of blood pressure in the body.**

    Your essay should contain some physiology, some reference to homeostasis and some physical science. Your essay should explain the concept of pressure and the units relevant to this essay. You should discuss the role of baroreceptors, rapid control of blood pressure and long-term control of blood pressure. You may wish to discuss mean arterial pressure, cardiac output, peripheral resistance and blood volume in relation to the regulation of blood pressure. (Do not write about the anatomy of the cardiovascular system.)

44. **Write an essay about the respiratory system.**

    Your essay should contain some physiology and some physical science. It should explain how the anatomical structures of the respiratory system are well suited to perform the functions that they do. Your essay could include a discussion of lung ventilation, the physiology of gas exchange and/or gas transport in the blood. It may also discuss one or two pathologies of the respiratory system and their explanation. (Do not write about the anatomy of the respiratory system.)

45. **Discuss the respiratory system and chronic obstructive pulmonary disease (COPD).**

    You may choose to discuss some or all of how the lungs are ventilated, what happens to the airways in disease, how spirometry is used in diagnosis. You could discuss the respiratory membrane and what effect partial pressures of the gases in air have on gas exchange and on the oxygen saturation of haemoglobin. You could include how the body monitors the amount of CO_2 in the body, the effects of too much CO_2 (or too little oxygen) and how your body tries to cope. (Do not merely describe the anatomy of the respiratory system in the manner found in your textbook.)

46. **Discuss the roles for carbon dioxide in the body.**

    Carbon dioxide is produced by, transported around and excreted from the body. Discuss how and where CO_2 is produced, how and in what form it is transported in the blood and how it is removed from the body. You may choose to detail how it is produced in the mitochondria, what effect it has on blood pH, how its dissolved concentration is measured or how it is handled in the alveoli. You could discuss how the body monitors CO_2 concentration, how it moves about in body fluids, what happens when the body has difficulty getting rid of CO_2 rapidly enough (and how such a situation might arise) and what could be done about it. (Do not merely describe the anatomy of the lungs in the manner found in your textbook.)

47. **Discuss pulmonary ventilation, gas exchanges in the body and transport of respiratory gases.**

    Your essay should contain some physiology, some reference to homeostasis and some physical science. Your essay should explain pulmonary ventilation, gas exchanges in the body and the role of blood. You may wish to discuss some (but not necessarily all) of oxygen transport, carbon dioxide transport, basic properties of gases, gas exchanges between blood, lungs and tissues, physical factors influencing pulmonary ventilation, the composition of alveolar gas. (Do not write about the anatomy of the respiratory system.)

48. **Describe and discuss the role (or the significance) of CO_2 in the body.**
    You may choose to discuss CO_2 in the context of respiration. You may wish to discuss the role of CO_2 in stimulating breathing or in the body's acid–base balance and blood buffer system. You should explain how CO_2 is produced by the body and eliminated from the body. Your essay should contain some physiology and (where appropriate) some physical science and chemistry (e.g. how CO_2 is transported in the blood and the role of carbonic anhydrase). You could include (among other things) appropriate discussions about the effect of having too much CO_2 dissolved in the body. (Do not merely describe the anatomical structure associated with the handling of CO_2 in the body.)
49. **Discuss oxygen in the body.**
    You may choose to discuss some or all of how the body obtains oxygen, how it is transported and what it is used for. You could discuss how the body's anatomy is adapted to handle oxygen. You may wish to discuss what happens to oxygen within cells and or how it moves in solution. You could include how the body monitors the amount of oxygen in the body, the effects of too little oxygen and how your body tries to cope with a deficiency. (Do not merely describe the anatomy of the respiratory system in the manner found in your textbook.)
50. **Discuss how the lungs and body are affected by COPD.**
    COPD is one of the most common causes of death worldwide. Discuss the changes from normal respiratory physiology that characterise COPD. You should discuss the effect of too much CO_2 in the blood. You may wish to discuss the respiratory membrane and how gas exchange occurs across it. You could discuss the causes, management, prognosis and how respirometry is used in its diagnosis. (Do not merely describe the anatomy of the respiratory system in the manner found in your textbook.)
51. **Write an essay on respiratory acidosis and metabolic acidosis.**
    Your discussion will probably refer to acids, bases, pH, neutralisation, buffers, carbonic acid, acidosis, alkalosis, electrolytes. Be sure to include the relevant chemical background for your essay. You may want to discuss how the conditions arise and how they are dealt with (compensated for) by the respiratory system and by the renal system.
52. **Discuss how the brain and body are affected by a CVA (stroke).**
    Stroke is one of the most common causes of death worldwide. Discuss the physiology and pathophysiology of ischemic and haemorrhagic stroke. You should discuss the blood supply to the brain and why its maintenance is important. You could discuss how (and why) the location of the CVA produces different disabilities in the person. You may choose to discuss some of the epidemiology, prevention strategies, diagnosis, prognosis or management etc. of the disease. (Do not merely describe the anatomy of the brain and spinal cord in the manner found in a textbook.)
53. **Discuss the blood–brain barrier and how it protects the brain.**
    The BBB restricts the entry to the brain of some blood-borne substances while allowing others to enter. You should discuss how the structure of the BBB prevents the entry of some molecules but allows others through. You should

discuss why this discrimination is necessary. You could discuss the entry of drugs, both those used as medicines and those sold illegally. You may choose to discuss the BBB in newborn babies, or how it makes the treatment of some neurological disorders difficult, or how the CSF is made and differs from blood. You could describe the blood supply to the brain. (Do not merely describe the brain anatomy in the manner found in a textbook.)

54. **Write an essay about the autonomic nervous system.**

    Your essay should describe the organisation of the ANS, the role of neurotransmitters and of the hypothalamus. You may wish to discuss some (but not necessarily all) of the division between sympathetic and parasympathetic, reflex arcs, receptors and the levels of control of the ANS. You could (in order to provide illustrative examples) focus on one body system or function and relate the influences that the ANS has on it. (Do not write in great detail about the anatomy of the autonomic nervous system.)

55. **Discuss the role of the autonomic nervous system in controlling the body's functions.**

    You may choose to discuss all of the organs that are influenced by the ANS or you may focus on one (or some)—the depth of your treatment will vary accordingly. You should discuss both the sympathetic and the parasympathetic divisions. You may wish to discuss the roles of neurotransmitters and receptors. You could include (among other things) appropriate discussions about how the ANS links with the endocrine organs, and visceral reflexes. (Do not merely describe the anatomy of the autonomic nervous system in the manner found in your textbook.)

56. **Write an essay on the differences and similarities between light and sound and the consequences for sight and hearing.**

    You may approach the topic from an anatomical perspective or from a waves perspective (in either case your essay should contain some anatomy and some physical science). Your essay should explain how the anatomical structures are well suited to detect sound or light waves. You may (if you wish) prefer to concentrate the majority of your essay on either vision or hearing. Your essay could include the physiology of sight and/or hearing and may also discuss one or two defects of sight and vision and their explanation.

57. **Discuss the special senses and the roles that they perform in the body.**

    The special senses are based in organs that gather information about our environment and provide it to our brain. You could discuss all our special senses or just some (or one) of them. You should describe what they do and how they do it. You could discuss what sort of information they gather, how they pass it to the brain and/or how our brain interprets it. You may choose to discuss some of the defects of the special senses or how they deteriorate with age. (Do not merely describe the anatomy of special sense organs in the way that a textbook does.)

58. **Discuss the reproductive system and its physiology.**

    The male and female reproductive systems together allow us to produce new humans. You could discuss how the reproductive structures are well suited to their functions. You should discuss the changes in the system as the child moves through puberty and the adult ages. You may choose to discuss the genetics of reproduction or the development of the foetus or the replication of cells. You could discuss the role of hormones in reproduction. You may wish to discuss some medical problems that occur when reproductive physiology does not work as it is intended. Do not merely describe the anatomy in the "sideways view" way that an anatomy text does. Do not describe birth.

59. **Write an essay on LASERs and their use in endoscopy**.

    Your essay should discuss at an appropriate level the relevant physical science. In your discussion you may want to refer to some (but not necessarily all) of the following: the difference between ionising and non-ionising radiation, fibre optics, the wavelength of the LASER and its property of being monochromatic, its collimation and the level of power delivered, the procedures required for safe operation of lasers, why LASERs are suitable/necessary for endoscopic procedures.

21.5 Challenging Essay Topics

60. **Discuss the physiology of zombies.**

    Zombie humans do not exist, yet we all seem to know how they work. They stumble around in crowds bleeding from open sores on their skin, they are badly dressed, they don't talk but are able to see and distinguish us (living humans) from them. They pass on the zombie infection (virus/bacteria/prion/brain worms) by biting people, are strong, seemingly able to tear people open in order to feed on them, are persistent but are "killed" by severing the head from the body, etc. What are the characteristics of zombies? Are there different types of zombies? (Fast zombies/slow zombies). Explain what has gone wrong with their physiology in order to produce the observed behaviours. Eschew "magical" or unscientific explanations, rather consider that zombie behaviour is constrained by the physiological characteristics of human anatomy and physiology and discuss what is not working and so would result in their characteristic behaviours. (Use your critical faculties scientifically. Do not cut and paste material from the Internet—it is mostly nonsense.)

    Zombie Movie Bibliography *(for Backgrounding Rather Than Citing!)*

    Juan of the Dead/Juan de los Muertos (Spanish language Cuban movie 2010, with subtitles, comedy).

    Shaun of the Dead (British 2004, set in London, comedy).

    Night of the Living Dead (USA 1968, remade 1990, horror).

    Dawn of the Dead (USA 1978, remade 2004, action, horror, thriller).

    Cargo (Australian, 2017, set in Flinders Ranges).

    http://zombieresearchsociety.com(but don't take this site very seriously ☺).

    Neuroscientists Explain the Undead (http://zombieresearchsociety.com/archives/27009).

61. **Design a better human anatomy**.

Human anatomy is rather good, thanks to evolution. However, can you think of ways to improve it?

For example: allow females to have four breasts instead of two—useful for breast-feeding multiple births. Append an extra pair of arms—useful for carrying things and performing tasks. Have an intake port for food that is separate from the orifice used for talking/breathing—we could eat and talk at the same time and be less likely to choke on our food. Add an extra eye—perhaps in the back of our head? Add a third kidney. Have a way to store oxygen for the brain so that we can withstand interruption to breathing for longer. Have a blood supply to the myocardium that has more anastomoses so that blood can navigate around blockages to the coronary arteries. A wider or more elastic birth canal so that birth can be accomplished more rapidly and with less drama. Eliminate the appendix. Have the ability to digest cellulose, etc. (Do not consider the imaginary powers of "superheroes".)

Describe some improvements you would make and justify them. What other changes to our supporting "anatomical infrastructure" would be necessary to ensure that your improvements are fully functional? What would be the physiological implications of your new design? Remember that there is probably a good reason for being built the way we are, so what would be the detriment/consequences of your changes?

Some websites with useful (?) discussions *(but not for citing!)*

The Most Unfortunate Design Flaws in the Human Body (http://io9.gizmodo.com/the-most-unfortunate-design-flaws-in-the-human-body-1518242787).

Top 10 Design Flaws in the Human Body (http://nautil.us/issue/24/error/top-10-design-flaws-in-the-human-body).

62. **Which organ in the body is the most important?**

Which organ or tissue, in your opinion is the (or one of the) most important organ(s) in the body? Make a case, using anatomical and physiological science, to support the paramount importance of your choice. That is, describe the physiological purpose of your chosen organ, and the consequences to human physiology if the organ was malfunctioning or not present to perform its functions. You may wish to scientifically assess the importance of your choice of organ against the importance of a few other likely contenders. Your assessment could be based on whether a human could survive without the organ, the detriment to the human organism of a malfunctioning organ, how such malfunction might be compensated for (e.g. kidney dialysis, hormone replacement therapy, coronary artery bypass surgery and use of medicine). (Do not merely describe the physiology of the organ in the way a physiology textbook would.)

63. **Will drinking your urine save you from dying of dehydration?**

Occasionally the media reports on people who have been rescued from dying of dehydration in the desert or at sea saying that they drank their urine to stay alive. Scientifically discuss whether this is indeed a life-saving strategy and make a conclusion. Consider why we produce urine, what its composition is and how its composition changes from when we are well hydrated (and hence need to get rid of excess water) compared to when we are dehydrated and hence conserving water and are producing urine of the highest concentration physiologically possible. You may wish to discuss the effects of dehydration on blood osmolarity and on blood pressure. You could discuss the effectiveness of our ability to lose heat by sweating when we are dehydrated. You may choose to discuss the hormonal signals that are used to regulate the production of urine.

64. **Discuss organ transplantation from a scientific view point** (avoid emotive and ethical issues and build an argument as to why or why not organ donation should occur).

    The need for more people to posthumously donate their organs for transplanting has been raised. However, the public are not well informed about such things as the processes involved, anti-rejection drug therapy, success rates and long-term survival rates. You could focus on the transplantation of particular organs, for example liver transplants, kidney transplants, heart transplantation, lung transplants, corneal grafts, bone marrow transplant. You may wish to discuss whether blood donation lies within the realm of transplantation.

Index

M. Caon, *Examination Questions and Answers in Basic Anatomy and Physiology*, https://doi.org/10.1007/978-3-030-47314-3

R

S

Printed in the United States
by Baker & Taylor Publisher Services